ADVANCES IN
HUMAN GENETICS **21**

CONTRIBUTORS TO THIS VOLUME

Audrey D. Goddard
Somatic Cell Genetics
Imperial Cancer Research Fund
London, England

Gregory A. Grabowski
Division of Human Genetics
Children's Hospital Medical Center
Cincinnati, Ohio

Dimitris Kardassis
Section of Medical Genetics
Cardiovascular Institute
Departments of Medicine and Biochemistry
Housman Medical Research Center
Boston University Medical Center
Boston, Massachusetts

Hugo W. Moser
Kennedy Institute
Baltimore, Maryland

Jennifer M. Puck
Department of Pediatrics
University of Pennsylvania School of Medicine
and Children's Hospital of Philadelphia
Philadelphia, Pennsylvania

Ellen Solomon
Somatic Cell Genetics
Imperial Cancer Research Fund
London, England

Eleni Economou Zanni
Section of Molecular Genetics
Cardiovascular Institute
Departments of Medicine and Biochemistry
Housman Medical Research Center
Boston University Medical Center
Boston, Massachusetts

Vassilis I. Zannis
Section of Molecular Genetics
Cardiovascular Institute
Departments of Medicine and Biochemistry
Housman Medical Research Center
Boston University Medical Center
Boston, Massachusetts

A Continuation Order Plan is available for this series. A continuation order will bring delivery of each new volume immediately upon publication. Volumes are billed only upon actual shipment. For further information please contact the publisher.

ADVANCES IN HUMAN GENETICS **21**

Edited by

Harry Harris

Harnwell Professor of Human Genetics
University of Pennsylvania, Philadelphia

and

Kurt Hirschhorn

Herbert H. Lehman Professor and Chairman of Pediatrics
Mount Sinai School of Medicine of The City University of New York

SPRINGER SCIENCE+BUSINESS MEDIA, LLC

The Library of Congress catalogued the first volume of this title as follows:

Advances in human genetics. 1—
  New York, Plenum Press, 1970—
    (1) v. illus. 24-cm.
    Editors: V. 1— H. Harris and K. Hirschhorn.
  1. Human genetics—Collected works. I. Harris, Harry, ed. II. Hirschhorn, Kurt, 1926—
joint ed.
QH431.A1A32   573.2'1   77-84583

ISBN 978-1-4613-6311-8   ISBN 978-1-4615-3010-7 (eBook)
DOI 10.1007/978-1-4615-3010-7

© 1993 Springer Science+Business Media New York
Originally published by Plenum Press in 1993
Softcover reprint of the hardcover 1st edition 1993

ARTICLES PLANNED FOR FUTURE VOLUMES

CONTENTS OF EARLIER VOLUMES

Preface to Volume 1

During the last few years the science of human genetics has been expanding almost explosively. Original papers dealing with different aspects of the subject are appearing at an increasingly rapid rate in a very wide range of journals, and it becomes more and more difficult for the geneticist and virtually impossible for the nongeneticist to keep track of the developments. Furthermore, new observations and discoveries relevant to an overall understanding of the subject result from investigations using very diverse techniques and methodologies and originating in a variety of different disciplines. Thus, investigations in such various fields as enzymology, immunology, protein chemistry, cytology, pediatrics, neurology, internal medicine, anthropology, and mathematical and statistical genetics, to name but a few, have each contributed results and ideas of general significance to the study of human genetics. Not surprisingly it is often difficult for workers in one branch of the subject to assess and assimilate findings made in another. This can be a serious limiting factor on the rate of progress.

Thus, there appears to be a real need for critical review which summarizes the positions reached in different areas, and it is hoped that *Advances in Human Genetics* will help to meet this requirement.

Each of the contributors has been asked to write an account of the position that has been reached in the investigations of a specific topic in one of the branches of human genetics. The reviews are intended to be critical and to deal with the topic in depth from the writer's own point of view. It is hoped that the articles will provide workers in other branches of the subject, and in related disciplines, with a detailed account of the results so far obtained in the particular area, and help them to assess the relevance of these discoveries to aspects of their own work, as well as to the science as a whole. The reviews are also intended to give the reader some idea of the nature of the technical and methodological problems involved, and to indicate new directions stemming from recent advances.

The contributors have not been restricted in the arrangement or organization of their material or in the manner of its presentation, so that the reader should be able to appreciate something of the individuality of approach which goes to make up the subject of human genetics, and which, indeed, gives it much of its fascination.

HARRY HARRIS
The Galton Laboratory
University College London

KURT HIRSCHHORN
Division of Medical Genetics
Department of Pediatrics
Mount Sinai School of Medicine

Preface to Volume 10

This is the tenth volume of *Advances in Human Genetics* and some fifty different reviews covering a very wide range of topics have now appeared. Many of the earlier articles still stand as valuable sources of reference. But the subject continues to move forward at an increasing speed and its vitality is indicated by its remarkable recruitment of young investigators. New areas of research which could hardly have been envisaged only a few years ago have emerged, and quite unexpectedly discoveries have been made in parts of the subject which only recently had come to be thought as fully explored. So there continues to be a need for authoritative and critical reviews intended to keep workers in the various branches of this seemingly ever-expanding subject fully informed about the progress that is being made and also, of course, to provide a ready and accessible account of new developments in human genetics for those whose primary interests are in other fields of biological and medical research.

We see no reason to alter the general policy which was outlined in the preface to the first volume. We believe that it has served our readers well. The subject seems to us to be just as exciting and intellectually stimulating and rewarding as it did when this series was first started. We expect the next decade of research in human genetics to be as innovative and productive as the last and our aim is to record its progress in *Advances in Human Genetics*.

HARRY HARRIS
University of Pennsylvania, Philadelphia

KURT HIRSCHHORN
Mount Sinai School of Medicine of the City University of New York

Note About Addenda

To make the volume as up-to-date as possible, each author was given the opportunity to write a short Addendum at the time he or she received the page proofs of that particular chapter. This allows for any important new material to be presented at the latest possible time in the publication process. The Addenda are presented at the end of the book, beginning on page 443.

Contents

Chapter 3

Genetic Mutations Affecting Human Lipoproteins, Their Receptors, and Their Enzymes

Vassilis I. Zannis, Dimitris Kardassis, and Eleni Economou Zanni

Chapter 4

Genetic Aspects of Cancer

Audrey D. Goddard and Ellen Solomon

Chapter 5

Gaucher Disease: Enzymology, Genetics, and Treatment

Gregory A. Grabowski

Chapter 1

Peroxisomal Diseases

Hugo W. Moser

Kennedy–Krieger Institute
Baltimore, Maryland 21205

HISTORICAL PERSPECTIVE

While knowledge about clinically diverse diseases that are now classified as peroxisomal disorders dates back nearly 70 years, the unifying concept that they could be traced to dysfunction of this subcellular organelle was not developed until 12 years ago. This delay is due in part to the fact that the organelle was not identified as a distinct structure until 1954 (Rhodin, 1954; de Duve and Baudhuin, 1966). It was not until 1960 and later that it was recognized that certain enzymes, namely urate oxidase, catalase, and D-amino acid oxidase, were localized in this organelle. The severe deficits associated with human peroxisomal disorders highlight that it is much more than the "fossil organelle" status proposed by de Duve and Baudhuin (1966), albeit with the caution that "it is difficult not to assume that at least some selective pressure has favored the retention [of the peroxisomes] and therefore that they fulfill a useful function wherever they are found." The disorders that are now assigned to the peroxisome disease category include X-linked adrenoleukodystrophy, first described in 1923 (Siemerling and Creutzfeldt, 1923), acatalasemia (Takahara and Mijamoto, 1948), and the Zellweger cerebrohepatorenal syndrome (Bowen *et al.*, 1964). The diversity of clinical manifestations of these disorders is consistent with the multiplicity of functions that are now known to reside in the peroxisome and readily accounts for the delay in recognizing that they are all related to malfunction of this organelle.

Advances in Human Genetics, Volume 21, edited by Henry Harris and Kurt Hirschhorn. Plenum Press, New York, 1993.

Two sets of observations were particularly important for the development of the concept that there exists a group of human disorders attributable to disturbed peroxisomal functions. First, Goldfischer *et al.* (1973) demonstrated that patients with the Zellweger cerebrohepatorenal syndrome lack demonstrable peroxisomes and indeed referred to a peroxisome disease category. Second, it was shown that mammalian peroxisomes perform functions beyond those originally assigned by de Duve and Baudhuin (1966), namely oxidation of fatty acids (Lazarow, 1978) and the synthesis of plasmalogens (Hajra and Bishop, 1982). Many other functions have been assigned since then. Most or all of the biochemical abnormalities associated with the Zellweger syndrome can be traced to peroxisomal dysfunction. The Zellweger syndrome, with its multiple and fatal abnormalities, thus can be viewed as a dramatic example of the loss of peroxisomal function in humans. The significance of this finding has been highlighted by Opitz (1985), who stated:

> Now, 20 years after the discovery of the Zellweger syndrome, it has undergone one of the most dramatic changes in modern medical history, namely, [reassignment] from that of a true multiple congenital anomaly (i.e., static nonmetabolic malformation) syndrome to the metabolic dysplasia-malformation category. Simply a semantic difference? No, rather a true one of such fundamental importance that the very concept "Zellweger syndrome" has become paradigmatic of a whole new class of genetic disorders.

Since the first reference to the peroxisome disease category by Goldfischer *et al.* (1973) many more peroxisomal functions and disease states have been recognized; the peroxisomal disorders are subdivided into three major categories (Schutgens *et al.*, 1986; H. W. Moser and Goldfischer, 1985; H. W. Moser, 1989; Zellweger, 1987; Wilson *et al.*, 1986; I. Singh *et al.*, 1988). At least 14 peroxisomal disorders have now been identified, and it is likely that this number will increase further. Their combined incidence is estimated to be 1:25,000 or greater. All of these disorders are genetically determined, cause serious disability, and can be identified by noninvasive diagnostic assays. Moreover, all can be identified prenatally, and therapy is becoming available for some.

THE NORMAL PEROXISOME

Peroxisome Structure

In the liver and kidney, peroxisomes appear in transmission electron microscopy as round or oval organelles, bounded by a single membrane, with an average diameter of 500 nm (Hruban and Rechcigl, 1969). Demonstration that this organelle stains with the alkaline 3,3'-diaminobenzidine (DAB) reaction for catalase is an important criterion for histological identification (Roels and Gold-fischer, 1979). Serial section analysis of mouse liver peroxisomes has shown that they are complex elongated tubular structures that vary in diameter and in intensity of DAB staining. They occupy various intracellular positions, and are in close apposition to other cell organelles, such as the endoplasmic reticulum, mito-chondrion, and the Golgi apparatus (Gorgas, 1985), consistent with the close metabolic interrelationships that exist between the peroxisome and the other organelles. Despite the proximity of the peroxisome tubular network and the endoplasmic reticulum, actual connections between these two networks have never been observed (Gorgas, 1985).

Peroxisomes probably are present in all human tissues except the mature red blood cells (Hruban et al., 1972), but show a great deal of variation in respect to size and number. They are prominent in liver and kidney, the intestinal epithelium, yolk sac, adrenal cortex of the pregnant rat, and in the sebaceous gland (Gorgas, 1987), and also in tissues that synthesize lipids. Peroxisomes are not demonstrable in rat liver before 14–15 days of gestation but are abundant by day 19 (Tsukada et al., 1968; Stefanini et al., 1985). Fetal peroxisomes can be increased in number and size by the administration of clofibrate, a peroxisome proliferator, to the mother (Wilson et al., 1991). Nervous system peroxisomes are smaller (140 nm) and fewer than those in the liver. They are most prominent in the first two postnatal weeks in the neurons of the cerebrum, cerebellum, locus coeruleus, and spinal cord, but are much rarer in the neurons of the adult animal. Peroxisomes in oligodendrocytes are most prominent during the period of myelin formation, and may be adjacent to the outer lamellae that form the myelin sheath. They are rare in oligodendrocytes of neonatal or adult animals (Arnold and Holtzman, 1978). Biochemical studies show a similar pattern. The activities of catalase and peroxi-somal acyl-CoA oxidase and oxidation of lignoceric acid in brain reach a peak 10–16 days postnatally and then diminish (Adamo et al., 1986; Lazo et al., 1991). Peroxisomes in normal cultured skin fibroblasts also are small (100 nm in diameter) and relatively rare (1–2 profiles/μm^3) (Beard et al., 1985b).

Peroxisome Functions

By 1981, 40 enzymes had been localized to the peroxisome (Tolbert, 1981) and the number has increased considerably since then. The enzyme composition varies among species, and possibly among tissues within a species, and enzyme activities vary with maturation, metabolic state, and environmental factors. While

many peroxisomal activities, such as fatty acid oxidation and cholesterol biosynthesis, are duplicated in other cellular compartments, at least four reactions
(oxidation of very long-chain fatty acids and of pipecolic acid, synthesis of
plasmalogens, and certain steps in bile acid synthesis) take place exclusively or
nearly so in the peroxisome. These latter reactions are disturbed most strikingly in
the peroxisomal disorders.

Hydrogen Peroxide-Based Cellular Respiration

Peroxisomes contain oxidases that reduce oxygen to hydrogen peroxide, and
a large amount of catalase, which decomposes the hydrogen peroxide (de Duve
and Baudhuin, 1966). This function was recognized early and led to the name of
the organelle. Substrates for the oxidases include D- and L-amino acids, L-α-
hydroxy acids, urate (except in humans), glutaryl-CoA (Vamecq et al., 1985),
oxalate (Beard et al., 1985a), and other compounds that will be discussed.
Catalase in humans and other mammals has been localized to the peroxisome with
certainty and indeed is used as the marker for the organelle. While catalase clearly
has a role in the defense against accumulation of hydrogen peroxide, it is only one
of a cluster of enzymes, including superoxide dismutase and glutathione peroxi-
dase, that protect against activated oxygen species. Probably because of the
multiplicity of these protective mechanisms, acatalasemia is a benign entity in
most instances (Eaton, 1989).

Fatty Acid Oxidation

In 1976 Lazarow and de Duve (1976) demonstrated that rat liver peroxisomes
could oxidize palmitoyl-CoA; in the following year Lazarow showed that acetyl-
CoA was produced and that enzymes of the β-oxidation system were present
(Lazarow, 1978). This observation was of seminal importance to the recognition
that peroxisomes have more than a "fossil" role in mammalian metabolism.

Figure 1 shows the reactions and enzymes involved in peroxisomal β-oxidation. As in the mitochondrion, each cycle shortens the fatty acids by a two-carbon
unit, but unlike in the mitochondrion, the second step is catalyzed by an oxidase
rather than a dehydrogenase. All of the peroxisomal β-oxidation enzymes have
been purified (Hashimoto, 1982) and shown to be entirely distinct from their
mitochondrial counterparts, and the complete cDNA sequence for the rat liver
enzymes has been determined. Defects involving each of these four enzymes have
now been linked to four distinct human diseases.

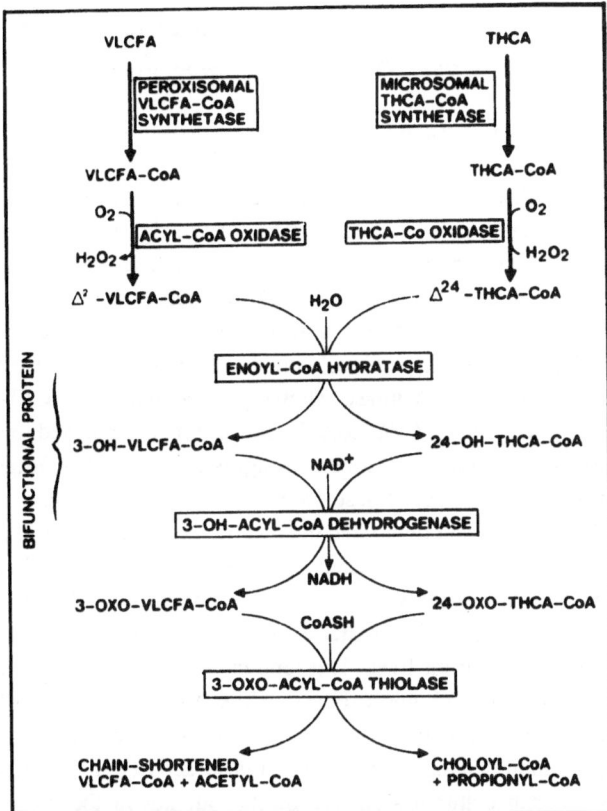

Fig. 1. The activation and subsequent β-oxidation of very long-chain fatty acids and trihydroxy-cholestanoic acid in peroxisomes. (From Wanders *et al.*, 1990c. Used by permission.)

Long-Chain Acyl-CoA Ligase

Long-chain acyl-CoA ligase has received a great deal of recent study because a closely related enzyme appears to be involved in X-linked adrenoleukodystrophy (ALD) (see p. 61). Tomoda *et al.* (1991) have proposed that the enzyme has an essential role in proliferation of animal cells. Initially it was thought that there was a single long-chain (C10–C18) acyl-CoA ligase in microsomes, mitochondria, and peroxisomes of rat liver (S. Miyazawa *et al.*, 1985). The nucleotide sequence that encodes this enzyme has been determined (H. Suzuki *et al.*, 1990). The peroxisomal enzyme has been localized to the cytoplasmic side of the peroxisomal membrane (Mannaerts *et al.*, 1982), unlike most other peroxisomal enzymes, which are located in the matrix. The human long-chain acyl-CoA ligase has recently been mapped to chromosome 3 (Stanczak *et al.*, 1991).

A Separate Lignoceroyl-CoA Ligase

Study of X-linked ALD (see p. 61) led to a series of investigations in three laboratories that establish the existence of a separate ligase for *very long-chain fatty acids* (VLCFA), referred to as lignoceroyl-CoA ligase, and it is this enzyme that is thought to be deficient in X-linked ALD (Wanders *et al.*, 1988c; Lazo *et al.*, 1988). This enzyme is most active toward C24 (lignoceric) and C26 (hexacosanoic) fatty acids, whereas the previously cited *long-chain ligase*, also referred to as palmitoyl- or stearoyl-CoA ligase, shows peak activity with C10–C18 fatty acids. Since the postulated VLCFA-CoA ligase has not been purified, evidence for the existence of separate enzymes, albeit strong, is circumstantial. Bhushan *et al.* (1986) resolved palmitoyl-CoA ligase from lignoceroyl-CoA ligase activity by column chromatography and found that the two fractions had different properties, but Nagamatsu *et al.* (1986) suggested that these distinctions could be due to differences in the aggregation state of the same enzyme. H. Singh and Poulos (1988) favored the existence of distinct enzymes because of differences in their subcellular distribution in rat liver. The long-chain ligase was present in mitochondria, peroxisomes, and microsomes, while VLCFA ligase was virtually absent in purified mitochondria. Lazo *et al.* (1990) found that an antibody to palmitoyl-ligase inhibited lignoceroyl-CoA ligase activity in the mitochondria isolated from human cultured skin fibroblasts, but not in the peroxisomes or endoplasmic reticulum. Lageweg *et al.* (1991b) reported more than a tenfold difference in the inhibition of LCFA and VLCFA ligase activity produced by 1-pyrenedecanoic acid, and cited this as additional evidence that the two enzymes are distinct. I. Singh *et al.* (1984a) had demonstrated that lignoceric acid was oxidized mainly, and perhaps exclusively, in the peroxisome.

The preponderance of evidence thus is that VLCFA and LCFA ligase are distinct enzymes, and that in liver and cultured skin fibroblasts the VLCFA are oxidized mainly in the peroxisome. Findings in brain are less clear, probably because of the difficulty of separating subcellular organelles in that tissue. Poulos and associates concluded that long-chain and very long-chain ligase activity are present in both rat brain peroxisomes and mitochondria (H. Singh *et al.*, 1990). The Charleston group, using a different fractionation procedure, concluded that the subcellular distribution in brain was similar to that in liver and fibroblasts, with the VLCFA oxidation being concentrated in peroxisomes and the long-chain ligase in mitochondria (Lazo *et al.*, 1991). Resolution of this discrepancy must await the development of better procedures for the separation of brain mito-chondria and peroxisomes.

There is dispute about the topographical localization of lignoceroyl-CoA ligase. Lazo *et al.* (1990) localize it to the luminal surface of the peroxisome, while Lageweg *et al.* (1991a) assign it to the cytoplasmic surface. This discrepancy, which has important implications in respect to physiological function and disease mechanism, may be related to the method in which the peroxisomal membrane was disrupted. Lageweg *et al.* (1991a) used ultrasonic disruption, whereas Lazo *et al.* (1991) used the detergent Triton X-100.

Acyl-CoA Oxidase

Rat liver acyl-CoA oxidase is a flavoprotein with a molecular weight of 139,000. It contains three polypeptide components, with subunits of molecular weight 71,900, 51,700, and 20,500. The smaller units result from the cleavage of the larger unit. The complete nucleotide sequence of the cDNA and the predicted amino acid sequence have been determined (W. Miyazawa *et al.*, 1987). Two species of acyl-CoA oxidase cDNA were identified. They differed in their coding sequence only within a small region, contain the same number of nucleotides, and can be translated in a common reading frame. Based upon their prediction of the properties of these two species, Schepers *et al.* (1990) isolated two species of rat liver acyl-CoA oxidase and made the interesting observation that one of them is induced by peroxisome proliferators (see p. 20), while the other is not.

In common with most peroxisomal proteins, acyl-CoA oxidase is synthesized in its mature form. It contains the carboxyl-terminal serine–lysine–leucine target-ing sequence (S. Miyazawa *et al.*, 1989), which is found in many peroxisomal matrix enzymes (Gould *et al.*, 1990). Acyl-CoA oxidase is most active toward saturated and unsaturated fatty acids with a 12- to 18-carbon chain length. The oxidase for bile acid intermediates, such as trihydroxycoprostanoyl-CoA, is a separate enzyme (Schepers *et al.*, 1990).

Bifunctional Enzyme

In the peroxisomal β-oxidation system, the hydration (enoyl-CoA hydratase) and the second dehydrogenation step (3-hydroxyacyl-CoA dehydrogenase) are catalyzed by a single enzyme, molecular weight 78,000, referred to as bifunc-tional enzyme. Its nucleotide sequence and gene structure have been determined (Ishii *et al.*, 1987). Comparison with the mitochondrial enzymes showed that the carboxy-terminal side of the peroxisomal bifunctional protein had sequence homology with the mitochondrial 3-hydroxyacyl-CoA dehydrogenase, while the amino-terminal side showed homology with mitochondrial enoyl-CoA hydratase, pointing to a common evolutionary origin between the peroxisomal and mitochondrial enzymes. Of interest is the recent demonstration that peroxisomal bifunctional enzyme also has Δ3, Δ2-enoyl-CoA isomerase activity (Palosaari and Hiltunen, 1990), so that the enzyme is in fact *trifunctional*. The isomerase is required for the oxidation of unsaturated fatty acids. The isomerase activity has been localized to the amino-terminal side of the trifunctional protein in proximity to the hydratase activity and has been found to have amino acid sequence similarities to the mitochondrial short-chain Δ3, Δ2-enoyl-CoA isomerase (Palosaari and Hiltunen, 1990). As discussed on p. 8, peroxisomes also play an important role in the metabolism of mono- and polyunsaturated fatty acids.

3-Ketoacyl-CoA Thiolase

Peroxisomal thiolase has a molecular weight of 80,000 and exists as a dimer. It is most active toward substrates with 6–12 carbon atoms (Hashimoto, 1982), and differs from the mitochondrial thiolases in molecular weight and substrate specificity. The gene structure for the rat liver enzyme has been determined (Hijikata *et al.*, 1990). Unlike most other peroxisomal enzymes, thiolase is synthesized as a precursor form, which is converted to the mature form by cleavage of a basic 26-amino-terminal leader sequence, which resembles those present in mitochondrial enzymes. The mature enzyme is composed of 398 amino acids and has a 34% homology with the mitochondrial thiolase (Hijikata *et al.*, 1987). Analysis of the gene structure gave the surprising result that there were two genes with 97% homology, but with the intriguing difference that one of the genes is activated by peroxisome proliferators while the other is not (Hijikata *et al.*, 1990). The complete cDNA of human peroxisomal thiolase has been determined (Bout *et al.*, 1988).

Oxidation of Unsaturated Fatty Acids

Recent studies have modified concepts about the pathway and enzymes involved in the oxidation of mono- and polyunsaturated fatty acids (Schulz and Kunau, 1987). In addition to the β-oxidation enzymes shown in Figure 1, the unsaturated fatty acids require three auxiliary enzymes, namely Δ3 *cis*-Δ2 *trans*-enoyl-CoA isomerase (EC 5.3.3.8), 2,4-dienoyl-CoA reductase (EC 1.3.1.34), and 3-hydroxyacyl-CoA epimerase (EC 5.1.2.3). As discussed above, the peroxisomal bifunctional enzyme also has Δ3 *cis*-Δ2 *trans* isomerase activity, but the bulk of this activity appears to be associated with another enzyme present both in peroxisomes and mitochondria, which is also induced by peroxisome proliferators (Tomioka *et al.*, 1991). The 2,4-dienoyl-CoA reductase is present in rat liver peroxisomes and has been shown to have a role in the peroxisomal oxidation of polyunsaturated fatty acids in isolated perfused rat liver (Hiltunen *et al.*, 1986). The capacity to shorten linolenic and docosohexanoic acid chains in rat liver peroxisomes is at least equal to that in mitochondria (Hiltunen *et al.*, 1986). The physiological role of the peroxisome in the oxidation of polyunsaturated fatty acids is supported further by the observation that these substances accumulate in the disorders of peroxisome biogenesis.

Prostaglandin Degradation

The chain-shortening reactions of prostaglandin F2 (Diczfalusy and Alexson, 1988) and prostaglandin E2 (Schepers *et al.*, 1986) take place in the peroxisome. These reactions, which are analogous to fatty acid β-oxidation, are part of the process of inactivating these biologically active compounds into excretable compounds. The pattern of metabolites in peroxisomal fractions suggested that other reactions that involve prostaglandins may also take place in this organelle (Schepers *et al.*, 1986). Prostaglandin metabolism has not been studied in detail in patients with peroxisomal disorders. Tiffany *et al.* (1991) found that basal and interleukin 1-stimulated synthesis of prostaglandin E2 was increased in cultured skin fibroblasts of patients with peroxisomal disorders.

Formation of Bile Acids

The formation of bile acids from cholesterol involves shortening the cholesterol side chain by a β-oxidation pathway analogous to that in the fatty acid oxidation. Shortening of the side chain takes place in the peroxisome and, as discussed later, this leads to biochemical alterations that are of diagnostic and pathogenetic importance in patients with peroxisomal disorders. Conversion of cholesterol to bile acids in most instances involves first the formation of $3\alpha,7\alpha,12\alpha$-trihydroxy-5β-cholestanoic acid (THCA), or of $3\alpha,7\alpha$-dihydroxy-5β-cholestanoic acid (DHCA). Pedersen and Gustafsson (1980) showed that this reaction took place in rat liver peroxisomes. Kase *et al.* (1986) showed that THCA was converted to cholic acid in liver peroxisomes, and that the conversion of

DHCA to chenodeoxycholic acid also took place in the peroxisome and that it was deficient in patients with Zellweger syndrome (Kase *et al.*, 1991). Shortening of the side chain can also take place before the hydroxylation of the cholesterol ring structure. In this reaction 26-hydrocholesterol is converted to 3β-hydroxy-5-cholenoic acid. This reaction also takes place in the peroxisome (Krisans *et al.*, 1985). The enzymes involved in the shortening of cholesterol side chains are similar to those in fatty acid oxidation. It has been shown, however, that the peroxisomal oxidase for THCA-CoA is separate from that for fatty acids (Casteels *et al.*, 1990), and these enzymes are affected differently in human disease states. An NADPH cytochrome P-450 that catalyzes the hydroxylation of the C26/27 carbon of cholesterol in the bile acid synthesis pathway has also been localized to the peroxisome (Gutierrez *et al.*, 1988). Peroxisomes are also involved in the formation of taurine conjugates of bile acids (Kase and Bjorkheim, 1989).

Cholesterol Biosynthesis

Recent studies, mainly from Dr. Skraidrite Krisans' laboratory, have yielded the surprising result that peroxisomes synthesize cholesterol, in apparent duplication of the well-known and predominant microsomal pathway. Peroxisomes contain 3-hydroxy-3-methylglutaryl-coenzyme A reductase (Keller *et al.*, 1986), the key regulatory enzyme for cholesterol biosynthesis that converts 3-hydroxy-3-methylglutaryl-CoA to mevalonic acid; the organelles can synthesize cholesterol from mevalonic acid (Thompson *et al.*, 1987), and also catalyze the initial step in cholesterol synthesis, the condensation of acetyl-CoA units into acetoacetyl-CoA (Thompson and Krisans, 1990). Sterol carrier protein is also present in high concentration in liver peroxisomes (van Amerongen *et al.*, 1989). The combined data provide strong evidence that peroxisomes play a role in cholesterol synthesis. The biological significance of this newly identified pathway is not known. In the rat, the peroxisomal cholesterol synthesis shows a pattern of diurnal variation that differs from that in the microsome (Rusnak and Krisans, 1987), suggesting that the pathways have distinct functions. The implications of this pathway for peroxisomal disorders have not yet been explored. It is of interest, however, that patients with disorders of peroxisome biogenesis have low levels of cholesterol in plasma.

Plasmalogen Biosynthesis

Plasmalogens belong to the class of ether-linked phospholipids. As shown in Fig. 2, these lipids differ from the more common ester-linked lipids in that the

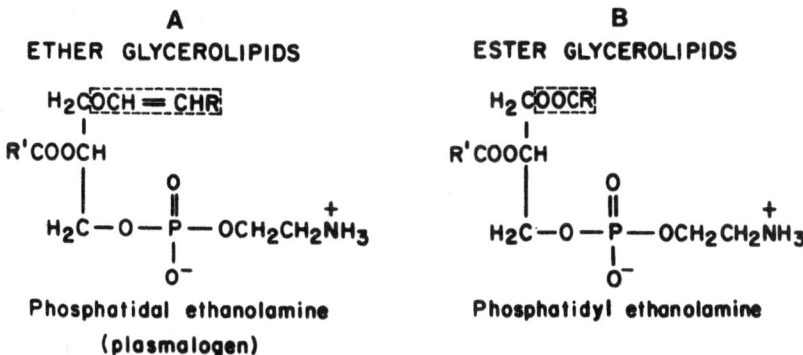

Fig. 2. Phospholipid structures. (A) Plasmalogen, (B) phosphatidyl ethanolamine. A is an ether lipid, whereas B is an ester-linked phospholipid. The plasmalogens (A) are distinguished from other ether lipids by the double bond between the first two carbons of the long-chain alcohol. R and R' represent the long tails of the fatty acids or fatty alcohols.

first carbon atom of the glycerol backbone is linked to a long-chain alcohol through a vinyl ether linkage. Ester lipids are referred to as plasmalogens when the long-chain alcohol is unsaturated in the 1–2 position. Plasmalogens constitute 5–20% of the phospholipids in most mammalian cell membranes (Snyder, 1972). They are especially abundant in nervous tissue; indeed, one-third of myelin phospholipids are plasmalogens (Norton and Autilio, 1966). Apart from a structural role, plasmalogens also appear to protect animal cells against photosensitized killing (Zoeller et al., 1988).

Plasmalogens are synthesized by a complex series of reactions (Hajra and Bishop, 1982) (Fig. 3). Plasmalogen synthesis begins with dihydroxyacetone phosphate (DHAP), which is acylated by DHAP acyltransferase. The acyl group is then replaced by a long-chain alcohol catalyzed by alkyl-DHAP synthase. Both of these enzymes are located (Hajra and Bishop, 1982) on the inner surface of the peroxisomal membrane (Hardeman and van den Bosch, 1989). The third enzyme is acyl/alkyl dihydroxyacetone-phosphate reductase, which reduces 1-alkyl-DHAP to 1-alkyl glycerol 3-phosphate. This enzyme, which has been purified recently (Datta et al., 1990) is located both in the peroxisome and the microsomes, and all subsequent steps of ether lipid synthesis are microsomal.

The participation of two organelles has permitted the development of a double-substrate/double-isotope method that determines the ratio of the activities of the peroxisomal and microsomal components of the pathway and has proven of value in the study of peroxisomal disorders (Roscher et al., 1985).

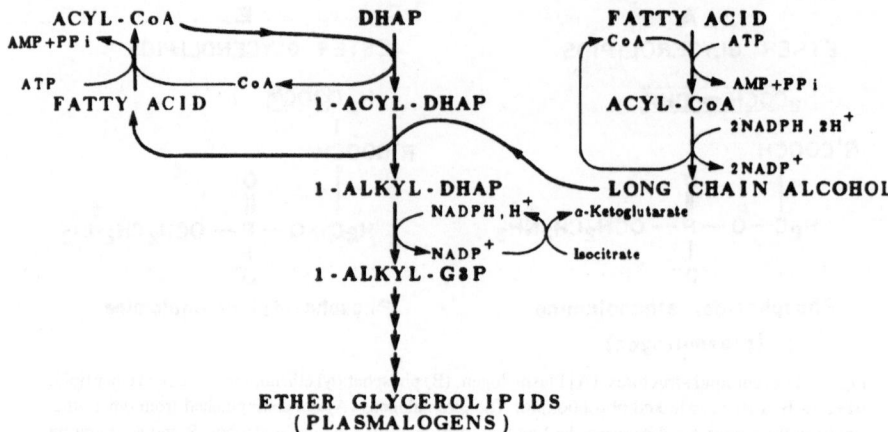

Fig. 3. The peroxisomal steps of plasmalogen synthesis. See text for discussion. (From Lazarow and Moser, 1989. Used by permission.)

The participation of two subcellular organelles in plasmalogen synthesis also requires that intermediate substrates can pass freely from the peroxisome to the microsome. Van Veldhoven *et al.* (1987) showed that isolated peroxisomes are permeable to molecules with molecular weight up to 800, and they concluded that the 22-kD integral membrane protein formed pores through which these molecules could pass. Recently, Wolvetang *et al.* (1990) used a model in which they exposed cultured skin fibroblasts to low levels of digitonin (20 μg/ml), which permeabilized the plasma membrane but left the integrity of intracellular membranes intact. In this system, unlike all previous studies with isolated peroxisomes, they did demonstrate latency of acyl-CoA acetyltransferase. The latency could be overcome by the addition of ATP. These studies are beginning to provide insight about the role of peroxisomal membranes in the transport of substrates and proteins, a topic that will be discussed in more detail in reference to the pathogenesis of the disorders of peroxisome biogenesis.

Amino Acid Metabolism

D-Amino Acid Oxidase. D-amino acid oxidase, along with uricase, catalase, and other oxidases, was among the original group of enzymes shown by de Duve and Baudhuin (1966) to be associated with the peroxisome. The enzyme is a flavoprotein with absolute stereospecificity for D-amino acids but is active with a broad range of these acids, with greatest activity toward D-proline. It is widely distributed in vertebrate species, with highest activity in kidney. Its physiological role is not known.

Alanine:Glyoxylate Aminotransferase. The subcellular localization of alanine:glyoxylate aminotransferase (AGT) is species-dependent. In human, rabbit, guinea pig, and macaque the enzyme is peroxisomal (Noguchi and Takada, 1979), while in cat and dog it is mitochondrial, and in rat, mouse, and hamster significant activities are present in both organelles (Noguchi *et al.*, 1978). The cDNA sequence of AGT has been determined and the subcellular localization appears to be related to the targeting signal at the amino-terminal region. The rat enzyme has a mitochondrial targeting signal that is not expressed in the human enzyme (Takada *et al.*, 1990). It is of great interest that alterations in this mitochondrial targeting signal occur in some patients with hyperoxaluria type 1 (see p. 75), causing the enzyme to be targeted to the mitochondrion instead of the peroxisome (Purdue *et al.*, 1990).

The metabolic role of AGT is shown in Figure 4. AGT catalyzes the transamination of glyoxylate to glycine, with alanine serving as the amino group

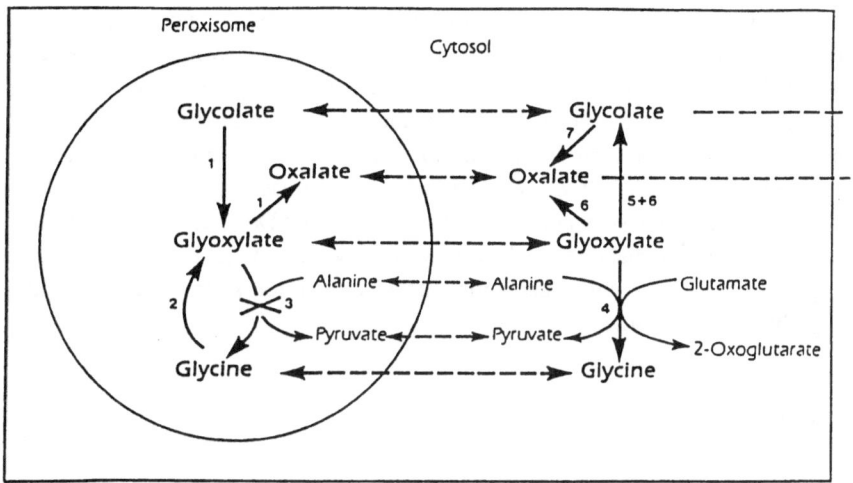

Fig. 4. Some of the main pathways involved in the metabolism of glyoxylate in human liver. Solid lines, metabolic pathways; borken lines, diffusion pathways; X, metabolic block in primary hyper-oxaluria type 1. The enzymes catalyzing the reactions are: 1, glycolate oxidase (Lα-hydroxyacid oxidase); 2, glycine oxidase (D-amino acid oxidase); 3, alanine:glyoxylate aminotransferase (serine: pyruvate aminotransferase); 4, glutamate:glyoxylate aminotransferase (alanine:2-oxoglutarate amino-transferase); 5, glyoxylate reductase (D-glycerate dehydrogenase); 6, lactate dehydrogenase; 7, glycolate dehydrogenase. (From Danpure, 1989. Used by permission.)

donor. The physiological role of this reaction has been reviewed (Danpure, 1989; Hillman, 1989). Glyoxylic acid has a key role in plants, where it is part of the glyoxylate cycle (Hogg, 1991) that leads to production of sugars. The glyoxylate cycle does not exist in humans and other mammals, and the main sources of glyoxylic acid are uncertain (Hillman, 1989), although glycine and ascorbic acid have been implicated. In hyperoxaluria type 1 (see p. 75) the defective AGT leads to conversion of glyoxylate to oxalic acid, which is produced in large quantities. It seems unlikely that there would be enough ascorbic acid to provide large quantities of the oxalic acid precursor, and while glycine could theoretically serve as a major source, the flux of the reaction appears to be in the direction of glycine (Hillman, 1989). Nevertheless, the severity of the metabolic disturbance associated with hyperoxaluria type 1 indicates the importance of this pathway.

Pipecolic Acid Oxidase. L-Pipecolic acid is an intermediate in one of the lysine degradation pathways in animals. The major pathway, which begins with the formation of saccharopine (e-N-glutaryl-2)-L-lysine, does not involve pipecolic acid, but it is an intermediate in a second pathway that appears to be the main route for lysine degradation in brain (Chang, 1982). L-Pipecolic acid is oxidized by L-pipecolic acid oxidase. Similar to AGT, the subcellular localization of L-pipecolic acid oxidase is species-dependent: In humans and monkeys the enzyme is peroxisomal, while in rabbits it is mitochondrial (Mihalik and Rhead, 1989). The oxidase, which contains flavin, catalyzes the conversion of pipecolic acid to α-aminoadipic acid with the concomitant evolution of H_2O_2 (Fig. 5). The enzyme has been purified and has a molecular weight of 46,000. An unusual feature is that flavin is bound covalently, a finding that has so far been observed in only five other enzymes in eukaryotes (Edmondson and De Francesco, 1991), and in no other peroxisomal enzyme. The enzyme shows no activity toward D-alanine, D-pipecolic acid, or any other amino acids except for slight reactivity with L-proline and sarcosine.

Purine Catabolism

Urate oxidase is a peroxisomal enzyme in the liver of most vertebrates, including several genera of New and Old world monkeys, but it is absent in hominoids and humans (Friedman *et al.*, 1985). Curiously, the enzyme is found exclusively within the "core" structure that is present in the liver peroxisomes of species other than humans and hominoids, and is not distributed throughout the remainder of the matrix as is catalase (Volkl *et al.*, 1988). Humans lack this peroxisomal enzyme, and hence the capacity to degrade uric acid.

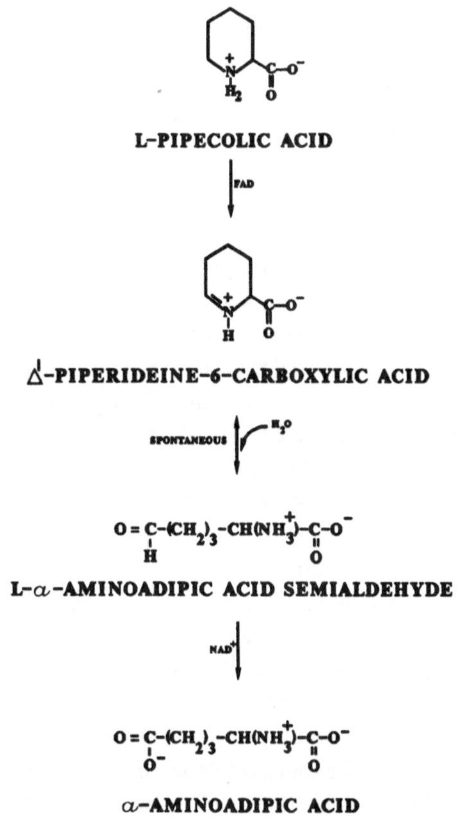

L-PIPECOLIC ACID

Δ^1**-PIPERIDEINE-6-CARBOXYLIC ACID**

L-α-AMINOADIPIC ACID SEMIALDEHYDE

$$O = C-(CH_2)_3-CH(NH_3^+)-C-O^-$$

α**-AMINOADIPIC ACID**

Fig. 5. The metabolism of pipecolic acid. L-Pipecolic oxidase, the enzyme that is deficient in the disorders of peroxisome biogenesis, catalyzes the first reaction. (From Mihalik, 1991. Used by permission.)

Polyamine Catabolism

The enzyme polyamine oxidase degrades spermine and spermidine, with putrescine, 3-aminobenzaldehyde, and H_2O_2 as end products (Holtta, 1977). This enzyme has been localized to the peroxisome (Beard *et al.*, 1985a).

Ethanol Oxidation

While most ethanol is oxidized by the cytosolic alcohol dehydrogenase, 5–25% of alcohol clearance is insensitive to inhibitors of alcohol dehydrogenase, and part of this remaining oxidation is due to the peroxisomal peroxidatic action of catalase (Khanna and Israel, 1980; Oshino *et al.*, 1973).

Peroxisomal Membrane Proteins

The study of peroxisomal membrane proteins (PMP) is of key importance to understanding the disorders of peroxisome biogenesis, since, as will be discussed later, these disorders appear to represent a failure of the mechanisms that normally import proteins into the peroxisome, and this import process clearly must involve the peroxisomal membrane. PMP may be prepared by osmotic shock (Huang and Beaver, 1973) or by a frequently used procedure in which peroxisomes are treated with alkaline sodium carbonate at an alkaline pH (Fujiki *et al.*, 1982). Proteins with molecular masses of 22, 26, 27, 41, 68, and 70 kD have been identified (Hashimoto *et al.*, 1986), with the 22-, 68-, and 70-kD PMPs being most prominent. Imanaka *et al.* (1991) have recently identified a novel 57-kD peroxisomal integral membrane protein that is located on the cytoplasmic surface and is induced by peroxisome proliferators.

In an important recent advance, Kamijo *et al.* (1990) determined the complete nucleotide cDNA sequence of the 70-kD PMP. They made the intriguing observation that this protein belongs to the superfamily of certain ATP-binding proteins (Higgins *et al.*, 1988), which are involved in bacterial transport systems (Ames, 1986), those that confer multiple-drug resistance (Gros *et al.*, 1986), and also appear to be involved in the transport defect in cystic fibrosis (Cutting *et al.*, 1990; Kerem *et al.*, 1990). This conclusion is based upon sequence homology between the carboxy-terminal portion of the 70-kD PMP and these ATP-binding proteins. Figure 6 shows a hypothetical model of the 70-kD PMP. Based upon the

Matrix

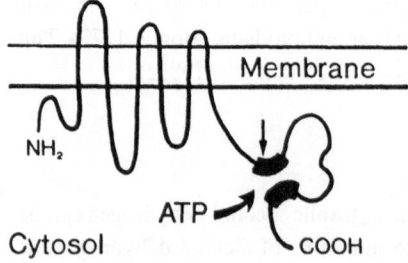

Fig. 6. Hypothetical model for the topology of PMP70 in the peroxisomal membrane. Consensus sequences for the ATP-binding fold are indicated by solid boxes. (From Kamijo *et al.*, Used by permission.)

hydropathy profile and analogy with other members of the ATP-binding super-family, there are six transmembrane segments and the hydrophilic carboxy-terminal portion is on the cytosolic surface, with the ATP-binding consensus sequence extending for 200–250 residues on the carboxy side, and the membrane-anchoring domain forms channels for ATP-dependent import. This model pro-poses that the 70-kD PMP is part of the import machinery for peroxisomal proteins, which, as already noted, is of such great importance to the understand-ing of peroxisomal disorders. While much work remains to be done, several other observations are consistent with this general concept. Van Veldhoven *et al.* (1987) had concluded that the permeability of peroxisomal membranes to small solutes was based on the presence of a nonselective pore-forming protein, and on the basis of reconstitution experiments with liposomes suggested that this depended on the 22-kD PMP. Imanaka *et al.* (1987) showed that translocation of acyl-CoA oxidase into the peroxisome required ATP. Wolvetang *et al.* (1990) noted recently that ATP was required to overcome the latency of DHAP acyltransferase activity in cultured skin fibroblasts that had been treated with low concentrations of digitonin that disrupted the plasma membrane, but left intracellular membranes intact. Since the DHAP acyltransferase is located on the matrix side of the peroxisomal membrane, this finding is compatible with an ATP-linked translocation of the substrate across the peroxisomal membrane. Nevertheless we note that most of the ATP-binding proteins reported so far have functioned in the transport of small molecules. There is, however, already one report of such a transport that involved the bacterial protein hemolysin A (Wandersman and Delepelaire, 1990).

Biogenesis of Peroxisomes

Concepts about the biogenesis of peroxisomes have undergone an interesting and controversial evolution, reviewed in Lazarow and Fujiki (1985). The early concept that peroxisomes were formed by budding of the endoplasmic reticulum was based on the biochemical finding that newly synthesized catalase (a peroxi-somal enzyme) was located in the microsomal fraction (Higashi and Peters, 1963) and on the demonstration of connections between peroxisomes and endoplasmic reticulum in electron micrographs (Novikoff and Shin, 1964). As discussed in Lazarow and Fujiki (1985), the current consensus is that both studies had technical limitations. Later investigations in several laboratories, particularly those of de Duve and Lazarow, are based upon experiments with several peroxisomal proteins, including catalase, the peroxisomal membrane proteins, and the fatty acid β-oxidation enzymes. These have led to a different theory. It was shown that the mRNA that encodes catalase is found in the free polyribosome fraction

(Goldman and Blobel, 1978). The cell-free translation product of the enzyme was identical in size to the mature enzyme and indistinguishable from it by one- and two-dimensional peptide mapping (Robbi and Lazarow, 1982). The newly synthesized apomonomer of catalase appears first in the cytosol, but is transferred into preexisting peroxisomes within 14 min (Lazarow and de Duve, 1973). Addition of heme and polymerization of catalase takes place mainly within the peroxisome. Analogous results have been obtained for peroxisomal membrane proteins (Fujiki *et al.*, 1984) and β-oxidation enzymes (Rachubinski *et al.*, 1984). These data have led to the scheme shown in Figure 7. According to this scheme, peroxisomal proteins are synthesized on free polyribosomes, enter the cytosol, and are transferred posttranslationally into preexisting peroxisomes. Peroxisomes then subdivide by fission or budding. The majority of electron microscopic studies are consonant with this scheme. As reviewed by Lazarow and Fujiki (1985), detailed studies in several laboratories have failed to confirm the earlier reports of continuities between peroxisomes and endoplasmic reticulum. The formation of peroxisomal proteins on free polyribosomes and their posttranslational entry into the organelle places peroxisomes in the same general category as mitochondria and chloroplasts, and has led to the interesting speculation that they may also be descendants of ancient endosymbionts (de Duve, 1983).

Targeting Mechanism for Peroxisomal Proteins

Except for 3-ketoacyl-CoA thiolase (Fujiki *et al.*, 1985), all of the peroxisomal proteins appear to be synthesized in their mature form and hence the topogenic information must be located in the mature polypeptide sequence. The topogenesis of peroxisomal enzymes has been reviewed recently (Osumi and

Posttranslational Import
Growth and Division

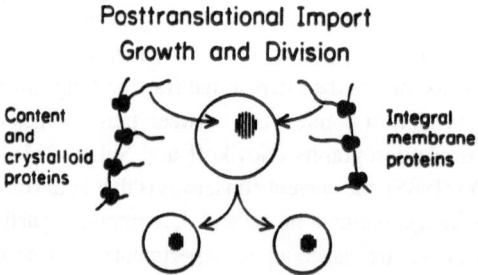

Content and crystalloid proteins

Integral membrane proteins

Fig. 7. Current scheme of peroxisome biogenesis. (From Lazarow and Fujiki, 1985. Used by permission.)

Fujiki, 1990). For many, but not all, matrix enzymes the topogenic signal has been shown to reside in a serine–lysine–leucine (SKL) amino acid sequence attached to the carboxy terminal. The first lead to a topogenic sequence was provided in studies with firefly luciferase conducted by Keller *et al.* (1987). These investigators had shown previously that in the firefly this enzyme was localized in a peroxisomelike organelle. They introduced cloned luciferase cDNA, placed under the control of a viral promoter, into CV-1 monkey kidney cells in culture, and found that the expressed luciferase colocalized with catalase and that it gave a punctate staining pattern by double immunofluorescence microscopy. Transport into the peroxisome was abolished when the 12–20 carboxy-terminal amino acids were deleted. Conversely, if this sequence was attached to two proteins that are not normally found in peroxisomes (mouse dihydrofolate reductase and *Escherichia coli* chloramphenicol acetyltransferase), these enzymes were then imported into the peroxisome (Gould *et al.*, 1987). Parallel studies of peroxisomal protein import were conducted with peroxisomal acyl-CoA oxidase by S. Miyazawa *et al.* (1989). These investigators found that the targeting signal was contained in the five carboxy-terminal amino acids of this enzyme. Comparison of the structure of these two enzymes revealed that they shared the carboxy-terminal SKL sequence and this same sequence was also present in three other peroxisomal proteins (Gould *et al.*, 1988). Gould prepared an antibody to a synthetic peptide that contained the carboxy-terminal SKL sequence, and showed that this antibody recognized 15–20 rat liver peroxisomal proteins, but reacted with only a few proteins from other subcellular compartments (Gould *et al.*, 1990). Substitution of any amino acid for the terminal leucine abolished the topogenic signal. The serine residue could be replaced by cysteine or alanine, and the lysine residue by histidine or arginine, both without loss of import activity. So far the SKL carboxy-terminal sequence has been identified only in matrix enzymes and it has not been found in peroxisomal membrane enzymes (Gould *et al.*, 1990). This finding is of great interest for the peroxisomal disorders, where the import defect affects proteins in the matrix, but not those in the membranes (see p. 44).

Not all matrix proteins contain the carboxy-terminal SKL sequence. Indeed it is lacking in catalase (Gould *et al.*, 1988). The topogenic signal for 3-ketoacyl-CoA thiolase probably resides in the positively charged 26-leader sequence at the amino-terminal end (Osumi and Fujiki, 1990). A cleavable amino-terminal topogenic signal has been identified for glyoxosomal malate dehydrogenase (Gietl, 1990). The acyl-CoA oxidase in yeast contains two internal regions that are necessary and sufficient for targeting (Small *et al.*, 1988b). Thus, while the carboxy-terminal SKL sequence is of great significance, other topogenic sequences that have not yet been identified undoubtedly exist, and it is likely that they will be important in the pathogenesis of peroxisomal disorders.

The mechanisms by which proteins and substrates are transported into the peroxisome have not yet been defined. Imanaka *et al*. (1987) showed that the import of rat acyl-CoA oxidase requires ATP, but was unaffected by ionophores and by the uncoupler carbonylcyanide-*m*-chlorophenylhydrazone (Imanaka *et al*., 1987). We have already referred to the recent important observation that the 70-kD PMP appears to be a member of the ATP-binding protein superfamily (Kamijo *et al*., 1990). This field is under active investigation, and study of human mutants (represented by the peroxisomal disorders) as well as those in yeast (Hansen and Roggenkamp, 1989) and mutant cell lines (Zoeller *et al*., 1989) will advance knowledge about the normal processes.

Induction of Peroxisome Proliferation

Hess *et al*. (1965) were the first to show that hypolipemic agents such as clofibrate and industrial phthalate plasticizers resulted in proliferation of liver peroxisomes. This proliferation accounts for the striking liver enlargement produced by these agents in rodents. The administration of peroxisome proliferators has facilitated the study of peroxisome function: The initial demonstration that peroxisomes played a role in fatty acid β-oxidation was conducted in clofibrate-treated rats (Lazarow, 1978). The phenomenon is also of public health significance, since in rodents peroxisome proliferation is associated with hepatocarcinogenesis (Reddy, 1990). Induction of proliferation of liver peroxisomes is associated with a striking 20- to 30-fold increase in the activity of the fatty acid β-oxidation system. Other peroxisomal enzymes, such as catalase, are induced to a much lesser extent (twofold). The increased activity of the β-oxidation enzymes is due to the rapid (within 1 hr) coordinated increase of the transcription of the genes for peroxisomal acyl-CoA oxidase and bifunctional enzyme (Reddy *et al*., 1986). Hijikata *et al*. (1987) have noted three sequences in the 5′-flanking region that are common to the inducible fatty acid oxidation enzymes. The effect is thought to be mediated by a binding protein (Lalwani *et al*., 1987). This receptor protein may be related to the steroid hormone receptor superfamily (Issemann and Green, 1990). Peroxisomes are also induced by high-fat diets (Neat *et al*., 1980), in the adrenal cortex by ACTH (Black and Russo, 1980), and by administration of thyroid hormone (Fringes and Reith, 1982). Peroxisome function thus is influenced by metabolic state, nutrition, and pharmaceutical agents in ways that have been studied only incompletely, but may prove relevant to the pathogenesis and treatment of peroxisomal disorders.

PEROXISOMAL DISORDERS

Table I lists the genetically determined peroxisomal disorders that have been identified so far. They are subdivided into two major categories: The disorders of peroxisome biogenesis and the single-enzyme defects. These two categories differ fundamentally. In the first category the organelle fails to form or be maintained normally and this leads to a multiplicity of peroxisomal dysfunctions. In the second category peroxisome structure remains intact and all of the deficits can be traced to malfunction of a single peroxisomal enzyme. One disorder, rhizomelic chondrodysplasia punctata (RCDP), has features of both categories, in that there is malfunction of three peroxisomal functions even though peroxisome structure appears to be intact. Finally, disorders have been described only recently that appear to be peroxisomal but have not yet been fully defined.

Disorders of Peroxisome Biogenesis

Historical Considerations

The four disorders assigned to this category were described before the peroxisomal disorders were recognized. For three of the disorders the names assigned were based upon one set of abnormalities that happened to be identified first. The recent realization that all four disorders were associated with a deficiency of peroxisomes and multiple dysfunctions of this organelle came as a surprise. The *Zellweger syndrome* was described in 1964 (Bowen *et al.*, 1964) and was assigned to the multiple congenital anomaly category. It represents the most severe proto-type of a human peroxisome deficiency disorder, and affected children only rarely survive beyond the fourth month. *Neonatal adrenoleukodystrophy* (NALD) was

TABLE I.   Peroxisomal Disorders

Disorders of peroxisome biogenesis	Defects of single peroxisomal enzyme	Multiple biochemical defects; peroxisome intact
Zellweger syndrome	X-linked ALD	Rhizomelic chondrodysplasia punctata
Neonatal ALD	Oxidase deficiency	
Infantile Refsum disease	Bifunctional enzyme deficiency	
Hyperpipecolic acidemia	Thiolase deficiency	
	Glutaryl CoA oxidase deficiency	
	Acatalasemia	
	Hyperoxaluria type 1	

described first by Ulrich *et al.* (1978) as a "connatal" form of adrenoleukodystrophy in a male infant with neonatal seizures and severe psychomotor retardation and who died at 20 months. Adrenal atrophy and the presence of cytoplasmic inclusions characteristic of X-linked ALD (see p. 62) were a surprising finding and led to the name. It was not until 1982 and later that the peroxisome deficiency and the multiplicity of peroxisomal biochemical defects were recognized (Mobley *et al.*, 1982; Partin and McAdams, 1983; Kelley and Moser, 1984; Kelley *et al.*, 1986; Wanders *et al.*, 1987a). *Hyperpipecolatemia* was first described in 1968 by Gatfield *et al.* (1968) in a boy who died at about 27 months of a neurodegenerative disorder associated with an enlarged liver; a marked excess of pipecolic acid was demonstrated in his body fluids and tissues. It was not until 1975 that pipecolic acid excess was also demonstrated in the Zellweger syndrome (Danks *et al.*, 1975), and not until 1988 that general peroxisomal dysfunction was described in this disorder (Wanders *et al.*, 1988d), including samples from Gatfield's original case. The name *infantile phytanic acid storage disease* or *infantile Refsum disease* was assigned in 1982 to three unrelated boys, ages 3–6 years old, who had enlarged livers, dysmorphic features, mental retardation, sensorineural hearing loss, and retinitis pigmentosa. Liver biopsy revealed inclusions similar to those in plant chloroplasts that contained phytol and this led to the measurement of phytanic acid levels and the demonstration that they were elevated as in Refsum syndrome (Scotto *et al.*, 1982). Generalized peroxisomal dysfunction was documented in 1984 (Poulos *et al.*, 1984) and later (Poll-The *et al.*, 1986a,b; Wanders *et al.*, 1986d). While classical Zellweger syndrome has a characteristic phenotype (Kelley, 1983), there is clinical overlap between the other three disorders, which are all slightly milder than the Zellweger syndrome. As discussed on p. 42, it is anticipated that complementation analysis followed by additional biochemical and genetic studies will lead to a more meaningful subclassification of the disorders of peroxisome biogenesis.

Clinical and Pathological Features

Zellweger syndrome has striking and characteristic abnormalities that are easy to recognize and have been described in detail (Wilson *et al.*, 1986; Kelley, 1983; Govaerts *et al.*, 1982). Of central importance are the typical face (high forehead, hypoplastic supraorbital ridges, epicanthal folds, midface hypoplasia); severe weakness and hypotonia; eye abnormalities (cataracts, glaucoma, corneal clouding, Brushfield spots, pigmentary retinopathy and optic nerve dysplasia); weakness and hypotonia; and neonatal seizures. Table II lists the most common

TABLE II.   Main Clinical Abnormalities
in Zellweger Syndrome[a]

Abnormal feature	Cases in which information about the feature was available		Cases in which the feature was present	
	Number	Percent	Number	Percent
High forehead	60	53	58	97
Flat occiput	16	14	13	81
Large fontanelle(s)	57	50	55	96
Shallow orbital ridges	33	29	33	100
Low/broad nasal bridge	23	20	23	100
Epicanthus	36	32	33	92
High arched palate	37	32	35	95
External ear deformity	40	35	39	97
Micrognathia	18	16	18	100
Redundant skin folds of neck	13	11	13	100
Brushfield spots	6	5	5	83
Cataract/cloudy cornea	35	31	30	86
Glaucoma	12	11	7	58
Abnormal retinal pigmentation	15	13	6	40
Optic disk pallor	23	20	17	74
Severe hypotonia	95	83	94	99
Abnormal Moro response	26	23	26	100
Hyporeflexia or areflexia	57	50	56	98
Poor sucking	77	68	74	96
Gavage feeding	26	23	26	100
Epileptic seizures	61	54	56	92
Psychomotor retardation	45	39	45	100
Impaired hearing	21	18	9	40
Nystagmus	37	32	30	81

[a]From Heymans (1984) survey of 114 patients with Zellweger syndrome reported in the literature. (From Lazarow and Moser, 1989. Used by permission.)

abnormalities and their relative frequency in a series of 114 patients compiled by Heymans (1984). In Wilson's literature survey of 90 patients, 79 had died at an average of 12.5 months (Wilson *et al.*, 1986). Most children show no signs of psychomotor development at all (Zellweger, 1987).

The brains of patients with Zellweger syndrome show striking and characteristic abnormalities of neuronal migration (Volpe and Adams, 1972; Evrard *et al.*, 1978). The *cerebral hemispheres* show some abnormally small convolutions (microgyria), others that are abnormally thick (pachygyria), and others that are normal. The abnormal gyral pattern is most prominent in the frontal, temporal, and parietal regions (the so-called opercular regions) that surround the insular region of the Sylvian fissure. In other regions, such as the anterior frontal, the gyral pattern is normal. The pachygyric and microgyric areas show abnormal cytoarchitecture (Fig. 8). The cells are not arranged in the normal six-layer pattern, in that those that normally are destined for the outer cortical layers are found in abnormal (heterotopic) positions in the deep layers of the cortex and in the underlying white matter. This type of abnormality is referred to as a defect in neuronal migration. For a general discussion of this topic see Barth (1937). In the *cerebellum* a large number of heterotopic Purkinje cells are scattered throughout the cerebellar cortex and in the granule cell layer. Finally, there is an abnormality of the *inferior olivary nucleus* that appears to be unique to Zellweger syndrome, namely laminar discontinuities that involve the principal olivary nucleus and are presumed to be secondary to the Purkinje cell abnormalities.

Evrard *et al*. (1978) emphasize two features that set the Zellweger malformation apart from other migrational defects. First, only a portion of neurons are affected, since some parts of cerebral cortex have normal cytoarchitectonic pattern. Second, detailed analysis of the pattern of abnormalities led them to the conclusion that this partial disturbance of migration occurred throughout the greater duration of the neuronal migration "epoch." They speculated that this

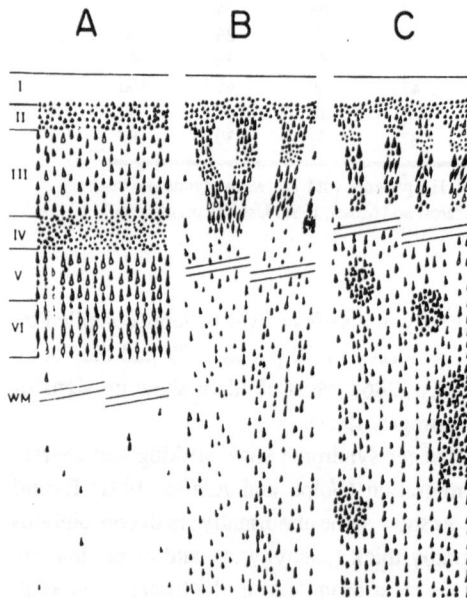

Fig. 8. Schematic representation of cell pattern in (A) normal, (B) Zellweger microgyric, and (C) Zellweger pachygyric neocortical regions. Roman numerals to left correspond to normal cortical layers. White matter and horizontal slashes mark junction of cortex and central white matter. The white matter in B and C contains numerous heterotopic neurons that failed to reach their normal location in the neocortex. In addition, in the abnormal cortex (B and C) the neurons are arranged in fascicles that contain three to six neurons abreast of each other. [Evard and Caviness (1982).] (From Lazarow and Moser, 1989. Used by permission.)

could be due to the presence of "as yet unidentified cytotoxic substances . . . that are not adequately cleared by the placenta from the circulation of fetuses with the Zellweger malformation." Consistent with this formulation is the later observation that the migrational abnormality was demonstrable in 14- to 21-week fetuses with Zellweger syndrome who had been identified prenatally (Powers *et al.*, 1985).

At the time of the Evrard *et al.* (1978) report, the only biochemical abnormality that had been associated with the Zellweger syndrome was the abnormally high level of pipecolic acid (Danks *et al.*, 1975). Table III shows the additional abnormalities that have been identified since then, and one of the challenges for the future is to determine which ones may be responsible for the migrational defect. It is likely that the neuronal migration defect is the principal cause of the seizures, hypotonia, and profound mental retardation associated with the Zellweger syndrome. Different degrees of myelin abnormalities (leukodystrophy) may also be present (Agamanolis *et al.*, 1976; Liu *et al.*, 1976), but appear to be less critical since leukodystrophy is not found consistently (Volpe and Adams, 1972; Danks *et al.*, 1975).

The *eyes* show multiple abnormalities: corneal clouding, congenital cataracts and glaucoma, loss of ganglion cells, optic atrophy, and changes resembling those of retinitis pigmentosa (Cohen *et al.*, 1983). The electroretinogram is extinguished (Hittner *et al.*, 1981). The *liver* is enlarged in 78% of patients and fibrotic in 76%. Cholestasis was present in 59% and micronodular cirrhosis in 37% (Heymans, 1984). Excessive iron deposits in the liver were reported in an early study (Vitale *et al.*, 1969), but these levels diminish with time and are normal in infants who live beyond 20 weeks (Gilchrist *et al.*, 1976). *Renal cysts* were observed in 78 of 80 patients who were studied pathologically (Heymans, 1984). The cysts vary from glomerular microcysts to large cortical cysts of glomerular and tubular origin (Bernstein *et al.*, 1981). Such cysts are also present in the fetus (Powers *et al.*, 1985). The changes in the *adrenal gland* are similar to those in X-linked ALD (see

TABLE III. **Diagnostically Significant Abnormalities in Zellweger Syndrome**

1. Peroxisomes absent or reduced in number
2. Catalase in cytosol
3. Deficient synthesis and reduced tissue levels of plasmalogens
4. Defective oxidation and abnormal accumulation of very long-chain fatty acids
5. Deficient oxidation and age-dependent accumulation of phytanic acid
6. Defects in certain steps of bile acid formation and accumulation of bile acid intermediates
7. Defect in oxidation and accumulation of L-pipecolic acid
8. Increased urinary excretion of dicarboxylic acids

From Lazarow and Moser, 1989. Used by permission.

p. 62). The reticularis-inner fasciculata zones of the cortex contain striated cells, some of which are ballooned, and have lamellar cytoplasmic inclusions (Gold-fischer *et al.*, 1983). *Calcific stippling* of the patella and synchondrosis of the acetabulum occur in 50% of Zellweger patients (Heymans, 1984). Powers *et al.* (1985) suggest that this represents a premature rather than a dysplastic mineralization.

Disorders of Peroxisome Biogenesis with Milder Involvement

As noted in the historical introduction, the faulty assembly of peroxisomes may also be associated with phenotypes that are somewhat less severe than classical Zellweger syndrome. The names and classification of these disorders are confusing and in a state of flux. The disorders range in severity from phenotypes that are almost as severe and as rapidly fatal as classical Zellweger syndrome to those that permit survival to the second or third decade and possibly longer, although it must be emphasized that even these milder forms are associated with mental retardation and serious handicaps. The severe category cases are differentiated from classical Zellweger syndrome in that they lack the full-blown facial dysmorphism and do not have renal cysts or punctate stippling of cartilage (chondrodysplasia punctata) on radiographs. The first reported case (Ulrich *et al.*, 1978) of what is referred to as *neonatal adrenoleukodystrophy* (NALD), a boy who died at 20 months and thus belonged to the severe phenotype category, was so named because of the demonstration of adrenal atrophy and lamellar inclusions similar to X-linked ALD. The first reported cases of *infantile Refsum disease* (IRD) (Scotto *et al.*, 1982) were so named because they had increased levels of phytanic acid in plasma, as in Refsum disease. These three moderately disabled boys are alive at ages 3–6 years, and thus represent the milder phenotype. Since both NALD and IRD now are known to have the same generalized peroxisomal dysfunctions listed in Table III, it seems likely that they represent two forms of the same entity that differ mainly in degree of severity. Figure 9 compares the life expectancy of patients with Zellweger syndrome, NALD, and IRD. Complementation analysis (see p. 42) indicates that IRD and NALD may occur within the same complementation groups. Subject to these caveats, the description of the clinical features of NALD and IRD that follows is based upon the historical designations from the literature.

Neonatal ALD.   Aubourg *et al.* (1986) provide an excellent description of the severe form of the disease. Patients are severely ill at birth, with severe seizures that usually begin in the neonatal period, mild to moderate dysmorphic features, hypotonia, poor feeding, pigmentary degeneration of the retina that becomes evident only at 4–6 months of age, and an enlarged liver with micronodular

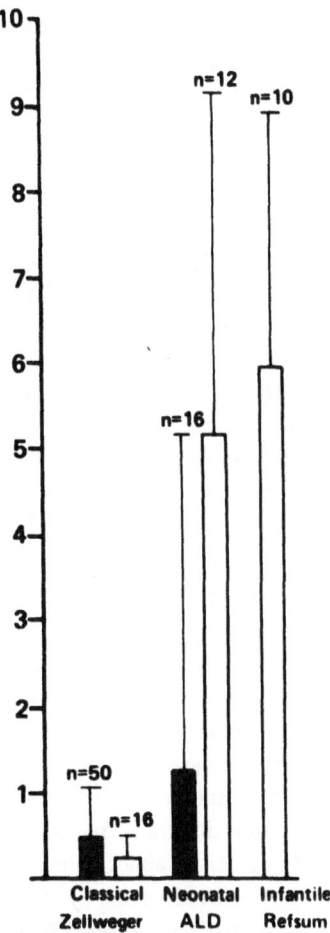

Fig. 9. Life expectancy of patients with Zellweger syndrome, NALD, and IRD. (From Lazarow and Moser, 1989. Used by permission.)

cirrhosis. Some of the children make minimal maturational gains, such as evidence of ocular pursuit, some degree of head control, and the ability to sit with support, but others display no recognizable psychomotor development. In this severely involved group, death occurred between 1 month and 3 years. The neuropathology in this group differs from that in classical Zellweger syndrome. Degeneration in cerebral and cerebellar white matter with preservation of axons was the main abnormality, and in some instances perivascular cuffs of lymphocytes were present as in X-linked ALD (see p. 63). Neuronal migrational defects were confined to the cerebellar Purkinje cells. The cytoarchitecture of the cerebral hemispheres and the olivary nucleus was normal. Thus, whereas in Zellweger syndrome the neuronal migrational defect is severe and the white matter abnormality is variable and relatively mild, in neonatal ALD the reverse is true. The adrenal glands showed cortical atrophy and lamellar cytoplasmic inclusions similar to those in X-linked ALD (see p. 62). The retinal pathology is similar to that in Zellweger syndrome, with changes resembling those of retinitis pigmentosa. There is loss of ganglion cells with gliosis of the nerve-fiber layer and distinctive bileaflet inclusions in pigment epithelium and macrophages (Javitt, 1990; Cohen et al., 1983).

Infantile Refsum Syndrome. The findings in the original report of Scotto et al. (1982) have been confirmed in several additional studies (Budden et al., 1986; Wanders et al., 1990a; Poll-The et al., 1987; Torvik et al., 1988; Weleber et al., 1984). Patients with this disorder resemble those with NALD in that they have mild dysmorphic features (high arched palate, epicanthal folds, anteverted nostrils, midface hypoplasia, enlarged livers, and pigmentary degeneration of the retina). Hypotonia may be present, but it is not as severe as in Zellweger syndrome. Patients differ from those with Zellweger syndrome and the severe forms of NALD in that seizures have not been noted, and early motor development does occur. Most patients were able to stand and walk, albeit with an ataxic component, but speech was not attained, and the children function in the profoundly retarded range, with mental ages estimated at 6–24 months. Vision is severely impaired because of pigmentary degeneration of the retina, with an absent or severely abnormal electroretinogram, and there is marked hearing loss. Liver function tests show only moderate abnormalities, but at least two of the patients had abnormal bleeding that was corrected by administration of vitamin K. The liver may show micronodular cirrhosis. Adrenal function was tested by Poll-The et al. (1987) in two patients and it was normal in one and severely impaired in the other. Torvik et al. (1988) reported the only complete postmortem study, performed in a 12-year-old boy. They demonstrated micronodular cirrhosis, hypoplastic adrenals without degenerative changes, and large groups of lipid macrophages in the liver, lymph nodes, and certain areas of the cerebral white matter. There were no malformations of cerebral hemispheres, but the cerebellum showed hypoplasia of the granule cell layer and ectopic locations of the Purkinje cells that were interpreted as representing a neuronal migrational defect. The olivary nucleus was normal. Thus, both in respect to clinical features and pathological findings, IRD can be viewed as a milder variant of Zellweger syndrome and of the severe form of NALD.

Patients Who Survive to Adulthood.   We have identified one 23-year-old man with a disorder of peroxisome biogenesis who is in stable condition (Naidu et al., 1990). He is profoundly retarded, wheelchair-bound, with hypertonia and increased deep tendon reflexes. Initial diagnosis had been the cerebral form of X-linked ALD, but was differentiated by the absence of early normal development, the presence of pigmentary degeneration, and the demonstration of generalized peroxisomal dysfunction (see p. 29). Others, such as patient 3 reported by Kelley et al. (1986) and the patient reported by Brown et al. (1982a), are now in stable condition in their late teens and are expected to reach adulthood.

Hyperpipecolic Acidemia.   Five patients with hyperpipecolic acidemia have been reported (Arneson et al., 1982; Burton et al., 1981; Challa et al., 1983; Gatfield et al., 1968; Thomas et al., 1975). Since the clinical manifestations and pathological findings resemble those of NALD and IRD, the biochemical abnormalities are similar (Wanders et al., 1988d), and complementation analysis (see p. 42) also places it into the large complementation group along with the other three disorders, there appears little justification to continue to list it as a distinct entity.

Abnormalities That Are Present in All Patients with Disorders of Peroxisome Biogenesis

Table III lists a panel of abnormalities that are shared by all patients with disorders of peroxisome biogenesis. These abnormalities are of key importance for the understanding of pathogenesis and for diagnosis.

Absence or Reduced Number of Catalase-Containing Subcellular Particles.   While in the past this feature was referred to as "absence or reduced number of *peroxisomes*," the recent demonstration that membranous structures containing the membrane proteins characteristic of peroxisomes (peroxisome "ghosts") (Santos et al., 1988b) *are* present makes it more appropriate to use the more restrictive terminology, "absence or reduced number of *catalase-containing subcellular particles*." Irrespective of the precise terminology, this finding rests upon the seminal observation made by Goldfischer et al. (1973), which provided the foundation for all our knowledge about peroxisomal disorders. These investigators studied liver and kidney biopsy specimens of two children with classical Zellweger syndrome and searched for peroxisomes by using the diaminobenzidine reaction described by Novikoff and Goldfischer (1969) which stains catalase, a procedure that has since been modified and made more specific by Roels and Goldfischer (1979). With this procedure no catalase-containing particles were demonstrable in hepatocytes or renal proximal tubules, although they were abundant in control tissues. They stated that, while "it is difficult, it not impossible to unequivocally demonstrate the absence of an organelle by morphological studies. . ., however, if any peroxisomes are present in the cerebro-hepato-renal syndrome, our observations indicate that they must be greatly reduced in number or altered in form and enzymatic activity." All subsequent studies have confirmed this finding (Versmold et al., 1977; Lazarow et al., 1985; Mooi et al., 1983). Indeed, the *presence* of a significant number of catalase-containing particles excludes the diagnosis of Zellweger syndrome or of the other disorders of peroxisome biogenesis. Peroxisomes are also markedly reduced in number or are absent in cultured skin fibroblasts of Zellweger patients (Santos et al., 1985; Arias et al., 1985). Reduction of the number of peroxisomes (judged on the basis of the size and number of catalase-containing particles) has also been demonstrated in NALD (Vamecq et al., 1986; Wanders et al., 1986d). Catalase is cytosolic rather than particulate in cultured skin fibroblasts of patients with hyperpipecolic acidemia (Wanders et al., 1988d), from which it can be inferred that peroxisomes are reduced also in this disorder. Challa et al. (1983) reported the presence of normal peroxisomes in a liver biopsy specimen of one patient with hyperpipecolic acidemia, but this identification appears to have been made without the use of the histochemical stain for catalase. It is of interest that both Arias et al. (1985) (see also Fig. 10) and Vamecq et al. (1986) noted that NALD patients had a smaller reduction in peroxisomes than did the patients with classical Zellweger syndrome. This finding suggests that patients with the most severe form of the disease also show the greatest reduction in peroxisome number.

In their 1973 report, Goldfischer et al. (1973) also demonstrated mitochondrial abnormalities in the Zellweger syndrome. While further study is required, it is likely that peroxisomal dysfunction is the primary abnormality and that mitochondrial changes are secondary and less significant than those in the peroxisome. As discussed on p. 25, most or all of the biochemical abnormalities can be traced to peroxisomal defects. Patients with Zellweger syndrome have normal lactate/pyruvate and β-hydroxybutarate/acetoacetate ratios in blood and their cultured skin fibroblasts metabolize [1-^{14}C]- and [2-^{14}C]pyruvate at a normal rate (Trijbels et al., 1983), and mitochondrial structure was normal in a liver biopsy specimen (Lazarow et al., 1985). Nevertheless, Kelley and Corkey (1983) showed increased sensitivity to the complex III inhibitor antimycin A in cultured skin fibroblasts of patients with Zellweger syndrome and Trijbels et al. (1983) suggested involvement of this same mitochondrial complex and proposed that there was an impairment in the interaction between ubiquinone and succinate dehydrogenase (Trijbels et al., 1983).

Recent studies of phytanic acid metabolism in the disorders of peroxisome

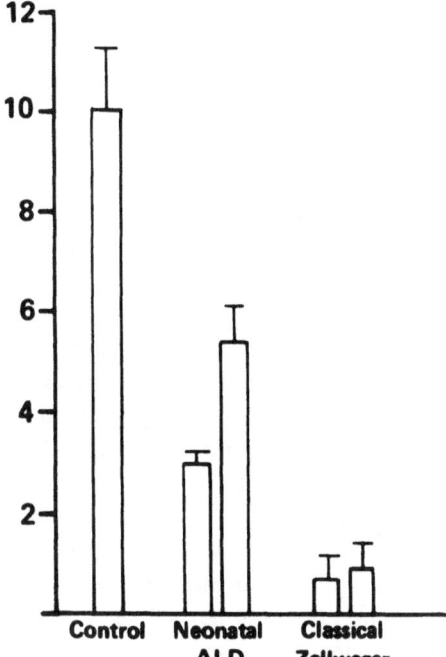

Fig. 10. Numbers of peroxisomes in NALD and Zellweger patients. These values were taken from Arias *et al.* (1985). While in this reference the cases labeled here as "neonatal ALD" were referred to as "atypical cases with features of both Zellweger syndrome and neonatal ALD," a more recent review of their case records places them in the neonatal ALD category. (From Lazarow and Moser, 1989. Used by permission.)

biogenesis (see p. 38) also suggest abnormalities in the function of both organelles. Mitochondrial dysfunction could result from changes in the mitochondrial membrane lipids that result from the primary peroxisomal defect (Trijbels *et al.*, 1983). Analysis of this phenomenon is of general interest and may be aided by analysis of these presumably secondary mitochondrial defects in the disorders of peroxisome biogenesis.

Demonstration of Peroxisome "Ghosts" and Their Implication for Pathogenesis.   Although failure to form peroxisomal membranes would be a plausible explanation for the apparent absence of this organelle in the disorders of peroxisome biogenesis, a series of studies demonstrated that postmortem liver tissue of Zellweger disease patients contained normal or near-normal amounts of the peroxisomal membrane proteins (Small *et al.*, 1988a; Gaertner *et al.*, 1991; Y. Suzuki *et al.*, 1989). The only exception is the study of Aikawa *et al.* (1987), who reported a deficiency of 22-, 26-, and 70-kD PMP in the postmortem liver of one patient with Zellweger syndrome. The reason for the discrepancy between this report and the others is not clear. While it may be due to technical factors, the complementation analyses discussed below show that there is genetic hetero-geneity, and it is possible that in some patients the failure to form the peroxisome is due to a defect in membrane protein synthesis. Nevertheless, the preponderance of evidence indicates that the peroxisomal membrane proteins *are* synthesized in the majority of patients.

Exciting new insights about the pathogenesis of these disorders are provided by the studies of Santos *et al*. (1988b), who showed that cultured skin fibroblasts of patients with Zellweger syndrome contained membranous structures that reacted with a polyspecific antibody to the 140-, 69-, and 53-kD PMPs and with another antibody against the 22-kD PMP (Santos *et al.*, 1988a). Fractionation studies in normal fibroblasts demonstrated that these antibodies reacted only with peroxi-somes. These fractionations were performed with a linear Nycodenz gradient in which peroxisomes from control fibroblasts had an equilibrium density of 1.17 g/cm^2. The peroxisomal membranes from Zellweger fibroblasts had a lighter equilibrium density (1.10 g/cm^2) and immunoelectron microscopy showed that the peroxisomal membrane proteins were located in membranous structures with a diameter 2–4 times that of normal peroxisomes (Santos *et al.*, 1988a). Studies that utilized antibody to catalase gave completely different results. In Zellweger syndrome fibroblasts the catalase was found exclusively in the cytosol, whereas in control samples it was present in the peroxisome fraction. Wiemer *et al*. (1989) obtained similar results and added the important information that these findings also applied to samples from the Brul *et al*. (1988a) complementation groups 2, 3, and 4 (Table IV), which also include samples from patients with NALD, infantile Refsum disease, and hyperpipecolic acidemia.

It is concluded that patients with disorders of peroxisome biogenesis do form membranous structures that contain peroxisome membrane proteins similar to the normal proteins in size and immunological reactivity, but that these structures do not contain the matrix enzyme catalase. These abnormal membrane structures appear to be "ghosts" that lack some or all of the normal matrix proteins. As discussed earlier, proteins destined for the peroxisome are synthesized in free polyribosomes, enter the cytoplasm, and then are transported into the peroxisome. Chase studies in cultured skin fibroblasts of patients with the Zellweger syndrome have shown that the peroxisomal β-oxidation enzymes, which normally are located in the peroxisome matrix, are formed at a normal rate but are degraded abnormally rapidly, presumably because they are not transported into the peroxi-somes (Schram *et al.*, 1986). These findings led to the hypothesis that the basic defect in the disorders of peroxisome biogenesis is the failure of import of matrix enzymes into the organelle (Fig. 11). Recent studies indicate that this scheme may be an oversimplification. Balfe *et al*. (1990) showed that the immature 3-ketoacyl-

TABLE IV.   Results of Complementation Analyses in Human Peroxisomal
Disorders. Complementation Groups

Brul	Roscher	Poll-The	McGuinness
		1 CREF	
1 RCDP (1)		2 RCDP (1)	
2 ZS, IR, HP (3)	1 ZS, IR, NALD, HP (33)	3 ZS, IR (5)	1 ZS (3)
3 ZS (1)			
4 NALD (1)			
5 ZS (1)	2 ZS, NALD (2)		2 ZS, NALD (2)
	3 ZS (2)		3 ZS (1)
	4 ZS (5)		4 ZS (3)
	5 ZS (1)		
	6 NALD (1)		6 NALD (2)
		4 NALD (3)	
		5 ACOX (2)	7 ACOX (1)
			8 BIF (1)
			9 X-ALD (5)

RCDP, Rhizomelic chondrodysplasia punctata; CREF, classical Refsum disease; ZS,
Zellweger syndrome; IR, infantile Refsum disease; NALD, neonatal adrenoleukodystrophy; HP, hyperpipecolic acidemia; ACOX, acyl-CoA oxidase deficiency; BIF,
bifunctional enzyme deficiency; X-ALD, X-linked adrenoleukodystrophy.

Note: Restoration of capacity to synthesize plasmalogens was the criterion for complementation in the studies of Brul *et al.* (1988a) and Roscher *et al.* (1989); Poll-The *et al.*
(1989) used phytanic acid oxidation as the criterion; and McGuinness *et al.* (1990)
showed the capacity ot oxidize very long-chain fatty acids. The unbracketed numbers
indicate the group assignments designated by each of the authors, followed by the
phenotypes assigned to each group. The numbers in parentheses indicate the number of
cell lines in each group. In several instances cell lines from the same patient were
studied by more than one investigator. Where this is the case, the groups are listed on the
same line. For instance, Brul group 2 included one or more cell lines that were assigned
to their group 1 by Roscher *et al.*; to their group 3 by Poll-The *et al.*; and to group 1 by
McGuinness *et al.*; indicating that these groups, even though they have been assigned
different numbers by the various authors, are probably identical. Similar equivalence
applied to Brul group 1 and Poll-The group 2; and to Brul group 5, Roscher group 2, and
McGuinness group 2. For the groups that are not listed on the same line, cell lines have
not yet been exchanged between the investigators.

CoA thiolase present in the liver and cultured skin fibroblasts of Zellweger disease
patients (see p. 48) is located in a subcellular fraction with a lower density than
that of peroxisomes or mitochondria, and similar to that of the peroxisome ghosts.
Possibly the "ghosts" are not entirely devoid of matrix enzymes or the formation
of peroxisomes may proceed via as yet unidentified precursor particles.

Recent advances in knowledge about the topogenesis of peroxisomal enzymes will undoubtedly contribute to the understanding of the pathogenesis of the

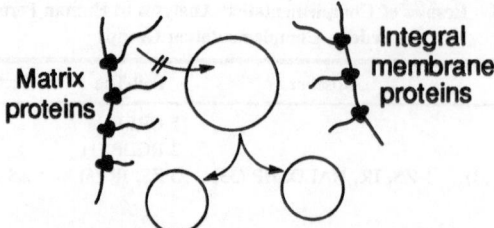

Fig. 11. Schematic representation of the basic defect in the disorders of peroxisome biogenesis. The double line indicates that the underlying defect is thought to involve the import of matrix enzymes. The demonstration of normal or near-normal levels of membrane proteins and the existence of peroxisome "ghosts" suggest that the import of integral membrane proteins is not disturbed. This scheme is an oversimplification, since there are at least six types of import defects.

disorders of peroxisome biogenesis. As noted above, many of the matrix enzymes contain the carboxy-terminal serine–lysine–leucine sequence, while the membrane proteins do not (Gould *et al.*, 1990). One possibility is that in the disorders of peroxisome biogenesis there is a defect in a receptor that recognizes this topogenic sequence and leaves the targeting of peroxisome membrane proteins intact since their import does not depend upon this sequence. Very recently we have demonstrated a mutation in one patient with Zellweger syndrome that affects a splice junction of the 70-kD PMP (J. Gaertner, H. W. Moser, and D. Valle, unpublished observation). The 70-kD PMP protein is of particular interest, since, as noted above, it contains an ATP-binding sequence that is postulated to be involved in transport mechanisms (Kamijo *et al.*, 1990). As will be discussed on p. 42, complementation analysis has demonstrated that there are at least six genetically distinct subgroups within the disorders of peroxisome biogenesis. This suggests that at least six distinct genetic defects may impair the import mechanisms. Application of molecular and cell biology techniques to these human mutants provides the opportunity to define the defects precisely.

Biochemical Abnormalities in the Disorders of Peroxisome Biogenesis

Increased Levels of Very Long-Chain Fatty Acids.
Very long-chain fatty acids (VLCFA) here are referred to as those with a carbon chain length of greater than 22. The chemistry, occurrence, and metabolism of these substances have been reviewed recently (Rezanka, 1989). Levels of saturated and unsaturated VLCFA are increased greatly in the tissues and body fluids of patients with

disorders of peroxisome biogenesis (Brown *et al.*, 1982a; A. B. Moser *et al.*, 1984; Tanaka *et al.*, 1990; Poulos *et al.*, 1986a,b, 1989; Molzer *et al.*, 1986; Bakkeren *et al.*, 1984; Govaerts *et al.*, 1985; Martinez, 1989) and this finding forms the basis of the most widely used diagnostic assay (see p. 46). Table V shows the changes in plasma levels observed in our clinic. The abnormality is greater in the Zellweger syndrome than in NALD and infantile Refsum disease. Poulos and associates have demonstrated the accumulation of polyenoic fatty acids with chain lengths up to 38 in the brain (Sharp *et al.*, 1987) and plasma (Poulos *et al.*, 1989) of patients with Zellweger syndrome. These species are increased markedly in a brain phosphatidylcholine fraction (Sharp *et al.*, 1991) and they may be of considerable physiological significance. In the retina, where the structure and function are severely disturbed in Zellweger syndrome, the fatty acid composition of phosphatidylcholine normally influences its binding to rhodopsin (Aveldano, 1988). Martinez has reported a striking reduction of the levels of docosohexacosanoic acid (22:6 ω3) and 22:5 ω6 in postmortem tissues of a Zellweger disease patient (Martinez, 1989). Docosohexacosanoic acid normally is present in high concentration in retina and brain (Aveldano, 1988) and a reduction of the levels of this substance could have profoundly deleterious effects. Martinez postulated that the reduced level of these fatty acids could be due to deficiency of Δ4 desaturase. However, the activity of this enzyme was not reduced in cultured skin fibroblasts of Zellweger disease patients (Gronn *et al.*, 1990). These fibroblasts did show greatly reduced capacity for the retroconversion of 22:6 ω3 to eicosapentanoic acid (20:5 ω3). This reaction requires the action of Δ4 enoyl-CoA reductase, which has been shown to be localized to the peroxisome. The impaired capacity to retroconvert C22:6 may cause it to be elongated to very long-chain polyunsaturated fatty acids that, as mentioned above, accumulate in Zellweger disease patients.

TABLE V.   Plasma Very Long-Chain Fatty Acids in Patients
with Peroxisomal Disorders (Mean ± SD, µg/ml)

Disorder	C26:0	C26:0/C22:0	C24:0/C22:0
X-ALD (males) (684)	1.34 ± 0.51	0.065 ± 0.044	1.55 ± 0.25
X-ALD (heterozygotes) (929)	0.74 ± 0.40	0.036 ± 0.07	1.10 ± 0.43
Infantile Refsum disease (14)	1.29 ± 0.79	0.150 ± 0.14	1.20 ± 0.39
Neonatal ALD (72)	2.52 ± 1.20	0.34 ± 0.29	1.76 ± 0.47
Zellweger syndrome (113)	3.42 ± 1.6	0.46 ± 0.16	1.90 ± 0.35
Rhizomelic chondrodysplasia punctata (22)	0.252 ± 0.13	0.017 ± 0.014	0.82 ± 0.16
Control and other disorders (1858)	0.24 ± 0.16	0.014 ± 0.0056	0.81 ± 0.09

The accumulation of VLCFA in the disorders of peroxisome biogenesis is almost certainly due to the great reduction of tissue levels of peroxisomal β-oxidation enzymes. Immunoblot studies have shown that the peroxisomal acyl-CoA oxidase and bifunctional enzyme are essentially absent from the liver of patients with Zellweger syndrome (Lazarow et al., 1985; Tager et al., 1985; Y. Suzuki et al., 1986). In NALD there was a marked deficiency of the bifunctional enzyme (Chen et al., 1987). As discussed above, the deficiency of these peroxisomal β-oxidation enzymes is due to their failure to be imported into the peroxisome and their rapid degradation in the cytosol.

Impaired Synthesis of Plasmalogens. The report of Hajra and Bishop (1982) that the initial steps of plasmalogen synthesis take place in the peroxisome led the Dutch investigators to examine this function in patients with the Zellweger syndrome. The first two enzymes in the plasmalogen synthesis pathway (see Fig. 3) are deficient in patients with disorders of peroxisome biogenesis. Acyl-CoA:dihydroxyacetone acyltransferase (DHAPAT) is deficient in tissues, cultured skin fibroblasts, cultured amniocytes and chorion villus cells, leukocytes, and thrombocytes (Datta et al., 1984; Besley and Broadhead, 1987; Schutgens et al., 1985; Wanders et al., 1985). Deficiency of dihydroxyacetone phosphate synthase has been demonstrated in tissues and cultured skin fibroblasts (Webber et al., 1987). Webber et al. (1987) showed that the Km values and other properties of the residual enzymes are normal, suggesting that the enzyme is not defective, but is present at a reduced level (Webber et al., 1987), a finding that is compatible with the concept that the protein is formed normally, but is abnormally labile because of failure of normal transport into the peroxisome. Plasmalogen levels in brain, liver, kidney, muscle, and heart of Zellweger patients are generally less than 5% of control (Heymans et al., 1983, 1984). Although plasmalogens normally are components of all cell membranes, their physiological role is not fully understood. They may play a role in protection against photosensitized killing (Zoeller et al., 1988), and also in the structure of myelin, where they are present in high concentration (Norton and Autilio, 1966). There is a moderate but significant age-dependent reduction of plasmalogen levels in red cell membranes in Zellweger patients. During the first 5 weeks of life they were reduced to 5–10% of control, while normal levels were observed in Zellweger patients aged 20 weeks or older (Heymans et al., 1984; Wanders et al., 1986a).

Accumulation and Impaired Catabolism of Pipecolic Acid. Pipecolic acid accumulation was the first biochemical abnormality to be identified in Zellweger syndrome (Wanders et al., 1986a). It is the L-isomer that accumulates (Lam et al., 1986). Pipecolic acid levels in blood and urine are age-dependent (Dancis and Hutzler, 1986; Kelley, 1991). Table VI shows that with the use of a

TABLE VI. Plasma and Urinary Pipecolic Acid Levels Determined by Isotope-Dilution Gas Chromatography/Mass Spectrometry

	Pipecolic acid concentration					
	Plasma (μmole/liter)			Urine (μmole/g creatinine)		
Diagnosis	Mean ± SD	n	Range	Mean ± SD	n	Range
Normal and nonperoxisomal neurological disease						
Newborn–7 days	1.9 ± 0.7	12	1.0 ± 3.2	34.6 ± 11.3	3	16.9 ± 48.0
1 week–1 month	2.3 ± 1.4	9	0.8 ± 5.3	28.8 ± 15.4	3	13.9 ± 44.7
1–6 months	1.8 ± 0.9	32	0.6 ± 4.1	26.6 ± 20.3	16	4.6 ± 62.0
7–12 months	1.9 ± 0.8	25	0.5 ± 3.8	9.8 ± 7.3	12	2.3 ± 20.4
1–5 years	1.9 ± 0.8	60	0.4 ± 4.9	3.9 ± 3.7	11	1.6 ± 13.4
>5 years	1.6 ± 0.6	23	0.7 ± 2.6	4.7 ± 3.7	4	0.9 ± 9.7
Nonperoxisomal hepatocellular disease	7.1 ± 3.9	6	2.6 ± 13.6	—	—	—
Peroxisomal disorders						
Zellweger syndrome, infantile Refsum disease, neonatal ALD						
Newborn–7 days	12.6 ± 7.7	8	5.9 ± 30.0	371.0 ± 267.5	4	75.4 ± 804.3
1 week–1 month	10.3 ± 7.2	10	4.4 ± 28.5	715.1 ± 36.3	2	678.8 ± 751.4
1–6 months	47.0 ± 42.1	11	6.7 ± 125.8	468.1 ± 169.4	4	178.0 ± 597.0
6–12 months	117.7 ± 116.8	5	10.0 ± 341.8	105.6 ± 110.2	3	30.0 ± 261.5
1–5 years	98.7 ± 81.6	20	11.5 ± 298.0	57.0 ± 46.0	4	14.5 ± 68.5
>5 years	21.1 ± 11.3	7	11.1 ± 43.3	—	—	—
"Pseudo-Zellweger" syndromes (single-enzyme defects of peroxisomal β-oxidation)	2.2 ± 0.8	7	0.9 ± 3.1	2.5	1	—
Rhizomelic chondrodysplasia	1.5 ± 0.7	6	0.7 ± 2.5	—	—	—

From Kelley, 1991. Used by permission.

sensitive isotope dilution method abnormally high levels can be demonstrated in all patients with disorders of peroxisome biogenesis irrespective of age. The accumulation is due to deficient activity of hepatic L-pipecolic acid oxidase (Wanders *et al.*, 1988a; Mihalik *et al.*, 1989), an enzyme that has been localized to the peroxisome in humans and primates (Mihalik and Rhead, 1989; Mihalik *et al.*, 1991).

Accumulation of Intermediates of Bile Acid Metabolism. Abnormally high plasma levels of $3\alpha,7\alpha,12\alpha$-trihydroxy-5β-cholestanoic acid (THCA) and of $3\alpha,7\alpha$-dihydroxy-5β-cholestanoic acid (DHCA) are present consistently in Zellweger syndrome patients and in those with the other disorders of peroxisome biogenesis (Hanson *et al.*, 1979; Monnens *et al.*, 1980; Mathis *et al.*, 1980; Eyssen *et al.*, 1985; Gustafsson *et al.*, 1983; Clayton *et al.*, 1987; Poulos and Whiting, 1985). Total bile acid levels are normal except when there is liver damage. In Zellweger syndrome, THCA, DHCA, varanic acid, and certain other metabolites may account for 30–50% of total plasma bile acids, while normally these metabolic intermediates are either absent or present in very low concentration. Defects in peroxisomal function can readily account for these abnormalities. Normally, conversion of THCA and of DHCA into deoxycholic acid takes place in the peroxisome (Pedersen and Gustafsson, 1980). Kase *et al.* (1985) have shown these reactions to be deficient in Zellweger syndrome both *in vivo* and *in vitro*.

Impaired Oxidation and Accumulation of Phytanic Acid. Until 1982, the accumulation and impaired oxidation of phytanic acid was thought to be pathognomonic of Refsum disease (Pettit *et al.*, 1986). In this section we will refer to this disorders as "classical Refsum syndrome," and emphasize that it must be distinguished sharply from infantile Refsum syndrome, a disorder of peroxisome biogenesis discussed above.

Phytanic acid is a branched fatty acid (3,7,11,15-tetramethylhexadecanoic acid). The methyl group at the 3 position makes it impossible for this compound to be oxidized by the usual β-oxidation pathway, and an α-oxidation system is utilized. The first step is α-hydroxylation, which is followed by decarboxylation to form pristanic acid (2,6,10,14-tetramethylpentadecanoic acid) (Fig. 12). When normal fibroblasts are incubated with labeled phytanic acid, the cells contain labeled α-hydroxyphytanic acid and pristanic acid as well as smaller degradation products, while none of these compounds are formed by fibroblasts from Refsum disease patients. It was concluded that the initial α-hydroxylation reaction is deficient in Refsum disease (Herndon *et al.*, 1969). This α-oxidation system has been reported to be localized to the mitochondrion both in rat (Skjeldal and Tokke, 1987) and in human and primate liver (Watkins *et al.*, 1990).

In 1982, it was shown that patients with infantile Refsum disease also had

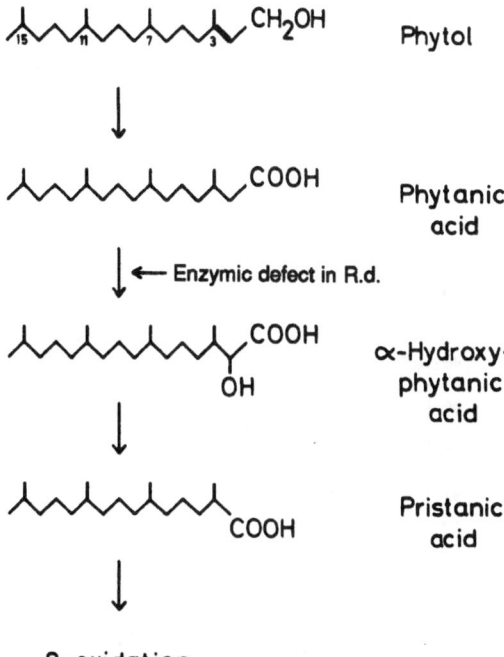

Fig. 12. Degradation of phytanic acid and pristanic acid. The initial step in the degradation of phytanic acid involves an α-oxidation pathway that is deficient in classical Refsum syndrome (CREF) and probably also in rhizomelic chondrodysplasia punctata. The primary defect in the disorders of peroxisome biogenesis appears to involve the β-oxidation of the branched fatty acids such as pristanic acid.

abnormal accumulation of phytanic acid (Scotto *et al.*, 1982), and in 1984 defective oxidation of phytanic acid was demonstrated both in this condition and in Zellweger syndrome (Poulos *et al.*, 1984, 1985). An age-related accumulation of phytanic acid is a consistent feature in the disorders of peroxisome biogenesis (Fig. 13) (Wanders *et al.*, 1987c). This finding was unexpected and paradoxical: Why should a reaction that appears to be localized to the mitochondrion consistently be abnormal in disorders that involve the peroxisome? Notably the plasma levels of phytanic acid in patients with disorders of peroxisome biogenesis are probably lower than those in patients with untreated classical Refsum disease. In 14 patients with classical Refsum disease these levels ranged between 317 and 2050 μg/ml (Britton *et al.*, 1989), while in patients with peroxisome biogenesis they tend not to exceed 100 μg/ml (Fig. 13).

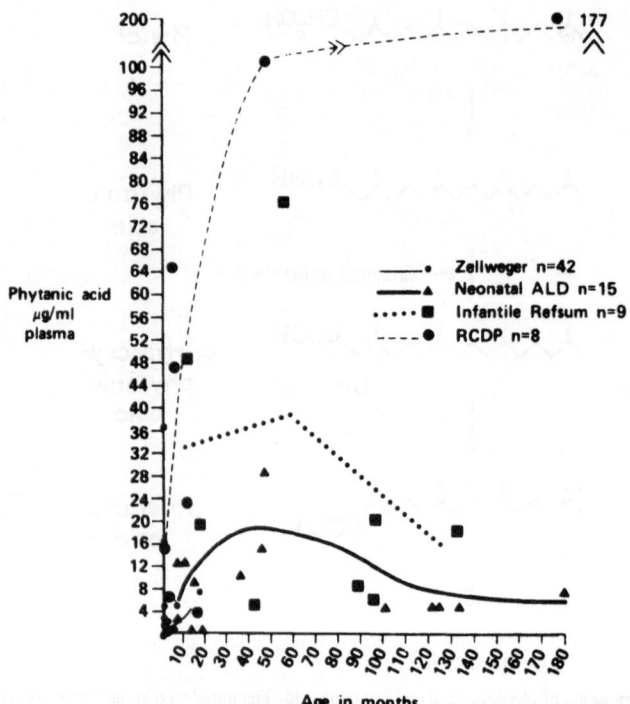

Fig. 13. Variation with age of plasma phytanic acid levels in patients tested at the Kennedy Institute. All points represent individual patients. Because of the scale, only a few of the 42 Zellweger patients are shown on this graph. Twenty-eight of these patients (ages 0–6 months) had normal plasma phytanic levels (below 3 μg/ml). While neonatal ALD and infantile Refsum patients had higher levels, this appears to be related to their longer survival, since the youngest neonatal ALD patients had levels similar to the Zellweger patients. For unknown reasons the neonatal ALD and infantile Refsum patients who are older than 80 months appear to have somewhat lower plasma levels than those aged 30–60 months, even though in all cases the phytanic acid oxidation in cultured skin fibroblasts was less than 2% of control (H. W. Moser, unpublished observation). In the RCDP patients plasma phytanic levels are higher and become elevated at an earlier age than in patients in the other disease groups. (From Lazarow and Moser, 1989. Used by permission.)

were the first to demonstrate a fundamental difference between classical Refsum syndrome and the disorders of peroxisome biogenesis. In 1988 they showed that pristanic acid levels were increased in patients with disorders of peroxisome biogenesis (Poulos *et al.*, 1988). Ten Brink *et al.* (1991, 1992) have developed highly specific isotope dilution assays of pristanic acid and phytanic acid and have shown that the ratio of pristanic acid to phytanic acid in plasma is significantly higher in patients with disorders of peroxisome biogenesis than in those with classical Refsum disease (ten Brink *et al.*, 1991, 1992). This difference was explained by H. Singh *et al.* (1990), who showed that patients with disorders of peroxisome biogenesis had an impaired capacity to oxidize pristanic acid, whereas patients with classical Refsum syndrome did not. This finding suggests that the oxidation of pristanic acid takes place in the peroxisome. In classical Refsum disease (Fig. 12) the defect is confined to the formation of pristanic acid, whereas in the disorders of peroxisome biogenesis the further metabolism of pristanic acid is impaired. Several questions about the subcellular localization and interrelationships between phytanic acid and pristanic acid metabolism are unresolved and are being investigated at this time. The α-oxidation of phytanic acid to pristanic acid involves two separate reactions, namely hydroxylation and decarboxylation. The previously cited reports that point to the mitochondrial localization of this reaction (Skjeldal and Tokke, 1987; Watkins *et al.*, 1990) measured [^{14}C]carbon dioxide from [^{14}C]phytanic acid, which provides information about the two steps combined. Use of [2,3-^{3}H]phytanic acid permits investigation of the α-hydroxylation reaction separately. Preliminary data (P. A. Watkins and S. J. Mihalik, unpublished observations) suggest that this hydroxylation may take place in the peroxisome. In any case, the data indicate that the overall degradation of phytanic acid involves both the mitochondrion and the peroxisome. The control, interaction, and transfer of metabolites between these two organelles require further study.

Until the subcellular localization of the α-hydroxylation has been determined more precisely, it is unclear whether or not classical Refsum disease is a peroxisomal disorder.

Medium- and Long-Chain Dicarboxylic Aciduria. Patients with disorders of peroxisome biogenesis excrete modestly elevated levels of medium-chain-length dicarboxylic acids in the urine (Bjorkhem *et al.*, 1984a; Rocchiccioli *et al.*, 1986). These acids include those with an even number of carbon atoms, such as adipic (C6), suberic (C8), and sebacic (C10) acids, as well as those with an odd number of carbon atoms, such as pimelic (C7) and azelaic (C9) acids, and 2-hydroxy odd- and even-chain dicarboxylic acids. The ratio of medium-chain (C10) to adipic (C6) acids and the accumulation of longer-chain (C12 and greater) acids indicate that in peroxisomal disorders there is at least a partial block in the degradation of long-chain dicarboxylic acids.

Reduced Levels of Cholesterol and Lipoproteins in Plasma. Relatively low levels of cholesterol in plasma (55–135 mg/dl) have been reported in ten patients with infantile Refsum disease (Budden *et al.*, 1986; Poll-The *et al.*,

1986b; Scotto *et al.*, 1982). LDL and HDL levels were both reduced (Scotto *et al.*, 1982). As noted above, peroxisomes have been shown to contain cholesterol biosynthesis enzymes (Keller *et al.*, 1986; Thompson *et al.*, 1987). Possibly the moderate hypocholesterolemia in patients with peroxisomal disorders is due to impaired function of this pathway.

Complementation Analysis and Disorders of Peroxisome Biogenesis

Complementation analysis has revolutionized the classification of peroxisomal disorders and, in conjunction with new techniques in cell and molecular biology, promises to lead to a better understanding of the basic defects. The general principle is that fibroblasts from two different patients, both deficient in a peroxisomal process, are induced to fuse (Gravel *et al.*, 1975); the resulting multinuclear cells are examined for their ability to carry out this metabolic process. Restoration of activity can only occur if each cell line provides the gene product defective in the other. Therefore, cell lines that complement each other in this way thus must represent distinct genotypes.

Complementation analysis of the disorders of peroxisome biogenesis has been aided by the availability of at least four entirely different techniques to test for complementation, and results have been congruous. In the earliest studies (Brul *et al.* (1988a) used restoration of the activity of acyl-CoA:dihydroxyacetone phosphate acyltransferase, which catalyzes the first step in plasmalogen synthesis, as the index of complementation. Roscher *et al.* (1989) used as the index a method that determines the ratio between the peroxisomal and microsomal steps of plasmalogen synthesis (Roscher *et al.*, 1985); a significant increase in this ratio indicates that complementation has occurred. Other biochemical indices are the restoration of the capacity to oxidize phytanic acid (Poll-The *et al.*, 1989) or very long-chain fatty acids (McGuinness *et al.*, 1990). An elegant method involves the formation of catalase-containing intracellular particles by immunofluorescence techniques (Santos *et al.*, 1988b; Zoeller *et al.*, 1989).

Table IV shows the results of complementation analyses that have been published. The phenotype distribution patterns established in different laboratories appear to be the same. Furthermore, as indicated in the table, in several instances the same cell line has been tested in several laboratories, thus permitting a direct comparison between the group designations in different laboratories. For example, Brul *et al.* (1988a) group 2 appears to be the equivalent of Roscher *et al.* (1989) and McGuinness *et al.* (1990) groups 1 and of Poll-The *et al.* (1989) group 3. While a more systematic exchange of cell lines is desirable, and indeed in progress, the limited data available permit the following conclusions: (1) Congruous results appear to be obtained with the various indices of complementation, thus supporting the general validity of the approach. (2) When the capacity to oxidize VLCFA is used as the index (McGuinness *et al.*, 1990), cell lines from patients with single-enzyme defects (acyl-CoA oxidase deficiency and bifunctional enzyme deficiency) complement each other and those from patients with disorders of peroxisome biogenesis. With phytanic acid oxidation as the index (Poll-The *et al.*, 1989), complementation is demonstrated between cell lines from patients with classical Refsum disease, rhizomelic chondrodysplasia punctata (RCDP), and the disorders of peroxisome biogenesis. These not unexpected findings are consistent with the concept that there is a fundamental difference between the disorders of peroxisome biogenesis, the single-enzyme defects, and RCDP. (3) Complementation analysis of the disorders of peroxisome biogenesis did yield unexpected results that have important implications for disease classification and provide the basis for correlative enzymology and molecular biology studies that can clarify disease mechanisms. Cell lines from patients with the four phenotypes established on an historical basis (Zellweger, NALD, IRD, and HP) are found within a single complementation group, namely Brul *et al.* (1988a) group 2, Roscher *et al.* (1989) group 1, Poll-The *et al.* (1989) group 3, and McGuinness *et al.* (1990) group 1. Furthermore, the most common phenotype, Zellweger syndrome, is represented in at least four other complementation groups. These was no correlation between phenotype and complementation group (Roscher *et al.*, 1989). These findings suggest strongly that the historically based phenotype designations that have been cited here, and are in general use in the literature, do not represent distinct genotypes. Six complementation groups have been identified so far, however, suggesting that there are at least six distinct genotypes. Systematic comparison of various indices of complementation and exchanges of cell lines from various laboratories are in progress. Completion of these studies should permit assignment of patients to specific genotype categories and it is likely that additional complementation groups will be identified.

The identification of the distinct complementation groups provides the opportunity to conduct studies of the genetics and molecular and cell biology in cell lines that are homogeneous in respect to the primary genetic defect. Conversely, the lack of correlation between phenotypes makes it unwise to conduct studies of basic disease mechanisms in cell lines of patients with disorders of peroxisome biogenesis without definition of their complementation group assignment.

A few studies have already used complementation analysis to clarify basic disease mechanisms. Brul *et al.* (1988b) studied the kinetics of complementation and found that in most instances it was relatively slow (8–10 hr) and was inhibited by cycloheximide, which inhibits protein synthesis. In contrast, with two cell lines, complementation was rapid (<3 hr) and was not inhibited by cycloheximide, "suggesting that the components necessary for the assembly of peroxisomes can occur in a stable form in the cytosol of certain fibroblast cell lines but not in others." Zoeller *et al.* (1989) studied a Chinese hamster ovary (CHO) mutant cell line that is deficient in peroxisomes. They showed restoration of normal peroxisome structure and function when these cells were fused with fibroblast cell lines from Zellweger patients. They used two Zellweger cell lines. One of these cell lines (GM 4340) had been assigned to complementation group 4 by Roscher *et al.* (1989). The complementation group of the other cell line (GM 228) has not yet been specified.

Tsukamoto *et al.* (1991) recently reported the exciting finding that addition of a 35-kD peroxisomal membrane protein restores the assembly of peroxisomes in mutant CHO cells that are deficient in peroxisomes. This protein contains 303 amino acids and its cDNA nucleotide sequence has been determined; it does not contain the Ser–Lys–Leu–COOH peroxisomal targeting signal (Gould *et al.*, 1988). This protein may be a "component of a putative peroxisomal signal receptor or of a transmembrane import machinery of newly synthesized proteins."

Diagnosis of Disorders of Peroxisome Biogenesis

Most commonly the disorders of peroxisome biogenesis present in the *neonatal* period with hypotonia, failure to feed, seizures, and varying degrees of dysmorphic features. They must be distinguished from other, more common causes of life-threatening illness in the neonatal period, such as infections, and from many other types of inborn errors of metabolism, 56 of which are listed in a review by Burton (1987). This is a formidable task. Experienced clinicians may recognize the characteristic facies of infants with classical Zellweger syndrome, but dysmorphic features are often mild or nonspecific and may be absent. A review of the clinical features of patients diagnosed at the Kennedy Institute indicates that the most common reason for seeking the diagnostic assays was the combination of hypotonia, failure to thrive, and seizures rather than the presence of characteristic dysmorphic features. The dysmorphic features, Brushfield spots, and hypotonia may lead to confusion with Down syndrome. Dysmorphic features, hypotonia, seizures, renal cysts, and neuronal migrational defects are also present in the most severe form of glutaric acidemia type 2 (Loehr *et al.*, 1990; Colevas *et al.*, 1988). Moreover, to complicate diagnosis, pipecolic acid levels in glutaric aciduria type

2 are elevated to the same extent as in the disorders of peroxisome biogenesis (R. Kelley, personal communication).

In *childhood* the disorders of peroxisome biogenesis must be distinguished from conditions that cause psychomotor retardation, eye disease, hearing deficits, liver disease, and other disorders associated with dysmorphic features. The patients who live beyond the first half-year show profound or severe psychomotor retardation. The more mildly involved children may acquire the ability to sit or stand. If they do walk independently, gait is unsteady or ataxic. Rarely are they able to say a few words. The hypotonia present in the neonatal period often is followed later by increased tone. Some children make slight psychomotor gains during the first 1–4 years, but then regress with quadriparesis and total loss of vision, possibly related to further demyelination. The severe neurological deficits in childhood thus require differentiation from static cerebral palsy and from other neurodegenerative disorders. The presence of an enlarged liver, dysmorphic features, and eye pathology would direct attention toward peroxisomal disorders, but these findings are rarely specific, and most patients in this category have been identified as part of a systematic survey for metabolic disorders. While some degree of progressive illness occurs, the overall course differs from most of the lysosomal disorders and X-linked ALD, where progressive neurological disability follows a period of normal or near-normal early development.

In some patients with disorders of peroxisome biogenesis, the enlarged liver and impaired liver function have dominated the clinical picture, thus suggesting other types of liver disease. Optic atrophy can be confused with Leber optic atrophy (Ek *et al.*, 1986), and the combination of pigmentary degeneration of the retina and deafness can lead to misdiagnosis as Usher syndrome (Noetzel *et al.*, 1983; Kelley *et al.*, 1986). While patients with Usher syndrome may have psychiatric or neurological disturbances, they function at a much higher mental level than do patients with disorders of peroxisome biogenesis (McVay, 1969). Some patients with neonatal ALD are in relatively stable condition in early adulthood and must be differentiated from those with X-linked ALD. The neonatal ALD patients do not show the normal early development characteristic of X-linked ALD, have pigmentary degeneration of the retina, and show laboratory evidence of peroxisomal dysfunctions over and above the defect of very long-chain fatty acid degradation that is present in both disorders.

As awareness of peroxisomal disorders has increased, there is now the risk of overdiagnosis. One patient who had Killian/Teschler-Nicola syndrome (Teschler-Nicola and Killian, 1981) was misdiagnosed as having a peroxisomal disorder because of the superficial resemblance of facial features to those of Zellweger syndrome.

Laboratory Diagnosis. A plethora of laboratory tests for the diagnosis of disorders of peroxisome biogenesis has been developed. The challenge is to select those that are informative, reliable, and available. We utilize the assay of plasma very long-chain fatty acids (VLCFA) as the initial diagnostic test, and, depending upon circumstances, follow this up with other assays, all of which are aimed at detecting the abnormalities listed in Table III. The plasma VLCFA assay in our laboratory utilizes capillary gas–liquid chromatography (A. B. Moser *et al.*, 1984; H. W. Moser *et al.*, 1981) as modified recently (H. W. Moser and Moser, 1990a). Other valid techniques for VLCFA in plasma (Aubourg *et al.*, 1985; Alberghina *et al.*, 1984; Kobayashi *et al.*, 1983), red blood cells (Tsuji *et al.*, 1981b; Antoku *et al.*, 1987), dried "Guthrie" blood spots (Nishio *et al.*, 1986), white blood cells (Molzer *et al.*, 1982), and cultured skin fibroblasts (A. B. Moser *et al.*, 1984; H. W. Moser *et al.*, 1980b) are available. Table V shows the results of plasma VLCFA assays in the peroxisomal disorders. The levels of saturated VLCFA (C26:0 and C24:0) are increased in all of the disorders of peroxisome biogenesis, as are levels of monounsaturated VLCFA (A. B. Moser *et al.*, 1984). Poulos *et al.* (1989) found that levels of polyunsaturated VLCFA were increased in the plasma of Zellweger syndrome patients but not in plasma in the less severe neonatal ALD and infantile Refsum syndrome. Table V shows also that the excess of saturated VLCFA is most severe in Zellweger syndrome and somewhat less so in neonatal ALD and infantile Refsum disease. Barth *et al.* (1987) reported two siblings with a relatively mild variant of Zellweger syndrome, with only subtle, although significant, elevation of plasma VLCFA levels. The combined experience in several laboratories indicates that some degree of elevation of VLCFA in plasma or other accessible tissues is present in all patients with disorders of peroxisome biogenesis and thus can serve as the initial diagnostic test.

Additional diagnostic assays are often advisable for the following reasons: (1) The phenotypes of defects of singe peroxisomal β-oxidation enzymes often mimic those of the disorders of peroxisome biogenesis, and they can be distinguished only through tests *other* than the VLCFA. (2) For the disorders of peroxisome biogenesis, demonstration of multiple peroxisomal defects confirms the diagnosis. (3) The patterns of peroxisomal defects are variable. These variations help to characterize disease states and provide leads to therapeutic approaches (see p. 52).

The biochemical assays that are particularly useful include measurement of plasma and urinary pipecolic acid levels by an isotope dilution method (Kelley, 1991). While pipecolic acid concentrations vary with age (Table VI), elevated levels are demonstrable at all ages. Moderate elevations occur in nonperoxisomal liver disease, and increases equivalent to that in Zellweger syndrome also occur in glutaric aciduria type 2 (R. Kelley, personal communication). Moderate age-dependent increases in the level of plasma phytanic acid are noted in the disorders of peroxisome biogenesis (Lazarow and Moser, 1989; H. W. Moser and Moser, 1990b; Wanders *et al.*, 1987c), but the degree of elevation is slight in comparison to classical Refsum disease and the levels may be normal. Abnormally high levels of 3α,7α,12α-trihydroxy-5β-cholestanoic acid and of 3α,7α-dihydroxy-5β-cholestanoic acid (DHCA) are consistently found in the disorders of peroxisome biogenesis (Hanson *et al.*, 1979; Monnens *et al.*, 1980; Mathis *et al.*, 1980; Eyssen *et al.*, 1985; Gustafsson *et al.*, 1983; Clayton *et al.*, 1987; Poulos and Whiting, 1985). These are highly sensitive indices of peroxisomal dysfunction, and help pinpoint the site of the metabolic lesion in the single-enzyme defects, but specialized equipment such as fast atom bombardment-mass spectrometry may be required for definitive identification (Lawson *et al.*, 1986; Setchell and Vestal, 1989). Plasmalogen levels in red blood cell membranes are markedly reduced in patients who are less than 20 weeks old (Wanders *et al.*, 1986a) but normalize thereafter (Fig. 14). We utilize the analytical procedure described by Bjorkhem *et al.* (1986) in which the plasmalogens are converted into dimethylacetals.

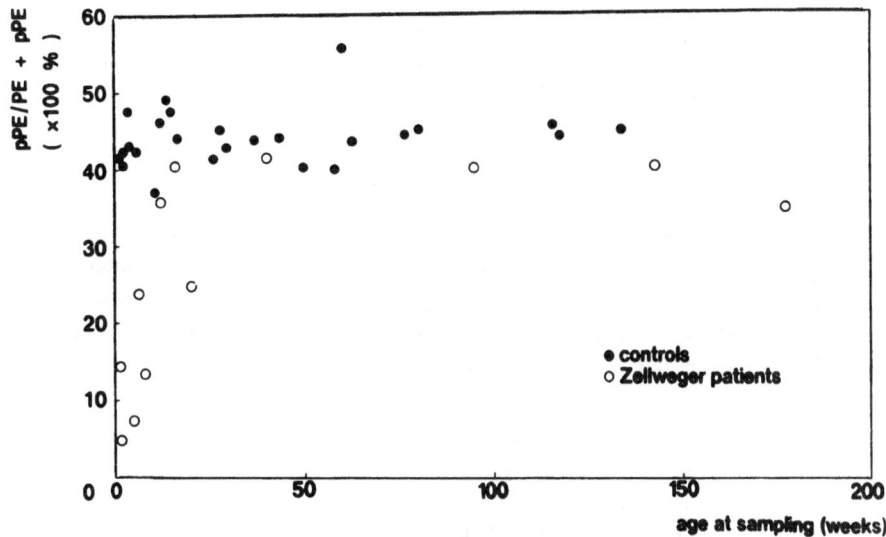

Fig. 14. Plasmalogen phosphatidylethanolamine (see Fig. 2) content of erythrocyte membranes from 25 controls and 12 Zellweger patients as a percentage of the total phosphatidylethanolamine fraction. Note that after 20 weeks of age this ratio becomes normal in Zellweger patients even without dietary supplementation of plasmalogens or their precursor, so that measurement of this ratio is of diagnostic value mainly in early infancy. (From Wanders *et al.*, 1986a. Used by permission.)

All of the above assays can be performed on 1- to 2-ml samples of venous blood, and in most instances permit specific diagnosis. Additional information can be obtained by study of cultured skin fibroblasts, including a variety of enzyme assays, assessment of peroxisome structure, and complementation group assignment, and they aid in the interpretation of prenatal studies that may be undertaken in the future. Apart from diagnosis, study of cultured skin fibroblasts is of central importance for understanding the basic defects in individual patients and the cell and molecular biology of peroxisomal disorders in general. We recommend, therefore, that such cultures be established.

Of particular diagnostic value are assays that cannot be performed in blood samples. These include the subcellular localization of catalase, plasmalogen synthesis, and phytanic acid oxidation and complementation analysis. The levels (A. B. Moser et al., 1984a) and oxidation (Wanders et al., 1987e) of VLCFA can also be determined in cultured skin fibroblasts. Immunoblot studies in cells from patients with disorders of peroxisome biogenesis demonstrate low levels or absence of peroxisomal acyl-CoA oxidase and bifunctional enzyme and the presence of immature 3-oxoacyl-CoA thiolase (Schram et al., 1986; Chen et al., 1987; Balfe et al., 1990). The subcellular localization of catalase in cultured skin provides the key information that sets the disorders of peroxisome biogenesis apart from the other peroxisomal disorders and from all other diseases states. In patients with disorders of peroxisome biogenesis catalase activity is in the cytosol, whereas in normal individuals and in all other disease states it is in the sedimentable fraction (Santos et al., 1985; Wanders et al., 1984; Arias et al., 1985). Deficient activity of the peroxisomal reactions involved in plasmalogen synthesis is a consistent and diagnostically important abnormality in the disorders of peroxisome biogenesis. The activity of acyl-CoA:dihydroxyacetone phosphate acyltransferase is diminished (Schutgens et al., 1984; Datta et al., 1984). An alternate procedure which we use in our laboratory is the double-label/double-isotope procedure devised by Roscher et al. (1985). The substrates are [1-^{14}C]hexadecanol and [9,10-^{3}H]sn-hexadecylglycerol. Radioactivity in the plasmalogen fraction is measured after incubation with these substrates. As shown in Figure 3, the ^{14}C-labeled substrate requires the peroxisomal components of the synthetic pathway, while tritium-labeled substrate requires the microsomal components and thus serves as an internal standard. Plasmalogen synthesis is impaired in the Zellweger syndrome (Schutgens et al., 1984; Datta et al., 1984; Roscher et al., 1985) and also in neonatal ALD (Wanders et al., 1987b), infantile Refsum disease (Wanders et al., 1986e), and hyperpipecolic acidemia (Wanders et al., 1988d). The abnormality is more severe in Zellweger syndrome than in neonatal ALD (Fig. 15). Plasmalogen synthesis is also impaired in RCDP (Heymans et al., 1986; Heymans

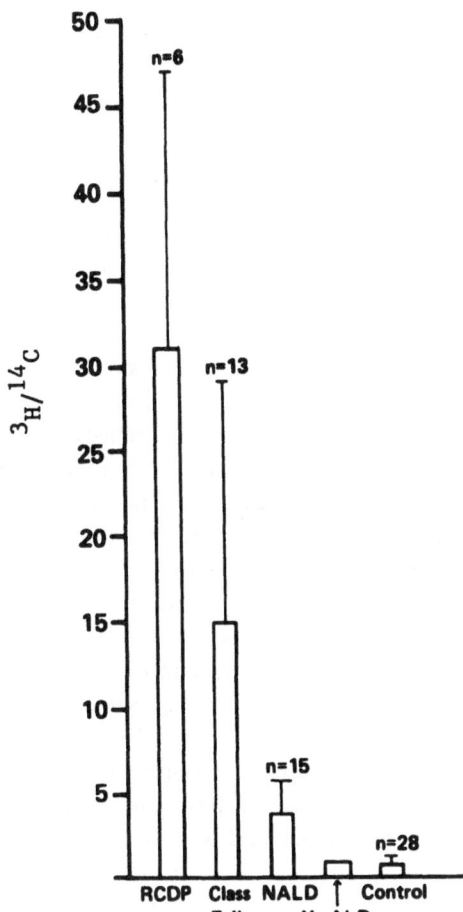

Fig. 15. The ratio of the peroxisomal to microsomal steps of plasmalogen synthesis, determined using the method of Roscher *et al.* (1985). The abnormality is most severe in RCDP (G. Hoefler *et al.*, 1988). In neonatal ALD the defect is much less severe, while the ratio is normal in X-linked ALD. (From Lazarow and Moser, 1989. Used by permission.)

et al., 1985; G. Hoefler *et al.*, 1988). In fact, plasmalogen synthesis in RCDP is compromised more severely than it is in Zellweger syndrome (G. Hoefler *et al.*, 1988). Phytanic acid oxidation is impaired severely in all of the disorders of peroxisome biogenesis (Poll-The *et al.*, 1989; Poulos *et al.*, 1985). The defect in phytanic acid oxidation is as severe as in classical Refsum syndrome even though, as already noted, plasma levels of phytanic acid are not increased as much. The importance of *complementation analyses* (which utilize cultured fibroblasts) for the classification and basic understanding of the disorders of peroxisome biogenesis has already been emphasized.

Studies in Other Tissues. Liver biopsy is used to determine the number and size of peroxisomes. Catalase cytochemistry and appropriate positive controls must be used (Roels and Goldfischer, 1979), which is of historical significance, since it provided the first clue to the very existence of the peroxisome disease category (Goldfischer *et al.*, 1973). It continues to be important for clinical diagnosis and the understanding of disease mechanisms. Morphometry has provided evidence that liver peroxisomes are abnormally large in peroxisomal acyl-CoA oxidase deficiency (Poll-The *et al.*, 1988). The pseudo-Zellweger disease entity was first identified by liver biopsy (Goldfischer *et al.*, 1986). Sensitive biochemical and morphological techniques have been used to confirm the diagnosis of Zellweger syndrome by rectal biopsy (Shimozawa *et al.*, 1988). Wanders *et al.* (1985) have demonstrated that the plasmalogen synthesis in thrombocytes of Zellweger disease patients is diminished.

Prenatal Diagnosis. All of the disorders of peroxisome biogenesis can be diagnosed prenatally by studies of cultured amniocytes or chorionic villus cells (Table VII). We have monitored more than 340 at-risk pregnancies and have identified 68 affected fetuses. The prenatal diagnosis was confirmed in all instances by follow-up of live births or postmortem study of the fetus when pregnancy was interrupted. In our laboratory we utilize two independent techniques, namely the measurement of VLCFA levels (A. B. Moser *et al.*, 1984; Solish *et al.*, 1985) and of plasmalogen synthesis with the double-substrate/double-label technique of Roscher *et al.* (1985) or the measurement of acyl-CoA:dihydroxyacetone phosphate acyltransferase activity (Schutgens *et al.*, 1985). Other techniques involve the assay of VLCFA β-oxidation (Wanders *et al.*, 1987f) or phytanic acid oxidation (Poulos *et al.*, 1986c). Since the above tech-

TABLE VII.   Kennedy Institute Experience
with Prenatal Diagnosis of Peroxisomal Disorders—
October 1990

	Test number and results		
Diagnosis at risk	Total	Positive	Percent positive
X-linked ALD	132	31	23
Zellweger syndrome	141	28	20
Neonatal ALD	31	3	10
RCDP	12	2	17
Other	24	4	17
Total	340	68	20

Note: 340 pregnancies monitored; 89 chorionic villus sample, 251 amniocentesis.

niques are performed with cultured cells, they impose a delay of 2 or more weeks before an answer can be obtained. Studies of cultured chorion cells carry the risk of a false-negative result due to maternal overgrowth (Carey *et al.*, 1986). Techniques based upon the demonstration of the lack of catalase-containing peroxisomes by immunofluorescence microscopy (Wanders *et al.*, 1986c, 1989; Roels *et al.*, 1987a,b) can avoid this delay, but require positive controls, and would not detect the single-enzyme defects that may mimic Zellweger syndrome (Gold-fischer *et al.*, 1986) or neonatal ALD (Poll-The *et al.*, 1988). Elevated levels of VLCFA (Rocchiccioli *et al.*, 1987) or impaired VLCFA degradation can be demonstrated by direct analysis of chorion villus biopsy samples, but in our experience (Ann Moser, unpublished observation) results may not be decisive and we recommend that they be confirmed by subsequent measurements in cultured samples.

Bjorkhem *et al.* (1984b) proposed that affected fetuses might be identified by measurement of VLCFA levels in amniotic fluid, but Jakobs *et al.* (1989) reported that this technique yielded uncertain results in one study. Stellaard *et al.* (1988) have reported a decisively elevated level of trihydroxycoprostanic acid (THCA) in the amniotic fluid of one fetus with Zellweger syndrome. If confirmed, this technique may prove valuable for more rapid prenatal identification, but it would be applicable only when elevated levels of bile acid intermediates had been demonstrated in previously affected family members.

Carrier Detection

Carrier identification has not been achieved. Presumably this is because all of the abnormalities detected in affected patients represent changes secondary to peroxisomal dysfunction and the underlying defect has not yet been defined for any of the disorders of peroxisome biogenesis.

Genetics

An autosomal recessive mode of inheritance is likely for all of the disorders of peroxisome biogenesis. Information is most complete for the Zellweger syndrome. In his review of 111 cases, Heymans (1984) found 55 males and 56 females. In 17 of 90 families, two or more sibs of either sex were affected. Parental consanguinity was observed in 17%.

Since the initial cases of infantile Refsum syndrome were male (Scotto *et al.*, 1982; Poll-The *et al.*, 1987), a sex-linked mode of inheritance was considered at first. Subsequently, however, female cases have been reported (Weleber *et al.*, 1984) and also observed in our clinic (H. W. Moser, unpublished observation). We have noted one instance of a woman who had two male children with neonatal ALD from different fathers. With these rare exceptions, the preponderance of experience points to an autosomal recessive mode of inheritance. Confirmation must await techniques for carrier identification. Consistent with an autosomal recessive inheritance is the finding of Naritami *et al.* (1989), who have demonstrated abnormalities in the proximal arm of chromosome 7 in two patients with Zellweger syndrome. One patient had a microdeletion [del(7)(q11.22q11.23)] and the other had a pericentric inversion of chromosome 7 (Naritami *et al.*, 1989). The findings led to a tentative gene assignment to 7q11.23. The complementation group assignment of these two cases is not known. It must also be kept in mind that the complementation analyses indicate genetic heterogeneity (Table IV), so that the possibility of X-linked recessive inheritance of a small subgroup cannot be excluded.

Disorders of peroxisome biogenesis have been noted in all races and countries. There are no precise figures about their incidence. Danks estimated that the incidence of Zellweger syndrome in Australia was 1:100,000 (Danks *et al.*, 1975). Heymans (1984) considered this to be an underestimate and placed the incidence in the Netherlands at 1:25,000 to 1:50,000 (Zellweger, 1987). Zung *et al.* (1990) proposed that the Zellweger syndrome may have a high incidence in the Karaite community in Israel. During the last 5 years our laboratory has identified on the average 45 new patients per year with disorders of peroxisome biogenesis in the United States (Table VIII). Based upon a U.S. birth rate of 3.6 million per year, this yields an incidence of 1.25:100,000. There is no doubt that this is an underestimate, since not all patients are recognized, and not all recognized cases are tested in our laboratory.

Therapy

The potential of postnatal therapy is limited because of the severe multiple malformations present at birth. For infants with classical Zellweger syndrome, therapy is confined to supportive care, with addition of vitamin K when prothrombin deficiency exists in association with liver disease. More active therapy may be considered for children with the somewhat milder phenotypes, such as neonatal ALD and infantile Refsum disease. In spite of the existence of profound or severe mental retardation, gratifying albeit limited progress has been achieved with multidisciplinary habilitative approaches, including hearing aids and communication training, ophthalmology, and physical and occupational therapy. It is pos-

TABLE VIII. Peroxisomal Disease Patients Identified
by Kennedy Institute Screening Program 1979–1991

Year	X-ALD Hemizygote	Other peroxisome disorders	Total
1979	20	1	21
1980	27	0	27
1981	52	5	57
1982	107	17	124
1983	62	33	95
1984	72	45	117
1985	81	39	120
1986	80	64	144
1987	87	49	136
1988	100	51	151
1989	82	53	135
1990	92	43	135
1991 (2 months)	8	8	16
Total	870	408	1278
Percent United States and Canada	67	87	

Note: The Kennedy Institute conducts a diagnostic program for peroxisomal disorders that tests a significant but unknown proportion of patients with these disorders mainly in the United States and Canada. The number of new cases identified each year thus can provide a minimum value for their incidence. The diagnostic procedure for the peroxisomal disorders other than X-linked ALD did not become available until 1981, and this accounts for the low figure in the early years. The proportion of "non-U.S. and Canada" cases is diminishing, since these tests are now conducted also in Europe, Australia, and Japan. The values for 1991 represent only the first 2 months.

sible to normalize, at least in part, some of the biochemical abnormalities, and several of the more mildly affected children have been placed on special diets. The regimens that are being employed include oral ether lipid therapy (Wilson *et al.*, 1986; Holmes *et al.*, 1987a; Poulos *et al.*, 1990) and the dietary restriction of very long-chain fatty acids (Van Duyn *et al.*, 1984) and phytanic acid (Steinberg, 1989; C. R. Greenberg *et al.*, 1987). Because of the phenotypic variability, it has not yet been possible to assess the clinical effectiveness of these interventions. These approaches have normalized plasma levels of phytanic acid, increased the levels of red blood cell plasmalogens, and reduced the plasma VLCFA levels. An additional potential approach is the administration of ursodeoxycholic acid (Colombo *et al.*, 1990), which may reduce the levels of bile acid intermediates. The administration of peroxisome proliferators such as clofibrate was tested, and did not induce the formation of catalase-containing peroxisomes or improve clinical status (Lazarow *et al.*, 1985).

Disorders Due to Deficient Function of a Single Peroxisomal Enzyme

At this time this category includes seven disorders (Table I) in each of which there is deficient function of an enzyme that has been localized to the peroxisome in human beings. The subcellular localization of the enzyme deficient in classical Refsum disease is uncertain at this time. Some authors include it in the peroxisome disease category, while others do not. The group identified so far includes four disorders that involve one of the peroxisomal β-oxidation enzymes, and a recently described form of riboflavin-responsive glutaric aciduria. The other members in this category are hyperoxaluria type 1, knowledge about which is advancing rapidly, and acatalasemia, a rare disorder first described in 1946, that surprisingly is usually benign, and involves the classic peroxisome marker enzyme catalase. It is likely that additional disorders in which there is a defect of a single peroxisomal enzyme will be identified in the future.

Defects of Peroxisomal β-Oxidation Enzymes

The peroxisomal β-oxidation pathway involves four sequential steps shown in Figure 1. X-linked ALD, in which the primary defect probably involves the enzyme lignoceroyl-CoA ligase, is by far the most common, and it is probably the most common peroxisomal disorder. In spite of the apparently close biochemical relationship, there is a striking clinical dichotomy. Patients with X-linked ALD always are normal at birth and during the early developmental period. There are no dysmorphic features and symptoms do not begin until age 4 years or later, and sometimes not until midlife. In contrast, patients with single-enzyme defects that involve the three other enzymes of the peroxisomal β-oxidation system already are severely abnormal at birth and show dysmorphic features and multiple organ involvement, so that their clinical features resemble those in the disorders of peroxisome biogenesis, and they are often misdiagnosed as Zellweger syndrome or neonatal ALD.

X-Linked Adrenoleukodystrophy (ALD). *Clinical Features.* What is now referred to as the childhood cerebral form of X-linked ALD (X-ALD) was first described by Siemerling and Creutzfeldt (1923). They reported a boy who developed normally until 3–4 years of age, when increasingly dark skin pigmentation was noted. At 6 years of age he became "wild" and hyperactive, running around without purpose, and would not sit still. His speech and gait deteriorated, he became incontinent and had difficulty swallowing. He developed a spastic paralysis of legs, became totally helpless, and died at age 7 years. Postmortem study showed widespread demyelination in the brain with perivascular accumulation of lymphocytes and adrenocortical atrophy, and this form of the disease represents the most common phenotype. In 1974 Powers and Schaumburg made the key observation that adrenal cortical cells and brain macrophages of patients with childhood ALD contained characteristic lamellar lipid-soluble cytoplasmic inclusions (Powers and Schaumburg, 1974). Budka *et al.* (1976) and J. W. Griffin *et al.* (1977) reported similar inclusions in men with adrenal insufficiency and spastic paraparesis, and named this condition adrenomyeloneuropathy (AMN). Because of similarities in the histopathology, both groups of authors considered AMN to be an adult variant of childhood ALD, and this supposition was strengthened by the observation that one of the AMN patients had a nephew with childhood ALD (J. W. Griffin *et al.*, 1977). The biochemical basis of ALD was defined in 1976 by Igarashi *et al.* (1976b), who demonstrated a striking excess of saturated very long-chain fatty acids in the postmortem adrenal cortex and cerebral white matter of ALD patients. Between 1978 and 1981 noninvasive diagnostic assays that utilize cultured skin fibroblasts, plasma, or red blood cells were developed (Kawamura *et al.*, 1978; H. W. Moser *et al.*, 1980b, 1981; Tsuji *et al.*, 1981b). Application of these assays has shown that X-ALD is more common than had been realized, and that the phenotype is varied.

Table IX lists the phenotypes of X-linked ALD as well as their relative frequency. Figure 16 shows the age of onset of neurological symptoms and Figure 17 the mean interval between the onset of neurological symptoms and vegetative state or death in untreated patients. The *childhood cerebral* form is the most serious and frequent phenotype. The boy almost always develops entirely normally until age 4–8 years, when he is noted to be hyperactive, withdrawn, or emotionally labile and school performance drops off. Difficulty in understanding speech in a noisy room or over the telephone are common early symptoms and

TABLE IX.   Phenotypes among X-Linked ALD
Hemizygotes

Type	Frequency (%)
1. Childhood cerebral	48
2. Adolescent cerebral	5
3. Adult cerebral	3
4. Adrenomyeloneuropathy	25
5. Addison only	10
6. Asymptomatic and presymptomatic	8

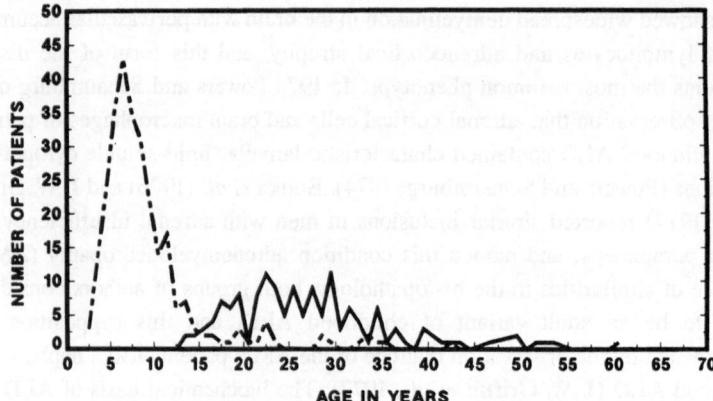

Fig. 16. Age of onset of neurological symptoms of cerebral forms of adrenoleukodystrophy (-----) and adrenomyeloneuropathy (—). Note that the cerebral forms of ALD are by far most common in childhood, with a peak at 7 years of age. The earliest onset of AMN, the spinal cord form of the disease, was at 12 years of age. The most common age of onset is in the third decade, but in a few men initial symptoms were delayed until their 50s. Note that there are a few instances in which cerebral symptoms presented in adulthood; Fig. 17 shows that this mode of presentation is nearly as rapidly progressive as the childhood cerebral form. (From Moser *et al.*, 1992. Used by permission.)

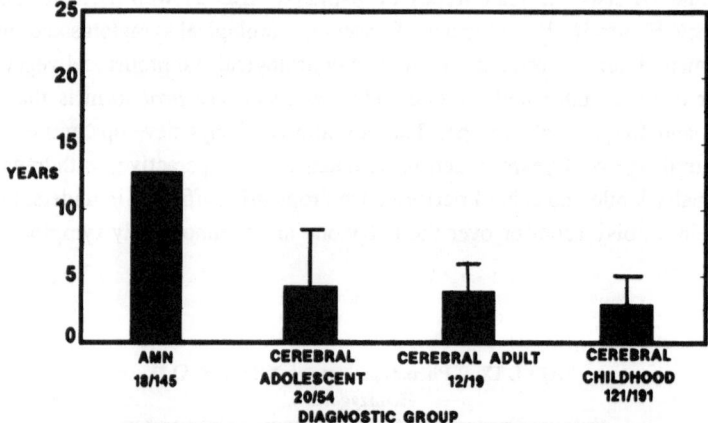

Fig. 17. Averaged time interval between the initial neurological symptom and an apparently vegetative state or death. For the childhood cerebral form, the mean interval between initial neurological symptoms and death was 2.29 ± 2.01 years, and the interval between first symptom and death was 3.48 ± 2.75 years. Note that the prognosis for the adult cerebral form is nearly the same as that for the childhood cerebral form, probably relating to the inflammatory response present in both. The bar shown for the AMN patients applies only to the 18 men who had died or reached a vegetative state. The remaining 127 patients were alive and only partially disabled at time of last follow-up, so that the difference in prognosis from the cerebral forms is even greater than indicated by the figure.

reflect impaired auditory discrimination. Visual impairment is an early symptom in about one-third of the cases, and may include field cuts, impaired visual acuity, and cortical blindness. Strabismus may be an early symptom. About one-third of the patients have focal or generalized seizures; indeed, a seizure may be the initial symptom. What at first may have been diagnosed as an attention deficit disorder evolves to include clear-cut signs of parietal lobe dysfunction, such as constructional or dressing apraxia, and disturbances in language. The illness tends to advance rapidly with spastic paresis, difficulty in swallowing, and loss of speech, vision, and hearing, leading to an apparently vegetative state within 1.9 ± 2 years from the time of the first definite neurological symptom. In 86% of cases the neurological manifestations precede the recognition of adrenal insufficiency, but 85% of the neurologically involved patients showed an impaired cortisol response to ACTH stimulation at the time of initial neurological study (Moser *et al.*, 1991b). The history often reveals earlier episodes of unexplained vomiting, dehydration, weakness, and hyperpigmentation that, in retrospect, were attributable to primary adrenal insufficiency.

The term *adolescent cerebral ALD* is applied to patients who present between ages of 10 and 21 years with symptoms that resemble those in the childhood cerebral form. The *adult cerebral* form presents with behavioral disturbances, dementia, seizures, or other adult cerebral neurological deficits after age 21 years (Esiri *et al.*, 1984; Bresnan and Richardson, 1979; Powers *et al.*, 1980a; Turpin *et al.*, 1985; James *et al.*, 1984; Kitchin *et al.*, 1987; Sereni *et al.*, 1987). The illness may be misdiagnosed as schizophrenia, multiple sclerosis, or brain tumor. Figure 17 shows that the rate of progression of the adult cerebral form is nearly as rapid as in the childhood cerebral form.

Adrenomyeloneuropathy (AMN) is the second most common form of ALD. It contrasts with the childhood cerebral form because of its later age of onset (Fig. 16), rate of progression (Fig. 17), and site of main neurological involvement. The neuropathology of AMN also differs from that of the childhood cerebral form (Schaumburg *et al.*, 1975). The spinal cord is involved mainly, Schaumburg *et al.*, 1977; Powers, 1985; Probst *et al.*, 1980; Budka *et al.*, 1976, 1983), with loss of myelinated axons in the long ascending and descending tracts. The pattern is consistent with a distal or "dying-back" axonopathy in that the greatest losses are observed in the (descending) lumbar corticospinal and the (ascending) cervical gracile and dorsal spinocerebellar tracts. The perivascular lymphocytic infiltration characteristic of the cerebral forms of the disease is much less prominent or lacking altogether. The mean age of onset of symptoms is 27.6 ± 8.7 years and the progression is over decades rather than years. The main neurological deficits are spasticity and weakness of the lower extremities, and impaired vibration sense most severe in the distal lower extremities. Bladder function is almost always impaired. Approximately one-half of the AMN patients also appear to have some degree of cerebral involvement. Mild to moderate abnormalities in brain MRI were noted in 46% of 76 AMN patients studied at Johns Hopkins Hospital, with lesions being most frequent in the parieto-occipital white matter and the optic radiations (A. J. Kumar, unpublished observations). About 60% of AMN patients showed relatively subtle neuropsychological abnormalities, most commonly a pattern of spared and impaired functions typical of subcortical dementia. Frontal executive functions and memory, particularly visual memory, appear to have been disproportionately affected (Edwin *et al.*, 1990). Adrenal function is impaired in most patients, and gonadal function in about one-third. In our series of 69 patients, adrenal insufficiency was present in 67%. Serum testosterone levels were abnormally low in 22% and luteinizing or follicle stimulating hormone levels were increased in 42% (H. W. Moser *et al.*, 1991b).

The *Addison-only* category is an important one that was recognized only recently. It includes patients who have Addison disease in the absence of neuro-logical signs or symptoms. This category is more common than had been recognized in the past. In a recent study conducted at the Tufts University Endocrine Clinic, five of eight male Addison disease patients had the biochemical defect of ALD (Sadeghi-Nejad and Senior, 1990), and we have identified another 25 patients in this category during the past year. It appears that in developed countries, where tuberculosis of the adrenal gland occurs infrequently, as many as 40% of male Addison patients may have the biochemical defect of ALD (H. W. Moser *et al.*, 1991b). Recognition of these patients is of great importance for genetic counseling. Detailed neurological examination and MRI may reveal subtle neurological abnormalities in about half the Addison-only patients who have the ALD biochemical defect. It is our impression that the majority will develop signs of AMN later in life.

In the Kennedy Institute series, 8% of males with the biochemical defect of ALD were *asymptomatic*. These persons were identified by measuring plasma VLCFA levels of at-risk relatives of symptomatic patients. It is possible that the percentage of asymptomatic persons is underestimated because of ascertainment bias, i.e., diagnostic assays are performed more frequently in persons with symptoms. About half the asymptomatic persons are boys who vary in age from newborn to 7 years and are still at serious risk of developing the childhood cerebral form of the disease, and the remainder are adolescents and men in their twenties, who have passed the age of greatest risk. In previous reports we referred to a 62-year-old man who had been tested for ALD because his brother had died of the adult cerebral form of the disease at age 48 years (H. W. Moser *et al.*, 1967). The

62-year-old was examined by a neurologist who reported that neurological examination was normal and that there was no evidence of adrenal insufficiency. When reexamined by the same neurologist 4 years later, he had abnormally increased deep tendon reflexes in the lower extremities and extensor plantar responses, findings compatible with mild AMN. We have demonstrated similar neurological findings as well as impaired cortisol response to ACTH in several young men with the biochemical defect of ALD who were free of symptoms. It is my impression that most asymptomatic adolescents and adults with the biochemical defect will later develop signs of AMN.

Symptomatic Heterozygotes.   A considerable number of reports describe neurological disability in women who are heterozygous for ALD (Heffungs *et al.*, 1980; Pilz and Schiener, 1973; Morariu *et al.*, 1982; H. W. Moser *et al.*, 1980a; Naidu *et al.*, 1986; O'Neill *et al.*, 1982, 1984; Penman, 1960; Dooley and Wright, 1985; Noetzel *et al.*, 1987). Progressive spastic paraparesis combined with sphincter disturbances, impaired vibration sense, and other sensory disturbances in the distal parts of the lower extremities are the most frequent findings, which resemble those in male AMN patients, but are somewhat milder and occur later (mean age of onset 37.8 ± 14.6 years, compared to 27.6 ± 8.7 in the male AMN patients). About 10% of the neurologically symptomatic women have dementia or visual disturbances. Exacerbations and remissions may occur (Cooley and Wright, 1985) and the diagnosis of multiple sclerosis was made in 11% of the symptomatic women.

We have used two approaches to estimate the incidence of neurological involvement in heterozygous women. Through our screening program we have identified 1032 carriers. A total of 165 women in this group had reported neurological involvement attributable to the heterozygote state. This is a minimum number, since the neurological status of many of these carriers is unknown. In the second approach, Dr. Sakkubai Naidu performed detailed neurological examinations of 60 heterozygotes who attended the annual meetings of the United Leukodystrophy Foundation. Almost all of the women came to the meeting because of concern about their sons health, and not because of concern about themselves. It is likely, therefore, that they are representative of the ALD carrier population as a whole. Fifty-three percent of the women had neurological signs or symptoms that varied in severity from mild hyperreflexia and vibratory sense impairment with little or no functional disability, to paraparesis requiring a wheelchair.

In contrast to affected males, adrenal insufficiency occurs rarely in heterozygotes; it has been reported only three times (Heffungs *et al.*, 1980; Penman, 1960; Pilz and Schiener, 1973), and we know of only three other instances in more than 1000 carriers (H. W. Moser *et al.*, 1991b). The rarity of adrenal insufficiency, which when present in men serves as a diagnostic alerting signal, combined with the still limited general awareness of this syndrome, probably accounts for the failure in diagnosis. Ninety-five percent of women identified as neurologically symptomatic carriers were relatives of symptomatic male ALD patients. It seems certain that there are additional symptomatic carriers who do not have an affected male relative. VLCFA assay of plasma or cultured skin fibroblasts would permit diagnosis in 85% of cases (Moser *et al.*, 1983).

The Biochemical Basis of X-Linked ALD. The main biochemical defect is the abnormal accumulation of very long-chain fatty acids (VLCFA). These fatty acids are saturated and unbranched, and vary in chain length from C24 (tetracosanoic or lignoceric acid) to C28 or longer. C26:0 (hexacosanoic acid) and C25:0 (pentacosanoic acid) are most abundant. The C26:0 excess is most striking in the cholesterol ester fraction of the adrenal gland and brain white matter, where it was first described by Igarashi *et al.* (1976b), and confirmed by other investigators (Molzer *et al.*, 1981; Menkes and Corbo, 1977; Ramsey *et al.*, 1979; Brown *et al.*, 1983; Reinecke *et al.*, 1985; Taketomi *et al.*, 1987). Smaller amounts of excess VLCFA are present in virtually all tissues and body fluids, including plasma (H. W. Moser *et al.*, 1981), red blood cells (Tsuji *et al.*, 1981b), and leukocytes (Molzer *et al.*, 1982). In brain gangliosides the VLCFA excess involves mainly C24:0 (Igarashi *et al.*, 1976a), while in brain tissue that is undergoing active demyelination the C26:0 excess is most striking in the cholesterol ester fraction. Sharp *et al.* (1991) have reported that monoenoic VLCFA are increased in the phosphatidylcholine fraction. Theda *et al.* (1987) assessed the temporal evolution of the VLCFA excess in ALD brain tissue, by comparing the degree of enrichment in areas of brain that were relatively intact with those that were undergoing active demyelination. They noted that in areas of brain where myelin was still intact, there was a 20-fold excess of C26:0 in the phosphatidylcholine fraction, whereas the cholesterol ester fraction had a normal fatty acid composition. The striking enrichment with C26:0 in the cholesterol ester fraction that has been reported by previous investigators was present only in the zones that were undergoing active demyelination and to a lesser extent in the zones where myelin had been replaced by glial tissue. They concluded that enrichment of VLCFA in phosphatidylcholine may be a primary event, and that the excess in the cholesterol ester fraction is a secondary event associated with myelin breakdown. Bizzozero *et al.* (1991) studied the composition of the fatty acids that are covalently bound to proteolipid. They noted that in normal brain tissue this fraction contains a relatively high proportion of VLCFA (8–21% unsaturated, 3.9–7.6% saturated.

The proportion is increased in X-linked ALD, but not in neonatal ALD or Zell-weger syndrome.

It is likely, but not yet fully established, that the accumulation of VLCFA is due to a defect in peroxisomal lignoceroyl-CoA ligase. I. Singh *et al.* (1984b) showed that the capacity to oxidize C24:0 and C26:0 was reduced to 17–37% of control in cultured skin fibroblasts, amniocytes, and leukocytes of patients with X-linked ALD, whereas the capacity to oxidize palmitate (C16:0) was unimpaired. Lazo *et al.* (1988) and Wanders *et al.* (1988c) demonstrated that the capacity to oxidize the lignoceroyl-CoA was intact, indicating that the defect involved the capacity to form the CoA derivative of VLCFA. Lignoceroyl-CoA formation was assayed in the peroxisomal, microsomal, and mitochondrial fractions of fibro-blasts from patients with X-linked ALD, and found to be reduced to 11.8–19.6% of control in the peroxisomal fraction only, whereas this activity was normal in the other fractions. We have cited previously the evidence that this enzyme is distinct from palmitoyl-CoA ligase (Bhushan *et al.*, 1986; H. Singh and Poulos, 1988; Lageweg *et al.*, 1991b). It is puzzling that the microsomal lignoceroyl-CoA ligase activity was unimpaired. Possibly microsomal lignoceroyl-CoA ligase differs from that in the peroxisome, or the defect involves an as yet undefined interaction between the enzyme and the peroxisome. More definitive conclusions must await purification and characterization of the enzyme.

Tsuji *et al.* (1981a, 1984) have shown that there is also increased synthesis of VLCFA by the microsomal elongating system. Street *et al.* (1990) have confirmed this finding.

These apparently discrepant results may reflect compartmentalization of very long-chain fatty acid metabolism, with the peroxisomal acyl-CoA ligase providing acyl-CoA esters for β-oxidation and the microsomal enzyme providing acyl-CoA esters for integration into cellular lipids. While the preponderance of evidence indicates that the VLCFA excess in X-linked ALD is due to impaired β-oxidation of these substances in the peroxisome, this β-oxidation system is not as severely impaired as in the disorders of peroxisome biogenesis or in the single-enzyme peroxisomal β-oxidation defects to be described below. This point has been emphasized in a recent paper by Street *et al.* (1990). It may account for the fact that in X-linked ALD the clinical abnormalities are more restricted and of later onset than in these other disorders.

Pathogenesis of X-Linked ALD. It is likely but not proven that the pathogenesis of X-linked ALD is related in some way to the accumulation of VLCFA. Substantial evidence establishes that VLCFA accumulation is toxic to the adrenal cortical cells. The abnormal cholesterol esters that contain the VLCFA are demonstrable ultrastructurally as cytoplasmic lamellar lipid profiles and clefts. The cells that contain these abnormal profiles show decreased amounts of endoplasmic reticulum, depression of several histochemically demonstrable enzymes, and a diminished capacity to adapt to changes in the microenvironment. Powers (1985) postulated that these lamellar profiles behave as solids at room temperature and cause a generalized impairment in the cell. Knazek *et al.* (1983) showed that the red cell membrane microviscosity is increased in X-linked ALD. That this increase may be of physiological significance is suggested by the studies in human adrenal gland cultures, where impaired cortisol response to ACTH was demonstrated when C26:0 fatty acid was added to the culture medium in concentrations equivalent to that in ALD plasma (Whitcomb *et al.*, 1988). Cholesterol esterified with VLCFA is a poor substrate for cholesterol ester hydrolase (Ogino and Suzuki, 1981). Since release of free cholesterol from cholesterol esters may be rate-limiting for steroidogenesis, the poorly hydrolyzable VLCFA containing cholesterol esters thus may trap cholesterol in the ester fraction and thus limit steroid production.

While there is considerable evidence that the VLCFA excess impairs adrenal function in a dose-related fashion, the pathogenesis of nervous system lesions is more complex. This issue is discussed in more detail in several recent reviews (Powers *et al.*, 1980b; H. W. Moser *et al.*, 1987, 1989). There is no correlation between the severity of neurological diseases and the levels of VLCFA in plasma (Table X) or with the VLCFA levels or metabolism in cultured skin fibroblasts (Boles *et al.*, 1991). The degree of VLCFA abnormality in boys with the rapidly progressive childhood cerebral ALD does not differ from that in men with AMN or those adolescents or adults who are neurologically intact. These findings have led to the hypothesis that the VLCFA excess is necessary but not sufficient for the development of neurological disease, i.e., that one or more additional factors are required. There is provocative evidence that at least one additional factor

TABLE X. Levels of Plasma Very Long-Chain Fatty Acids in X-Linked
Adrenoleukodystrophy Phenotypes (Mean ± SD µg/ml)

Disorder	C26:0	C26:0/C22:0	C24:0/C22:0
Childhood cerebral (239)	1.36 ± 0.48	0.066 ± 0.025	1.57 ± 0.25
AMN (191)	1.39 ± 0.59	0.061 ± 0.025	1.58 ± 0.23
Adolescent cerebral (43)	1.19 ± 0.45	0.054 ± 0.016	1.51 ± 0.21
Adult cerebral (22)	1.36 ± 0.52	0.067 ± 0.033	1.57 ± 0.35
Addison only (61)	1.18 ± 0.40	0.062 ± 0.023	1.52 ± 0.24
Presymptomatic (31)	1.41 ± 0.60	0.074 ± 0.033	1.58 ± 0.21
Asymptomatic (51)	1.23 ± 0.045	0.063 ± 0.024	1.53 ± 0.25
Control (1858)	0.24 ± 0.16	0.014 ± 0.0056	0.81 ± 0.09

may be immunological. We have already referred to the striking perivascular lymphocytic infiltration that is a characteristic feature of the neurological lesions in the cerebral forms of ALD (Schaumburg *et al.*, 1975; Powers and Schaumburg, 1974) and is not seen in the adrenal gland of ALD patients (Powers *et al.*, 1980b) or the metachromatic or globoid forms of leukodystrophy. D. E. Griffin *et al.* (1985) have typed these cells in autopsy material from four ALD patients. On the average they contained 59% T cells, 34% T4 cells, 16% T8 cells, 24% B cells, and 11% monocyte/macrophages. The pattern of cells is similar to that found in the central nervous system during a cellular immune response. Bernheimer *et al.* (1983) have demonstrated increased levels of IgG and IgA in ALD brain tissue, comparable to those found in brain tissue of patients with multiple sclerosis and 2– 10 times higher than control levels, and accumulation of cells staining for IgG, IgA, and IgM has been observed. Unlike the finding in multiple sclerosis, bound immunoglobulins were not detected cytochemically. Boutin *et al.* (1989) studied the macrophages in the perivascular cuffs with a panel of monoclonal antibodies directed against antigenic determinants at the surface of mononuclear cells. They concluded that these cells have "various phenotypes, and very likely different states of activation, and [the results] suggest that interleukin-2 plays an important role during the activation." Taken together, these findings suggest that immuno- pathogenetic mechanisms are involved in the development of rapidly progressive white matter lesions of the childhood cerebral form of ALD. Possibly the altered brain lipid composition secondary to the enzymatic defect provokes an immuno- logical response in some ALD patients, such as the 48% with the childhood cerebral form, while others, such as those with "pure" AMN or the "Addison- only" form, escape it. These immunological aspects are under investigation in several centers. As will be discussed in the genetics section, the childhood cerebral form and AMN frequently cooccur within the same kindreds and also in nuclear families. Segregation analysis suggests the existence of a modifier gene (Moser *et al.*, 1992). Further study of this putative gene may lead to the identification of the additional factor or factors that we presume to be involved in the pathogenesis of the nervous system lesion.

   Diagnosis.   Because of the wide range of phenotypic variation, X-linked ALD can be mistaken for many other conditions. *The early psychological mani- festations* of the childhood cerebral form are often mistaken for hyperactivity or attention deficit disorder. Once *progressive neurological deterioration* has oc- curred in the childhood form, ALD has been misdiagnosed as metachromatic leukodystrophy, subacute sclerosing leukoencephalopathy, and Batten disease. The childhood cerebral form may present with *increased intracranial pressure* (Chaves-Carballo *et al.*, 1984) or as an asymmetric brain lesion on magnetic resonance imaging (Young *et al.*, 1985) and hence be mistaken for a brain tumor. Differential diagnosis from other disorders causing *adrenal insufficiency without neurological involvement* has become of increasing importance since the recent recognition that ALD is a common cause of "Addison disease only" (Sadeghi-Nejad and Senior, 1990; H. W. Moser *et al.*, 1991b). It must be differentiated from autoimmune Addison disease. Patients with Addison disease due to ALD do not have antibodies to adrenal tissue (H. W. Moser, unpublished observation) and have increased levels of VLCFA. When *adrenal insufficiency is combined with neurological deficit in childhood* it must be distinguished from glycerol kinase deficiency (Renier *et al.*, 1983; Seltzer *et al.*, 1985); from the syndrome of familial glucocorticoid deficiency with achalasia of the cardia and deficient tear production (Allgrove *et al.*, 1978; Lanes *et al.*, 1980), in which plasma VLCFA levels are normal (H. W. Moser, unpublished observation); from central pontine myelinolysis (Kandt *et al.*, 1983); and from brain damage in "non-ALD" Addison disease caused by hypoglycemia or hypotension associated with an Addisonian crisis. *Adrenomyeloneuropathy* is often misdiagnosed as chronic multiple sclerosis and the *adult cerebral form* as psychiatric disease (Kitchin *et al.*, 1987). *Neurologically symptomatic heterozygotes* are misdiagnosed frequently as having multiple sclerosis or a chronic nonprogressive spinal cord disorder (Noetzel *et al.*, 1987). A hereditary adult-onset leukodystrophy with an autosomal dominant inheritance and normal VLCFA has been described by Eldridge *et al.* (1984). The laboratory diagnosis of X-linked ALD depends upon the demonstration of elevated levels of saturated VLCFA in plasma (Alberghina *et al.*, 1984; Aubourg *et al.*, 1985; Koyabashi *et al.*, 1983; H. W. Moser *et al.*, 1981), red cell membranes (Antoku *et al.*, 1984, 1985a,b; Tanaka *et al.*, 1986; Tsuji *et al.*, 1981b), leukocytes (Molzer *et al.*, 1982), or cultured skin fibroblasts (H. W. Moser *et al.*, 1980b). In our laboratory we depend mainly upon the plasma assay (H. W. Moser and Moser, 1990a). Table X shows that the VLCFA levels are increased to an equal extent in all the male phenotypes. The levels are already elevated on the day of birth (H. W. Moser, unpublished observation). False-negative results are difficult to exclude, but in our own experience with assays from 25,000 persons we are not aware of false-negative results in male ALD patients. However, Dr. Joseph Tager (personal communication) noted that the Dutch investigators saw a normal plasma VLCFA assay in one male ALD patient in whom the VLCFA levels were increased in cultured skin fibroblasts.

Radiological investigations are of great diagnostic value. In the cerebral forms of the disease, computer tomography or magnetic resonance imaging studies show striking white matter lesions in the parieto-occipital region with accumulation of contrast material at the advancing edges of the lesions (Duda and

Huttenlocher, 1976; Eiben and DiChiro, 1977; H. S. Greenberg et al., 1977; Kumar et al., 1987). These lesions are characteristic and occur early in the course of the disease so that abnormal MRI or CT scans often provide the first clue to the diagnosis. In about 15% of patients the initial lesions are frontal (MacDonald et al., 1984). Other patterns of evolution are also observed (DiChiro et al., 1980). About 85% of heterozygotes show increased levels of VLCFA in plasma and cultured skin fibroblasts (H. W. Moser et al., 1983). When affected and unaffected male relative are available, heterozygote identification can be achieved by linkage analysis with the DXS52 recombinant DNA probe (Aubourg et al., 1987). Affected male fetuses can be identified reliably by demonstration of increased VLCFA levels in cultured amniocytes (H. W. Moser et al., 1982) and cultured chorion villus cells (Table VII). The DSX52 probe has also been used successfully (Boue et al., 1985).

Genetics.   Fanconi et al. (1963) suggested X-linked recessive inheritance on the basis of pedigree analysis of ten cases. Analysis of the pattern of inheritance in more than 600 kindreds is consistent with this proposal (H. W. Moser, unpublished observation). There are no exact figures on the incidence of X-linked ALD. During the last 5 years our laboratory has identified each year an average of 57 new ALD patients in the United States. With yearly U.S. births of 3.6 million, this yields a minimum incidence of 1.58:100,000. This is almost certainly an underestimate since our laboratory does not test all cases.

X-Linkage was confirmed by Migeon et al. (1981), who demonstrated that clones from heterozygotes were of two types: one with normal fatty acid pattern, the other showing a pattern similar to that of affected males. Linkage to glucose-6-phosphate dehydrogenase (G6PD) was established by studies of families with women heterozygous for both ALD and electrophoretic variants of G6PD. No recombinants between these genes were observed in the offspring of 18 heterozygous women. Since G6PD has been mapped to Xq28 (Martin-deLeon et al., 1985), the ALD gene must also be located there. Confirmation for this finding is provided by the demonstration of linkage with a logarithm of odds (lod) score of 13.766 at a theta value of 0 between ALD and the DXS52 probe (Aubourg et al., 1987) which recognizes polymorphic loci in that region (Oberle et al., 1985). The Xq28 region also contains the genes that control the formation of the red-green color vision pigments (Nathans et al., 1986a,b). Therefore, it was of considerable interest when we demonstrated an increased frequency of color-vision defects (Sack et al., 1989) and frequent alterations in red-green color vision genes (Aubourg et al., 1988) in American patients with ALD. A subsequent study did not demonstrate an abnormally high incidence of abnormalities of red-green color vision genes in French ALD patients (Aubourg et al., 1990b), and additional studies are required to rule out the possibility that the defects in red-green color vision in AMN patients are related to nervous system lesions secondary to the disease process, rather than to a primary defect of the ALD gene. Recent and ongoing studies of the American "O" family provide additional support for the hypothesis that the ALD gene is in close proximity to the red-green color vision genes. The proband in this family has AMN and is a blue monochromat (lacks the capacity to distinguish red and green pigments); he has been shown to have deletions and rearrangements of the red-green color vision genes (Aubourg *et al.*, 1988, 1990b). This patient has a deletion that involves the red-green color vision genes and that extends beyond the 5' region of the red pigment gene (Aubourg *et al.*, 1990b). We have shown recently that in the "O" family this gene cosegregates with the ALD gene with a lod score of 3.19 at theta of zero (G. Sack, unpublished observation). Intense efforts to identify the ALD gene are in progress.

Table XI shows that the AMN and the childhood cerebral form of ALD commonly cooccur within the same kindred and nuclear families. As noted earlier and illustrated in Figures 16 and 17 and Table XII, these two phenotypes show a bimodal distribution in respect to age of onset, site and nature of pathology, and length of survival. Since the primary defect in ALD involves only a single gene, compound heterozygosity is excluded as an explanation, and the frequent cooccurrence of the two phenotypes within the same family implies that a single primary

TABLE XI. Distributions of Childhood Cerebral and AMN Forms of ALD within Kindreds and Nuclear Families

| | Phenotype distribution | | | |
| | Among multiplex kindred | | Among multiplex nuclear families | |
	Kindred	Percent	Number of families	Percent
ALD	54	30.3	66	49.3
ALD/AMN	90	50.6	35	26.1
AMN	34	19.1	33	24.6
Total	178		134	

Note: These are the two most common types of ALD (Table IX) and are easily distinguished on the basis of age of onset (Fig. 16), prognosis (Fig. 17), and main site of neurological involvement (Table XII). Note that in the extended kindred it is more common for the mild and severe form to cooccur within an extended kindred than for either form to occur alone. Even in the nuclear families there is a high frequency of cooccurrence of the two forms. Preliminary results of segregation analysis are consistent with the existence of a modifier gene.

TABLE XII. Comparison between Childhood Cerebral Adrenoleukodystrophy (CCER) and Adrenomyeloneuropathy (AMN)

	CCER	AMN
Age of onset of neurological symptoms (years)	7.1 ± 1.7	27.6 ± 8.7
Nature of pathology	Inflammatory	Tract "dying-back"
Main site of pathology	Cerebral white	Spinal core, peripheral nerve
Interval to vegetative state or death (years)	3.0 ± 2.0	> 13

gene defect is associated with a wide range of phenotypic expression. Epidemiological studies have not identified environmental factors that can account for this. Preliminary genetic analysis is most compatible with the existence of a single modifier gene with either a dominant or recessive protective allele (K. D. Smith, unpublished observation); this finding has led our laboratory to initiate a search for this putative gene by linkage analysis.

At this time we still lack laboratory assays that distinguish between the severe and milder phenotypes of ALD. The levels of VLCFA in plasma (Table X) and cultured skin fibroblasts (Boles *et al.*, 1991) are identical, as is the capacity to metabolize VLCFA in cultured fibroblasts. These data suggest that the phenotypic variation cannot be attributed to variation in the severity of the primary metabolic defect. This conclusion may be drawn into question, however, by a recent report by Antoku *et al.* (1991). These authors reported that patients with childhood cerebral ALD had higher levels of VLCFA in blood mononuclear cells than did patients with AMN, while in concordance with the previous literature the VLCFA levels in plasma and red blood cell membranes showed no difference. The study involved four patients in each group, the p value was <0.05, and although the differences were small, there was no overlap. This intriguing result should be followed up in a larger series of patients. Tiffany *et al.* (1991) have found that cultured skin fibroblasts of patients with AMN have lower levels of interleukin 1-stimulated prostaglandin E2 synthesis than do those with the childhood cerebral ALD. Interestingly, both of these studies point to possible differences in immunological or inflammatory responses, differences that are consistent with the postmortem histological finding that in the childhood cerebral form there is an inflammatory response, while in AMN this is absent or less prominent (Powers, 1985). The ability to distinguish prospectively whether an asymptomatic child with the biochemical defect of ALD is "destined" for the mild or severe form of the disease would be of great value for counseling and the selection of therapeutic modalities.

Therapy. The therapy of X-linked ALD is under intense investigation in several clinics, and the results and issues have been reviewed recently (H. W. Moser *et al.*, 1991a). Briefly, two forms of therapy are under active consideration: dietary therapy and bone marrow transplantation.

Recent modifications of dietary therapy now make it possible to normalize plasma levels of VLCFA within 4 weeks. The initial thrust toward dietary therapy was based upon the observation of Kishimoto *et al.* (1980) that the VLCFA that accumulated in the brain of ALD patients were of dietary origin at least in part. This was followed by the development (Van Duyn *et al.*, 1984) and evaluation (Brown *et al.*, 1982b) of a diet that restricts VLCFA intake. This diet did not alter plasma VLCFA levels and had no effect on clinical course. The explanation of this finding and the lead to a new approach were provided by the studies of Rizzo *et al.* (1986), who showed that cultured skin fibroblasts synthesized substantial amounts of saturated VLCFA including C26:0, and that the rate of synthesis could be reduced substantially by the addition of monounsaturated fatty acids such as oleic acid. As already noted, saturated and monounsaturated fatty acids are elongated to VLCFA by the same microsomal elongating system (Bourre *et al.*, 1976). Addition of monounsaturated acids presumably reduces synthesis of saturated VLCFA by competition for the same enzyme system. This results also in increased levels of monosaturated VLCFA, but it has been shown that the LD_{50} of monounsaturated VLCFA is at least 25 times higher than that of saturated VLCFA (W. Rizzo, personal communication). These findings led to clinical trials in which oral glyceryl trioleate administration was combined with dietary restriction of VLCFA.

Such a regimen achieved a 50% reduction of the levels of saturated VLCFA in plasma (Rizzo *et al.*, 1987; A. B. Moser *et al.*, 1987a) as well as a statistically significant improvement in peripheral nerve function (H. W. Moser *et al.*, 1991a).

Subsequently it was shown that erucic acid (C22:1) has a still more powerful effect on the synthesis of saturated VLCFA, and administration of a diet that includes both glyceryl trioleate and glyceryl trierucate normalizes the level of C26:0 in plasma within 1 month (Rizzo *et al.*, 1989; H. W. Moser *et al.*, 1991a) (Fig. 18). This regimen also normalizes the microviscosity of red cell membranes (H. W. Moser *et al.*, 1991a). Red cell membrane microviscosity is increased in untreated ALD patients (Knazek *et al.*, 1983) and may be of pathogenetic significance (Whitcomb *et al.*, 1988).

The clinical effectiveness of this new dietary regimen is now being tested in our clinic in a 5-year trial involving more than 170 patients. The regimen does not appear to alter the course of the rapidly progressive childhood cerebral form of ALD (Uziel *et al.*, 1990). The main questions under investigation are whether it can arrest the progression of AMN and, perhaps more important, whether it can

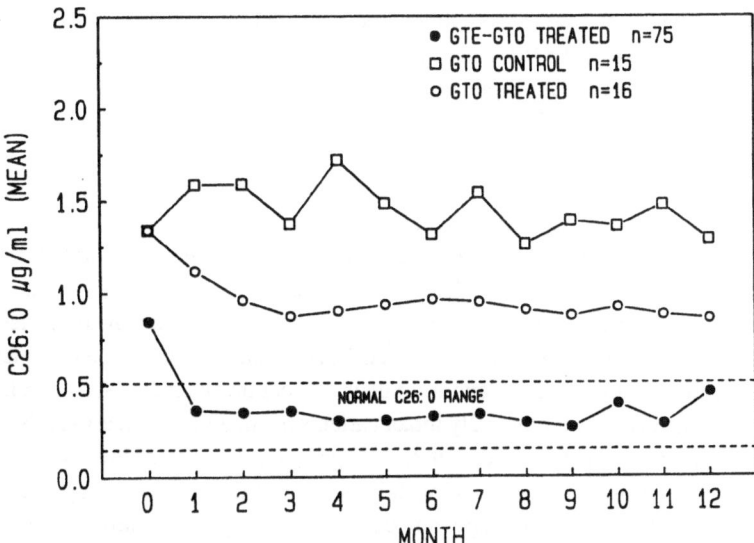

Fig. 18. Effect of dietary regimens on the plasma level of hexasonoic acid, C26:O. Accumulation of saturated very long-chain fatty acids is the principal biochemical abnormality in X-linked ALD. GTO, Glyceryl trioleate; GTE, glyceryl trierucate oil. The "GTO-control" group were AMN patients who continued their customary diet and did not receive any of the oils. The "GTO-treated" men received glyceryl trioleate in a daily dosage of 2–3 ml per kg of body weight and also restricted their dietary intake of VLCFA (Van Duyn *et al.*, 1984). The GTE–GTO group received a mixture of 1 part of GTE to 4 parts of GTO (Rizzo *et al.*, 1989) along with dietary restriction of VLCFA. Note that the GTE–GTO mixture normalized the plasma C26:O levels within 1 month.

prevent the onset of neurological involvement in persons who have the biochemical abnormality of ALD but are neurologically intact. At this time the main source of encouragement is the improved peripheral nerve function in those glyceryl trioleate patients whose plasma C26:0 levels were reduced (H. W. Moser *et al.*, 1991a). Longer follow-up is required to determine whether the diet can prevent or retard the onset of neurological disability.

The hope that bone marrow transplantation may be of benefit has been raised by a spectacular success in one 8-year-old patient (Aubourg *et al.*, 1990a). This patient had mild neurological disability and MRI change at the time of transplant. His nonaffected, nonidentical twin served as donor. Now, 3 years after transplant, the neurological disability and MRI abnormalities have cleared, and cognitive function is the same as that of the unaffected twin. Transplants in patients with more advanced disease have been associated with worsening of neurological

status in the immediate period following transplantation and are not advised
(H. W. Moser et al., 1984; Weinberg et al., 1988). A total of 14 transplants have
now been performed in ALD patients. Early results in mildly affected patients are
encouraging (Krivit et al., 1991), but additional follow-up is required for evaluation.

Present recommendations for therapy depend upon the stage and type of
illness. For *neurologically intact* males and men with *AMN* and *symptomatic
heterozygotes*, we recommend participation in the dietary therapeutic trial with the
glyceryl trierucate–glyceryl trioleate oils. For boys and adolescents with *mild
cerebral* involvement, we recommend consideration of bone marrow transplantation if a matched donor is available. We emphasize that this should be considered
only for those who have early and mild neurological involvement. As a corollary
we recommend that neurologically intact patients be monitored at least yearly, and
possibly every 6 months, to detect early signs of neurological involvement by
neurological examination, psychological tests, and MRI so that the opportunity
for successful transplant therapy not be lost. For boys who have moderately
advanced, rapidly progressive childhood form of the disease, none of the above
therapies appears to be effective. Dietary therapy does not appear to alter clinical
course, and bone marrow transplant appears to accelerate neurological progression. Novel therapeutic approaches are needed for this form of the illness, possibly
involving those that can alter the inflammatory response that appears to be
responsible for the rapid progression. Immunosuppression with cyclophospha-
mide did not arrest the rapid progression of the illness (Naidu et al., 1988a; Stumpf
et al., 1981). At this time we do not recommend any therapeutic interventions
for asymptomatic heterozygotes.

Acyl-CoA Oxidase Deficiency.   This entity has also been referred to as
pseudo-neonatal ALD, in analogy to the term pseudo-Zellweger syndrome (see
below under 3-oxo-CoA thiolase deficiency). Three documented case of acyl-CoA
oxidase deficiency have been reported: a brother and sister (Poll-The et al., 1988)
and one other girl who had three healthy older siblings (Kyllerman et al., 1990).
Both sets of parents were consanguineous. All three patients were hypotonic at
birth, with feeding difficulties and neonatal seizures that continued and were
difficult to control. Dysmorphic features were slight or absent. During the first 2
years, psychomotor development was severely impaired, but not totally absent:
one of the patients learned to crawl and stand with support. None developed
speech. After 2 years, development appeared to regress further, so that the
children became totally helpless. Ocular involvement was evidenced by extinguished electroretinogram response. Adrenal insufficiency was not evident clinically, but one patient had elevated serum ACTH and low cortisol levels. Liver and

spleen were not enlarged. Brain CT scan showed white matter hypodensities and abnormal contrast enhancement in the centrum semiovale bilaterally. MRI in one of the patients showed a thin corpus callosum, underdeveloped vermis of the cerebellum, and increased signal density in the white matter.

While analysis of the clinical features bears out the resemblance to neonatal ALD (Kelley *et al.*, 1986), laboratory tests define the fundamental difference. Peroxisomes in the liver are abundant, and morphometric analysis demonstrates that they are larger than normal. Plasma levels of VLCFA are increased markedly, equivalent to those observed in Zellweger syndrome and neonatal ALD. In contrast, all other measures of peroxisomal function, such as levels of phytanic acid, pipecolic acid, bile acid intermediates, and plasmalogen and their metabolism, are normal. Immunoblot studies in one patient revealed that acyl-CoA oxidase was absent in the liver, while the other peroxisomal β-oxidation enzymes were normal (Poll-The *et al.*, 1988). Cultured skin fibroblasts had a greatly reduced capacity to oxidize lignoceric acid, and the activity of peroxisomal palmitoyl-CoA oxidase was reduced by 95%.

The occurrence of acyl-CoA oxidase deficiency in siblings of both sexes with healthy and consanguineous parents suggests that it is a rare autosomal recessive disorder. It may be somewhat more frequent than recognized so far. The experience at the Kennedy Institute indicates that about 5% of patients with phenotypes that resemble the disorders of peroxisome biogenesis in fact have single defects of peroxisomal β-oxidation enzymes. The precise defect has not yet been defined in most of these patients, and it is likely that this group includes others with acyl-CoA deficiency. It is remarkable and surprising that a defect of VLCFA oxidation alone appears to cause profound brain and eye malfunction equivalent to that in the disorders of peroxisome biogenesis and indicates that abnormalities in these substances cause severe but as yet undefined disturbances in the early development of these organs.

Bifunctional Enzyme Deficiency.   Watkins *et al.* (1988) reported a male patient whose clinical presentation resembled that of the acyl-CoA oxidase deficiency cases discussed above: hypotonia and neonatal seizures in the absence of dysmorphic features and organomegaly, no psychomotor development, and death at 5½ months. Autopsy showed polymicrogyria and focal areas of cortical heterotopias, indicating neuropathological abnormalities similar to those in disorders of peroxisome biogenesis. Liver peroxisomes were present but had not been subjected to quantitative morphometric analysis. Biochemical studies showed accumulation of very long-chain fatty acids. All other peroxisomal functions were normal, *except* that the level of 3α,7α,12α-trihydroxy-5β-cholestanoic acid (THCA) was elevated.

Immunoblot studies of postmortem liver showed that bifunctional protein was absent, while the other peroxisomal β-oxidation enzymes were present in normal amounts. mRNA for the bifunctional protein also was normal, indicating that the defect involves either the defective translation of the mRNA or abnormal posttranslational processing, such as a defect in the peroxisomal import mechanism. The clinical manifestations and pathogenesis of bifunctional enzyme deficiency resemble those of acyl-CoA deficiency. Measurement of the levels of bile acid intermediates presents a convenient method of distinction: THCA levels are normal in acyl-CoA deficiency but are increased in bifunctional enzyme deficiency. Figure 1 provides the probable explanation for this difference. The initial oxidation step for bile acid intermediates and VLCFA is carried out by separate enzymes, while subsequent steps are carried out by the same enzymes. Wanders *et al.* (1990b) recently described a second patient with bifunctional enzyme deficiency. Immunologically reactive bifunctional enzyme protein was present in this patient, who thus is presumed to have a mutation that differs from that described by Watkins *et al.* (1988).

3-Oxoacyl-CoA Thiolase Deficiency. In 1986 Goldfischer *et al.* (1986) described a girl with clinical features compatible with classical Zellweger syndrome, who died at age 11 months. It came as a surprise when a liver biopsy sample showed the presence of abundant peroxisomes. The condition was therefore named "pseudo-Zellweger syndrome." Biochemical studies of plasma revealed that levels of VLCFA were elevated to the same degree as in Zellweger syndrome, and the bile acid intermediate THCA was 3–9% of total bile acids, compared to less than 0.03% in controls. The phytanic acid level was normal, but there was an (even now) unexplained moderate increase in the serum pipecolic acid level (17.5 μmole/liter compared to 5–10 μmole/liter in the control). Neither plasmalogen level nor synthesis was measured, but in light of later knowledge would have been expected to be normal. Postmortem examination revealed a slight degree of stellate portal fibrosis in the liver, microscopic renal cysts, and adrenal atrophy; the adrenal cortical cells contained the cytoplasmic inclusions characteristic of Zellweger syndrome or ALD. The brain showed severe demyelination, gliosis, and foci of heterotopic nuclei most evident in the cerebellum. These anatomical findings are consistent with those found in the disorders of peroxisome biogenesis.

Schram *et al.* (1987) showed that peroxisomal β-oxidation was diminished to less than 10% of control in the postmortem liver of this patient. Immunoblot studies showed that 3-oxoacyl-CoA thiolase was diminished, while the other peroxisomal β-oxidation enzymes were present in normal amounts. That the thiolase deficiency indeed represented the primary defect was corroborated by the finding that addition of rat 3-oxoacyl-CoA thiolase restored peroxisomal β-oxidation to normal in a dose-dependent manner.

The 3-oxoacyl-CoA thiolase deficiency was the first demonstration that clinical syndromes similar to the disorders of peroxisome biogenesis can be caused by defects that involve a single enzyme in the peroxisomal β-oxidation system. The thiolase deficiency can be distinguished from the oxidase deficiency because levels of bile acid intermediates are elevated in the thiolase deficiency but not in the oxidase deficiency. This criterion cannot be used to distinguished it from bifunctional enzyme deficiency, because bile acid intermediate levels are also increased in that condition. The elevation of the pipecolic acid level in the thiolase deficiency is unexpected and unexplained and should be reassessed in other cases.

Undefined Defects of a Single Peroxisomal β-Oxidation Enzyme. Naidu et al. (1988b), Barth et al. (1990), and Kaufmann et al. (1991) presented patients with presumed defects of a single peroxisomal β-oxidation enzyme in whom the precise biochemical defect was not defined. These patients were severely involved with neonatal seizures, hypotonia, and severe psychomotor retardation. Dysmorphic features were absent or mild. Cerebral malformations, including neuronal migration defects, varied from mild to severe. Diagnosis rests on the demonstration of increased levels and impaired oxidation of VLCFA, and depending upon the locus of the defect, normal or elevated levels of bile acid intermediates. Peroxisome structure and other peroxisomal functions are intact. These disorders are probably more frequent than is realized at present. Since dysmorphic features may be slight or absent, patients may not be diagnosed at all, or they may be mistaken for those with disorders of peroxisome biogenesis.

Diagnosis, Genetics, and Therapy of Peroxisomal β-Oxidation Single-Enzyme Defects Other Than X-Linked ALD. *Diagnosis.* Clinically these disorders resemble the disorders of peroxisome biogenesis and differentiation from them depends upon laboratory tests. The essential difference is that the biochemical abnormality is confined to VLCFA and, depending upon the location of the defect, to bile acid intermediates. Peroxisome structure is intact, catalase is sedimentable, and pipecolic acid, plasmalogen, and phytanic acid levels and metabolism are normal.

While the laboratory tests described above place a patient into the "single-enzyme-defect" category, differentiation among the three disorders may be difficult. Levels of bile acid intermediates are normal in oxidase deficiency and elevated in the bifunctional enzyme and thiolase deficiencies. Immunoblot studies in samples of liver or cultured skin fibroblasts may reveal the absence of the specific enzyme (Poll-The et al., 1988; Schram et al., 1987; Watkins et al., 1988), but this would not distinguish patients in whom immunologically reactive but catalytically inactive cross-reacting material is present. Specific enzyme assays require substrates that are not generally available, so that it may not be possible to pinpoint the exact defect. Complementation analysis with well-characterized cell lines may aid in the identification of the enzyme defect (McGuinness *et al.*, 1990), but this procedure has not been fully evaluated.

Genetics. Patterns of inheritance suggest that these disorders have an autosomal recessive mode of inheritance. The genes have not yet been mapped.

Therapy. Therapeutic options are limited because of the malformations present at birth. It may be possible to normalize the levels of VLCFA and bile acid intermediates, as described for the disorders of peroxisome biogenesis, but these approaches have not been tested.

Peroxisomal Glutaryl-CoA Oxidase Deficiency

Bennett *et al.* (1991) have recently reported the first patient with peroxisomal glutaryl-CoA oxidase deficiency. The patient was 5 years and 8 months old at last follow-up. She had been investigated at age 11 months because of failure to thrive and postprandial vomiting. Urine glutaric acid was 500 mmole per mole of creatinine compared to the normal of <2 mmole per mole. She also was homozygous for β-thalassemia. The glutaric aciduria was shown to be responsive to riboflavin 200 mg twice daily, and she has been maintained on this vitamin dosage until the time of her last follow-up. At that time her development was judged to be normal. It is not known whether the normal development is attributable to the vitamin therapy. Detailed metabolic investigations at this time ruled out glutaric aciduria types 1 and 2. The peroxisomal hydrogen peroxide-producing oxidases in cultured skin fibroblasts were measured by a fluorometric assay (Vamecq, 1990). The activity of peroxisomal glutaryl-CoA oxidase was reduced to <10% of control, while lauroyl-CoA and dodecanedoyl-CoA oxidase activities were normal. Loading studies suggested that the glutaric acid was derived from lysine, presumably via the pathway in which pipecolic acid is an intermediate, and not from leucine. This report is of interest for several reasons. It demonstrates that there is a third cause of glutaric aciduria, in addition to the well-known mitochondrial disorders glutaric aciduria types 1 and 2. Since mitochondrial metabolism of glutaric acid was normal, the data suggest that there is little interchange between the mitochondrial and peroxisomal glutaric acid pools, and since the activities of the other peroxisomal oxidases were normal, it appears likely that the peroxisomal glutaryl-CoA oxidase is a separate enzyme. It will be of great interest to determine the incidence of peroxisomal glutaryl-CoA oxidase deficiency, and to test whether other cases also are responsive to riboflavin.

Acatalasemia

Acatalasemia has been reviewed recently (Eaton, 1989). The majority of persons with acatalasemia are asymptomatic. The disorder appears to be heterogeneous. The most severe Japanese variant may be associated with oral ulcerations thought to be due to infection by peroxide-generating bacteria. These patients have greatly reduced catalase activity, but the electrophoretic mobility of the small amount of enzyme is normal. In contrast, persons with the Swiss variant are free of symptoms and have a larger amount of residual enzyme activity. However, their enzyme can be separated electrophoretically from the normal catalase and shows greater than normal heat lability. Crawford *et al.* (1988) studied enzyme activity and mRNA levels and concluded that the Japanese variant involves a regulatory mutation in which the gene is not transcribed, while the Swiss variant involves a structural mutation that renders the enzyme unstable. The lack of clinical symptoms associated with mutations that affect such a ubiquitous enzyme must reflect the efficiency of alternate mechanisms for the disposal of hydrogen peroxide (de Duve and Baudhuin, 1966).

Hyperoxaluria Type 1

Clinical Features and Biochemical Basis. Primary hyperoxaluria type 1 is an autosomal recessive disease characterized biochemically by the increased urinary excretion of oxalate, glycolate, and sometimes glyoxylate (Danpure, 1989). The disorder presents in children or young adults with recurrent urolithiasis combined with progressive renal failure. In a recent German survey it accounted for 2–2.7% of children with end-stage renal disease (Latta and Brodhel, 1990). It is firmly established that the disorder is due to deficient function of peroxisomal alanine:glyoxylate aminotransferase (AGT) (Danpure *et al.*, 1987; Danpure and Jennings, 1986, 1988; Wanders *et al.*, 1987d, 1988b). There is heterogeneity in respect to the severity of the enzyme deficiency (Danpure and Jennings, 1988). The patients with the milder defect are responsive to pyridoxine (Wanders *et al.*, 1988b). Peroxisomes are present in the liver (Iancu and Danpure, 1987; Cooper *et al.*, 1988) and the function of other peroxisomal enzymes such as acyl-CoA:dihydroxyacetone phosphate dehydrogenase, catalase, D-amino acid oxidase, and L α-hydroxyacid oxidase is normal (Wanders *et al.*, 1987d). Danpure and Jennings (1988) made the intriguing observation that in the approximately one-third of patients who have diminished but not absent AGT activity, the enzyme is erroneously routed to the mitochondrion instead of its normal (in humans) localization in the peroxisome (Danpure *et al.*, 1989a).

Takada *et al.* (1990) have recently determined the nucleotide sequence of normal human AGT and compared it to that in the rat. This comparison is of particular interest since rat AGT is localized to the mitochondrion. There was 74% sequence identity within the region encoding the rat amino-terminal mitochondrial targeting sequence, but this sequence is not expressed as part of the human protein because of a coding difference at the site corresponding to the rat protein translation start site. Recently the same group of investigators sequenced AGT cDNA from eight patients with hyperoxaluria type 1 in whom AGT was localized in the mitochondrion (Purdue *et al.*, 1990). They identified three mutations: one that encoded a glycine-to-arginine substitution at residue 170 was not present in any of the control individuals, while two other mutations were also present in 5–10% of the control population. Studies are in progress to define the structural alteration of the protein that is responsible for the misdirection of the enzyme to the mitochondrion. Nishiyama *et al.* (1991) cloned the cDNA of a patient with hyperoxaluria type 1 who had nearly absent AGT activity, and thus differed from the patients with mitochondrial AGT activity studied by Purdue *et al.* (1990). Nishiyama's patient was shown to have a point mutation that encoded a serine-to-proline substitution at residue 205.

Diagnosis of Hyperoxaluria Type 1.   The diagnosis of hyperoxaluria type 1 should be suspected in children or young adults who present with renal failure combined with a history of urolithiasis and a family history of urolithiasis, consanguinity, and nephrocalcinosis (Latta and Brodhel, 1990). The next step is the demonstration in the urine of abnormally high levels of oxalate (>50 mg/day per 1.73 m^2 of body surface) and of glycolate (>70 mg/day per 1.73 m^2). Definitive diagnosis depends upon demonstration of deficient activity of alanine: glyoxylate aminotransferase in a liver biopsy specimen (Danpure *et al.*, 1987; Danpure and Jennings, 1986). After the 17th week of gestation the enzyme normally is present in fetal liver at 30% or more of the adult level, and prenatal diagnosis has been achieved by second-trimester fetal liver biopsy (Danpure *et al.*, 1989b). It is anticipated that in the future prenatal diagnosis will be achieved by DNA analysis of amniocytes or chorion villus samples.

Genetics.   The defect is inherited as an autosomal recessive trait. Latta and Brodehl (1990) refer to a study of 330 patients, with family data available in 231. There were affected family members in 90 and parental consanguinity in 18. They estimated the incidence of the disorder to be 1 in 5–15 million children, but cautioned that this was probably an underestimate. The molecular biology of the disorder has been discussed above.

Therapy.   The most exciting therapeutic success has been achieved with combined liver and kidney transplants (Ruder *et al.*, 1990; McDonald *et al.*, 1989;

Morgan and Watts, 1989; Watts *et al.*, 1987). Twenty transplants had been performed in Europe by 1990 and preliminary results appear favorable. Ruder *et al.* (1990) report a 4-year, 8-months-old boy who was in terminal renal failure at time of transplant. Nine months later both transplants are functioning well and the metabolic defect is cured. Some of the more mildly affected patients are responsive to pyridoxine (Wanders *et al.*, 1988b; Watts *et al.*, 1979).

Multiple Biochemical Defects: Peroxisome Intact

Rhizomelic Chondrodysplasia Punctata

Clinical Features and Biochemical Basis. The term chondrodysplasia punctate is applied to infant patients with punctate epiphyseal and extraepiphyseal calcifications on roentgenograms. It is a nonspecific finding that can be caused by toxins such as in warfarin embryopathy (Hall *et al.*, 1980). The majority of cases are genetically determined. The mode of inheritance may be autosomal dominant (Spranger *et al.*, 1971), X-linked dominant (Happle, 1979), X-linked recessive (Curry *et al.*, 1984), or autosomal recessive (Spranger *et al.*, 1971). In 1971 Spranger *et al.* (1971) carried out a detailed clinical analysis of a large series of chondrodysplasia patients and separated out two distinct forms. A relatively mild form, referred to as the Conradi–Hunermann type, has an autosomal dominant mode of inheritance; intelligence is normal or nearly so, cataracts are relatively uncommon (17%), and patients often survive to adulthood. This contrasts with a severe form, which has an autosomal recessive mode of inheritance, and is associated with severe shortening of limbs, severe psychomotor retardation, and usually death during the first year of life. This form is referred to as rhizomelic chondrodysplasia punctata (RCDP).

In 1985 Heymans and associates made the important observation that plasmalogen synthesis was deficient in patients with RCDP (Heymans *et al.*, 1985, 1986) and classified it as a peroxisomal disorder. These and subsequent studies have shown that all RCDP patients have a triad of biochemical defects: (1) marked reduction of plasmalogen levels (Heymans *et al.*, 1983, 1984) and a profound impairment of plasmalogen synthesis (G. Hoefler *et al.*, 1988); (2) increased plasma levels of phytanic acid and impaired capacity to oxidize this substance (G. Hoefler *et al.*, 1988); and (3) presence of 3-oxoacyl-CoA thiolase in the immature form.

The clinical features of classical RCDP are short stature with disproportionate shortening of the proximal parts of the extremities, microcephaly, peculiar facial appearance, severe mental retardation, cataracts (72%), and ichthyosis.

Radiological features include severe shortening, metaphyseal cupping, splaying, and disturbed ossification of the humeri and femora. Lateral views of the spine show a universal coronal cleft of the vertebral bodies. Stippling of the epiphyses involves mainly the knee, hip, elbow, and shoulder. Stippling in the vertebral column is uncommon in RCDP, but is a prominent feature in the Conradi–Hunermann syndrome (Spranger *et al.*, 1971). Some patients with the classical form of RCDP survive beyond the first year (Wardinksy *et al.*, 1990). The oldest known patient is 16 years old (Oorthhuys *et al.*, 1987).

Dr. B. T. Poll-The (personal communication) observed a 9-month-old girl whose plasma and cultured skin fibroblasts showed the triad of biochemical abnormalities referred to above, but who is much less disabled. While radiographs show prominent chondrodysplasia punctata, her limbs are of normal size, there are no other radiological abnormalities, and psychomotor development appears normal. This causes a semantic problem, since one cannot comfortably use the designation of *rhizomelic* chondrodysplasia punctata when the limbs are not short! Two groups of investigators have reported a moderate impairment of acyl-CoA:dihydroxyacetone phosphate acyltransferase in a patient with Conradi–Hunermann syndrome (Clayton *et al.*, 1989; Holmes *et al.*, 1987b). We have not found such an abnormality (H. W. Moser, unpublished observation) and the other biochemical abnormalities characteristic of RCDP have not been reported. There is need for additional studies to assess peroxisomal function in the Conradi–Hunermann syndrome. We also failed to detect abnormalities of peroxisomal function in the patient with X-linked recessive chondrodysplasia punctata reported by Curry *et al.* (1984). We recommend that only those chondrodysplasia punctata patients with the triad of biochemical abnormalities be included in the peroxisomal disease category. Most of these patients have the RCDP phenotype. I believe that at this time there is not sufficient evidence to warrant inclusion of Conradi–Hunermann syndrome in the peroxisomal disease category.

RCDP differs from the disorders of peroxisome biogenesis in that peroxisomes are present (Heymans *et al.*, 1986) and catalase is in the particulate fraction rather than the cytosol (G. Hoefler *et al.*, 1988). It differs from the single-enzyme defects since the activities of at least two unrelated enzymes are deficient, namely acyl-CoA:dihydroxyacetone phosphate acyltransferase (Heymans *et al.*, 1985) and the oxidation of phytanic acid (G. Hoefler *et al.*, 1988). In addition, the enzyme 3-oxoacyl-CoA thiolase is present in its unprocessed form. While peroxisomes are present, their structure may differ from normal. H. S. A. Heymans (personal communication) noted that in a liver biopsy specimen some hepatocytes lacked peroxisomes while other liver cells contained an increased number of irregularly shaped, huge peroxisomes.

Recent studies by ten Brink *et al.* (1991) have clarified the mechanism of the phytanic acid excess. These investigators showed that patients with disorders of peroxisome biogenesis accumulate both pristanic acid and phytanic acid, whereas in classical Refsum disease and in RCDP only phytanic acid accumulates. This finding suggests that the metabolic block in RCDP, as in classic Refsum disease, involves the phytanic acid α-oxidation step. Plasma phytanic acid levels in RCDP are higher than in the disorders of peroxisome biogenesis (Fig. 13), and may be as high as those in patients with classic Refsum disease. In spite of these points of resemblance, the basic defects in these disorders must differ since the phytanic acid oxidation defect is corrected when cell lines from RCDP and classical Refsum disease patients are fused (Poll-The *et al.*, 1989; Wanders *et al.*, 1986b).

The third biochemical abnormality in RCDP, namely the presence of the immature form of peroxisomal thiolase, is also found in patients with Zellweger syndrome (Schram *et al.*, 1986) and in neonatal ALD (Chen *et al.*, 1987). However, VLCFA levels and VLCFA oxidation are normal in RCDP, whereas they are severely impaired in the disorders of peroxisome biogenesis (Wanders *et al.*, 1987b). Possibly the immature thiolase is enzymatically active. Alternatively, a small amount of mature enzyme that was not detected by the immunoblot technique may be present in sufficient quantity to permit complete β-oxidation of the fatty acids. Balfe *et al.* (1990) reported recently that the immature thiolase in RCDP and Zellweger patients is present in a subcellular particle that has a lower density than normal peroxisomes and mitochondria. They speculated that this particle may be related to the peroxisome "ghosts" described by Santos *et al.* (1988b).

The nature of the basic defect in RCDP presents a challenge to the cell biologist. Presumably in RCDP there is a defect in peroxisomal import mechanisms that differs and is more restricted than those presumed to exist in the disorders of peroxisome biogenesis. The peroxisomal thiolase enzyme does not contain the carboxy-terminal serine–lysine–leucine sequence characteristic of many peroxisomal proteins and the topogenic sequence is located in the amino-terminal leader sequence (S. Gould, personal communication). The topogenic sequences for the phytanic and plasmalogen degradation enzymes have not been determined. An understanding of the pathogenesis requires more knowledge about the biogenesis of normal peroxisomes. Further study of the RCDP mutation may speed the acquisition of this knowledge.

Diagnosis of Rhizomelic Chondrodysplasia Punctata (RCDP).

RCDP must be differentiated from other genetic and acquired causes of chondrodysplasia punctata. The three characteristic defects of peroxisomal function demonstrable by laboratory studies are present only in RCDP. The most easily performed are the demonstration of abnormally low levels of plasmalogens in red blood cells (Bjorkhem et al., 1986) and elevated levels of phytanic acid in plasma (H. W. Moser and Moser, 1990b). We recommend that the diagnosis be confirmed by demonstration of impaired plasmalogen synthesis (Schutgens et al., 1984; Datta et al., 1984; Roscher et al., 1985) and phytanic acid oxidation (Poll-The et al., 1989; Poulos et al., 1985) in cultured skin fibroblasts. Prenatal diagnosis utilizes the same assays in cultured amniocytes or chorion villus samples (S. Hoefler et al., 1988). Biochemical confirmation of the diagnosis of RCDP is of great importance for genetic counseling since none of the other types of chondrodysplasia punctata have defined biochemical abnormalities that permit prenatal identification. Dr. B. T. Poll-The (personal communication) has recently observed the triad of peroxisomal biochemical defects characteristic of RCDP in a patient with chondrodysplasia punctata but who did not have shortened limbs or impaired intelligence. This finding suggests that the phenotype of RCDP may be more varied than so far recognized, and may require a reappraisal of the indications for the performance of these biochemical assays. Duff et al. (1990) have reported the prenatal diagnosis of classical RCDP by ultrasound.

Genetics. RCDP has an autosomal recessive mode of inheritance. It is a rare disorder and incidence figures are not available.

Treatment. Therapeutic options are limited because of the severe abnormalities present at birth. The existence of patients with longer survival (Wardinsky et al., 1990) indicates consideration of habilitative therapy including orthopedic and ophthalmological care, as well as dietary measures that may correct at least in part the plasmalogen deficiency and the phytanic acid accumulation. (These dietary measures have been described in the section on the treatment of the disorders of peroxisome biogenesis.)

Peroxisomal Disorders That Are Incompletely Characterized

1. *Zellweger-like syndrome with detectable peroxisomes.* T. Suzuki et al. (1988) described a male infant with the Zellweger phenotype in whom liver peroxisomes were present. The patient differed from the pseudo-Zellweger syndrome [3-oxoacyl-CoA thiolase deficiency (Schram et al., 1987; Goldfischer et al., 1986)] since multiple peroxisomal functions, including plasmalogen synthesis, were deficient.

2. *Congenital syndrome associated with calcific epiphyseal stippling and peroxisomal dysfunction* (Pike et al., 1990). In this patient, phytanic acid levels in serum were greatly increased and there was a slight increase in plasma VLCFA levels and a slight impairment of plasmalogen synthesis.

3. *Ataxia and peripheral neuropathy: A benign variant of peroxisome biogenesis* (MacCollin *et al.*, 1991). This patient presented with calcium oxalate renal stones, ataxia, and peripheral neuropathy, but normal intellect. VLCFA and pipecolic acid levels were increased greatly and catalase in cultured skin fibroblasts was in the cytosolic fraction. This is the only known patient in whom hyperoxaluria is combined with defects in VLCFA oxidation and abnormalities of peroxisome structure. Studies are in progress to characterize the defects more completely.

4. *Di- and trihydroxycholestanoic acidemia in twin sisters.* Wanders *et al.* (R. J. A. Wanders, personal communication) have identified mentally retarded twin sisters in whom the only biochemical abnormality was the accumulation of di- and trihydroxycholestanoic acid in a pattern similar to that seen in the disorders of peroxisome biogenesis. The biochemical defect has not yet been defined. The patients are of particular interest because they may permit assessment of the degree to which accumulation of bile acid intermediates contributes to the pathogenesis of the peroxisomal disorders.

Classical Refsum Disease

Since recent studies of the subcellular localization of phytanic acid oxidation suggest that this takes place in the mitochondrion (Watkins *et al.*, 1990; Skjeldal and Tokke, 1987), classical Refsum disease (Steinberg, 1989) is not classified as a peroxisomal disorder at this time.

CONCLUDING REMARKS

The peroxisomal disorders have attracted increasing interest because they are more frequent and varied than had been realized in the past. Fourteen different types are known and it is certain that still others will be identified. A conservative estimate of their combined incidence is 1:25,000. Their phenotype ranges from multiple congenital anomaly syndromes that lead to death in the first few months of life (Zellweger syndrome) to essentially asymptomatic states (acatalasemia). Between these two extremes are disorders that may be misdiagnosed as isolated adrenal insufficiency, brain tumor, or multiple sclerosis (adrenoleukodystrophy), deafness and blindness (infantile Refsum), renal failure (hyperoxaluria type 1), nonspecific mental retardation (neonatal ALD), or progressive neurodegenerative disorders (ALD).

All the peroxisomal disorders can be identified by noninvasive biochemical tests and prenatal diagnosis is available for all. Consideration of peroxisomal diseases has now become part of the standard differential diagnosis of many disease states. While many of the peroxisomal disorders are serious and even fatal, therapeutic approaches are becoming available for some, such as hyperoxaluria type 1 and possibly X-linked adrenoleukodystrophy.

The very existence and the serious consequence of peroxisomal disorders led to the recognition that this organelle has important physiological functions even in humans, and that the organelle is much more than an evolutionary vestige. New peroxisomal functions, such as its role in metabolism of cholesterol, polyunsaturated fatty acids, and prostaglandins and in the control of neuronal migration, continue to be identified. Very recent studies have provided new insights about the biogenesis of the organelle and the mechanisms that control the import of proteins into the peroxisome. It is these mechanisms that are at fault in the disorders of peroxisome biogenesis. Study of the cell and molecular biology of these human mutants is contributing in an exciting way to our understanding of normal cell biology and the pathogenesis and therapy of these newly recognized human disease states.

REFERENCES

Adamo, A. M., Aloise, P. A., and Pasquini, J. M, 1986, A possible relationship between concentration of microperoxisomes and myelination, *Int. J. Dev. Neurosci.* **4**:513–517.

Agamanolis, D. P., Robinson, H. B., and Timmons, G. D, 1976, Cerebro-hepato-renal syndrome: Report of a case with histochemical and ultrastructural observations, *J. Neuropathol. Exp. Neurol.* **35**:226–246.

Aikawa, J., Ishizawa, S., Narisawa, K., Tada, K., Yokota, S., and Hashimoto, T, 1987, The abnormality of peroxisomal membrane proteins in Zellweger syndrome, *J. Inherited Metab. Dis.* **10**(2):211–213.

Alberghina, M., Fiumara, A., Pavone, L., and Giuffrida, A. M, 1984, Determination of C20–C30 fatty acids by reversed-phase chromatographic techniques: An efficient method to quantitate minor fatty acids in serum of patients with adrenoleukodystrophy, *Neurochem. Res.* **9**(12):1719–1727.

Allgrove, J., Clayden, G. S., and Grant, D. B, 1978, Familial glucocorticoid deficiency with achalasia of the cardia and deficient tear production, *Lancet* **1**:1284–1286.

Ames, G. F. L, 1986, Bacterial periplasmic transport systems: Structure, mechanism, and evolution, *Annu. Rev. Biochem.* **55**:397–425.

Antoku, Y., Sakai, T., Goto, I., Iwashita, H., and Kuroiwa, Y, 1984, Adrenoleukodystrophy: Abnormality of very long-chain fatty acids in the erythrocyte membrane phospholipids, *Neurology* **34**:1499–1501.

Antoku, Y., Sakai, T., Goto, I., Miyoshino, S., Iwashita, H., and Kuroiwa, Y, 1985a, Adrenoleukodystrophy: Abnormality of "tightly bound" fatty acids in the erythrocyte membrane proteins, *Neurology* **35**:1512–1514.

Antoku, Y., Sakai, T., Goto, I., Katafuchi, Y., Sato, H., Iwashita, H., and Kuroiwa, Y., 1985b, Adrenoleukodystrophy: Fatty acid analysis of total glycerophospholipids in erythrocyte membranes, *Acta Neurol. Scand.* **72:**193–197.

Antoku, Y., Sakai, T., Tsukamoto, K., Imanishi, I., Ohtsuka, Y., Iwashita, H., and Goto, I., 1987, A simple diagnostic method of adrenoleukodystrophy: Total fatty acid analysis of erythrocyte membranes, *Clin. Chim. Acta* **169:**121–126.

Antoku, Y., Koike, F., Ohtsuka, Y., Sakai, T., Tsukamtot, K., Nagara, H., Iwashita, H., and Goto, I., 1991, Adrenoleukodystrophy: A correlation between saturated very long chain fatty acids in mononuclear cells and phenotype, *Ann. Neurol.* **30:**101–103.

Arias, J. A., Moser, A. B., and Goldfischer, S. L., 1985, Ultrastructural and cytochemical demonstration of peroxisomes in cultured fibroblasts from patients with peroxisomal deficiency disorders, *J. Cell Biol.* **100:**1789–1792.

Arneson, D. W., Tipton, R. E., and Ward, J. C., 1982, Hyperpipecolic acidemia: Occurrence in an infant with clinical findings of the cerebrohepatorenal (Zellweger) syndrome, *Arch. Neurol.* **39:**713–716.

Arnold, G., and Holtzman, E., 1978, Microperoxisomes in the central nervous system of the postnatal rat, *Brain Res.* **155:**1–17.

Aubourg, P., Bougneres, P. F., and Rocchiccioli, F., 1985, Capillary gas-liquid chromatographic-mass spectrometric measurement of very long chain (C22 to C26) fatty acids in microliter samples of plasma, *J. Lipid Res.* **26:**263–267.

Aubourg, P., Scotto, J., Rocchiccioli, F., Feldman-Pautrat, D., and Robain, O., 1986, Neonatal adrenoleukosdystrophy, *J. Neurol. Neurosurg. Psychiatry* **49:**77–86.

Aubourg, P., Sack, G. H., Meyers, D. A., Lease, J. J., and Moser, H. W., 1987, Linkage of adrenoleukodystrophy to a polymorphic DNA probe, *Ann. Neurol.* **21:**240–249.

Aubourg, P., Sack, G. H., and Moser, H. W., 1988, Frequent alteration of visual pigment genes in adrenoleukodystrophy, *Am. J. Hum. Genet.* **42:**408–413.

Aubourg, P., Blanche, S., Jambaque, I., Rocchiccioli, F., Kalifa, G., Naud-Saudreau, C., Rolland, M.-O., Debre, M., Chaussain, J. L., Griscelli, C., Fischer, A., and Bougneres, P.-F., 1990a, Reversal of early neurologic and neuroradiologic manifestations of X-linked adrenoleukodystrophy by bone marrow transplantation, *N. Engl. J. Med.* **322:**1860–1866.

Aubourg, P. B., Feil, R., Guidoux, S., Kaplan, J.-C., Moser, H. W., Kahn, A., and Mandel, J.-L., 1990b, The red-green visual pigment gene region in adrenoleukodystrophy, *Am. J. Hum. Genet.* **46:**459–469.

Aveldano, M. I., 1988, Phospholipid species containing long and very long polyenoic fatty acids remain with rhodopsin after hexane extraction of photoreceptor membranes, *Biochemistry* **27:**1229–1239.

Bakkeren, J. A., Monnens, L. A. H., Trijbels, J. M. F., and Maas, J. M., 1984, Serum very long chain fatty acid pattern in Zellweger syndrome, *Clin. Chim. Acta.* **138:**325–331.

Balfe, A., Hoefler, G., Chen, W. W., and Watkins, P. A., 1990, Aberrant subcellular localization of peroxisomal 3-ketoacyl-CoA thiolase in the Zellweger syndrome and rhizomelic chondrodysplasia punctata, *Pediatr. Res.* **27(3):**304–310.

Barth, P. G., 1987, Disorders of neuronal migration, *Can. J. Neurol. Sci.* **14:**1–16.

Barth, P. G., Schutgens, R. B. H., Wanders, R. J. A., Heymans, H. S. A., Moser, A. B., Moser, H. W., Bleeker-Wagemakers, E. M., Jansoniun-Schultheiss, K., Derix, M., and Melck, G. F., 1987, A sibship with a mild variant of Zellweger syndrome, *J. Inherited Metab. Dis.* **10:**253–259.

Barth, P. G., Wanders, R. J. A., Schutgens, R. B. H., Bleekerwagemakers, E. M., and Vanheemstra, D., 1990, Peroxisomal beta-oxidation defect with detectable peroxisomes—A case with neonatal onset and progressive course, *Eur. J. Pediatr.* **149(10):**722–726.

Beard, M. E., Baker, R., Conomos, P., Pugatch, D., and Holtzman, E., 1985a, Oxidation of oxalate and polyamines by rat peroxisomes, *J. Histochem. Cytochem.* **33(5):**460–464.

Beard, M. E., Sapirstein, V., Kolodny, E. H., and Holtzman, E., 1985b, Peroxisomes in fibroblasts from skin of Refsum's disease patients, *J. Histochem. Cytochem.* **33**(5):480–484.

Bennett, M. J., Pollit, R. J., Goodman, S. I., Hall, D. E., and Vamecq, J., 1991, Atypical riboflavin-responsive glutaric aciduria, and deficient peroxisomal glutaryl-CoA oxidase activity: A new peroxisomal disorder, *J. Inherited Metab. Dis.* **14**:165–173.

Bernheimer, H., Budka, H., and Muller, P., 1983, Brain tissue immunoglobulins in adrenoleukodystrophy: A comparison with multiple sclerosis and systemic lupus erythematosis, *Acta Neuropathol.* (Berl.) **59**:95–102.

Bernstein, J., Brough, A. J., and McAdams, A. J., 1981, The renal lesion in syndromes of multiple congenital malformations: Cerebrohepatorenal syndrome; Jeune asphyxiating thoracic dystrophy; tuberous sclerosis; Meckel syndrome, in: *The Clinical Delineation of Birth Defects: Part XVI. Urinary Systems and Others*, Alan R. Liss/National Foundation—March of Dimes, New York, pp. 35–43.

Besley, G. T. N., and Broadhead, D. M., 1987, Dihydoxyacetone phosphate acyltransferase deficiency in peroxisomal disorders, *J. Inherited Metab. Dis.* **10**(2):236–238.

Bhushan, A., Singh, R., and Singh, I., 1986, Characterization of rat brain microsomal acyl-coenzyme A ligases: Different enzymes for the synthesis of palmitoyl-coenzyme A and lignoceroyl-coenzyme A, *Arch. Biochem. Biophys.* **246**(1):374–380.

Bizzozero, O. A., Zuniga, G., and Lees, M. B., 1991, Fatty acid composition of human myelin proteolipid protein in peroxisomal disorders, *J. Neurochem.* **56**:872–878.

Björkhem, I., Blomstrand, S., Haga, P., Kase, B. F., Palonek, E., Pedersen, J. I., Standvik, B., and Wikstrom, S.-A., 1984a, Urinary excretion of dicarboxylic acids from patients with the Zellweger syndrome, *Biochim. Biophys. Acta* **795**:15–19.

Björkhem, I., Sisfontes, K. L., Bostrom, B., Kase, F., Hagenfeldt, L., and Blomstrand, R., 1984b, Possibility of prenatal diagnosis of Zellweger syndrome, *Lancet* **2**:1234–1235.

Björkhem, I., Sisfontes, L., Bostrom, B., Kase, B. F., and Blomstrand, R., 1986, Simple diagnosis of the Zellweger syndrome by gas-liquid chromatography of dimethylacetals, *J. Lipid Res.* **27**: 786–791.

Black, V. H., and Russo, J. J., 1980, Stereological analysis of the guinea pig adrenal: Effects of dexamethasone and ACTH treatment with emphasis on the inner cortex, *Am. J. Anta.* **159**:85–120.

Boles, D. J., Craft, D. A., Padgett, D. A., Loria, R. M., and Rizzo, W. B., 1991, Clinical variation in X-linked adrenoleukodystrophy—Fatty acid and lipid metabolism in cultured fibroblasts, *Biochem. Med. Metab. Biol.* **45**(1):74–91.

Boue, J., Oberle, I., Mandel, J. L., Moser, A., Moser, H. W., Larsen, Jr., J. W., Dumez, Y., and Boue, A., 1985, First trimester prenatal diagnosis of adrenoleukodystrophy by determination of very long chain fatty acid levels and by linkage analysis to a DNA probe, *Hum. Genet.* **69**:272–274.

Bourre, J.-M., Daudu, O., and Baumann, N., 1976, Nervonic acid biosynthesis by erucyl-CoA elongation in normal and quaking mouse brain microsomes. Elongation of other unsaturated fatty acyl-CoAs (mono and polyunsaturated), *Biochim. Biophys. Acta* **424**:1–7.

Bout, A., Teunissen, Y., Hashimoto, T., Benne, R., and Tager, J. M., 1988, Nucleotide sequence of human peroxisomal 3-oxoacyl CoA thiolase, *Nucleic Acids Res.* **16**:10369.

Boutin, B., Matsuguchi, L., Lebon, P., Ponsot, G., and Arthuis, C., 1989, Immunohistochemical analysis of brain macropages in adrenoleukodystrophy, *Neuropediatrics* **20**(4):202–206.

Bowen, P., Lee, C. S. N., Zellweger, H., and Lindenberg, R., 1964, A familial syndrome of multiple congenital defects, *Bull. Johns Hopkins Hosp.* **114**:402–414.

Bresnan, M. J., and Richardson, E. P., Jr.., 1979, Case records of the Massachusetts General Hospital Case 18-1979, *N. Engl. J. Med.* **300**:1037–1045.

Britton, T. C., Gibber, F. B., Clemens, M. E., Billimoria, J. D., and Sidey, M. C.., 1989, The significance of plasma phytanic acid levels in adults, *J. Neurol. Neurosurg. Psychiatry* **52**: 891–894.

Brown III, F. R., McAdams, A. J., Cummins, J. W., Konkol, R., Singh, I., Moser, A. B., and Moser, H. W., 1982a, Cerebro-hepato-renal (Zellweger) syndrome and neonatal adrenoleukodystrophy. Similarities in phenotype and accumulation of very long chain fatty acids, *Johns Hopkins Med. J.* **151**:344–351.

Brown III, F. R., Van Duyn, M. A., Moser, A. B., Schulman, J. D., Rizzo, W. B., Snyder, R. D., Murphy, J. V., Kamoshita, S., Migeon, C. J., and Moser, H. W., 1982b, Adrenoleukodystrophy: Effects of dietary restriction of very long chain fatty acids and of administration of carnitine and clofibrate on clinical status and plasma fatty acids, *Johns Hopkins Med. J.* **151**:164–172.

Brown III, F., Chen, W. W., Kirschner, D. A., Frayer, K. I., Powers, J. M., Moser, A. B., and Moser, H. W., 1983, Myelin membrane from adrenoleukodystrophy brain white matter—Biochemical properties, *J. Neurochem.* **41**(2):341–347.

Brul, S., Westerveld, A., Strijland, A., Wanders, R. J. A., Schram, A. W., Heymans, H. S. A., Schutgens, R. B. H., van den Bosch, H., and Tager, J. M., 1988a, Genetic heterogeneity in the cerebrohepatorenal (Zellweger) syndrome and other inherited disorders with a generalized impairment of peroxisomal functions—A study using complementation analysis, *J. Clin. Invest.* **81**(5):1710–1715.

Brul, S., Wiemer, E. A., Westerveld, A., Strijland, A., Wanders, R. J., Schram, A. W., Heymans, H. S., Schutgens, R. B., Van den Bosch, H., and Tager, J. M., 1988b, Kinetics of the assembly of peroxisomes after fusion of complementary cell lines from patients with the cerebro-hepato-renal (Zellweger) syndrome and related disorders, *Biochem. Biophys. Res. Commun.* **152**(3):1083–1089.

Budden, S. S., Kennaway, N. G., Buist, N. R. M., Poulos, A., and Weleber, R. G., 1986, Dysmorphic syndrome with phytanic acid oxidase deficiency, abnormal very long chain fatty acids, and pipecolic acidemia: Studies in four children, *J. Pediatr.* **108**:33–39.

Budka, H., Sluga, E., and Heiss, W. D., 1976, Spastic paraplegia associated with Addison's disease: Adult variant of adrenoleukodystrophy, *J. Neurol.* **213**:237–250.

Budka, H., Molzer, B., Bernheimer, H., Lassmann, H., Pilz, P., and Toifl, K., 1983, Clinical, morphological and neurochemical findings in adrenoleukodystrophy and its variants, *Neuropathology* (Tokyo) **1**:209–224.

Burton, B., 1987, Inborn errors of metabolism: The clinical diagnosis in early infancy, *Pediatrics* **79**:359–369.

Burton, B. K., Reed, S. P., and Reny, W. T., 1981, Hyperpipecolic acidemia: Clinical and biochemical observations in two male siblings, *J. Pediatr.* **99**(5):729–734.

Carey, W. F., Robertson, E. F., Van Crugten, C., Poulos, A., and Nelson, P. V., 1986, Prenatal diagnosis of Zellweger's syndrome by chorionic villus sampling—and a *caveat*, *Prenat. Diagn.* **6**:227–229.

Casal, H. L., and McElhaney, R. N., 1990, Quantitative determination of hydrocarbon chain conformational order in bilayers of saturated phosphatidylcholines of various chain lengths by Fourier transform infrared spectroscopy, *Biochemistry* **29**(23):5423–5427.

Casteels, M., Schepers, L., Van Veldhoven, P., Eyssen, H. J., and Mannaerts, G. P., 1990, Separate peroxisomal oxidases for fatty acyl-CoAs and trihydroxycoprostanoyl-CoA in human liver, *J. Lipid Res.* **31**:1865–1872.

Challa, V. R., Geisinger, K. R., and Burton, B. K., 1983, Pathologic alterations in the brain and liver in hyperpipecolic acidemia, *J. Neuropathol. Exp. Neurol.* **42**(6):627–638.

Chang, Y.-F., 1982, Lysine metabolism in the human and the monkey: Demonstration of pipecolic acid formation in the brain and other organs, *Neurochem. Res.* **7**(5):577–587.

Chaves-Carballo, E., Frank, M., and Chrenka, B. A., 1984, Increased intracranial pressure in adrenoleukodystrophy, *Arch. Neurol.* **41**:339–340.

Chen, W. W., Watkins, P. A., Osumit, T., Hashimoto, T., and Moser, H. W., 1987, Peroxisomal beta-oxidation enzyme proteins in adrenoleukodystrophy: Distinction between X-linked adrenoleukodystrophy and neonatal adrenoleukodystrophy, *Proc. Natl. Acad. Sci. USA* **84**:1425–1428.

Clayton, P. T., Lake, B. D., Shortland, D. B., Carruthers, R. A., and Lawson, A. M., 1987, Plasma bile acids in patients with peroxisomal dysfunction syndromes: Analysis of capillary gas chromatography–mass spectrometry, *Eur. J. Pediatr.* **146:**166–173.

Clayton, P. T., Kalter, D. Chester, Atherton, D. J., Besley, G. T. N., and Broadhead, D. M., 1989, Peroxisomal enzyme deficiency in X-linked dominant Conradi–Hunermann syndrome, *J. Inherited Metab. Dis.* **12(2):**358–360.

Cohen, S. M. Z., Martyn, L., Brown III, F. R., Moser, H. W., Chen, W. W., Kistenmacher, M., Punnett, H., Grover, W., de la Cruz, Z. C., Chan, N. R., and Green, W. R., 1983, Ocular histopathologic and biochemical studies of the cerebro-hepato-renal (Zellweger) syndrome and its relationship to neonatal adrenoleukodystrophy, *Am. J. Ophthalmol.* **96:**448–501.

Colevas, A. D., Edwards, J. L., Hruban, R. H., Mitchell, G. A., Valle, D., and Hutchins, G. M., 1988, Glutaric acidemia type 2. Comparison of pathological features in two infants, *Arch. Pathol. Lab. Med.* **112:**1133–1139.

Colombo, C., Setchell, K. D. R., Padda, M., Crosingnani, A., Roda, A., Curcio, L., Ronchi, M., and Giunta, A., 1990, Effects of ursodeoxycholic acid therapy for liver disease associated with cystic fibrosis, *J. Pediatr.* **117:**482–489.

Cooper, P. J., Danpure, C. J., Wise, P. J., and Guttridge, K. M., 1988, Immunocytochemical localization of human hepatic alanine; glyoxylate aminotransferase in control subject and patients with primary hyperoxaluria type 1, *J. Histochem. Cytochem.* **36:**1285–1294.

Crawford, D. R., Mirault, M.-E., Moret, R., Zbinden, I., and Cerutti, P. A., 1988, Molecular defect in human acatalasia fibroblasts, *Biochem. Biophys. Res. Commun.* **153(1):**59–66.

Curry, C. J. R., Magenis, R. E., Brown, M., Lanman, J. T., Tsai, J., O'Lague, P., Goodfellow, P., Mohandas, T., Bergner, E. A., and Shapiro, L. J., 1984, Inherited chondrodysplasia punctata due to a deletion of the terminal short arm of an X-chromosome, *N. Engl. J. Med.* **311:**1010–1015.

Cutting, G. R., Kasch, L. M., Rosenstein, B. J., Zielenski, J., Tsui, L.-C., Antonarakis, S. E., and Kazazian, H. H., 1990, A cluster of cystic fibrosis mutations in the first nucleotide-binding fold of the cystic fibrosis conductance regulator protein, *Nature* **346:**366–369.

Dancis, J., and Hutzler, J., 1986, The significance of hyperpipecoliatemia in Zellweger syndrome, *Am. J. Hum. Genet.* **38:**707–711.

Danks, D. M., Tippett, P., Adams, C., and Campbell, P., 1975, Cerebro-hepato-renal syndrome of Zellweger: A report of eight cases with comments upon the incidence, the liver lesion, and a fault in pipecolic acid metabolism, *J. Pediatr.* **86(3):**382–387.

Danpure, C. J., 1989, Recent advance in the understanding, diagnosis and treatment of primary hyperoxaluria type 1, *J. Inherited Metab. Dis.* **12:**210–224.

Danpure, C. J., and Jennings, P. R., 1986, Peroxisomal alanine:glyoxylate aminotransferase deficiency in primary hyperoxaluria type 1, *FEBS Lett.* **201(1):**20–24.

Danpure, C. J., and Jennings, P. R., 1988, Enzymatic heterogeneity in primary hyperoxaluria type 1 (hepatic peroxisomal alanin:glyoxylate aminotransferase deficiency), *J. Inherited Metab. Dis.* **11(2):**205–207.

Danpure, C. J., Jennings, P. R., and Watts, R. W. E., 1987, Enzymological diagnosis of primary hyperoxaluria type 1 by measurement of hepatic alanine glyoxylate aminotransferase activity, *Lancet* **1:**289–291.

Danpure, C. J., Cooper, P. J., Wise, P. J., and Jennings, P. R., 1989a, An enzyme trafficking defect in two patients with primary hyperoxaluria type 1: Peroxisomal alanine:glyoxalate aminotransferase rerouted to mitochondria, *J. Cell Biol.* **108:**1345–1352.

Danpure, C. J., Jennings, P. R., Penketh, R. J., Wise, P. J., Cooper, P. J., and Rodeck, C., 1989b, Fetal liver alanine:glyoxylate aminotransferase and the prenatal diagnosis of primary hyperoxaluria type 1, *Prenat. Diagn.* **9:**271–281.

Datta, N. S., Wilson, G. N., and Hajra, A. K., 1984, Deficiency of enzyme catalyzing the biosynthesis of glycerol-ether lipids in Zellweger syndrome: A new category of metabolic disease involving the absence of peroxisomes, *N. Engl. J. Med.* **31**:1080–1083.

Datta, S. C., Ghosh, M. K., and Hajra, A. K., 1990, Purification and properties of acyl/alkyl dihydroxyacetone-phosphate reductase from guinea pig liver peroxisomes, *J. Biol. Chem.* **265**(14):8268–8274.

de Duve, C., 1983, Microbodies in the living cell, *Sci. Am.* **248**:74–84.

de Duve, C., and Baudhuin, P., 1966, Peroxisomes (microbodies and related particles), *Physiol. Rev.* **46**:323–357.

DiChiro, G., Eiben, R. M., Manz, H. J., Jacobs, I. B., and Schellinger, D., 1980, A new CT pattern in adrenoleukodystrophy, *Radiology* **137**(3):687–692.

Diczfalusy, U., and Alexson, S. E. H., 1988, Peroxisomal chain-shortening of prostaglandin $F(2)\alpha$, *J. Lipid Res.* **29**:1629–1636.

Dooley, J. M., and Wright, B. A., 1985, Adrenoleukodystrophy mimicking multiple sclerosis, *Can. J. Neurol. Sci.* **12**:73–74.

Duda, E. E., and Huttenlocher, P. R., 1976, Computed tomography in adrenoleukodystrophy: Correlation of radiological and histological findings, *Radiology* **120**:349–350.

Duff, P., Harlass, F. E., and Milligan, D. A., 1990, Prenatal diagnosis of chondrodysplasia punctata by sonography, *Obstet. Gynecol.* **76**:497–500.

Eaton, J. W., 1989, Actalasemia, in: *The Metabolic Basis of Inherited Disease* (C. R. Scriver, A. L. Beaudet, W. S. Sly, and D. E. Valle, eds.), McGraw-Hill, New York, pp. 1551–1561.

Edmondson, D. E., and De Francesco, R., 1991 Structure, synthesis, and physical properties of covalently bound flavins and 6- and 8-hydroxyflavins, in: *Chemistry and Biochemistry of Flavoenzymes*, CRC Press, Boca Raton, Florida, pp. 73–103.

Edwin, D., Speedie, L., Naidu, S., and Moser, H. W., 1990, Cognitive impairment in adult-onset adrenoleukodystrophy, *Mol. Chem. Neuropathol.* **12**:167–176.

Eiben, R. M., and DiChiro, G., 1977, Computer assisted tomography in adrenoleukodystrophy, *J. Computer. Assisted Tomogr.* **1**(3):308–314.

Ek, J., Kase, B. F., Reith, A., Bjorkhem, I., and Pedersen, J. I., 1986, Peroxisomal dysfunction in a boy with neurologic symptoms and amaurosis (Leber disease): Clinical and biochemical findings similar to those observed in Zellweger syndrome, *J. Pediatr.* **108**:19–24.

Eldridge, R., Anayiotos, C. P., Schlesinger, S., Cowen, D., Bever, C., Patronas, N., and McFarland, H., 1984, Hereditary adult-onset leukodystrophy simulating chronic progressive multiple sclerosis, *N. Engl. J. Med.* **311**:948–953.

Esiri, M. M., Hyman, N. M., Horton, W. L., and Lindenbaum, R. H., 1984, Adrenoleukodystrophy: Clinical, pathological and biochemical findings in two brothers with the onset of cerebral disease in adult life, *Neuropathol. Appl. Neurobiol.* **10**:429–445.

Evrard, P., Caviness, V. S., Prats-Vinas, J., and Lyon, G., 1978, The mechanism of arrest of neuronal migration in the Zellweger malformation: An hypothesis based upon cytoarchitectonic analysis, *Acta Neuropathol.* (Berl.) **41**:109–117.

Eyssen, H., Eggermont, E., Van Eldere, J., Jaeken, J., Parmenter, G., and Janssen, G., 1985, Bile acid abnormalities and the diagnosis of cerebro-hepato-renal syndrome (Zellweger syndrome), *Acta Pediatr. Scand.* **74**:538–544.

Fanconi, Von A., Prader, A., Isler, W., Luthy, F., and Siebenmann, R., 1963, Morbus Addison mit hirnsklerose im kindesalter—Ein hereditares syndrom mit X-chromosomaler vererbung? *Helv. Paed. Acta* **18**:480–501.

Frank, K., Schrecker, O., Brosi, K., Krause, K. H., Krause, P., Moser, H. W., and Ziegler, R., 1986, Adrenomyeloneuropathy, a rare cause of primary adrenal cortical insufficiency, *Dtsch. Med. Wochenschr.* **40**:1519–1522.

Friedman, T. B., Polanco, G. E., Appold, J. C., and Mayle, J. E., 1985, On the loss of uricolytic activity during primate evolution—I. Silencing of urate oxidase in a hominoid ancestor, *Comp. Biochem. Physiol.* **81B**(3):653–659.

Fringes, B., and Reith, A., 1982, A time course of peroxisome biogenesis during adaptation to mild hyperthyroidism in rat liver, *Lab. Invest.* **47**:19–26.

Fujiki, Y., Hubbard, A. L., Fowler, S., and Lazarow, P. B., 1982, Isolation of intracellular membranes by means of sodium carbonate treatment. Application to endoplasmic reticulum, *J. Cell Biol.* **93**:97–102.

Fujiki, Y., Rachubinski, R. A., and Lazarow, P. B., 1984, Synthesis of a major integral membrane polypeptide of rat liver peroxisomes on free polysomes, *Proc. Natl. Acad. Sci. USA* **81**:7127–7131.

Fujiki, Y., Rachubinski, R. A., Mortenseen, R. M., and Lazarow, P. A., 1985, Synthesis of 3-ketoacyl-CoA thiolase of rat liver peroxisomes on free polyribosomes as a larger precursor, *Biochem. J.* **226**:697–704.

Gaertner, J., Chen, W. W., Kelley, R. I., Mihalik, S., and Moser, H. W., 1991, The 22-kD peroxisomal integral membrane protein in Zellweger syndrome—Presence, abundance, and association with a peroxisomal thiolase precursor protein, *Pediatr. Res.* **29**:141–146.

Gatfield, P. D., Taller, E., Hinton, G. G., Wallace, A. C., Abdelnour, G. M., and Haust, M. D., 1968, Hyperpipecolatemia: A new metabolic disorder associated with neuropathy and hepatomegaly, *Can. Med. Assoc. J.* **99**(25):1215–1233.

Gietl, C., 1990, Glyoxysomal malate dehydrogenase from watermelon is synthesized with an amino-terminal transit peptide, *Proc. Natl. Acad. Sci. USA* **87**:5773–5777.

Gilchrist, K. W., Gilbert, E. F., Goldfarb, S., Goll, U., Spranger, J. W., and Opitz, J. M., 1976, Studies of malformation syndromes of man XIB: The cerebro-hepato-renal syndrome of Zellweger. Comparative pathology, *Eur. J. Pediatr.* **121**:99–118.

Goldfischer, S., Moore, C. L., Johnson, A. B., Spiro, A. J., Valsamis, M. P., Ritch, R. H., Wisniewski, H. K., Norton, W. T., Rapin, I., and Gartner, L. M., 1973, Peroxisomal and mitochondrial defects in the cerebro-hepato-renal syndrome, *Science* **182**:62–64.

Goldfischer, S., Powers, J. M., Johnson, A. B., Axe, S., Brown III, F. R., and Moser, H. W., 1983, Striated adrenocortical cells in cerebro-hepato-renal (Zellweger) syndrome, *Virchows Arch. A* **401**:355–361.

Goldfischer, S., Collins, J., Rapin, I., Neumann, P., Neglia, W., Spiro, A. J., Ishii, T., Roels, F., Vamecq, J., and Van Hoof, F., 1986, Pseudo-Zellweger syndrome: Deficiencies in several peroxisomal oxidative activities, *J. Pediatr.* **108**:25–32.

Goldman, B. M., and Blobel, G., 1978, Biogenesis of peroxisomes: Intracellular site of synthesis of catalase and uricase, *Proc. Natl. Acad. Sci. USA* **75**:5066–5070.

Gorgas, K., 1985, Serial section analysis of mouse hepatic peroxisomes, *Anat. Embryol.* **172**:21–32.

Gorgas, K., 1987, Morphogenesis of peroxisomes in lipid synthesizing epithelia, in: *Peroxisomes in Biology and Medicine* (H. D. Fahimi and H. Sies, eds.), Springer, Berlin, pp. 4–17.

Gould, S. J., Keller, G. A., and Subramani, S., 1987, Identification of a peroxisomal targeting signal at the carboxy terminus of firefly luciferase, *J. Cell Biol.* **105**:2923–2931.

Gould, S. J., Keller, G.-A., and Subramani, S., 1988, Identification of peroxisomal targeting signals located at the carboxy terminus of four peroxisomal proteins, *J. Cell Biol.* **107**:897–905.

Gould, S. J., Krisans, S., Keller, G. A., and Subramani, S., 1990, Antibodies directed against the peroxisomal targeting signal of firefly luciferase recognize multiple mammalian peroxisomal proteins, *J. Cell Biol.* **110**(1):27–34.

Govaerts, L., Monnens, L., Tegelaers, W., Trijbels, F., and van Raay-Selten, A., 1982, Cerebro-hepato-renal syndrome of Zellweger: Clinical symptoms and relevant laboratory findings in 16 patients, *Eur. J. Pediatr.* **139**:125–128.

Govaerts, L., Bakkeren, J., Monnens, L., Maas, J., and Trijbels, F., 1985, Disturbed very long chain (C24–C26) fatty acid pattern in fibroblasts of patients with Zellweger's syndrome, *J. Inherited Metab. Dis.* **8**:5–8.

Gravel, R. A., Mohoney, M. J., Ruddle, F. H., and Rosenberg, L. E., 1975, Genetic complementation in heterokaryons of human fibroblasts defective in cobalamin metabolism, *Proc. Natl. Acad. Sci. USA* **72**:3181–3185.

Greenberg, C. R., Hajra, A. K., and Moser, A. B., 1987, Triple therapy of a patient with a generalized peroxisomal disorder, *Am. J. Hum. Genet.* **41**(Suppl):A–64.

Greenberg, H. S., Halverson, D., and Lane, B., 1977, CT scanning and diagnosis of adrenoleukodystrophy, *Neurology* **27**:884–886.

Griffin, D. E., Moser, H. W., Mendoza, Q., Moench, T., O'Toole, S., and Moser, A. B., 1985, Identification of the inflammatory cells in the nervous system of patients with adrenoleukodystrophy, *Ann. Neurol.* **18**:660–664.

Griffin, J. W., Goren, E., Schaumburg, H., Engel, W. K., and Loriaux, L., 1977, Adrenomyeloneuropathy: A probable variant of adrenoleukodystrophy, *Neurology* **27**(12):1107–1113.

Gronn, M., Christensen, E., Hagve, T.-A., and Christophersen, B. O., 1990, The Zellweger syndrome: Deficient conversion of docosahexaenoic acid (22:6(n-3)) and normal delta desaturase activity in cultured skin fibroblasts, *Biochim. Biophys. Acta* **1044**(2):249–254.

Gros, P., Coop, J., and Housman, D., 1986, Mammalian multidrug resistance gene: Complete cDNA sequence indicates strong homology to bacterial transport proteins, *Cell* **47**:371–380.

Gustafsson, J., Gustavson, K. H., Karlaganis, G., and Sjövall, J., 1983, Zellweger's cerebro-hepatorenal syndrome—Variations in expressivity and in defects of bile and synthesis, *Clin. Genet.* **24**:313–319.

Gutierrez, C., Okita, R., and Krisans, S., 1988, Demonstration of cytochrome reductase in rat liver peroxisomes: Biochemical and immunochemical analysis, *J. Lipid Res.* **29**:613–628.

Hajra, A. K., and Bishop, J. E., 1982, Glycerolipid biosynthesis in peroxisomes via the acyl dihydroxyacetone phosphate pathway, *Ann. N. Y. Acad. Sci.* **386**:170–181.

Hall, J. G., Pauli, R. M., and Wilson, K. M., 1980, Maternal and fetal sequelae of anticoagulants during pregnancy, *Am. J. Med.* **68**:122–140.

Hansen, H., and Roggenkamp, R., 1989, Functional complementation of catalase-defective peroxisomes in a methylotrophic yeast by import of the catalase A from *Saccharomyces cerevisiae*, *Eur. J. Biochem.* **184**:173–179.

Hanson, R. F., Szczepanick-van Leeuwen, P., Williams, G. C., Grabowski, G., and Sharp, H. L., 1979, Defects of bile acid synthesis in Zellweger's syndrome, *Science* **203**:1107–1108.

Happle, R., 1979, X-linked dominant chondrodysplasia punctata. Review of literature and report of a case, *Hum. Genet.* **63**:65–73.

Hardeman, D., and van den Bosch, H., 1989, Topography of ether phospholipid biosynthesis, *Biochim. Biophys. Acta* **1006**:1–8.

Hashimoto, T., 1982, Individual peroxisomal beta-oxidation enzymes, *Ann. N. Y. Acad. Sci.* **386**:5–12.

Hashimoto, T., Kuwabara, T., Usuda, N., and Nagata, T., 1986, Purification of membrane polypeptides of rat liver peroxisomes, *J. Biochem.* **100**:301–310.

Heffungs, W., Hameisier, H., and Ropers, H. H., 1980, Addison disease and cerebral sclerosis in an apparently heterozygous girl: Evidence of inactivation of the adrenoleukodystrophy locus, *Clin. Genet.* **18**:184–188.

Herndon, J. H., Steinberg, D., Uhlendorf, B. W., and Fales, H., 1969, Refsum's disease: Characterization of the enzyme defect in cell culture, *J. Clin. Invest.* **48**:1017–1031.

Hess, R., Staubli, W., and Riess, W., 1965, Nature of the hepatomegalic effect produced by ethylchlorophenoxyisobutyrate in the rat, *Nature* **208**:856–858.

Heymans, H. S. A., 1984, Cerebro-hepato-renal (Zellweger) syndrome. Clinical and biochemical consequences of peroxisomal dysfunction, Thesis, University of Amsterdam, Amsterdam.

Heymans, H. S. A., Schutgens, R. B. H., Tan, R., van den Bosch, H., and Borst, P., 1983, Severe plasmalogen deficiency in tissues of infants without peroxisomes (Zellweger syndrome), *Nature* **306**(5938):69–70.

Heymans, H. S. A., van den Bosch, H., Schutgens, R. B. H., Tegelaers, W. H. H., Walther, J.-U., Muller-Hocker, J., and Borst, P., 1984, Deficiency of plasmalogens in the cerebro-hepato-renal (Zellweger) syndrome, *Eur. J. Pediatr.* **142**:10–15.

Heymans, H. S. A., Oorthuys, J. W. E., Nelck, G., Wanders, R. J. A., and Schutgens, R. B. H., 1985, Rhizomelic chondrodysplasia punctata: Another peroxisomal disorder, *N. Engl. J. Med.* **313**:(3):187–188.

Heymans, H. S. A., Oorthuys, J. W. E., Nelck, G., Wanders, R. J. A., Dingemans, K. P., and Schutgens, R. B. H., 1986, Peroxisomal abnormalities in rhizomelic chondrodysplasia punctata, *J. Inherited Metab. Dis.* **9**(2):328–331.

Higashi, T., and Peters, T., 1963, Studies on rat liver catalase. Incorporation of C14-leucine into catalase of liver cell fractions *in vivo*, *J. Biol. Chem.* **238**:3952–3954.

Higgins, C. F., Gallagher, M. P., Mimmack, M. L., and Pearce, S. R., 1988, A family of closely related ATP-binding subunits from prokaryotic and eukaryotic cells, *Bioessays* **8**:111–116.

Hijikata, M., Ishii, N., Kagamiyama, H., Osumi, T., and Hashimoto, T., 1987, Structural analysis of cDNA for rat peroxisomal 3-ketoacyl-CoA thiolase, *J. Biol. Chem.* **262**(17):8151–8158.

Hijikata, M., Wen, J. K., Osumi, T., and Hashimoto, T., 1990, Rat peroxisomal 3-ketoacyl-CoA thiolase gene—occurrence of 2 closely related but differentially regulated genes, *J. Biol. Chem.* **265**(8):4600–4606.

Hillman, R. E., 1989, Primary hyperoxalurias, in: *The Metabolic Basis of Inherited Disease* (C. R. Scriver, A. L. Beaudet, W. S. Sly, and D. Valle, eds.), McGraw-Hill, New York, pp. 933–944.

Hiltunen, J. K., Karki, T., Hassinen, I. E., and Osmundsen, H., 1986, β-Oxidation of polyunsaturated fatty acids by rat liver peroxisomes, *J. Biol. Chem.* **261**(35):16484–16493.

Hittner, H. M., Dretzer, F. L., and Mehta, R. S., 1981, Zellweger syndrome: Lenticular opacities indicating carrier status and lens abnormalities characteristic of homozygotes, *Arch. Ophthalmol.* **99**:1977–1982.

Hoefler, G., Hoefler, S. Watkins, P., Chen, W., Moser, A. B., Baldwin, V., McGillivray, B., Charrow, J., Friedman, J. M., Rutledge, L., and Moser, H. W., 1988, Biochemical abnormalities in rhizomelic chondrodysplasia punctata, *J. Pediatr.* **112**:726–733.

Hoefler, S., Hoefler, G., Moser, A. B., Watkins, P. A., Chen, W. W., and Moser, H. W., 1988, Prenatal diagnosis of rhizomelic chondrodysplasia punctata, *Prenat. Diagn.* **8**:571–576.

Hogg, J. F., 1991, Peroxisomes in *Tetrahymena* and their relation to gluconeogenesis, *Ann. N. Y. Acad. Sci.* **168**:281–291.

Holmes, R. D., Wilson, G. N., and Hajra, A., 1987a, Oral ether lipid therapy in patients with peroxisomal disorders, *J. Inherited Metab. Dis.* **10**(2):239–241.

Holmes, R. D., Wilson, G. N., and Hajra, A. K., 1987b, Peroxisomal enzyme deficiency in the Conradi–Hunerman form of chondrodysplasia punctata, *N. Engl. J. Med.* **316**:1608.

Holtta, E., 1977, Oxidation of spermidine and spermine in rat liver. Purification and properties of polyamine oxidase, *Biochemistry* **16**(1):91–100.

Hruban, Z., and Rechcigl, M., 1969, *Microbodies and Related Particles*, Academic Press, New York.

Hruban, Z., Vigil, E. L., Slesers, A., and Hopkins, E., 1972, Microbodies: Constituent organelles of animal cells, *Lab. Invest.* **27**:184–191.

Huang, A. H. C., and Beaver, H., 1973, Localization of enzymes within microbodies, *J. Cell Biol.* **58**:379–389.

Iancu, T. C., and Danpure, C. J., 1987, Primary hyperoxaluria type 1: Ultrastructural observations in liver biopsies, *J. Inherited Metab. Dis.* **10**:330–338.

Igarashi, M., Belchis, D., and Suzuki, K., 1976a, Brain gangliosides in adrenoleukodystrophy, *J. Neurochem.* **27**:327–328.

Igarashi, M., Schaumburg, H. H., Powers, J., Kishimoto, Y., Kolodny, E., and Suzuki, K., 1976b, Fatty acid abnormality in adrenoleukodystrophy, *J. Neurochem.* **26**:851–860.

Imanaka, T., Small, G. M., and Lazarow, P. B., 1987, Translocation of acyl-CoA oxidase into peroxisomes requires ATP hydrolysis but not a membrane potential, *J. Cell Biol.* **105**(6):2915–2922.

Imanaka, T., Lazarow, P., and Takano, T., 1991, A novel 57 kDa peroxisomal membrane polypeptide detected by monoclonal antibody (PXCM1a/207B), *Biochim. Biophys. Acta* **1062**:264–270.

Ishii, N., Hijikata, M., Osumi, T., and Hashimoto, T., 1987, Structural organization of the gene for rat enoyl-CoA hydratase:3-hydroxyacyl-CoA dehydrogenase bifunctional enzyme, *J. Biol. Chem.* **262**(17):8144–8150.

Issemann, I., and Green, S., 1990, Activation of a member of the steroid hormone receptor superfamily by peroxisome proliferators, *Nature* **347**:645–650.

Jakobs, C., ten Brink, H., Kok, R. M., Stellaard, F., Kleijer, W. J., Wanders, R. J. A., and Schutgens, R. B. H., 1989, Very long chain fatty acids in amniotic fluid from a fetus affected with Zellweger syndrome, *Eur. J. Pediatr.* **148**:581.

James, A. C. D., Kaplan, P., Lees, A., and Bradley, J. J., 1984, Schizophreniform psychosis and adrenomyeloneuropathy, *J. R. Soc. Med.* **77**:882–884.

Javitt, N. B., 1990, HEP-G2 cells as a resource for metabolic studies—Lipoprotein, cholesterol, and bile acids, *FASEB J.* **4**(2):161–168.

Kamijo, K., Taketani, S., Yokota, S., Osumi, T., and Hashimoto, T., 1990, The 70-kDa peroxisomal membrane protein is a member of the Mdr (P-glycoprotein)-related ATP-binding protein superfamily, *J. Biol. Chem.* **265**(8):4534–4540.

Kandt, R. S., Heldrich, F. S., and Moser, H. W., 1983, Recovery from probable central pontine myelinolysis associated with Addison's disease, *Arch. Neurol.* **40**:118–119.

Kase, B. F., and Björkhem, I., 1989, Peroxisomal bile acid-CoA: Amino acid *N*-acyltransferase in rat liver, *J. Biol. Chem.* **264**(16):9220–9223.

Kase, B. F., Pedersen, J. I., Standvik, B., and Björkhem, I., 1985, *In vivo* and *in vitro* studies on formation of bile acids in patients with Zellweger syndrome, *J. Clin. Invest.* **76**:2393–2402.

Kase, B. F., Prydz, K., Björkhem, I., and Pedersen, J. I., 1986, *In vitro* formation of bile acids form di- and trihydroxy-5β-cholestanoic acid in human liver peroxisomes, *Biochim. Biophys. Acta* **877**:37–42.

Kase, B. F., Pedersen, J. I., Wathne, K. O., Gustafsson, J., and Björkhem, I., 1991, Importance of peroxisomes in the formation of chenodeoxycholic acid in human liver. Metabolism of 3a,7a-dihydroxy-5B-cholestanoic acid in Zellweger syndrome, *Pediatr. Res.* **29**(1):64–69.

Kaufmann, W. E., Theda, C., Naidu, S., Moser, A. B., and Moser, H. W., 1991, Zellweger-like neuronal migration abnormality in a patient with an isolated defect of peroxisomal fatty acid oxidation, *Ann. Neurol.* **30**:497a–498a.

Kawamura, N., Moser, A. B., Moser, H. W., Ogino, T., Suzuki, K., Schaumburg, H., Milunsky, A., Murphy, J., and Kishimoto, Y., 1978, High concentration of hexacosanoate in cultured skin fibroblast lipids from adrenoleukodystrophy patients, *Biochem. Biophys. Res. Commun.* **82**:114–120.

Keller, G. A., Pazirandeh, M., and Krisans, S., 1986, 3-Hydroxy-3-methylglutaryl coenzyme A reductase localization in rat liver peroxisomes and microsomes of control and cholestyramine-treated animals: Quantitative biochemical and immunoelectron microscopical analyses, *J. Cell Biol.* **103**:875–886.

Keller, G. A., Gould, S., DeLuca, M., and Subramani, S., 1987, Firefly luciferase is targeted to peroxisomes in mammalian cells, *Proc. Natl. Acad. Sci. USA* **84**:3264–3268.

Kelley, R. I., 1983, The cerebrohepatorenal syndrome of Zellweger: Morphologic and metabolic aspects, *Am. J. Med. Genet.* **16**:503–517.

Kelley, R. I., 1991, Quantification of pipecolic acid in plasma and urine by isotope-dilution gas chromatography/mass spectrometry, in: *Techniques in Diagnostic Human Biochemical Genetics* (F. A. Hommes, ed.), Wiley-Liss, New York, pp. 205–218.

Kelley, R. I., and Corkey, B. E., 1983, Increased sensitivity of cerebrohepatorenal syndrome fibroblasts to antimycin A, *J. Inherited Metab. Dis.* **6**:158–162.

Kelley, R. I., and Moser, H. W., 1984, Hyperpipecolic acidemia in neonatal adrenoleukodystrophy, *Am. J. Med. Genet.* **19**:791–795.

Kelley, R. I., Datta, N. S., Dobyns, W. B., Hajra, A. K., Moser, A. B., Noetzel, M. J., Zackal, E. H., and Moser, H. W., 1986, Neonatal adrenoleukodystrophy: New cases, biochemical studies, and differentiation from Zellweger and related peroxisomal polydystrophy syndromes, *Am. J. Med. Genet.* **23**:869–901.

Kerem, B.-S., Zielenski, J., Markiewicz, D., Bozon, D., Gazit, E., Yahav, J., Kennedy, D., Riordan, J. R., Collins, F. S., Rommens, J. M., and Tsui, L.-C., 1990, Identification of mutations in regions corresponding to the two putative nucleotide (ATP)-binding folds of the cystic fibrosis gene, *Proc. Natl. Acad. Sci. USA* **87**:8447–8451.

Khanna, J. M., and Israel, Y., 1980, Ethanol metabolism, in: *Liver and Biliary Tract Physiology*, University Park Press, Baltimore, Maryland, pp. 275–315.

Kishimoto, Y., Moser, H. W., Kawamura, N., Platt, M., Pallante, B., and Fenselau, C., 1980, Evidence that abnormal very long chain fatty acids of brain cholesterol esters are of exogenous origin, *Biochem. Biophys. Res. Commun.* **96**:69–76.

Kitchin, W., Cohen-Cole, S. A., and Mickel, S. F., 1987, Adrenoleukodystrophy: Frequency of presentation as a psychiatric disorder, *Biol. Psychiatry* **22**:1375–1387.

Knazek, R. A., Rizzo, W. B., Schulman, J. D., and Dave, J. R., 1983, Membrane microviscosity is increased in the erythrocytes of patients with adrenoleukodystrophy and adrenomyeloneuropathy, *J. Clin. Invest.* **72**:245–248.

Kobayashi, T., Katayama, M., Suzuki, S., Tomoda, H., Goto, I., and Kuroiwa, Y., 1983, Adrenoleuko-dystrophy: Detection of increased very long chain fatty acids by high-performance liquid chromatography, *J. Neurol.* **230**:209–215.

Krisans, S. K., Thompson, S. L., Pena, L. A., Kok, E., and Javitt, N. B., 1985, Bile acid synthesis in rat liver peroxisomes: Metabolism of 26-hydroxycholesterol to 3β-hydroxy-5-cholenoic acid, *J. Lipid Res.* **26**(11):1324–1332.

Krivit, W., Shapiro, E., Lockman, L., Torres, F., Stillman, A., Moser, A., and Moser, H., 1992, Recommendations for treatment of childhood cerebral form of adrenoleukodystrophy, in: *Correction of Certain Genetic Diseases by Transplantation 1991* (J. R. Hobbs and P. G. Riches, eds.), Cogent Trust, London, pp. 38–49.

Kumar, A. J., Rosenbaum, A. E., Naidu, S., Wenger, L., Citrin, C. M., Lindenberg, R., Kim, W. S., Ziureich, S. J., Molliver, M. E., Mayber, H. S., and Moser, H. W., 1987, Adrenoleukodystrophy: Correlating MR imaging with CT, *Radiology* **165**:496–504.

Kyllerman, M., Blomstrand, S., Mansson, J. E., and Conradt, N. G., 1990, Central nervous system malformations and white matter changes in pseudo-neonatal adrenoleukodystrophy, *Neuropediatrics* **21**:199–201.

Lageweg, W. Tager, J. M., and Wanders, J. A., 1991a, Topography of very long chain fatty acid activating activity in peroxisomes from rat liver, *Biochem. J.* **276**:53–56.

Lageweg, W., Wanders, R. J. A., and Tager, J. M., 1991b, Long chain acyl-CoA synthetase and very long chain acyl-CoA synthetase activities in peroxisomes and microsomes from rat liver, *Eur. J. Biochem.* **196**:519–523.

Lalwani, N. D., Alvarez, K., Reddy, M. K., Reddy, M. N., Parikh, I., and Reddy, J. K., 1987, Peroxisome proliferator binding protein: Identification and partial characterization of nafenopin, clofibric acid and ciprofibrate-binding protein from rat liver, *Proc. Natl. Acad. Sci. USA* **84**:5342–5346.

Lam, S., Hutzler, J., and Dancis, J., 1986, L-Pipecolaturia in Zellweger syndrome, *Biochim. Biophys. Acta* **882**:254–257.

Lanes, R., Plotnick, L. P., Bynum, T. E., Lee, P. A., Casella, J. F., Fox, C. E., Kowarski, A. A., and Migeon, C. J., 1980, Glucocorticoid and partial mineralocorticoid deficiency associated with achalasia, *J. Clin. Endocrinol. Metab.* **50**:268–270.

Latta, K., and Brodhel, J., 1990, Primary hyperoxaluria type 1, *Eur. J. Pediatr.* **149**(8):518.

Lawson, A. M., Madigan, M. J., Shortland, D., and Clayton, P. T., 1986, Rapid diagnosis of Zellweger syndrome and infantile Refsum's disease by fast atom bombardment-mass spectrometry of urine bile salts, *Clin. Chim. Acta* **161**:221–231.

Lazarow, P. B., 1978, Rat liver peroxisomes catalyze the β-oxidation of fatty acids, *J. Biol. Chem.* **253**(5):1522–1528.

Lazarow, P. B., and de Duve, C., 1973, The synthesis and turnover of rat liver peroxisomes. V. Intracellular pathway of catalase synthesis, *J. Cell Biol.* **59**:507–524.

Lazarow, P. B., and de Duve, C., 1976, A fatty acyl-CoA oxidizing system in rat liver peroxisomes: Enhancement by clofibrate, a hypolipidemic drug, *Proc. Natl. Acad. Sci. USA* **73**:(6):2043–2046.

Lazarow, P. B., and Fujiki, Y., 1985, Biogenesis of peroxisomes, *Annu. Rev. Cell. Biol.* **1**:489–530.

Lazarow, P., and Moser, H. W., 1989, Disorders of peroxisomal biogenesis, in: *The Metabolic Basis of Inherited Disease* (C. R. Scriver, A. L. Beaudet, W. S. Sly, and D. Valle, eds.), McGraw-Hill, New York, pp. 1479–1509.

Lazarow, P. B., Black, V., Shio, H., Fujiki, Y., Hajra, A., Datta, N., Bangaru, B. S., and Dancis, J., 1985, Zellweger syndrome: Biochemical and morphological studies on two patients treated with clofibrate, *Pediatr. Res.* **19**(12):1356–1364.

Lazo, O., Contreras, M., Hashmi, M., Stanley, W., Irazu, C., and Singh, I., 1988, Peroxisomal lignoceroyl-CoA ligase deficiency in childhood adrenoleukodystrophy and adrenomyeloneuropathy, *Proc. Natl. Acad. Sci. USA* **85**:7647–7651.

Lazo, O., Contreras, M., and Singh, I., 1990, Topographical localization of peroxisomal Acyl-CoA ligases—Differential localization of palmitoyl-CoA and lignoceroyl-CoA ligases, *Biochemistry* **29**(16):3981–3986.

Lazo, O., Singh, A. K., and Singh, I., 1991, Postnatal development and isolation of peroxisomes from brain, *J. Neurochem.* **56**:1343–1353.

Liu, H. M., Bangaru, B. S., Kidd, J., and Boggs, J., 1976, Neuropathological considerations in cerebro-hepato-renal syndrome (Zellweger's syndrome), *Acta Neuropathol. (Berl.)* **34**:115–123.

Loehr, J. P., Goodman, S. I., and Freeman, F. E., 1990, Glutaric acidemia type 2: Heterogeneity of clinical and biochemical phenotypes, *Pediatr. Res.* **27**:311–315.

MacCollin, M., DeVivo, D. C., Moser, A. B., and Beard, M., 1991, Ataxia and peripheral neuropathy: A benign variant of peroxisome dysgenesis, *Ann. Neurol.* **28**:833–836.

MacDonald, J. T., Stauffer, A. E., and Heitoff, K., 1984, Adrenoleukodystrophy: Early frontal lobe involvement on computed tomography, *J. Computer Assisted Tomogr.* **8**(1):128–130.

Mannaerts, G. P., van Veldhoven, P., Van Broekhoven, A., Vandebroek, G,. and Debeer, L. J., 1982, Evidence that peroxisomal acyl-CoA synthetase is located at the cytoplasmic side of the peroxisomal membrane, *Biochem. J.* **204**:17–23.

Martin-deLeon, P. A., Wolf, S. F., Persico, G., Toniolo, D. Martini, G., and Migeon, B. R., 1985, Localization of glucose-6-phosphate dehydrogenase in mouse and man by *in-situ* hybridization: Evidence for a single locus and transposition of homologous X-linked genes, *Cytogenet. Cell Genet.* **39**:87.

Martinez, M., 1989, Polyunsaturated fatty acid changes suggesting a new enzymatic defect in Zellweger syndrome, *Lipids* **24**:261–265.

Mathis, R. K., Watkins, J. B., Szczepanik-Van Leeuwen, P., and Lott, I. T., 1980, Liver in the cerebro-hepato-renal syndrome: Defective bile acid synthesis and abnormal mitochondria, *Gastroenterology* **79**:1311–1317.

McDonald, J. C., Landrenaeau, M. D., Rohr, M. S., and DeVault, G. A., 1989, Reversal by liver transplantation of the complications of primary hyperoxaluria as well as the metabolic defect, *N. Engl. J. Med.* **321**:1100–1103.

McGuinness, M. C., Moser, A. B., Moser, H. W., and Watkins, P. A., 1990, Peroxisomal disorders:

Complementation analysis using beta-oxidation of very long chain fatty acids, *Biochem. Biophys. Res. Commun.* **172**(1):364–369.

McVay, V., 1969, Usher syndrome: Deafness and progressive blindness. Clinical cases, prevention, theory and literature survey, *J. Chronic Dis.* **22**:133–151.

Menkes, J. H., and Corbo, L. M., 1977, Adrenoleukodystrophy: Accumulation of cholesterol esters with very long chain fatty acids, *Neurology* **27**:928–932.

Migeon, B. R., Moser, H. W., Moser, A. B., Axelman, J., Sillence, D., and Norum, R. A., 1981, Adrenoleukodystrophy: Evidence for X-linkage, inactivation and selection favoring the mutant allele in heterozygous cells, *Proc. Natl. Acad. Sci. USA* **78**:5066–5070.

Mihalik, S., and Rhead, W. J., 1989, L-Pipecolic acid oxidation in the rabbit and cynomolgous monkey, *J. Biol. Chem.* **264**(5):2509–2517.

Mihalik, S. J., Moser, H. W., Watkins, P. A., Dans, D. M., Poulos, A., and Rhead, W. J., 1989, Peroxisomal L-pipecolic acid oxidation is deficient in liver from Zellweger syndrome patients, *Pediatr. Res.* **25**:548–552.

Mihalik, S. J., McGuiness, M., and Watkins, P. A., 1991, Purification and characterization of peroxisomal L-pipecolic acid oxidase from monkey liver, *J. Biol. Chem.* **266**:4822–4830.

Mihalik, S., and Rhead, W. J., 1989, L-Pipecolic acid oxidation in the rabbit and cynomolgous monkey, *J. Biol. Chem.* **264**(5):2509–2517.

Miyazawa, S., Hashimoto, T., and Yokota, S., 1985, Identity of long-chain acyl-coenzyme A synthetase of microsomes, mitochondria, and peroxisomes in rat liver, *J. Biochem.* **98**:723–733.

Miyazawa, S., Osumi, T., Hashimoto, T., Ohno, K., Miura, S., and Fujiki, Y., 1989, Peroxisome targeting signal of rat liver acyl-coenzyme A oxidase resides at the carboxy terminus, *Mol. Cell. Biol.* **9**(1):83–91.

Miyazawa, W., Hayashi, H., Hijikata, M., Ishii, N., Furuta, S., Kagamiyama, H., Osumi, T., and Hashimoto, T., 1987, Complete nucleotide sequence of cDNA and predicted amino acid sequence of rat acyl-CoA oxidase, *J. Biol. Chem.* **262**(17):8131–8137.

Mobley, W. C., White, C. L., Tennekoon, G., Clark, A. W., Cohen, S. R., Green, R., and Moser, H. W., 1982, Neonatal adrenoleukodystrophy, *Ann. Neurol.* **12**(2):204–205.

Molzer, B., Bernheimer, H., Budka, H., Pilz, P., and Toifl, K., 1981, Accumulation of very long chain fatty acids is common to 3 variants of adrenoleukodystrophy, *J. Neurol. Sci.* **51**:301–310.

Molzer, B., Bernheimer, H., Heller, R., Toifl, K., and Vetterlein, M., 1982, Detection of adrenoleuko-dystrophy by increased C26:O fatty acid levels in leukocytes, *Clin. Chim. Acta* **125**:299–305.

Molzer, B., Korschinsky, M., Bernheimer, H., Schid, R., Wolf, C., and Roscher, A., 1986, Very long chain fatty acids in genetic peroxisomal disease fibroblasts: differences between the cerebro-hepato-renal (Zellweger) syndrome and adrenoleukodystrophy variants, *Clin. Chim. Acta* **161**: 81–90.

Monnens, L., Bakkeren, J., Parmentier, G., Janssen, G., van Haelst, U., Trijbels, F., and Eyssen, H., 1980, Disturbances in bile acid metabolism of infants with the Zellweger (cerebro-hepato-renal) syndrome, *Eur. J. Pediatr.* **133**:31–35.

Mooi, W. J., Dingemans, K. P., van den Berg, H., Weerman, M. A., Jobsis, A. C., Heymans, H. S. A., and Barth, P. G., 1983, Ultrastructure of the liver in the cerebro-hepato-renal syndrome of Zellweger, *Ultrastruct. Pathol.* **5**:135–144.

Morariu, M. A., Chasan, J. L., Norum, R. A., Moser, H. W., and Migeon, B. R., 1982, Adrenoleuko-dystrophy variant in a heterozygous female, *Neurology* **32**:A81.

Morgan, S. H., and Watts, R. W. E., 1989, Perspectives in the assessment and management of patients with primary hyperoxaluria type 1, *Adv. Nephrol.* **18**:95–106.

Moser, A. B., Singh, I., Brown III, F. R., Solish, G. I., Kelly, R. I., Benke, P. J., and Moser, H. W., 1984, The cerebro-hepato-renal (Zellweger) syndrome: Increased levels and impaired degradation of very long chain fatty acids, and prenatal diagnosis, *N. Engl. J. Med.* **310**:1141–1145.

Moser, A. B., Borel, J., Odone, A., Naidu, S., Cornblath, D., Sanders, D. B., and Moser, H. W., 1987,

A new dietary therapy for adrenoleukodystrophy: Biochemical and preliminary clinical results in 36 patients, *Ann. Neurol.* **21**:240–249.

Moser, H. W., 1989, Peroxisomal disorders, *Curr. Opinion Pediatr.* **1**:284–289.

Moser, H. W., and Goldfischer, S., 1985, The peroxisomal disorders, *Hosp. Pract.* **20**(9):61–70.

Moser, H. W., and Moser, A. B., 1990a, Measurements of saturated very long chain fatty acids in plasma, in: *Techniques in Diagnostic Human Biochemical Genetics* (F. A. Hommes, ed), Wiley-Liss, New York, pp. 117–191.

Moser, H. W., and Moser, A. B., 1990b), Measurement of phytanic acid levels, in: *Techniques in Diagnostic Human Biochemical Genetics* (F. A. Hommes, ed), Wiley-Liss, New York, pp. 193–203.

Moser, H. W., Moser, A. B., Kawamura, N., Migion, B., O'Neill, B. P., Fenselau, C., and Kishimoto, Y., 1980a, Adrenoleukodystrophy: Studies of the phenotype, genetics and biochemistry, *Johns Hopkins Med. J.* **147**:217–224.

Moser, H. W., Moser, A. B., Kawamura, N., Murphy, J., Milunsky, A., Suzuki, K., Schaumburg, H., and Kishimoto, Y., 1980b, Adrenoleukodystrophy: Elevated C-26 fatty acid in cultured skin fibroblasts, *Ann. Neurol.* **7**:542–549.

Moser, H. W., Moser, A. B., Frayer, K. K., Chen, W. W., Schulman, J. D., O'Neill, B. P., and Kishimoto, Y., 1981, Adrenoleukodystrophy: Increased plasma content of saturated very long chain fatty acids, *Neurology* **31**:1241–1249.

Moser, H. W., Moser, A. B., Powers, J. M., Nitowsky, H. M., Schaumburg, H. H., Norum, R. A., and Migeon, B. R., 1982, The prenatal diagnosis of adrenoleukodystrophy. Demonstration of increased hexacosanoic acid in cultured amniocytes and fetal adrenal gland, *Pediatr. Res.* **16**:172–175.

Moser, H. W., Moser, A. B., Trojak, J. E., and Supplee, S. W., 1983, The identification of female carriers for adrenoleukodystrophy, *J. Pediatr.* **103**:54–59.

Moser, H. W., Tutschka, P. J., Brown III, F. R., Moser, A. B., Yeager, A. M., Singh, I., Mark, S. A., Kumar, A. J., McDonnell, J. M., White, C. L., Maumenee, I. H., Green, W. R., Powers, J. M., and Santos, G. W., 1984, Bone marrow transplant in adrenoleukodystrophy, *Neurology* **34**:1410–1417.

Moser, H. W., Naidu, S., Kumar, A. J., and Rosenbaum, A. E., 1987, The adrenoleukodystrophies, *CRC Crit. Rev. Neurobiol.* **3**:29–88.

Moser, H. W., Aubourg, P., Cornblath, D., Borel, J., Wu, Y.-W., Bergin, A., Naidu, S., and Moser, A. B., 1991a, The therapy for X-linked adrenoleukodystrophy, in: *Treatment of Genetic Diseases* (R. J. Desnick, ed.), Churchill Livingstone, New York, pp. 111–129.

Moser, H. W., Bergin, A., Naidu, S., and Ladenson, P. W., 1991b, Adrenoleukodystrophy: New aspects of adrenal cortical disease, in: *Endocrinology and Metabolism Clinics of North America: New Aspects of Adrenal Cortical Disease* (D. H. Nelson, ed.), W. B. Saunders, Philadelphia, pp. 297–318.

Moser, H. W., Moser, A. B., Smith, K. D., Bergin, A., Borel, J., Shankroff, J., Stine, O. C., Merette, C., Ott, J., Krivit, W., Shapiro, E., 1992, Adrenoleukodystrophy: Phenotypic variability and implications for therapy, *J. Inher. Metab. Dis.* **15**:645–664.

Nagamatsu, K., Soeda, S., and Kishimoto, Y., 1986, Change of substrate specificity of rat liver microsomal fatty acyl-CoA synthetase activity by triton X-100, *Lipids* **21**:328–332.

Naidu, S., Cornblath, D., Moser, A., and Moser, H., 1986, Neurological abnormalities in ALD heterozygotes, *Muscle Nerve* **9**:129.

Naidu, S., Bresnan, M. J., Griffin, D., O'Toole, S., and Moser, H. W., 1988a, Childhood adrenoleukodystrophy: Failure of intensive immunosuppression to arrest neurologic progression in childhood adrenoleukodystrophy, *Arch. Neurol.* **45**:846–848.

Naidu, S., Hoefler, G., Hoefler, S., Watkins, P., Chen, W., Rance, N., Powers, J. M., Beard, M., and Moser, H. W., 1988b, Neonatal seizures and retardation in a female with biochemical changes resembling X-linked ALD: A probable new peroxisomal disease entity, *Neurology* **38**:1100–1107.

Naidu, S., Moser, A., Moser, H., and Abern, S., 1990, Neonatal adrenoleukodystrophy: Survival to adulthood, in: *Vth International Congress. Inborn Errors of Metabolism*, Asilomar, CA, OC3.4.

Naritami, K., Izumikawa, Y., Ohshiro, S., Yoshida, K., Shimozawa, N., Suzuki, Y., Orii, T., and Hirayama, K., 1989, Gene assignment of Zellweger syndrome to 7q11.23: Report of the second case associated with a pericentric inversion of chromosome 7, *Hum. Genet.* **84**:79–80.

Nathans, J., Piantianida, T. P., Eddy, R. L., Shows, T. B., and Hogness, D. S., 1986a, Molecular genetics of inherited variation in human color vision, *Science* **232**:203–210.

Nathans, J., Thomas, D., and Hogness, D. S., 1986b, Molecular genetics of human color vision: The genes encoding blue, green and red pigments, *Science* **232**:193–202.

Neat, C. E., Thomassen, M. S., and Osmundsen, H., 1980, Induction of peroxisomal β-oxidation in rat liver by high-fat diets, *Biochem. J.* **186**:369–371.

Nishio, H., Kodama, S., Yokoyama, S., Matsuo, T., Mio, T., and Sumino, K., 1986, A simple method to diagnose adrenoleukodystrophy using a dried blood spot on filter paper, *Clin. Chim. Acta* **159**:77–82.

Nishiyama, K., Funai, T., Katakfuchi, R., Hatton, F., Onoyama, K., and Ichiyama, A., 1991, Primary hyperoxaluria type 1 due to a point mutation of T to C in the coding region of the serine:pyruvate aminotransferase gene, *Biochem. Biophys. Res. Commun.* **176**:1093–1099.

Noetzel, M. J., Clark, H. B., and Moser, H. W., 1983, Neonatal adrenoleukodystrophy with prolonged survival, *Ann. Neurol.* **14**:380.

Noetzel, M. J., Landau, W. M., and Moser, H. W., 1987, Adrenoleukodystrophy carrier state presenting as a chronic nonprogressive spinal cord disorder, *Arch. Neurol.* **44**:566–567.

Noguchi, T., and Takada, Y., 1979, Peroxisomal localization of alanine-glyoxalate aminotransferase in human liver, *Arch. Biochem. Biophys.* **196**:645–647.

Noguchi, T., Okuno, S., Takada, Y., Minatogawa, Y., Okai, K., and Kido, R., 1978, Characteristics of hepatic alanine:glyoxylate aminotransferase in different mammalian species, *Biochem. J.* **169**:113–122.

Norton, W., and Autilio, L. A., 1966, The lipid composition of purified bovine brain myelin, *J. Neurochem.* **13**:213–222.

Novikoff, A. B., and Goldfischer, S., 1969, Visualization of peroxisomes (microbodies) and mitochondria with diaminobenzidine, *J. Histochem. Cytochem.* **17**:675–680.

Novikoff, A. B., and Shin, W. Y., 1964, The endoplasmic reticulum in the Golgi zone and its relation to microbodies, Golgi apparatus and autophagic vacuoles in rat liver, *J. Microsc.* **3**:186–206.

Oberle, I., Drayna, D., Camerino, G., White, R., and Mandel, J. L., 1985, The telomere of the human X-chromosome long arm: Presence of a highly polymorphic DNA marker and analysis of recombination frequency, *Proc. Natl. Acad. Sci. USA* **82**:2824–2828.

Ogino, T. and Suzuki, K., 1981, Specificities of human and rat brain enzymes of cholesterol ester metabolism toward very long chain fatty acids: Implications for biochemical pathogenesis of adrenoleukodystrophy, *J. Neurochem.* **36**:776–779.

O'Neill, B. P., Moser, H. W., Saxena, K. M., and Marmion, L. C., 1982, Adrenoleukodystrophy (ALD): Neurological disease in carriers and correlation with very long chain fatty acids (VLCFA) concentrations in plasma and cultured skin fibroblasts, *Neurology* **32**:A216.

O'Neill, B. P., Moser, H. W., Saxena, K. M., and Marmion, L. C., 1984, Adrenoleukodystrophy: Clinical and biochemical manifestations in carriers, *Neurology* **34**:789–801.

Oorthhuys, J. W. E., Loewe-Sieger, D. H., Schutgens, R. B. J., Wanders, R. J. A., Heymans, E. M., and Bleeker-Wagermakers, E. M., 1987, Peroxisomal dysfunction in chondrodysplasia punctata, rhizomelic type, *Ophthalm. Paediatr. Genet.* **8**:183–185.

Opitz, J. M., 1985, The Zellweger syndrome: Book review and bibliography, *Am. J. Med. Genet.* **22**:419–426.

Oshino, N., Oshino, R., and Chance, B., 1973, The characteristics of the "peroxidatic" reaction of catalase in ethanol oxidation, *Biochem. J.* **131**:555–567.

Osumi, T., and Fujiki, Y., 1990, Topogenesis of peroxisomal proteins, *Bioessays* **13**:217–222.

Palosaari, P. M., and Hiltunen, J. K., 1990, Peroxisomal bifunctional protein from rat liver is a trifunctional enzyme possessing s-eonyl-CoA hydratase, 3-hydroxyacyl-CoA dehydrogenase, and delta-3, delta-2-enoyl-CoA isomerase activities, *J. Biol. Chem.* **265**(5):2446-2449.

Partin, J., and McAdams, A. J., 1983, Absence of hepatic peroxisomes in neonatal onset adrenoleuko-dystrophy, *Pediatr. Res.* **17**:294a.

Pedersen, J. I., and Gustafsson, J., 1980, Conversion of $3\alpha,7\alpha,12\alpha$-trihydroxy-5β-cholestanoic acid into cholic acid by rat liver peroxisomes, *FEBS Lett.* **121**(2):345–348.

Penman, R. W. B., 1960, Addison's disease in association with spastic paraplegia, *Br. Med. J.* **1**:402.

Pettit, H., Leys, D., Skjeldal, O. H., Caron, J. C., Lambert, P., Lehembre, P., and Hache, J. C., 1986, La maladie de Refsum: Correlations epidemiologiques, cliniques et biologiques, *Rev. Neurol.* (Paris) **142**(5):500–508.

Pike, M. G., Applegarth, D. A., Dunn, H. G., Bamforth, S. J., Tingle, A. J., Wood, B. J., Dimmick, J. E., Harris, H., Chantler, J. K., and Hall, J. G., 1990, Congenital rubella syndrome associated with calcific epiphyseal stippling and peroxisomal dysfunction, *J. Pediatr.* **116**(1):88–94.

Pilz, P., and Schiener, P., 1973, Kombination von morbus Addison und morbus Schilder bei einer 43 Jahrigen Frau, *Acta Neuropathol.* **26**:357–360.

Poll-The, B. T., Ogier, H., Saudubray, J. M., Schutgens, R. B. H., Wanders, R. J. A., van den Bosch, H., and Schrakamp, G., 1986a, Impaired plasmalogen metabolism in infantile Refsum's disease, *Eur. J. Pediatr.* **144**:513–514.

Poll-The, B. T., Saudubray, J. M., Ogier, H., Schutgens, R. B. H., Wanders, R. J. A., Schrakamp, G., Van den Bosch, H., Trybels, J. M. F., Poulos, A., Moser, H. W., van Eldere, J., and Eyssen, H. J., 1986b, Infantile Refsum's disease: Biochemical findings suggesting multiple peroxisomal dys-function, *J. Inherited Metab. Dis.* **9**:169–174.

Poll-The, B. T., Saudubray, J. M., Ogier, H., Schutgens, R. B. H., Wanders, R. J. A., Schramkamp, G., Van den Bosch, H., Trybels, J. M. F., Poulos, A., Moser, H. W., van Eldere, J., and Eyssen, H. J., 1987, Infantile Refsum's disease: An inherited peroxisomal disorder. Comparison with Zellweger syndrome and neonatal adrenoleukodystrophy, *Eur. J. Pediatr.* **146**:477–483.

Poll-The, B. T., Roels, F., Ogier, H., Scotto, J., Vamecq, J., Schutgens, R. B. H., Wanders, R. J. A., van Roermund, C. W. T., van Wijland, M. J. A., Schram, A. J., Tager, J. M., and Saudubray, J.-M., 1988, A new peroxisomal disorder with enlarged peroxisomes and a specific deficiency of Acyl-CoA oxidase (pseudo-neonatal adrenoleukodystrophy), *Am. J. Hum. Genet.* **42**:422–434.

Poll-The, B.-T., Skjeldal, O. H., Stokke, O., Poulos, A., Demaugre, F., and Saudubray, J.-M., 1989, Phytanic acid alpha-oxidation and complementation analysis of classical Refsum and peroxisomal disorders, *Hum. Genet.* **81**:175–181.

Poulos, A., and Whiting, M. J., 1985, Identification of $3\alpha,7\alpha,12\alpha$-trihydroxy-5β-cholestan-26-oic acid, an intermediate in cholic acid synthesis, in the plasma of patients with infantile Refsum's disease, *J. Inherited Metab. Dis.* **8**:13–17.

Poulos, A., Sharp, P., and Whiting, M., 1984, Infantile Refsum's disease (phytanic acid storage disease): A variant of Zellweger's syndrome? *Clin.Genet.* **26**:579–586.

Poulos, A., Sharp, P., Fellenberg, A. J., and Danks, D. M., 1985, Cerebro-hepato-renal (Zellweger) syndrome, adrenoleukodystrophy, and Refsum's disease: Plasma changes and skin fibroblast phytanic acid oxidase, *Hum. Genet.* **70**:172–177.

Poulos, A., Sharp, P., Singh, H., Johnson, D., Fellenberg, A., and Pollard, A., 1986a, Detection of a homologous series of C26–C38 polyenoic fatty acids in the brain of patients without peroxisomes (Zellweger's syndrome), *Biochem. J.* **235**:607–610.

Poulos, A., Singh, H., Paton, B., Sharp, P., and Derwas, N., 1986b, Accumulation and defective β-oxidation of very long chain fatty acids in Zellweger's syndrome, adrenoleukodystrophy and Refsum's disease variants, *Clin. Genet.* **29**:397–408.

Poulos, A., van Crugten, C., Sharp, P., Robertson, E., Becroft, D. M. O., Saudubray, J. M., Poll-The,

B. T., Christensen, E., and Brandt, N., 1986c, Prenatal diagnosis of Zellweger syndrome and related disorders: Impaired degradation of phytanic acid, *Eur. J. Pediatr.* **145**:507–510.

Poulos, A., Sharp, P., Fellenberg, A. J., and Johnson, D. W., 1988, Accumulation of pristanic acid (2,6,10,14 tetramethylpentadecanoic acid) in the plasma of patients with generalized peroxisomal dysfunction, *Eur. J. Pediatr.* **147**:143–147.

Poulos, A., Sharp, P., and Johnson, D., 1989, Plasma polyenoic very-long-chain fatty acids in peroxisomal disease: Biochemical discrimination of Zellweger's syndrome from other phenotypes, *Neurology* **39**:44–47.

Poulos, A., Robertson, E. F., Johnson, D. W., Beckman, K., Fellenberg, A. J., and Sharp, P., 1990, Use of stable-isotope-labelled-tracer to monitor ether lipid dietary supplementation in peroxisomal disease patients, in: *Stable Isotopes in Paediatric Nutritional and Metabolic Research*, Intercept, Hampshire, England, pp. 263–269.

Powers, J. M., 1985, Adrenoleukodystrophy (adreno-testiculo-leuko-myelo-neuropathic-complex), *Clin. Neuropathol.* **4**(5):181–199.

Powers, J. M., and Schaumburg, H. H., 1974, Adrenoleukodystrophy (sex-linked Schilder's disease). A pathogenetic hypothesis based on ultrastructural lesions in adrenal cortex, peripheral nerve and testis, *Am. J. Pathol.* **76**(3):481–492.

Powers, J. M., Schaumburg, H. H., and Gaffney, C. L., 1980a, Kluver–Bucy syndrome caused by adrenoleukodystrophy, *Neurology* **30**:1131–1132.

Powers, J. M., Schaumburg, H. H., Johnson, A. B., and Raine, C. S., 1980b, A correlative study of the adrenal cortex in adrenoleukodystrophy: Evidence for a fatal intoxication with very long chain fatty acids, *Invest. Cell Pathol.* **3**:353–376.

Powers, J. M., Moser, H. W., Moser, A. B., Upshur, J. K., Bradford, B. F., Pai, S. G., Kohn, P. H., Frias, J., and Tiffany, C., 1985, Fetal cerebro-hepato-renal (Zellweger) syndrome. Dysmorphic, radiologic, biochemical and pathologic findings in four affected fetuses, *Hum. Pathol.* **16**: 610–620.

Powers, J. M., Tummons, R. C., Moser, A. B., Moser, H. W., Huff, D. S., and Kelley, R. I., 1987, Neuronal lipidosis and neuroaxonal dystrophy in cerebro-hepato-renal (Zellweger) syndrome, *Acta Neuropathol.* **73**:333–343.

Probst, A., Ulrich, J., Heitz, P. U., and Herschkowitz, N., 1980, Adrenomyeloneuropathy: A protracted, pseudosystematic variant of adrenoleukodystrophy, *Acta Neuropathol.* **49**:105–115.

Purdue, P. E., Takada, T., and Danpure, C. J., 1990, Identification of mutations associated with peroxisome-to-mitochondrion mistargeting of alanine/glyoxylate aminotransferase in primary hyperoxaluria type 1, *J. Cell Biol.* **111**(6):2341–2351.

Rachubinski, R. A., Fujiki, Y., Mortensen, R. M., and Lazarow, P. B., 1984, Acyl-CoA oxidase and hydratase-dehydrogenase, two enzymes of the peroxisomal β-oxidation system, are synthesized on free polysomes of clofibrate-treated rat liver, *J. Cell Biol.* **99**:2241–2246.

Ramsey, R. B., Banik, N. L., and Davison, A. N., 1979, Adrenoleukodystrophy brain cholesteryl esters and other neutral lipids, *J. Neurol. Sci.* **40**:189–196.

Reddy, J. K., 1990, Carcinogenicity of peroxisome proliferators—Evaluation and mechanisms, *Biochem. Soc. Trans.* **18**(1):92–94.

Reddy, J. K., Goel, S. K., Nemali, M. R., Carrino, J. J., Laffler, T. G., Reddy, M. K., Sperbeck, S. J., Osumi, T., Hashimoto, T., Lalwani, N. D., and Rao, M. S., 1986, Transcriptional regulation of peroxisomal fatty acyl-CoA oxidase and enoyl-CoA hydratase/3 hydroxyacyl-CoA dehydrogenase in rat liver by peroxisome proliferators, *Proc. Natl. Acad. Sci. USA* **83**:1747–1751.

Reinecke, C. J., Knoll, D. P., Pretorius, P. J., Steyn, H. S., and Simpson, R. H. W., 1985, The correlation between biochemical and histopathological findings in adrenoleukodystrophy, *J. Neurol. Sci.* **70**:21–38.

Renier, W. O., Nabben, F. A. E., Hustinx, T. W. J., Veerkamp, J. H., Otten, B. J., Ter Laak, H. J., Ter Haar, B. G. A., and Gabreels, F. J. M., 1983, Congenital adrenal hypoplasia, progressive muscular dystrophy, and severe mental retardation, in association with glycerol kinase deficiency, in male sibs, *Clin. Genet.* **24**:243–251.

Rezanka, T., 1989, Very long chain fatty acids from the animal and plant kingdoms, *Prog. Lipid Res.* **28**:147–187.

Rhodin, J., 1954, Correlation of ultrastructural organization and function in normal and experimentally changed proximal convoluted tubule cells of the mouse kidney, Ph.D. thesis, Actiebolaget Godvil, Stockholm.

Rizzo, W. B., Watkins, P. A., Phillips, M. W., Cranin, D., Campbell, B., and Avigan, J., 1986, Adrenoleukodystrophy: Oleic acid lowers fibroblast saturated C22–C26 fatty acids, *Neurology* **36**:357–361.

Rizzo, W. B., Phillips, M. W., Dammann, A. L., Leshner, R. Y., and Jennings, S. V. K., 1987, Adrenoleukodystrophy: Dietary oleic acid lowers hexacosanoate levels, *Ann. Neurol.* **21:** 232–239.

Rizzo, W. B., Leshner, R. T., Odone, A., Dammann, A. L., Craft, D. A., Jensone, M. E., Jennings, S. S., Davis, S., Jaitly, R., and Sgro, J. A.., 1989, Dietary erucic acid therapy for X-linked adrenoleukodystrophy, *Neurology* **39**(11):1415–1422.

Robbi, M., and Lazarow, P. B., 1982, Peptide mapping of peroxisomal catalase and its precursor, *J. Biol. Chem.* **257**(2):964–970.

Rocchiccioli, F., Aubourg, P., and Bougneres, P. F., 1986, Medium- and long-chain dicarboxylic aciduria in patients with Zellweger syndrome and neonatal adrenoleukodystrophy, *Pediatr. Res.* **20**(1):62–66.

Rocchiccioli, F., Aubourg, P., and Choiset, A., 1987, Immediate prenatal diagnosis of Zellweger syndrome by direct measurement of very long chain fatty acids in chorionic villus cells, *Prenat. Diagn.* **7**:349–354.

Roels, F., and Goldfischer, S., 1979, Cytochemistry of human catalase. The demonstration of hepatic and renal peroxisomes by a high temperature procedure, *J. Histochem. Cytochem.* **27**(11):1471–1477.

Roels, F., Verdonck, V. Pauwels, M., De Catte, L., Lissens, W., Liebaers, I., and Elleder, M., 1987a, Light microscopic visualization of peroxisomes and plasmalogens in first trimester chorionic villi, *Prenat. Diagn.* **7**:525–530.

Roels, F., Verdonck, V., Pauwels, M., Foulon, W., Lissens, W., and Liebaers, I., 1987b, Visualization of peroxisomes and plasmalogens in first trimester chorionic villus, *J. Inherited Metab. Dis.* **10**(2):233–235.

Roscher, A., Molzer, B., Bernheimer, H., Stockler, S., Mutz, I., and Paltauf, F., 1985, The cerebro-hepatorenal (Zellweger) syndrome: An improved method for the biochemical diagnosis and its potential value of prenatal detection, *Pediatr. Res.* **19**(9):930–933.

Roscher, A. A., Hoefler, S., Hoefler, G., Paschke, E., Paltauf, F., Moser, A., and Moser, H. W., 1989, Genetic and phenotypic heterogeneity in disorders of peroxisome biogenesis—A complementation study involving cell lines from 19 patients, *Pediatr. Res.* **26**:67–72.

Ruder, H., Otto, G., Schutgens, R. B. H., Querfeld, U., Wanders, R. J. A., Herzog, K. H., Wolfel, P., Pomer, S., Scharer, K., and Rose, G. A., 1990, Excessive urinary oxalate excretion after combined renal and hepatic transplantation for correction of hyperoxaluria type 1, *Eur. J. Pediatr.* **150**:56–58.

Rusnak, N., and Krisans, S. K., 1987, Diurnal variation of HMG-CoA reductase activity in rat liver peroxisomes, *Biochem. Biophys. Res. Commun.* **148**(2):890–895.

Sack, G. H., Raven, M. B., and Moser, H. W., 1989, Color vision defects in adrenoleukodystrophy, *Am. J. Hum. Genet.* **44**:794–798.

Sadeghi-Nejad, A., and Senior, B., 1990, Adrenomyeloneuropathy presenting as Addison's disease in childhood, *N. Engl. J. Med.* **322**:13–16.

Santos, M. J., Ojeda, J. M., Garrido, J., and Leighton, F., 1985, Peroxisomal organization in normal and cerebrohepatorenal (Zellweger) syndrome fibroblasts, *Proc. Natl. Acad. Sci. USA* **82**:6556–6560.

Santos, M. J., Imanaka, T., Shio, H., and Lazarow, P. B., 1988a, Peroxisomal integral membrane proteins in control and Zellweger fibroblasts, *J. Biol. Chem.* **263**(21):10502–10509.

Santos, M. J., Imanaka, T., Shio, H., Small, G. M., and Lazarow, P. B., 1988b, Peroxisomal membrane ghosts in Zellweger syndrome—Aberrant organelle assembly, *Science* **239**:1536–1538.

Schaumburg, H. H., Powers, J. M., Raine, C. S., Suzuki, K., and Richardson, E. P., 1975, Adrenoleukodystrophy: A clinical and pathological study of 17 cases, *Arch. Neurol.* **33**:577–591.

Schaumberg, H., Powers, J. M., Raine, C. S., Spencer, P. S., Griffin, J. W., Prineas, J. W., and Boehme, D. M., 1977, Adrenomyeloneuropathy: A probable variant of adrenoleukodystrophy. II. General pathologic, neuropathologic, and biochemical aspects, *Neurology* **27**(12):1114–1119.

Schepers, L., Casteels, M., Vamecq, J., Parmentier, G., Van Veldhoven, P. P., and Mannaerts, G. P., 1986, β-Oxidation of the carboxyl side chain of prostaglandin E(2) in rat liver peroxisomes and mitochondria, *J. Biol. Chem.* **263**(6):2724–2731.

Schepers, L., Van Veldhoven, P. P., Casteels, M., Eyssen, H., and Mannaerts, G. P., 1990, Presence of three Acyl-CoA oxidases in rat liver peroxisomes. An inducible fatty acyl-CoA oxidase, a noninducible fatty acyl-CoA oxidase, and a noninducible trihydroxycoprostanoyl-CoA oxidase, *J. Biol. Chem.* **265**(9):5242–5246.

Schram, A., Strijland, A., Hashimoto, T., Wanders, R. J. A., Schutgens, R. B. H., van den Bosch, H., and Tager, J. M., 1986, Biosynthesis and maturation of peroxisomal β-oxidation enzymes in fibroblasts in relation to the Zellweger syndrome and infantile Refsum disease, *Proc. Natl. Acad. Sci. USA* **83**:6156–658.

Schram, A. W., Goldfischer, S., van Roermund, C. T., Brouwer-Kelder, E. M., Collins, J., Hashimoto, T., Heymans, H. S. A., van den Bosch, H., Schutgens, R. B. H., Tager, J. M., and Wanders, R. J. A., 1987, Human peroxisomal 3-oxoacyl-coenzyme A thiolase deficiency, *Proc. Natl. Acad. Sci. USA* **84**:2494–2496.

Schulz, H., and Kunau, W.-H., 1987, Beta-oxidation of unsaturated fatty acids: A revised pathway, *TIBS* **12**:403–407.

Schutgens, R. B. H., Romeyn, G. J., Wanders, R. J. A., van den Bosch, H., Schrakamp, G., and Heymans, H. S. A., 1984, Deficiency of acyl-CoA: Dihydroxyacetone phosphate acyltransferase in patients with Zellweger (cerebro-hepato-renal) syndrome, *Biochem. Biophys. Res. Commun.* **120**:179–184.

Schutgens, R. B. H., Heymans, H. S. A., and Wanders, R. J. A., 1985, Prenatal diagnosis of the cerebro-hepato-renal (Zellweger) syndrome by detection of an impaired plasmalogen bio-synthesis, *J. Inherited Metab. Dis.* **8**(2):153–154.

Schutgens, R. B. H., Heymans, H. S. A., Wanders, R. J. A., van den Bosch, H., and Tager, J. M., 1986, Peroxisomal disorders: A newly recognized group of genetic diseases, *Eur. J. Pediatr.* **144**:430–440.

Scotto, J. M., Hadchouel, M., Odievre, M., Laudat, M. H., Saudubray, J. M., Dulac, O., Beucler, I., and Beaune, P., 1982, Infantile phytanic acid storage disease: A possible variant of Refsum's disease: Three cases, including ultrastructural studies of the liver, *J. Inherited Metab. Dis.* **5**:83–90.

Seltzer, W. K., Firminger, H., Klein, J., Pike, A., Fennessey, P., and McCabe, E. R. B., 1985, Adrenal dysfunction in glycerol kinase deficiency, *Biochem. Med.* **33**:189–199.

Sereni, C., Ruel, M., Iba-Zizen, T., Baumann, N., Marteau, R., and Paturneau-Jouas, M., 1987, Adult adrenoleukodystrophy: A sporadic case? *J. Neurol. Sci.* **80**:121–128.

Setchell, K. D. R., and Vestal, C. H., 1989, Thermospray ionization liquid chromatograph-mass spectrometry—A new and highly specific technique for the analysis of bile acids, *J. Lipid Res.* **30**(9):1459–1469.

Sharp, P., Poulos, A., Fellenberg, A., and Johnson, D., 1987, Structure and lipid distribution of polyenoic very-long-chain fatty acids in the brain of peroxisome-deficient patients (Zellweger syndrome), *Biochem. J.* **248**:61–67.

Sharp, P., Johnson, D., and Poulos, A., 1991, Molecular species of phosphatidylcholine containing very long chain fatty acids in human brain: Enrichment in X-linked adrenoleukodystrophy brain and diseases of peroxisome biogenesis brain, *J. Neurochem.* **56**:30–37.

Shimozawa, N., Suzuki, Y., Orii, T., Yokota, S., and Hashimoto, T., 1988, Biochemical and morphologic aspects of peroxisomes in the human rectal mucosa: Diagnosis of Zellweger syndrome simplified by rectal biopsy, *Pediatr. Res.* **24**(6):723–727.

Siemerling, E., and Creutzfeldt, H. G., 1923, Bronzekrankheit und sklerosierende encephalomyelitis, *Arch. Psychiatr. Nervenkr.* **68**:217–244.

Singh, H., and Poulos, A., 1988, Distinct long chain and very long chain fatty acyl CoA synthetases in rat liver peroxisomes and mitochondria, *Arch. Biochem. Biophys.* **266**:486–495.

Singh, H., Usher, S., Johnson, D., and Poulos, A., 1990, A comparative study of straight chain and branched chain fatty acid oxidation in skin fibroblasts from patients with peroxisomal disorders, *J. Lipid Res.* **31**(2):217–225.

Singh, I., Moser, A. B., Goldfischer, S., and Moser, H. W., 1984a), Lignoceric acid is oxidized in the peroxisomes: Implications for the Zellweger cerebro-hepato-renal syndrome and adrenoleuko-dystrophy, *Proc. Natl. Acad. Sci. USA* **81**:4203–4207.

Singh, I., Moser, A. B., Moser, H. W., and Kishimoto, Y., 1984b, Adrenoleukodystrophy: Impaired oxidation of very long chain fatty acids in white blood cells, cultured skin fibroblasts and amniocytes, *Pediatr. Res.* **18**(3):286–290.

Singh, I., Johnson, G. H., and Brown III, F. R., 1988, Peroxisomal disorders: Biochemical and clinical diagnostic considerations, *Am. J. Dis. Child.* **142**:1297–1301.

Skjeldal, O. H., and Stokke, O., 1987, The subcellular localization of phytanic acid oxidase in rat liver, *Biochim. Biophys. Acta* **921**:38–42.

Small, G. M., Santos, M. J., Imanaka, T., Poulos, A., Dans, D. M., Moser, H. W., and Lazarow, P. B., 1988a, Peroxisomal integral membrane proteins in livers of patients with Zellweger syndrome, infantile Refsum's disease and X-linked adrenoleukodystrophy, *J. Inherited Metab. Dis.* **11**:358–371.

Small, G. M., Szabo, L. J., and Lazarow, P. B., 1988b, Acyl-CoA oxidase contains two targeting sequences each of which can mediate protein import into peroxisomes, *EMBO J.* **7**(4):1167–1173.

Snyder, F., 1972, Ed. *Ether Lipids*, Academic Press, New York.

Solish, G. I., Moser, H. W., Ringer, L. D., Moser, A. E., Tiffany, C., and Schutta, E., 1985, The prenatal diagnosis of the cerebro-hepato-renal syndrome of Zellweger, *Prenat. Diagn.* **5**:27–34.

Spranger, J. W., Opitz, J. M., and Bidder, U., 1971, Heterogeneity of chondrodysplasia punctata, *Humangenetik* **11**:190–212.

Stanczak, H., Stanczak, J. J., and Singh, I., 1991, *In-situ* localization of gene for palmitoyl-CoA ligase, *Trans. Am. Soc. Neurochem.* **22**(1):116.

Stefanini, S., Farrace, M. G., and Ceruargento, M. P., 1985, Differentiation of liver peroxisomes in the fetal and newborn rat, cytochemistry of catalase and D-amino acid oxidase, *J. Embryol. Exp. Morphol.* **88**:151–163.

Steinberg, D., 1989, Refsum disease, in: *The Metabolic Basis of Inherited Disease* (C. R. Scriver, A. L. Beaudet, W. Sly, and D. Valle, eds.), McGraw-Hill, New York, pp. 1533–1550.

Stellaard, F., Langelaar, S. A., Kok, R. M., Kleijer, W. J., Schutgens, R. B. H., and Jakobs, C., 1988, Prenatal diagnosis of Zellweger syndrome by determination of trihydroxycoprostanic acid in amniotic fluid, *Eur. J. Pediatr.* **148**(2):175–176.

Street, J. M., Singh, H., and Poulos, A., 1990, Metabolism of saturated and polyunsaturated very long chain fatty acids in fibroblasts from patients with defects in peroxisome beta-oxidation, *Biochem. J.* **269**:671–677.

Stumpf, D. A., Hayward, A., Haas, R., and Schaumburg, H. H., 1981, Adrenoleukodystrophy. Failure of immunosuppression to prevent neurological progression, *Arch. Neurol.* **38**:48–49.

Suzuki, H., Kawarabayasi, Y., Kondo, J., Abe, T., Nishikawa, K., Kimura, S., Hashimoto, T., and Yamamoto, T., 1990, Structure and regulation of rat long-chain acyl-CoA synthetase, *J. Biol. Chem.* **265**(15):8681–8685.

Suzuki, Y., Orii, T., and Hashimoto, T., 1986, Biosynthesis of peroxisomal β-oxidation enzymes in infants with Zellweger syndrome, *J. Inherited Metab. Dis.* **9**:292–296.

Suzuki, Y., Shimozawa, N., Orii, T., Igarashi, N., Kono, N., Matsui, A., Inoue, Y., Yokota, S., and Hashimoto, T., 1988, Zellweger-like syndrome with detectable hepatic peroxisomes: A variant form of peroxisomal disorder, *J. Pediatr.* **113**:841–845.

Suzuki, Y., Shimozawa, N., Orii, T., and Hashimoto, T., 1989, Major peroxisomal membrane polypeptides are synthesized in cultured skin fibroblasts from patients with Zellweger syndrome, *Pediatr. Res.* **26**(2):150–153.

Tager, J. M., Harmsen, T., van der Beek, W. A., Wanders, R. J. A., Hashimoto, T., Heymans, H. S. A., van den Bosch, H., Schutgens, R. B. H., and Schram, A. W., 1985, Peroxisomal beta-oxidation enzyme proteins in the Zellweger syndrome, *Biochem. Biophys. Res. Commun.* **126**:1269–1275.

Takada, Y., Kaneko, N., Esumi, H., Purdue, P. E., and Danpure, C. J., 1990, Human peroxisomal L-alanine-glyoxylate aminotransferase—Evolutionary loss of a mitochondrial targeting signal by point mutation of the initiation site, *Biochem. J.* **268**(2):517–520.

Takahara, S., and Miyamoto, H., 1948, Three cases of progressive oral gangrene due to lack of catalase, *J. Otorhinolaryngol. Soc.* **51**:165.

Taketomi, T., Hara, A., Kitazawa, N., Takada, K., and Nakamura, H., 1987, An adult case of adrenoleukodystrophy with features of olivo-ponto-cerebellar atrophy: 2. Lipid abnormalities, *Jpn. J. Exp. Med.* **57**:59–70.

Tanaka, K., Shimada, M., Naruto, T., Yamamoto, H., Saeki, Y., Sai, H., and Hirose, G., 1986, Very long-chain fatty acids in erythrocyte membrane sphingomyelin: Detection of ALD hemizygotes and heterozygotes, *Neurology* **36**:791–795.

Tanaka, K., Nishizawa, K., Yamamoto, H., Naruto, T., Izeki, E., Taga, T., Shimada, M., and Saeki, Y., 1990, Analysis of very long chain fatty acids and plasmalogen in the erythrocyte membrane—A simple method for the detection of peroxisomal disorders and discrimination between adrenoleukodystrophy and Zellweger syndrome, *Neuropediatrics* **21**(3):119–123.

ten Brink, H. J., Wanders, R. J. A., Stellaard, F., Schutgens, R. B. H., and Jacobs, C., 1991, Pristanic and phytanic acid in plasma from patients with a single peroxisomal enzyme deficiency, *J. Inherited Metab. Dis.* **14**:345–348.

ten Brink, H. J., Stellaard, F., van den Heuvel, C. M. M., Kok, R. M., Schor, D. S. M., Wanders, R. J. A., and Jakobs, C., 1992, Pristanic acid and phytanic acid in plasma from patients with peroxisomal disorders: Stable isotope dilution analysis with electron capture negative ion mass fragmentography, *J. Lipid Res.* **33**:41–47.

Teschler-Nicola, M., and Killian, W., 1981, Case report 72: Mental retardation, unusual facial appearance, abnormal hair, *Synd. Ident.* **7**:6.

Theda, C., Moser, A., Moser, H., and Debuch, H., 1987, Temporal evolution of brain biochemical changes in adrenoleukodystrophy, *J. Neurochem.* **48**(1):s35.

Thomas, G. H., Haslam, R. H. A., Batshaw, M. L., Capute, A. J., Neidengard, L., and Ransom, J. L., 1975, Hyperpipecolic acidemia associated with hepatomegaly, mental retardation, optic nerve dysplasia and progressive neurological disease, *Clin. Genet.* **8**:376–382.

Thompson, S. L., Burrows, R., Laub, R. J., and Krisans, S. K., 1987, Cholesterol synthesis in rat liver peroxisomes: Conversion of mevalonic acid to cholesterol, *J. Biol. Chem.* **262**(36):17420–17425.

Thompson, S. L., and Krisans, S. K., 1990, Rat liver peroxisomes catalyze the initial step in cholesterol synthesis—The condensation of acetyl-CoA units into acetoacetyl-CoA, *J. Biol. Chem.* **265**(10):5731–5735.

Tiffany, C. W., Hoefler, S. Moser, H. W., and Burch, R. M., 1991, Arachidonic acid metabolism in fibroblasts from patients with peroxisomal diseases, *Biochim. Biophys. Acta* **1096**:41–46.

Tolbert, N. E., 1981, Metabolic pathways in peroxisomes and glyoxysomes, *Annu. Rev. Biochem.* **50**:133–157.

Tomioka, Y., Aihara, K., Hirose, A., Hisinuma, T., and Mizugaki, M., 1991, Detection of heat-stable delta 3, delta 2-enoyl-CoA isomerase in rat liver mitochondria and peroxisomes by immunochemical study using specific antibody, *J. Biochem.* **109**:394–398.

Tomoda, H., Kgarashi, K., Cyong, J.-C., and Omura, S., 1991, Evidence for an essential role of long chain acyl-CoA synthetase in animal cell proliferation, *J. Biol. Chem.* **266**(7):4214–4219.

Torvik, A., Torp, S., Kase, B. F., Ek, J., Skjeldal, O., and Stokke, O., 1988, Infantile Refsum's disease: A generalized peroxisomal disorder. Case report with postmortem examination, *J. Neurol. Sci.* **85**:39–53.

Trijbels, J. M. F., Berden, J. A., Monnens, L. A. H., Willems, J. L., Janssen, A. J. M., Schutgens, R. B. H., and van den Broek-van Essen, M., 1983, Biochemical studies in the liver and muscle of patients with Zellweger syndrome, *Pediatr. Res.* **17**:514–517.

Tsuji, S., Sano, T., Ariga, T., and Miyatake, T., 1981a, Increased synthesis of hexacosanoic acid (C26:0) by cultured skin fibroblasts from patients with adrenoleukodystrophy (ALD) and adrenomyeloneuropathy (AMN), *Biochem. J.* (Tokyo) **90**:1233–1236.

Tsuji, S., Suzuki, M., Ariga, T., Sekine, M., Kuriyama, M., and Miyatake, T., 1981b, Abnormality of long-chain fatty acids in erythrocyte membrane sphingomyelin from patients with adrenoleukodystrophy, *J. Neurochem.* **36**:1046–1049.

Tsuji, S., Ohno, T., Miyatake, T., Suzuki, A., and Yamakawa, T., 1984, Fatty acid elongation activity in fibroblasts from patients with adrenoleukodystrophy (ALD), *J. Biochem.* (Tokyo) **96**:1241–1247.

Tsukada, H., Mochizuki, Y., and Konishi, T., 1968, Morphogenesis and development of microbodies of hepatocytes of rats during pre- and postnatal growth, *J. Cell Biol.* **37**:231–243.

Tsukamoto, T., Miura, S., and Fujiki, Y., 1991, Restoration by a 35K membrane protein of peroxisome assembly in a peroxisome-deficient mammalian cell mutant, *Nature* **350**:77–81.

Turpin, J. C., Paturneau-Jouas, M., Sereni, C., Pluot, M., and Baumann, N., 1985, Révélation à l'age adulte d'un cas d'adrénoleucodystrohie familiale, *Rev. Neurol.* (Paris) **141**(4):289–295.

Ulrich, J., Hershkowitz, N., Heits, P., Sigrist, T., and Baerlocher, P., 1978, Adrenoleukodystrophy: Preliminary report of a connatal case, light- and electron microscopical, immunohistochemical and biochemical findings, *Acta Neuropathol.* (Berl.) **43**:77–83.

Uziel, G., Bertini, E., Rimoldi, M., and Gambetti, M., 1990, Italian multicentric dietary therapeutical trial in adrenoleukodystrophy, in: *Adrenoleukodystrophy and Other Peroxisomal Disorders: Clinical, Biochemical, Genetic and Therapeutic Aspects*, Excerpta Medica, pp. 163–180.

Vamecq, J., 1990, Fluorometric assay of peroxisomal oxidases, *Analyt. Biochem.* **186**(2):340–349.

Vamecq, J., de Hoffman, E., and Van Hoof, F., 1985, Mitochondrial and peroxisomal metabolism of glutaryl-CoA, *Eur. J. Biochem.* **105**:1–7.

Vamecq, J., Draye, J.-P., Van Hoof, F., Misson, J.-P., Evradrd, P., Verellen, G., Eyssen, H., Van Eldere, J., Schutgens, R. B. H., Wanders, R. J. A., Roels, F., and Goldfischer, S., 1986, Multiple peroxisomal enzymatic deficiency disorders. A comprehensive biochemical and morphological study of Zellweger cerebrohepatorenal syndrome and neonatal adrenoleukodystrophy, *Am. J. Pathol.* **125**:524–535.

Van Amerongen, A., van Noort, M., van Beckhoven, J., Rommerts, F., Orly, J., and Swirtz, K., 1989, The subcellular distribution of the nonspecific lipid transfer protein (sterol carrier protein 2) in rat liver and adrenal gland, *Biochim. Biophys. Acta* **1001**:243–248.

Van Duyn, M. A., Moser, A. B., Brown III, F. R., Sacktor, N., Liu, A., and Moser, H. W., 1984, The design of a diet restricted in saturated very long chain fatty acids: Therapeutic application in adrenoleukodystrophy, *Am. J. Clin. Nutr.* **40**:277–284.

Van Veldhoven, P. P., Just, W. W., and Mannaerts, G. P., 1987, Permeability of the peroxisomal membrane to cofactors of β-oxidation: Evidence for the presence of a pore-forming protein, *J. Biol. Chem.* **262**(9):4310–4318.

Versmold, H. T., Bremer, H. J., Herzog, V., Siegel, G. V., Bassewitz, D. B., Irle, U. V., Voss, H.,
   Lombeck, and Brauser, B., 1977, A metabolic disorder similar to Zellweger syndrome with
   hepatic acatalasia and absence of peroxisomes, altered content and redox state of cytochromes,
   and infantile cirrhosis with hemosiderosis, *Eur. J. Pediatr.* **124**:261–275.
Vitale, L., Opitz, J. M., and Shahidi, N. T., 1969, Congenital and familial iron overload, *N. Engl.
   J. Med.* **280**:642–645.
Volkl, A., Baumgart, E., and Fahimi, H. G., 1988, Localization of urate oxidase in the crystalline core
   of rat liver peroxisomes by immunocytochemistry and immunoblotting, *J. Histochem. Cytochem.*
   36:329–336.
Volpe, J. J., and Adams, R. D., 1972, Cerebro-hepato-renal syndrome of Zellweger: An inherited
   disorder of neuronal migration, *Acta Neuropathol.* **20**:175–198.
Wanders, R. J. A., Kos, M., Roest, B., Meijer, A. J., Schrakamp, G., Heymans, H. S. A., Tegelaers, W.
   H. H., van den Bosch, H., Schutgens, R. B. H., and Tager, J. M., 1984, Activity of peroxisomal
   enzymes and intracellular distribution of catalase in Zellweger syndrome, *Biochem. Biophys. Res.
   Commun.* **123**(3):1054–1061.
Wanders, R. J. A., van Weringh, G., Schrakamp, G., Tager, J. M., van den Bosch, H., and Schutgens,
   R. B. H., 1985, Deficiency of acyl-CoA: Dihydroxyacetone phosphate acyltransferase in throm-
   bocytes of Zellweger patients: A simple postnatal diagnostic test, *Clin. Chim. Acta* **151**:217–221.
Wanders, R. J. A., Purvis, Y. R., Heymans, H. S. A., Bakkeren, J. A., Parmentier, G. G., van Eldere,
   J., Eyssen, H., van denBosch, H., Tager, J. M., and Schutgens, R. B. H., 1986a, Age-related
   differences in plasmalogen content of erythrocytes from patients with the cerebro-hepato-renal
   (Zellweger) syndrome: Implications for postnatal detection of the disease, *J. Inherited Metab. Dis.*
   9:335–342.
Wanders, R. J. A., Saelman, D., Heymans, H. S. A., Schutgens, R. B. H., Westerveld, A., Poll-The,
   B. T., Saudubray, J. M., van den Bosch, H., Strijland, A., Schram, A. W., and Tager, J. M., 1986b,
   Genetic relation between the Zellweger syndrome, infantile Refsum's disease, and rhizomelic
   chondrodysplasia punctata, *N. Engl. J. Med.* **314**:(12):787–788.
Wanders, R. J. A., Schrakamp, G., van den Bosch, H., Tager, J. M. Moser, H. W., Moser, A. E.,
   Aubourg, P., Kleijer, W. J., and Schutgens, R. B. H., 1986c, Pre- and postnatal diagnosis of the
   cerebro-hepato-renal (Zellweger) syndrome via a simple method directly demonstrating the
   presence or absence of peroxisomes in cultured skin fibroblasts, amniocytes of chorionic villi
   fibroblasts, *J. Inherited Metab. Dis.* **9**(2):317–320.
Wanders, R. J. A., Schutgens, R. B. H., Schrakamp, G., van den Bosch, H., Tager, J. M., Schram, A.
   W., Hashimoto, T., Poll-The, B. T., and Saudubray, J. M., 1986d, Infantile Refsum disease:
   Deficiency of catalase-containing particles (peroxisomes), alkyldihydroxyacetone phosphate
   synthase and peroxisomal β-oxidation enzyme proteins, *Eur. J. Pediatr.* **145**:172–175.
Wanders, R. J. A., Schutgens, R. B. H., Schrakamp, G., van den Bosch, H., Tager, J. M., Poll-The, B.
   T., and Saudubray, J. M., 1986e, Deficiency of dihydroxyacetonephosphate acyltransferase and
   catalase-containing particles in patients with infantile Refsum's disease, *J. Inherited Metab. Dis.*
   9(2):325–328.
Wanders, R. J. A., Schutgens, R. B. H., Schrakamp, G., Tager, J. M., Van Den Bosch, H., Moser,
   A. B., and Moser, H. W., 1987a, Neonatal adrenoleukodystrophy: Impaired plasmalogen
   biosynthesis and peroxisomal β-oxidation due to a deficiency of catalase-containing particles
   (peroxisomes) in cultured skin fibroblasts, *J. Neurol. Sci.* **77**:331–340.
Wanders, R. J. A., Schutgens, R. B. H., Schrakamp, G., van den Bosch, H., Tager, J. M., Moser, A.
   B., and Moser, H. W., 1987b, Generalized loss of peroxisomal function of neonatal adrenoleuko-
   dystrophy: Implications for pre- and postnatal detection and relationship to X-linked adrenoleukodystrophy, *J. Inherited Metab. Dis.* **10**(2):225–228.
Wanders, R. J. A., Smit, W., Heymans, H. S. A., Schutgens, R. B. H., Barth, P. G., Schierbeek, H.,
   Smit, G. P. A., Berger, R., Przyrembel, H., Eggelte, T. A., Tager, J. M., Maaswinkel-Mooy, P. D.,

Peters, A. C. B., Monnens, L. A. H., Bakkeren, J. A. J. M., Trijbels, J. M. F., Lommen, E. J. P., and Beganovic, N., 1987c, Age-related accumulation of phytanic acid in plasma from patients with the cerebro-hepato-renal (Zellweger) syndrome, *Clin. Chim. Acta* **166**:45–56.

Wanders, R. J. A., van Roermund, C. W. T., Westra, R., Schutgens, R. B. H., van der Ende, M. A., Tager, J. M., Monnens, L. A. H., Baadenhuysen, H., Govaerts, L., Przyrembel, H., Wolff, E. D., Blom, W., Huijmans, J. G. M., and van Laerhoven, F. G. M., 1987d, Alanine glyoxylate aminotransferase and the urinary excretion of oxalate and glycolate in hyperoxaluria type 1 and the Zellweger syndrome, *Clin. Chim. Acta* **165**:311–319.

Wanders, R. J. A., van Roermund, C. W. T., van Wijland, M. J. A., Schutgens, R. B. H., Heikoops, J., van den Bosch, H., Schram, A. W., and Tager, J. M., 1987e, Peroxisomal fatty acid β-oxidation in relation to the accumulation of very long chain fatty acids in cultured skin fibroblasts from patients with Zellweger syndrome and other peroxisomal disorders, *J. Clin. Invest.* **80**:1778–1783.

Wanders, R. J. A., van Wijland, M. J. A., van Roermund, C. W. T., Schutgens, R. B. H., van den Bosch, H., Tager, J. M., Nijenhuis, A., and Tromp, A., 1987f, Prenatal diagnosis of Zellweger syndrome by measurement of very long chain fatty acid (C26:O) β-oxidation in cultured chorionic villus fibroblasts: Implications for early diagnosis of other peroxisomal disorders, *Clin. Chim. Acta* **165**:303–310.

Wanders, R. J. A., Romeyn, G. J., van Roermund, C. W. T., Schutgens, R. B. H., van den Bosch, H., and Tager, J. M., 1988a, Identification of L-pipecolate oxidase in human liver and its deficiency in the Zellweger syndrome, *Biochem. Biophys. Res. Commun.* **154**:33–38.

Wanders, R. J. A., van Roermund, C. W. T., Jurrians, S., Schutgens, R. B. H., Tager, J. M., van den Bosch, H., Wolff, E. D., Przyrembel, H., Berger, R., Schaaphok, F. G., Reitsma, W., and van Luyk, W. H. J., 1988b, Diversity of residual alanine glyoxylate aminotransferase activity in hyperoxaluria type 1: Correlation with pyridoxine responsiveness, *J. Inherited Metab. Dis.* **11**(2):208–211.

Wanders, R. J. A., van Roermund, C. W. T., van Wijland, M. J. A., Schutgens, R. B. H., van den Bosch, H., Schram, A. W., and Tager, J. M., 1988c, Direct evidence that the deficient oxidation of very long chain fatty acids in X-linked adrenoleukodystrophy is due to an impaired ability of peroxisomes to activate very long chain fatty acids, *Biochem. Biophys. Res. Commun.* **153**:618–624.

Wanders, R. J. A., van Roermund, C. W. T., van Wijland, M. J. A., Schutgens, R. B. H., Tager, J. M., van den Bosch, H., and Thomas, G. H., 1988d, Peroxisomes and peroxisomal functions in hyperpipecolic acidemia, *J. Inherited Metab. Dis.* **11**(2):161–164.

Wanders, R. J. A., Wiemer, E. A. C., Brul, S., Schutgens, R. B. H., van den Bosch, H., and Tager, J. M., 1989, Prenatal diagnosis of Zellweger syndrome by direct visualization of peroxisomes in chorionic villus fibroblasts by immunofluorescence microscopy, *J. Inherited Metab. Dis.* **12**(2):301–304.

Wanders, R. J. A., Boltshauser, E., Steinmann, B., Spycher, M. A., Schutgens, R. B. H., van den Bosch, H., and Tager, J. M., 1990a, Infantile phytanic acid storage disease, a disorder of peroxisome biogenesis: A case report, *J. Neurol. Sci.* **98**:1–11.

Wanders, R. J. A., van Roermund, C. W. T., Schelen, A., Schutgens, R. B. H., Tager, J. M., Stephenson, J. B. P., and Clayton, P. T., 1990b, A bifunctional protein with deficient enzyme activity: Identification of a new peroxisomal disorder using novel methods to measure the peroxisomal beta-oxidation enzyme activities, *J. Inherited Metab. Dis.* **13**:375–379.

Wandersman, C., Delepelaire, P., 1990, TolC, an *Escherichia coli* outer membrane protein required for hemolysin secretion, *Proc. Natl. Acad. Sci. USA* **87**:4776–4780.

Wardinsky, T. D., Pagon, R. A., Powell, B. R., McGillivray, B. Stephan, M., Zonana, J., and Moser, A., 1990, Rhizomelic chondrodysplasia punctata and survival beyond one year: A review of the literature and five case reports, *Clin. Genet.* **38**:84–93.

Watkins, P. A., Chen, W. W., Harris, C. J., Hoefler, G., Hoefler, S., Blake, Jr., D. C., Balfe, A., Kelley,

R., Moser, A. B., Beard, M. E., and Moser, H. W., 1988, Peroxisomal bifunctional enzyme deficiency, *J. Clin. Invest.* **83**:771–777.

Watkins, P. A., Mihalik, S. J., and Skjeldal, O. H., 1990, Mitochondrial oxidation of phytanic acid in human and monkey liver—Implication that Refsum's disease is not a peroxisomal disorder, *Biochem. Biophys. Res. Commun.* **167**(2):580–586.

Watts, R. W., Chalmers, R. A., Gibbs, D. A., Lawson, A. M., Prukiss, P., and Spellacy, E., 1979, Studies on some biochemical treatments of primary hyperoxaluria, *O. J. Med.* **48**:259.

Watts, R. W., Calne, R. Y., Rolles, K., Danpure, C. J., Morgan, S. H., Mansell, M. A., Williams, R., and Purkiss, P., 1987, Successful treatment of primary hyperoxaluria type 1 by combined hepatic and renal transplantation, *Lancet* **2**:474.

Webber, K. O., Datta, N. S., and Hajra, A. K., 1987, Properties of the enzymes catalyzing the biosynthesis of lysophophatidate and its ether analog in cultured fibroblasts from Zellweger syndrome patients and normal controls, *Arch. Biochem. Biophys.* **254**(2):611–620.

Weinberg, K., Moser, A., Watkins, P., Lenarsky, C., Winter, S., Moser, H., and Parkman, R., 1988, Bone marrow transplantation (BMT) for adrenoleukodystrophy (ALD), *Pediatr. Res.* **23**:334A.

Weleber, R. G., Tongue, A. C., Kennaway, N. G., Budden, S. S., and Buist, N. R. M., 1984, Ophthalmic manifestations of infantile phytanic acid storage disease, *Arch. Ophthalmol.* **102**:1317–1320.

Whitcomb, R. W., Linehan, W. R., and Knazek, R. A., 1988, Effects of long-chain, saturated fatty acids on membrane microviscosity and adrenocorticotropin responsiveness of human adrenocortical cells *in vitro*, *J. Clin. Invest.* **81**:185–188.

Wiemer, E. A. C., Brul, S., Just, W. W., Vandriel, R., Brouwerkelder, E., Vandenberg, H., Weijers, P. J., Schutgens, R. B. H., Vandenbosch, H., Schram, A., Wanders, R. J. A., and Tager, J. M., 1989, Presence of peroxisomal membrane proteins in liver and fibroblasts from patients with the Zellweger syndrome and related disorders—Evidence for the existence of peroxisomal ghosts, *Eur. J. Cell Biol.* **50**(2):407–417.

Wilson, G. N., Holmes, R. G., Custer, J., Lipkowitz, J. L., Stover, J., Datta, N., and Majra, A., 1986, Zellweger syndrome: Diagnostic assays, syndrome delineation, and potential therapy, *Am. J. Med. Genet.* **24**:69–82.

Wilson, G. N., King, T., Argyle, J. C., and Garcia, R. F., 1991, Maternal clofibrate administration amplifies fetal peroxisomes, *Pediatr. Res.* **29**(3):256–262.

Wolvetang, E. J., Tager, J. M., and Wanders, R. J., 1990, Latency of the peroxisomal enzyme acyl-CoA:dihydroxyhacetonephosphate acyltransferase in digitonin-permeabilized fibroblasts: The effect of ATP and ATPase inhibitors, *Biochem. Biophys. Res. Commun.* **170**:1135–1142.

Young, R. S. K., Osbakken, M. D., Alger, P. M., Ramer, J. C., Weidner, W. A., and Daigh, J. D., 1985, Magnetic resonance imaging in leukodystrophies of childhood, *Pediatr. Neurol.* **1**:15–19.

Zellweger, H., 1987, The cerebro-hepato-renal (Zellweger) syndrome and other peroxisomal disorders, *Dev. Med. Child. Neurol.* **29**:821–829.

Zoeller, R. A., Morand, O. H., and Raetz, C. R. H., 1988, A possible role for plasmalogens in protecting animal cells against photosensitized killing, *J. Biol. Chem.* **263**:11590–11596.

Zoeller, R. A., Allen, L. A. H., Santos, N. J., Lazarow, P. B., Hashimoto, T., Tartakoff, A. M., and Raetz, C. R. H., 1989, Chinese hamster ovary cell mutants defective in peroxisome biogenesis—Comparison to Zellweger syndrome, *J. Biol. Chem.* **264**(36):21872–21878.

Zung, A., Mogilner, B. M., Nissani, R., Appelman, Z., and Gelman de Kohan, Z., 1990, Occurrence of cerebrohepatorenal (Zellweger) syndrome in the Karaite community in Israel: A genetic hypothesis, *Isr. J. Med. Sci.* **26**:570–572.

Chapter 2

X-Linked Immunodeficiencies

Jennifer M. Puck

Department of Pediatrics
University of Pennsylvania School of Medicine
and Children's Hospital of Philadelphia
Philadelphia, Pennsylvania 19104

INTRODUCTION

Advances in the molecular biology of diseases of the immune system are rapidly unfolding in many areas. Immunologists have become aware that many of the rare congenital immunodeficiency patients they encounter have specific genetic defects, many of them X-linked. Two X-linked genes, those for properdin and cytochrome b-245 beta chain, have been cloned and proven to cause, respectively, a complement system disorder, properdin deficiency, and the phagocyte-killing defect chronic granulomatous disease. There are at least five other genetic loci responsible for X-linked diseases involving lymphocytes. These appear to be disorders of distinct gene products which may be required for development, survival, or function of particular lymphocyte lineages. None of these genes have been cloned, and there are no biochemical or immunological indicators as to the primary pathogenesis. Linkage analysis has been used to establish a regional localization for each gene defect. Further genetic clues, such as affected individuals with chromosomal translocations or deletions, have not been detected except in chronic granulomatous disease and, recently, X-linked lymphoproliferative syndrome.

Fetal sex selection and fetal blood sampling for leukocyte functional studies

Advances in Human Genetics, Volume 21, edited by Henry Harris and Kurt Hirschhorn. Plenum Press, New York, 1993.

have been used occasionally in the past, but more specific tools have now been developed for carrier detection and prenatal diagnosis of several of these conditions (Puck, 1991). Even as the use of genetic testing by linkage analysis and X-chromosome-inactivation analysis becomes more widespread, further new technologies for dissecting the molecular lesions in immunodeficiency genes offer exciting possibilities for the future. The genes responsible for many of these diseases will soon be identified and cloned. Moreover, specific treatments, such as interferon-gamma in chronic granulomatous disease, and even gene therapy, currently in preliminary human trials for the autosomally inherited immunodeficiency caused by adenosine deaminase deficiency (Anderson, 1990), promise to change the attitudes of doctors and families about the outlook for patients with these gene defects.

HUMAN X-LINKED IMMUNODEFICIENCY SYNDROMES

In males, X-linked recessive mutations are uncompensated and the resulting phenotypes are thus readily recognized. In a system as complex as the immune system many X-encoded genes might be expected to be involved in the pathways of immune cell development, and indeed several distinct X-linked syndromes have been observed. Figure 1 shows a simplified version of the differentiation of totipotent bone marrow stem cells into the lymphoid and myeloid cell lineages, which upon further differentiation become the B and T lymphocytes; phagocytes, including neutrophils and monocytes; erythrocytes; and platelet-producing megakaryocytes. As a first step toward understanding the pathogenesis of the X-linked immunodeficiencies, one can identify one or more branches of the differentiation pathway scheme which appear to be affected in each disorder.

X-Linked Agammaglobulinemia

Agammaglobulinemia was the first recognized congenital immunodeficiency disease, described by Bruton (1952). The great majority of patients are males with X-linked defects of the B-lymphocytes compartment or humoral immune system, responsible for production of antibodies, as shown in the top branch of Fig. 1. The disease is now known as X-linked agammaglobulinemia (XLA). Its MIM number is 300300 (Online Mendelian Inheritance in Man, OMIM, 1991a), and its locus name, assigned by the Human Gene Mapping Workshop, is now AGMX1 (Mandel *et al.*, 1989), although it was formerly designated IMD1. XLA is characterized by

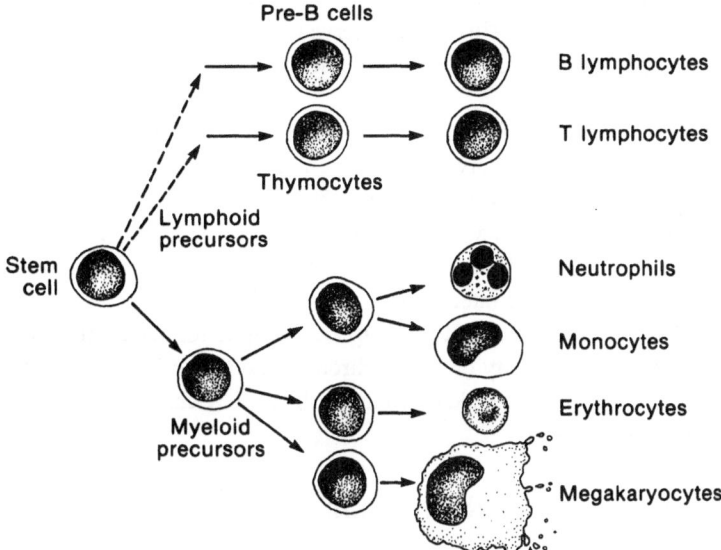

Fig. 1. Differentiation pathways of bone-marrow-derived stem cells. A small number of totipotent, self-renewing stem cells give rise to committed lymphoid and myeloid cell precursors. Lymphoid precursors further differentiate into cells of the T- and B-lymphocyte lineages, while myeloid precursors give rise to phagocyte and erythrocyte lineages and platelet-producing megakaryocytes. The X-linked agammaglobulinemia gene defect interferes with cell differentiation or survival between pre-B cells and B cells, while the X-linked SCID defect affects both the T- and B-lymphoid lineages, but not myeloid-derived lineages. The Wiskott–Aldrich syndrome defect appears to affect all bone-marrow-derived lineages.

recurrent invasive pyogenic bacterial infections and persistent infections with certain viruses from infancy onward. Fifty percent of the affected males have a family history of similarly affected maternal relatives. Clinical findings include a paucity of lymph node tissue, and immunological testing in the laboratory shows very low numbers of circulating B lymphocytes and, with few exceptions, very low serum concentrations of antibodies (WHO Scientific Group on Immunodeficiency, 1983; Lederman and Winkelstein, 1985; Conley and Puck, 1988).

Multiple lines of evidence indicate that the defect in XLA is specific to the B-lymphocyte lineage and is likely to be a developmental block in the normal progression of pre-B cells to mature B cells (Fig. 1). The numbers and function of T lymphocytes and other bone-marrow-derived cells are generally normal in affected patients. Although mature circulating B cells are rare or absent, XLA males have been found to have pre-B cells in their bone marrow (Pearl *et al.*,

1978). Confirming evidence for an intrinsic B-cell defect has been subsequently inferred by Conley *et al.* (1986), who observed that mature B cells of obligate female carriers of XLA were limited to cells with an active non-XLA-bearing X chromosome.

Replacement therapy for XLA with pooled human gammaglobulin has proven life-saving; but morbidity from recurrent and chronic respiratory, gastro-intestinal, and central nervous system infections is substantial. Furthermore, 16 of the 96 patients reviewed by Lederman and Winkelstein (1985) had died between 4 and 28 years of age from chronic pulmonary disease, chronic and disseminated viral diseases, and bacterial infections.

XLA has been localized to Xq21.3–q22.2. Despite former confusion about possible genetic heterogeneity on the X chromosome (Hendriks *et al.*, 1989), it is now agreed that there is no evidence for more than one X-chromosomal locus for XLA (Davies *et al.*, 1990).

X-Linked Severe Combined Immunodeficiency

Clinical Features

Severe combined immunodeficiency (SCID) is a term which includes a wide spectrum of clinical disease presentations caused by genetic defects in both the cellular, or T-cell-mediated, and humoral, or B-cell- and antibody-mediated, arms of the immune system, as shown in the top two branches of Fig. 1. Both males and females can have SCID, but the fact that approximately 60–80% of all SCID patients are male suggests that X-linked mutations account for a substantial proportion of cases. Patients with SCID appear normal at birth, but generally come to medical attention within the first 6 months of life with failure to thrive, chronic diarrhea, and recurrent and persistent infections with bacteria, viruses, and opportunistic pathogens such as *Pneumocystis carinii* and *Candida* spp., which are rarely seen in normal hosts. They lack a thymic outline on chest radiograms and have low numbers of T cells, which fail to proliferate upon stimulation with mitogens. Immunoglobulin concentrations are also low, and functional antibody responses, such as anti-tetanus antibody following immuniza-tion, are absent. The disease is invariably fatal early in life unless successfully treated by bone marrow transplantation.

The X-linked form of SCID has been given MIM number 300400 (OMIM, 1991b), and the locus designation is SCIDX1, formerly IMD4 (Mandel *et al.*, 1989). This entity was at one time referred to as Swiss-type agammaglobulinemia before the T-cell component was recognized. It was also sometimes referred to as SCID with B cells, because the majority of patients have normal or even elevated numbers of cells with B-cell surface markers in the peripheral blood upon presentation (WHO Scientific Group on Immunodeficiency, 1983). The presence or absence of B cells is not sufficiently consistent or specific, however, to assign the correct mode of inheritance (Conley et al., 1990a).

Distinguishing X-Linked from Autosomal SCID

Recognition and response to foreign antigens by lymphocytes requires a complex cascade of interrelated events involving products of a large number of genes. Defects in the expression of any of these genes might be expected to produce immune dysfunction. Moreover, the requirements for the differentiation of lymphoid progenitor cells into mature lymphocytes before and after arrival in the thymus have not yet been elucidated. Indeed, the wide clinical spectrum of diseases with combined cellular and humoral immune defects (Hong, 1990; Parkman, 1991) attests to the complexity of these developmental pathways. Although the etiology of most SCID cases is undetermined, there are several recognized autosomal recessive defects which result in a SCID phenotype with immunological impairment similar to the X-linked form of SCID. These have been defined to various degrees, as recently reviewed by Parkman (1991) and Weinberg (1991). Adenosine deaminase (ADA) and purine nucleoside phosphorylase are cloned purine pathway enzymes whose deficiency results in SCID, as reviewed by Hirschhorn (1986). Other etiologies recently recognized to cause SCID include failure to express class I (De Preval et al., 1988) or class II (Reith et al., 1988; Kara and Glimcher, 1991) histocompatibility antigens, known as bare lymphocyte syndrome; and lack of production of, or receptors for, lymphokines such as deficiency of the T-cell growth hormone interleukin 2 (Weinberg and Parkman, 1990). An autosomal recessively inherited SCID has been seen among the inbred Navajo tribe in the southwestern United States (Anthony Hayward, personal communication). Distinctive syndromes exist which include immunodeficiency similar to SCID accompanied by other congenital abnormalities. The best known is DiGeorge syndrome, in which absence of the thymus is associated with embryologically related defects of the parathyroids and heart, often in the presence of deletions of a small region of chromosome 22 (Fibison et al., 1990). A new form of SCID with multiple gastrointestinal atresias has also been recognized recently (Moreno et al., 1990).

The above recognizable forms of SCID remain the minority seen in clinical practice. Assignment of SCID genotype in the remaining cases, which are diagnosed on the basis of clinical and immunological evaluation, has been impossible due to a combination of factors. First, immunological studies are usually performed only on lymphocytes from the peripheral circulation, which are the most mature. It is generally not possible to enumerate and test human thymocytes, let alone pre-thymic T-cell precursors, to determine at which stage of differentiation a particular child's lymphocytes are blocked. Second, variability in the presentation of infants with SCID has been shown to be influenced by superimposed infections. For example, infections with cytomegalovirus tend to exaggerate the depression in lymphocyte numbers, particularly of T cells. On the other hand, Epstein–Barr virus, which infects B lymphocytes, can induce B-cell proliferation, producing elevated numbers of circulating B cells. Third, spontaneous engraftment of maternal lymphocytes transfused into the infant before or at the time of delivery has also been described (Conley et al., 1984), resulting in a presentation of SCID as graft-versus-host disease; circulating T cells are found in these infants, but upon karyotype or molecular analysis they prove to be of maternal origin. Fourth, allelic differences in SCID due to ADA deficiency have been correlated with a wide range of phenotypic severity (Hirschhorn et al., 1990; Wilson et al., 1991), and variation in severity as a result of distinct genetic lesions can be presumed to occur as well in SCID of other etiologies. Finally, factors such as age at diagnosis and effects of unknown modifying genes compound the difficulty of assigning a genotype based on clinical presentation.

A careful family history is important diagnostically when X-linked inheritance is found; similarly, affected females are likely to have autosomal recessive forms of SCID. Affected males with a negative family history, however, may represent the first manifestation of a new X-linked SCID mutation.

Wiskott–Aldrich Syndrome

Wiskott–Aldrich syndrome (WAS, MIM number 301000), formerly designated IMD2 (Mandel et al., 1989; OMIM, 1992a), is an X-linked condition classically characterized as a triad of clinical manifestations: thrombocytopenia, eczema, and immunodeficiency. Early in life affected males come to medical attention with bleeding disorders, bloody diarrhea, and petechiae. They are found to have thrombocytopenia with small platelets which function poorly. Later in infancy and childhood, eczema and other allergic and autoimmune phenomena may develop, and recurrent otitis and more invasive infections frequently occur. Among survivors of hemorrhagic and infectious complications in early life a very high incidence of malignancies, especially lymphoreticular tumors and leukemias, has been documented.

The nature of the immune defect in WAS has been intensely studied. Clinical tests show variable degrees of T- and B-cell dysfunction, often including decreased T-cell proliferation to mitogens and poor antibody responses to carbohydrate antigens, such as isohemagglutinins and bacterial cell-wall antigens. A lymphocyte surface glycoprotein, CD43 (also referred to as gp115 and sialoglyco-phorin), was found to be deficient or abnormally glycosylated on WAS lympho-cytes (Remold-O'Donnell et al., 1984; Reisinger and Parkman, 1987). Purification of the CD43 protein and cloning of the gene encoding it have been accomplished; the structural gene resides on chromosome 16, and not the X chromosome (Pallant et al., 1988). However, cross-linking of surface CD43 molecules, either by periodate binding of sialic acid residues or by anti-CD43 monoclonal antibody, has been shown to induce proliferation of normal, but not WAS, lymphocytes (Greer et al., 1989a; Puck et al., 1990a). Thus the WAS gene defect may interfere with normal glycosylation, transport or cell surface expression of CD43, and possibly other leukocyte and platelet cell-surface glycoproteins.

X-Linked Lymphoproliferative Syndrome

X-Linked lymphoproliferative syndrome (XLP), originally called Duncan syndrome after the first reported family (Purtilo et al., 1975) and designated by MIM number 308240 (OMIM, 1988), is characterized by abnormally severe outcomes following infection with the Epstein–Barr virus (EBV), which causes asymptomatic infection or mononucleosis in normal children and adults, respec-tively. The genetic locus, previously IMD5, has been renamed LYP (Mandel et al., 1989). EBV-induced disease states in a large number of affected individuals collected in the XLP registry of Purtilo include four distinct types. Most common, occurring in about 60% of patients, is fatal infectious mononucleosis, often with liver failure. Other outcomes include hypogammaglobulinemia, aplastic anemia, and lymphoma, especially of the American Burkitt type, presenting in the terminal ileum (Purtilo et al., 1982; Grierson and Purtilo, 1987).

The precise nature of the immunological defect in XLP remains obscure. In known XLP pedigrees, males who subsequently became affected had normal immunological evaluations prior to their exposure to EBV (Sullivan et al., 1983). Immune system abnormalities after EBV infection are variable and include defective anti-EBV antibody responses and depressed natural killer cell re-sponses; however, these may be secondary phenomena and can also be seen in individuals without XLP. Subtle abnormalities in anti-EBV immune responses and IgG subclasses have been reported in premorbid males and some female obligate carriers of XLP (Purtilo and Grierson, 1991), but the specificity and sensitivity of these have not been systematically studied.

Linkage studies have placed XLP near restriction fragment length polymor-phism (RFLP) markers in Xq25–q26 (Skare et al., 1989), and a recent family has been identified in which an interstitial deletion involving Xq25 was found in a mother and sister of a male who died of XLP (Sanger et al., 1990).

Hyper-IgM Syndrome

Immunodeficiency with increased serum concentrations of immunoglobulin M (IgM), but low concentrations of IgA and IgG, has been noted to occur in pedigrees with an X-linked pattern of inheritance (WHO Scientific Group on Immunodeficiency, 1983), but also in females (Gleich *et al.*, 1965) and in a father and two daughters (Brahmi *et al.*, 1983). Thus, not only X-linked, but also autosomal mutations, both recessive and dominant, may produce this phenotype. In addition, acquired forms have been reported (Geha *et al.*, 1979). The X-linked form, formerly called IMD3, has been given the locus name HIGM1, MIM number 308230 (OMIM, 1992b). The clinical picture of hyper-IgM syndrome (HIGM) is similar to that of XLA. However, in addition to recurrent bacterial infections due to poor antibody responses, HIGM patients commonly have autoimmune diseases and neutropenia. Moreover, T-lymphocyte dysfunction is suggested by the occurrence of opportunistic infections with *Candida* spp. and *Pneumocystis*. It has been suggested that the failure of isotype switching from IgM to IgG or IgA, whether in the X-linked HIGM or in other syndromes with similar phenotype, is due to lack of a soluble B-cell helper factor normally produced by T lymphocytes (Geha *et al.*, 1979; Levitt *et al.*, 1983; Mayer *et al.*, 1986; Hendricks *et al.*, 1990). That X-linked HIGM is even more rare than the other X-linked immunodeficiencies is suggested by the fact that only one linkage study of a single kindred has been published (Mensink *et al.*, 1987). This report found HIGM1 to be a distinct locus from the more proximal XLA locus; HIGM1 was found to be linked to DXS42 in Xq24–q27.

Chronic Granulomatous Disease

Defects in the cytochrome b-245 complex make up a group of genetic disorders called chronic granulomatous disease (CGD). Affected patients have recurrent infections, particularly with catalase-positive organisms, due to the failure of their phagocytic cells to generate a respiratory burst of the NADPH oxidase system, which normally produces the highly reactive products of oxygen metabolism responsible for killing microorganisms (Baehner, 1990). At least two membrane and three cytosolic protein components of this system have been identified through the study of patients with CGD. Two thirds of CGD patients have the X-chromosome-linked form, MIM number 306400 (OMIM, 1992c), due to mutations in the 91-kD chain of cytochrome b-559, the gene for which, designated CYBB (Mandel *et al.*, 1989), lies in Xp21.

Although CGD patients display a wide spectrum in disease severity, early death due to infection is the rule and has prompted many clinical trials of prophylactic antibiotics and immune mediators. Most recently, an international collaborative study found that prophylactic treatment of CGD patients with recombinant interferon-gamma was surprisingly effective in preventing serious infections in all genetic forms of the disease (International Chronic Granulomatous Disease Cooperative Study Group, 1991). Ezekowitz *et al.* (1990) have shown that one mechanism of action of interferon-gamma may be to induce newly developing myeloid progenitor cells to overexpress NADPH oxidase system components, enabling even cells from CGD patients to generate near-normal superoxide levels. Other, nonoxidative killing systems may also be induced (Baehner, 1990). Thus, for at least some CGD patients, especially those with mutations that do not completely delete or disable the involved genes, this disease may now have a more specific therapy in addition to trimethoprim–sulfamethoxazole or other prophylactic antibiotic regimens.

Properdin Deficiency

Properdin factor P is one of the serum proteins needed to activate complement factor C3 via the alternate pathway. Individuals who lack properdin P are unable to mount complement-mediated lysis of such bacteria as *Neisseria meningitidis* by the alternate pathway. Although cell-mediated immunity and antibody production are normal, these persons are susceptible to overwhelming infections with this and related organisms. Following recognition of kindreds in which properdin deficiency and fulminant neisserial infections were inherited in an X-linked pattern [MIM number 312060 (OMIM, 1991c)], the properdin P-deficient phenotype, and more recently the structural gene, have been localized to Xp11.23–21.1 (Goonewardena *et al.*, 1988; Goundis *et al.*, 1989).

New Syndromes

Nonlethal Combined Immunodeficiency

A newly defined X-linked combined immunodeficiency has been reported by Brooks *et al.* (1990), characterized by chronic and recurrent infections in males from 2.5 to 34 years of age, low numbers of T lymphocytes, and poor specific IgG responses. The disease has been assigned MIM number 312863 (OMIM, 1991d) and given the locus designation SCIDX2. Linkage analysis has not yet been performed on this single, large pedigree, and it will be interesting to see whether this turns out to be a new disease locus or an allelic variant of one of the others, such as X-linked SCID, XLP, or HIGM.

Such variation has been postulated to account for the range of immunodeficiency in WAS by Puck *et al.* (1990a), who reported two unrelated male infants presenting with isolated thrombocytopenia without demonstrable immune defects; in each case a variant of WAS was suspected upon finding that each of their mothers had T cells with nonrandom X-chromosome inactivation, as do carriers of typical WAS.

XLA Accompanied by Isolated Growth Hormone Deficiency

Two unrelated pedigrees have been reported in which the XLA phenotype was coinherited with deficiency of growth hormone without other endocrine abnormalities (Fleisher *et al.*, 1980; Sitz *et al.*, 1990). This entity, XLA with growth hormone deficiency (XLA/GHD), has thus been established as an X-linked disease rather than just a chance association. It has been given MIM number 307200 (OMIM, 1991e). Genetic evaluation of the second and a third family (Conley *et al.*, 1991) revealed the same female carrier X-inactivation pattern previously noted in mothers of boys with typical and atypical XLA. Moreover, linkage studies in these families were consistent with a regional localization for XLA/GHD at or near the XLA locus. Interestingly, X-linked GHD alone has also been described (Phillips and Vnencak-Jones, 1989; Najiar *et al.*, 1990), even though the growth hormone locus has been mapped to chromosome 17. It is possible that XLA and XLA/GHD will turn out to be distinct linked loci or allelic disorders at a single locus; alternatively, XLA/GHD may result from deletion or other disruption of contiguous genetic loci, the XLA locus and a locus affecting the expression of the growth hormone gene.

ANIMAL MODELS

Mouse Mutations with X-Linked Immunodeficiency

XID

Mouse strain CBA/N has been known for over 20 years to carry an X-linked immune deficiency (XID, locus *xid*) characterized by abnormal B-lymphocyte responses to polysaccharide antigens (Wicker and Scher, 1986). Affected mice lack a subset of B cells identified by Lyb-5 antiserum, and female carriers of *xid* manifest skewed X-chromosome inactivation in their B cells, but not other cell lineages (Nahm *et al.*, 1983), suggesting a developmental defect in B cells as the pathogenesis of the immunodeficiency. The mouse gene at the *xid* locus has not been cloned. Its map position is close, but distal to that for mouse phosphoglycerate kinase. This gene could therefore correspond to human XLA, which is also a B-cell-limited defect with a similar genetic localization. Human X-linked SCID, also linked to human phosphoglycerate kinase (PGK1), could by the same argument be the human analog of *xid*, although both B- and T-lymphocyte lineages, rather than B cells alone, are affected in X-linked SCID in humans.

XLR

XLR was originally identified as an X-linked multigene family cloned from a lymphocyte-specific library prepared by subtractive hybridization (Cohen *et al.*, 1985; Siegel *et al.*, 1987). Most of the transcripts appeared to be pseudogenes, but a cDNA, pM1, was found to encode a 30-kD protein expressed in the nucleus of T and B cells, but not other cell types (Garchon and Davis, 1989). The protein has distant homology to lamins, has a short half-life, and is stabilized in the nucleus by zinc ions, but its function is unknown. Studies of expression in tumor cell lines representing various stages of B-cell differentiation suggested that XLR is expressed in the late stages of B-cell development. There are plasmacytomas, however, in which it is not found. The possibility that mutations in XLR cause XID has been ruled out by mouse X-chromosome linkage studies (Mullins *et al.*, 1990), which have identified two genetic loci, *Xlr-1* and *Xlr-2*, the former close to *Otc*, *Timp*, and *Syn-1*, while the latter is near *Hprt*. Neither is genetically close to *xid*. It remains a matter of speculation whether the XLR genes are required for immune cell development or are involved in any human immunodeficiency.

Scurfy

The scurfy (*sf*) mouse strain is characterized by scaliness of the skin and hypogonadism. The locus has been found to map to the region of the mouse chromosome analogous to the human locus for Wiskott–Aldrich syndrome, prompting a search for hematological and immunological abnormalities in this strain (Lyon *et al.*, 1990). Indeed, low platelet counts and progressive anemia, as well as gastrointestinal bleeding and diarrhea, were found to be a feature of scurfy mice. The skin condition may not be the same as eczema, however, and hypogonadism is not seen in WAS. Moreover, on genetic grounds alone, *Xlr-1* could equally well be a WAS candidate in the mouse. It maps to the same region of the mouse X chromosome, as determined by backcross analysis in interspecies matings between *Mus spretus* and common laboratory mouse strains of *Mus musculus* (Mullins *et al.*, 1990).

Canine X-Linked Severe Combined Immunodeficiency

An X-linked form of immunodeficiency was recognized as a spontaneous mutation in a Basset hound (Patterson *et al.*, 1982; Jezyk *et al.*, 1989). The affected male pups were normal at birth, but developed growth retardation and subsequently died with pyogenic and viral infections in the first weeks to months of life. Immunological evaluations of the original dogs and subsequent litters revealed no thymic shadow on radiograms, lack of normal lymphoid tissue, and extremely poor B- and T-cell function, with only immature lymphoid forms present in the circulation and thymus. Similar to most humans with X-linked SCID, these pups had normal numbers of circulating B lymphocytes, but lacked specific antibody responses and had very low IgA and IgG concentrations. The dog disease thus seems to be very similar to human X-linked SCID; the likelihood that these diseases are caused by mutations in the same gene may prove valuable in evaluating gene candidates for human X-linked SCID, studying the pathogenesis of the disease, and developing strategies for gene therapy.

X-CHROMOSOME INACTIVATION PATTERNS AS AN INDICATOR OF PHENOTYPE OF X-LINKED IMMUNODEFICIENCIES

Basis for X-Inactivation Assays

Early in female embryogenesis one of the two X chromosomes in every cell becomes inactive and remains inactive in that cell's progeny as they divide and differentiate (Lyon, 1966). Because X inactivation is a random process, the mature tissues of normal women are composed of a mixture of cells with each X active. In a female in whom one X chromosome carries a deleterious mutation in a gene that is needed during development of a particular cell lineage, however, a secondary phenomenon occurs. Cells with an active X chromosome which does not bear such a mutation will have a normal gene product and will mature normally, but cells with the mutant X active will lack the needed gene product and will be selected against. As a result, the X-inactivation pattern in the mature cell lineage will be skewed in favor of the nonmutant X chromosome.

Methods for Assessing X-Inactivation Patterns

G6PD Isozymes

Nonrandom X inactivation was first demonstrated in T lymphocytes and platelets of female carriers of WAS (Gealy et al., 1980; Prchal et al., 1980), and then in B cells of carriers of XLA (Conley et al., 1986), who happened to be heterozygous for electrophoretic variants of the X-linked enzyme glucose-6-phosphate dehydrogenase (G6PD). It was recognized that skewed X inactivation might be useful as a carrier test for these diseases if general methods could be found to distinguish the active from inactive X chromosomes of women at risk.

Human/Rodent Hybrids Isolating the Active Human X

A subsequent method to determine the X-inactivation pattern in cell lineages from all women was based on forming hybrids between human lymphocytes and a hamster cell line deficient in another X-encoded enzyme, hypoxanthine phosphoribosyl transferase (HPRT) (Puck et al., 1987). Growth in selective medium resulted in a set of hybrid cell lines retaining the human active X chromosome to supply HPRT activity. This retained active human X chromosome could be identified by any X-linked polymorphic marker for which the donor woman was heterozygous. By this assay technique carriers of X-linked SCID were shown to have skewed X inactivation in T and B lineages, but not in monocytes (Puck et al., 1987; Conley et al., 1988). These results indicated that the X-linked SCID gene is required for lymphoid cell development, as shown in the top two branches of Fig. 1, but not for myeloid cell development, depicted in the lower part of the figure.

Methylation Differences

A third method for X-inactivation analysis took advantage of differences in methylation of specific cytosine bases in the DNA of active vs. inactive X chromosomes, as reviewed by Winkelstein and Fearon (1990). One such cytosine is located near the HPRT gene RFLP, and another is located near the PGK1 RFLP; only about half of women are heterozygotes for at least one of these RFLPs and thus eligible to have X inactivation assayed by this technique. DNA is subjected to double digestion with restriction enzymes, using one enzyme to reveal the RFLP alleles and the other, a methylation-sensitive enzyme, to cut into smaller fragments the DNA derived from the nonmethylated chromosomes. Interestingly, the active X chromosome is methylated at the HPRT locus, while the inactive X is methylated at PGK1, emphasizing that the role of methylation at these particular sites, while an apparently consistent phenomenon from one woman to another, is not yet understood.

A third locus with differential methylation, DXS255, is a variable number tandem repeat (VNTR) polymorphism in Xp with a heterozygosity of over 80% (Boyd and Fraser, 1990). It has an *Hpa*II restriction site which is methylated on active X chromosomes and unmethylated (partially or totally) on inactive X chromosomes. Although potentially very useful for demonstrating X-inactivation patterns because of the high proportion of women who are informative, DXS255 methylation patterns varied either relatively or absolutely in 12 of 119 samples (10%) as compared to the patterns observed with HPRT or PGK1 (unpublished data, Report of M27b Methylation Analysis Workshop, Oxford, England, January 1991). This locus is therefore not considered a reliable enough indicator of X inactivation to be used for carrier detection in X-linked immunodeficiencies.

Specificity of Skewed X Inactivation as a Function of Precursor Pool Size

The practical application of X-inactivation analysis for carrier detection depends on the specificity of skewed X inactivation to reflect carrier status. Therefore T-cell X-inactivation patterns have been evaluated quantitatively by the hybrid method both in X-linked SCID carriers and a large sample of normal women not at risk for immunodeficiency (Puck *et al.*, 1992). X-Inactivation ratios, defined as percent paternal X active, were clearly skewed in a group of 22 obligate carriers, as shown in Fig. 2, left. The range of normal X inactivation in 29 phase-known women not at risk for carrying X-linked SCID was very broad, however, from 20% to 86%, as shown in the right of the figure. The degree to which X inactivation was observed to be unbalanced in a few normal women raised the possibility that a normal woman with X inactivation skewed by chance could be confused with a carrier of X-linked SCID.

X inactivation in the bone marrow progenitors of T cells was modeled as a binomial distribution, with N progenitors each having an equal likelihood of a maternally derived or paternally derived active X chromosome. A small N would permit the occurrence of normal women with all maternal or all paternal active X's much more frequently than would a large N, just as it is more likely to get all

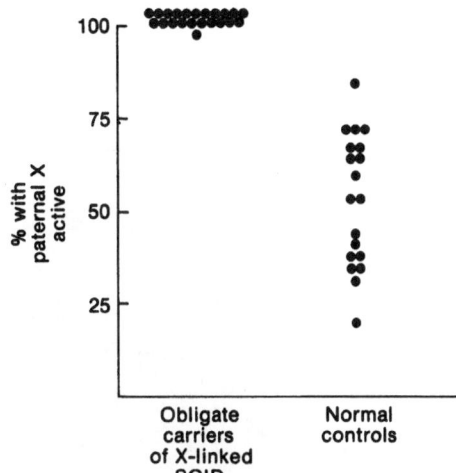

Fig. 2. Comparison of T-cell X-inactivation ratios, expressed as percent paternal X active, in obligate carriers of the X-linked form of SCID (left) and control women (right).

"heads" when flipping three or four coins than when flipping 30 or 40. Data from hybrid analysis of a total of 37 normal women was used in a maximum likelihood analysis to determine the likelihood of different values of N. As shown in Fig. 3, 10 is the most likely number of T-cell progenitors in the bone marrow stem cells of the normal women in this study.

This result is interesting from the standpoint of embryology and may be more accurate than estimates based on other methods (Failkow, 1973); it also may have implications for bone marrow transplantation and gene therapy. As applied to carrier testing for immunodeficiencies, it has made possible a quantitative assessment of carrier status as an odds ratio, for a given set of hybrids tested, of being a carrier vs. being a noncarrier (Puck *et al.*, 1992). Figure 4 is an evaluation of the odds ratio carrier test for X-linked SCID. The *y* axis shows, on a logarithmic scale, the odds ratios for at-risk women in pedigrees with known X-linked SCID whose carrier status had been independently arrived at by linkage analysis. All except 1 of 19 carriers by linkage had odds ratios exceeding 40, while only 2 of 22 noncarriers by linkage had odds ratios above 0.003. The test was therefore successful at diagnosing carriers and noncarriers in 38 of 41 cases, and gave indeterminate values in three women in whom 91.2–93.8% of hybrids scored had an active paternal X.

Without a family history to confirm X-linked inheritance and provide linkage data, skewed X inactivation is the only possible means of carrier detection. Application of the odds ratio test to the more common clinical problem of sporadic

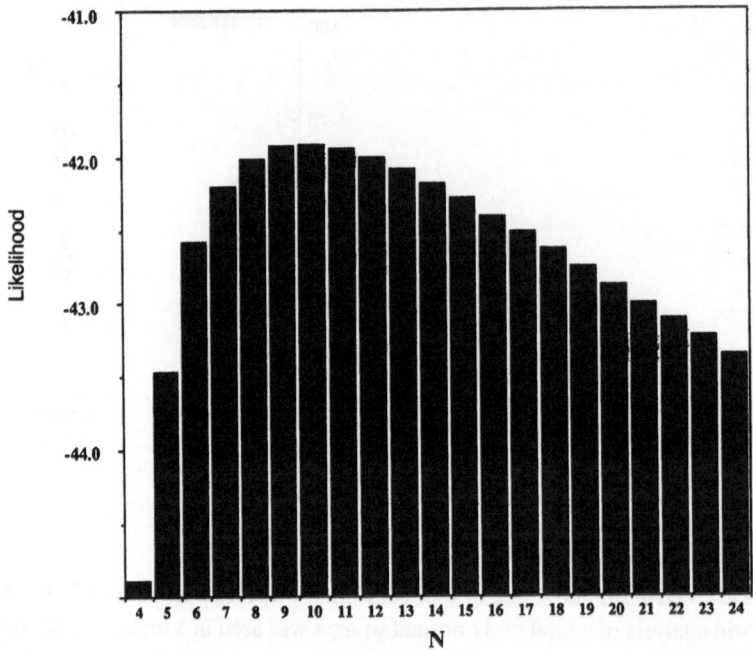

Fig. 3. Graph of overall likelihood (*y* axis) that the T-cell hybrids obtained from 37 control women reflected having been derived from *N* progenitors of T cells in the bone marrow. The logarithmic likelihood function was calculated for values of *N* from 4 to 24. The maximal likelihood was observed at *N* = 10.

cases of males with SCID, including autosomal and X-linked forms, is shown in Fig. 4B. The odds ratios clearly separated the mothers of sporadic SCID boys into two groups: the upper points, representing women with greater than 100:1 odds of having skewed T-cell X inactivation because they carry X-linked SCID; and the lower points, representing women with T-cell X inactivation in the normal range, with less than 1:100 odds of being carriers. By this test 10 of the 22 women evaluated were found to be carriers of the X-linked form of SCID.

Use of X-Inactivation Pattern as a Carrier Test

Both hybrid and differential methylation assays have been used to evaluate known carriers and women at risk of being carriers in families with X-linked SCID (Puck *et al.*, 1987; Conley *et al.*, 1988; Goodship *et al.*, 1988), XLA (Fearon *et al.*, 1987; Conley and Puck, 1988), and WAS (Fearon *et al.*, 1988, Greer *et al.*, 1989b);

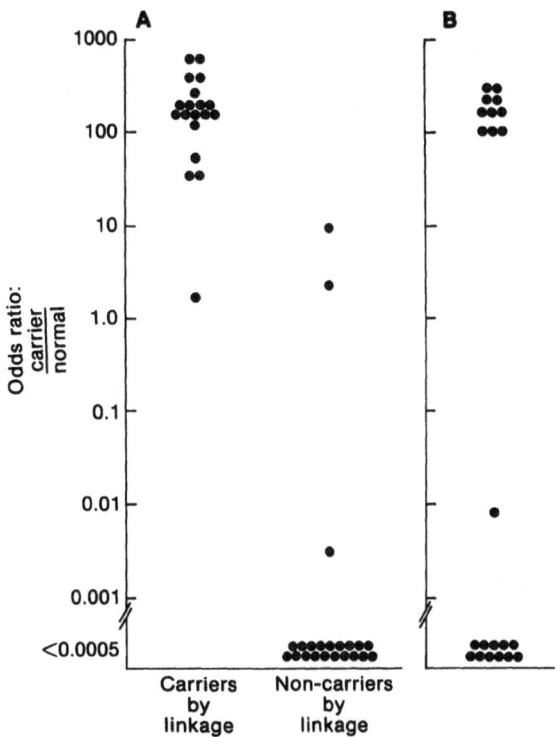

Fig. 4. Odds ratios expressing the likelihood of being a carrier of X-linked SCID, calculated from X-chromosome inactivation ratios and plotted on a logarithmic scale. (A) Odds ratios for women in multigeneration X-linked SCID pedigrees who were at 50% risk of having inherited the X-linked SCID mutation from their mothers. Carriers (left) and noncarriers (right) by linkage inherited flanking, tightly linked RFLP alleles from their mothers' SCID-bearing and normal X chromosomes, respectively. (B) Odds ratios for mothers of male infants with sporadic SCID of unknown inheritance pattern.

Puck *et al.*, 1990a). Skewed cell lineages include T and B cells in SCID, only B cells in XLA, and B and T cells plus phagocytes and platelets in WAS.

Other X-linked immunodeficiencies have not proven amenable to carrier detection by X inactivation. It has been demonstrated by both methylation and hybrids that skewed X inactivation is not a feature of T or B cells of carriers of X-linked lymphoproliferative syndrome (Conley *et al.*, 1990b). Similarly, random X-inactivation patterns as tested by DXS255 methylation were seen in not only T cells, but B cells of IgG, IgA, and IgM isotypes prepared from obligate carriers of X-linked hyper-IgM syndrome (Hendriks *et al.*, 1990).

It must be noted that only highly purified cell lineages that are polyclonal are appropriate for X-inactivation assays; in particular, B cells transformed with Epstein–Barr virus in bulk culture should not be used, as these may rapidly become oligoclonal or monoclonal. Important goals to be met in making X-inactivation testing available in a clinical setting include: (1) standardization, with a quantitative result for the likelihood that a woman is a carrier; (2) sensitivity and specificity; (3) applicability for all women; and (4) low cost and ease of performance. Currently, while the hybrid technique meets the first three objectives, production and testing of hybrids is expensive and takes 6 weeks or more. On the other hand, the methylation method is only possible in about half the population and has not been made quantitative. While a clearly random pattern on a qualitative methylation test may be sufficient to diagnose a noncarrier, difficulty arises in interpreting a predominantly skewed methylation pattern. This could be seen in a carrier with lineage-specific negative selection, but also in a small proportion of noncarrier women whose X inactivation may be unbalanced simply by chance.

LINKAGE ANALYSIS

Summarized Linkage Information

Chronic granulomatous disease (CGD) and properdin factor P deficiency, the only two X-linked immunodeficiencies for which the genes have been cloned, and Wiskott–Aldrich syndrome (WAS), have been mapped to Xp, the short arm of the X chromosome; the X-linked forms of SCID, agammaglobulinemia (XLA), lymphoproliferative syndrome (XLP), and hyper-IgM syndrome (HIGM) are on Xq, the long arm. Figure 5 shows the current map, based on the Human Gene Mapping Workshop (Davies *et al.*, 1990, 1991), of X-linked immunodeficiencies related to the standardized bands of a Giemsa-stained cytogenetic preparation of a human X chromosome. There is no evidence for more than one X-linked locus (genetic heterogeneity) for any of these diseases.

CYBB and Properdin, Cloned Loci

CYBB, encoding the beta chain of cytochrome b-245, lies in Xp21.1. Genetic studies of CYBB, the gene whose deficiency causes X-linked CGD, have detected restriction fragment length polymorphisms (RFLPs) in the cDNA for diagnosing female carriers in families with X-linked CGD and for prenatal diagnosis (Franke *et al.*, 1990; Pelham *et al.*, 1990). The many genetic polymorphisms now defined

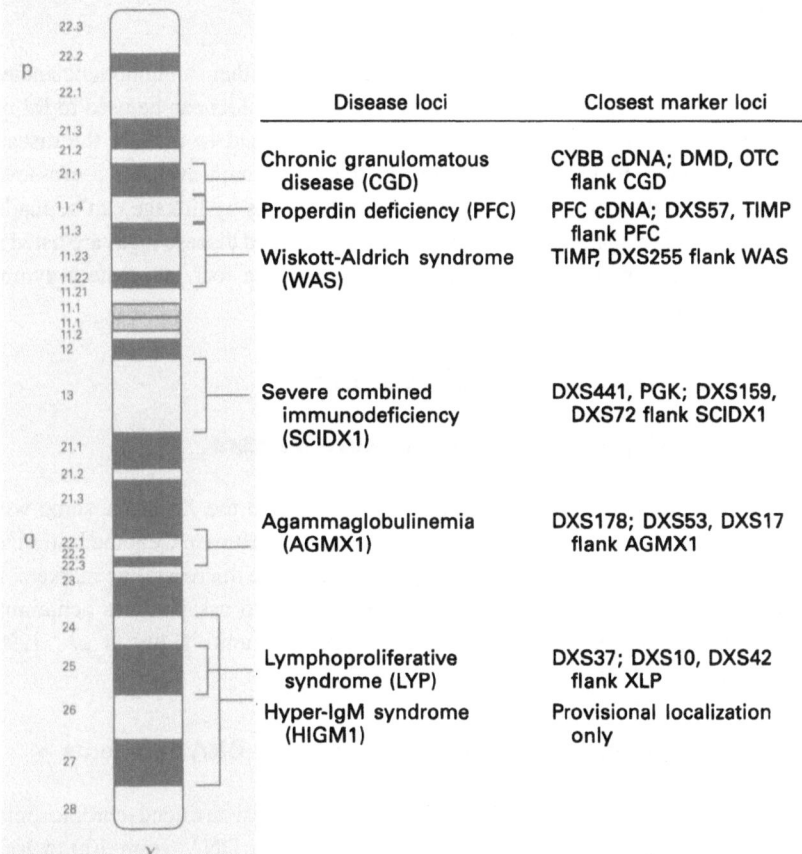

Disease loci	Closest marker loci
Chronic granulomatous disease (CGD)	CYBB cDNA; DMD, OTC flank CGD
Properdin deficiency (PFC)	PFC cDNA; DXS57, TIMP flank PFC
Wiskott-Aldrich syndrome (WAS)	TIMP, DXS255 flank WAS
Severe combined immunodeficiency (SCIDX1)	DXS441, PGK; DXS159, DXS72 flank SCIDX1
Agammaglobulinemia (AGMX1)	DXS178; DXS53, DXS17 flank AGMX1
Lymphoproliferative syndrome (LYP)	DXS37; DXS10, DXS42 flank XLP
Hyper-IgM syndrome (HIGM1)	Provisional localization only

Fig. 5. Regional localization on the X chromosome of immunodeficiency syndromes. Nonrecombinant loci are listed for each disease, followed by a pair of closely linked, flanking marker loci, which are polymorphic and can therefore be used for diagnosis by linkage. The flanking marker loci are indicated in the order Xpter-Xqter (Davies *et al.*, 1991).

in the dystrophin gene (DMD), just distal to CYBB and OTC, and just proximal to CYBB, are also available for family linkage studies.

The structural gene for properdin has been localized (Goundis *et al.*, 1989) to Xp11.23–p21.1, proximal to CYBB, by *in situ* hybridization and Southern blot studies using a panel of cell lines with well-defined X-chromosomal fragments. This position coincides with linkage studies, which have placed the disease locus between the RFLPs DXS7 in Xp11.3 and OTC in Xp21.1 (Goonewardena *et al.*, 1988; Davies *et al.*, 1991).

Immunodeficiencies Not Yet Cloned

Until the actual genes are identified for the other immunodeficiencies, inheritance of alleles at closely linked flanking marker loci can be used to follow the inheritance of a chromosomal subregion, presumed to include the disease locus, contained between the markers. In the uncommon event of a crossover between close flanking markers, however, no diagnosis by linkage can be made. The closest RFLP markers known to flank each uncloned disease locus are listed in Fig. 5; in cases with maternal homozygosity at these loci, alternate polymorphisms in the same region can be substituted.

Goals of Subregional Localization

Developing Closer and More Informative Markers

Rapid progress is occurring in efforts to saturate the X chromosome with markers very close to each disease locus as part of the Human Genome Initiative. Increasingly, the new marker loci are highly informative microsatellite markers, in which the polymerase chain reaction (PCR) is used to assay alleles containing different numbers of very short tandemly repeated units (Luty *et al.*, 1990; Edwards *et al.*, 1991; Davies *et al.*, 1991).

Spanning Chromosomal Regions with Cloned DNA Segments

Within the last 5 years the development of yeast artificial chromosomes (YACs) as cloning vectors for very large segments of DNA, from 200 to 1000 kilobases (kb), has revolutionized the process of genetic mapping. Chromosomal regions spanning several million bases of DNA have been cloned in overlapping YACs, including the Xp22.3 region containing the steroid sulfatase locus (Carrozzo *et al.*, 1992), and much of the Xq24–q28 region (Schlessinger *et al.*, 1991). The region Xq24–q25, within the latter region, encompasses the genetic locus for X-linked lymphoproliferative syndrome (XLP). Given the intensity of X-chromosome mapping efforts, it is likely that YAC clones will soon be available to span each of the immunodeficiency loci. This technology will make it possible to look for candidate genes in a precisely targeted section of the X chromosome.

Finding and Characterizing Candidate Genes

Aside from the genes for CYBB and properdin, the X-linked genes whose derangement causes immunodeficiency remain obscure for at least four reasons.

First, the diseases are so rare and, until recently, lethal early in life that it has been difficult to find enough informative meioses to perform linkage analysis. Second, sufficient numbers of highly informative markers have not been available. Third, XSCID and WAS reside in pericentromeric locations, known for decreased meiotic recombination frequency. This circumstance may be responsible for the paucity of crossover events, not only in SCID and WAS pedigrees to help localize the disease loci, but also in reference pedigrees used to order marker loci (Mahtani *et al.*, 1991). Finally, chromosomal alterations, such as translocations or small deletions, have not been found for X-linked immunodeficiencies, with the exception of a recent report of an interstitial deletion associated with XLP (Sanger *et al.*, 1990). Such chromosomal breakpoints have been successfully exploited in the identification of numerous other X-chromosomal and autosomal disease genes.

Progress on the Subregional Localization of SCIDX1

X-linked SCID was first mapped to proximal Xq in a European patient population by De Saint Basile *et al.* (1987). A further linkage study using six kindreds from the United States, A–F, confirmed this localization and placed SCIDX1 distal to DXS159 and proximal to DXS72 (Puck *et al.*, 1989). In that study random X-chromosome inactivation of T cells was used to assign noncarrier status to females without affected offspring. Females with apparently skewed X inactivation were not included. Since that time, additional polymorphic markers have become available, and the quantitation of X-inactivation analysis has made possible the inclusion of all females with odds ratios clearly in the carrier or noncarrier range, as discussed above. Furthermore, additional kindreds reported elsewhere (Conley *et al.*, 1990a; Puck *et al.*, 1990b), and newly presented, have been analyzed.

As shown in Fig. 6, Families G–M had on presentation multiple males affected with SCID in more than one generation, clearly inherited in an X-linked manner. Unfortunately, the high lethality of SCID, very obvious in Fig. 6, severely reduced the number of informative meioses available for study. Indeed, without the carrier test for females based on T-cell X inactivation, little linkage information could be gleaned from these pedigrees. For example, in Family G, only two phase-known meioses are available without carrier test data: the affected boy in Generation III and his mother, an obligate carrier. The X-inactivation data determined for each woman tested are summarized as S for skewed and R for random inside each symbol, with ratios of paternal to maternal X chromosomes in T-cell hybrids shown below. With carrier status thus assigned, the number of informative meioses in Family G increased from two to eight. Overall, carrier

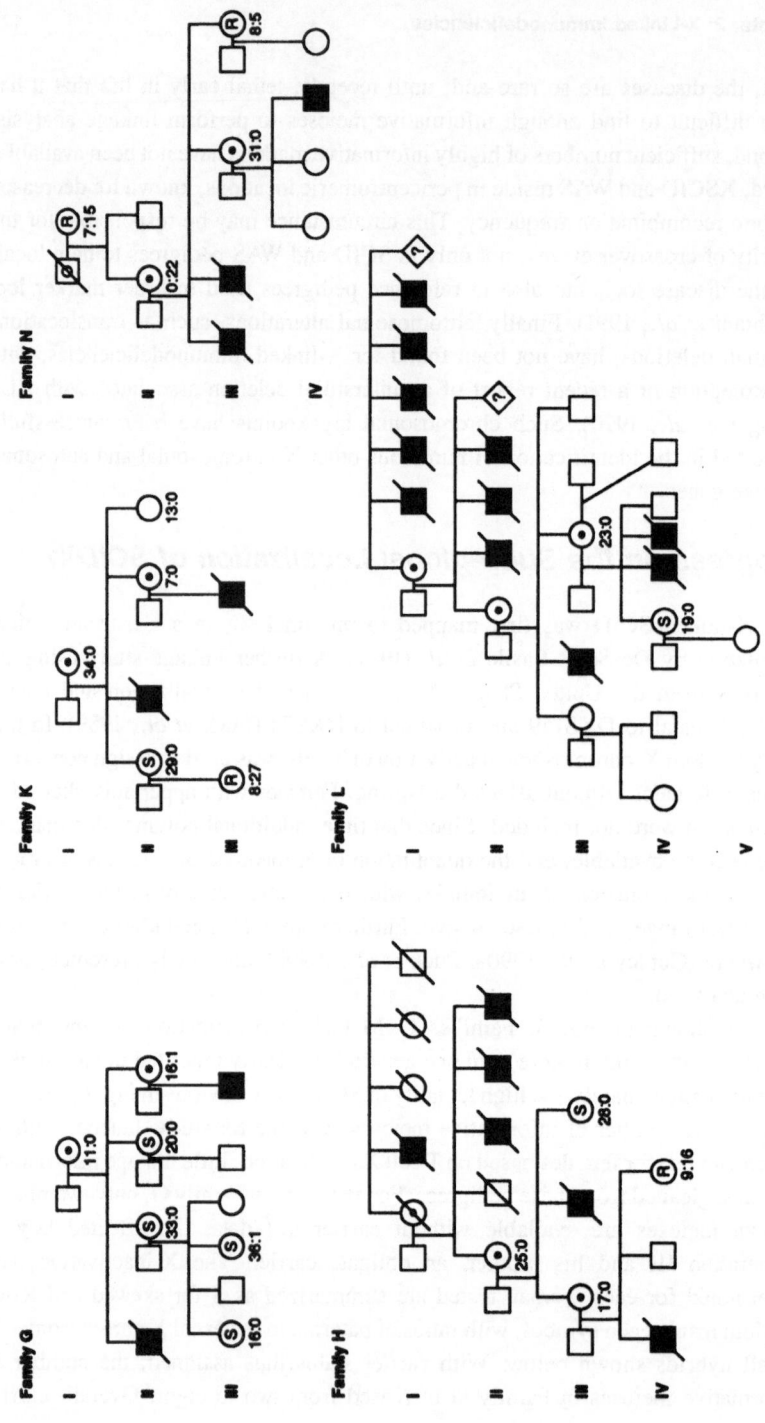

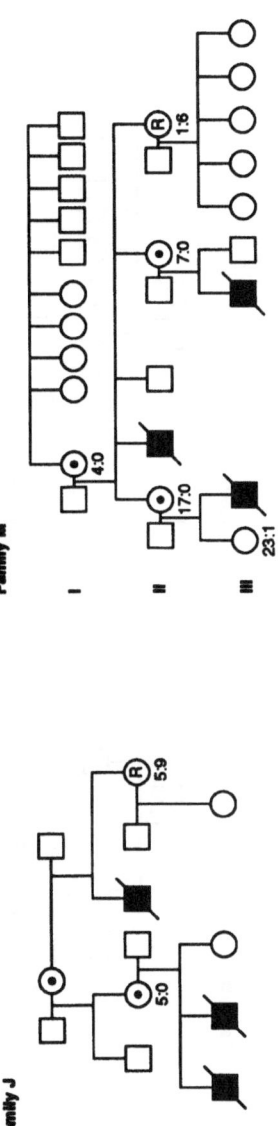

Fig. 6. Previously unpublished pedigrees of families with X-linked severe combined immunodeficiency. Dark squares indicate affected males, open squares indicate unaffected males, and circles with a black dot indicate obligate carrier females. Slash marks indicate deceased individuals. Letters R and S within circles designate females at risk for carrying SCID who have, respectively, random (noncarrier) or skewed (carrier) T-cell X inactivation by hybrid assay. The paternal to maternal ratios of hybrids are shown below the circles of females who were tested. In Family N and Family S, the squares containing open dots indicate the chromosome of origin of the SCID mutation, as determined by linkage studies. In Families G–M, the X-linked form of SCID was diagnosed by pedigree inspection, while in Families P–S, SCID was thought to be sporadic before the completion of T-cell X-inactivation studies, and in Family P and Family T, the birth of an affected son in Generation III post-genetic prediction. Families P, Q, and S were originally reported by Conley *et al.* (1990a).

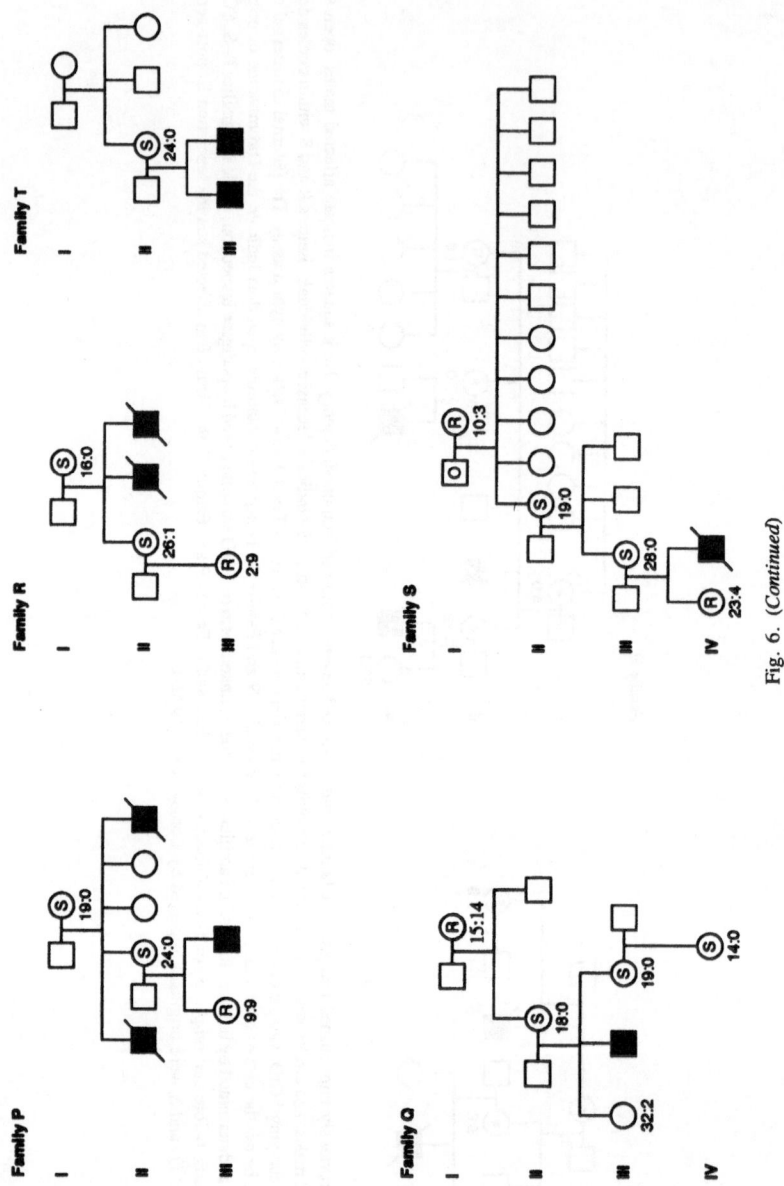

Fig. 6. (*Continued*)

testing completed to date has increased the number of informative meioses in Families G–M from 17 to 38.

Even more dependent on X-inactivation testing were Families P–S, which were ascertained through one or two males with SCID, but without documented X-linked inheritance (the affected male in Family P, Generation III, and one son in Family T, Generation III, were born after enrollment). Mothers of probands in Families P–S were among those with high odds ratios of being carriers in Fig. 4B. These families have contributed 15 informative meioses to date.

Linkage analysis has been performed with the original and new kindreds, A–S, using new RFLPs and carrier status assignment. Lod scores for two-point linkage analysis between SCIDX1 and pericentromeric and proximal Xq markers are shown in Table I, as calculated using LINKAGE v4.9 (Lathrop et al., 1985). The marker loci are listed in order based on genetic linkage data and translocation and deletion breakpoints (Mahtani et al., 1991; Lafreniere et al., 1991; Verga et al., 1991; and J. M. Puck, unpublished data). No recombination was observed between SCIDX1 and DXS325, DXS347, DXS441, PGK1, or DXS72. However, SCIDX1 is known to be proximal to DXS447 and DXS72 because of the existence of males with complex choroideremia who have interstitial deletions including these loci, but who do not have impaired immune function (Puck et al., 1989).

Multipoint ILINK analysis on this set of pedigrees gave the following most likely orders and distances in morgans (the distance between DXS447 and the immediately proximal locus group was set at 0.001 morgan): DXS1–0.07–DXS159–0.10–DXS106–0.02–DXS132–0.085–(DXS325, DXS347, DXS441, PGK1)–0.001–DXS447–0.048–DXS72–0.174–DXYS1X–0.051–DXS3. These values are consistent with previous reports. Multipoint analysis indicated that (1) SCIDX1 is over 3 logs more likely to be distal than proximal to DXS106 and DXS132; and (2) SCIDX1 is 10- to 24-fold more likely to be proximal than distal to DXS447 by linkage.

The abundance of very tightly linked markers will assist prenatal diagnosis and genetic management of SCIDX1 families. In order to increase the informativeness of markers very closely linked to SCIDX1, microsatellite polymorphisms assayed by PCR are being developed at these loci. For example, at DXS441, where a two-allele TaqI RFLP has shown no recombination with SCIDX1 with a lod score of 13.72 (Table I), a poly-TG dinucleotide repeat polymorphism has recently been detected (Ram et al., 1992). With eight different alleles consisting of from 14 to 22 repeated (TG) units this new polymorphism has a heterozygosity of 76%, and when combined with the original RFLP the overall informativeness of the locus is 82%.

The markers which have not recombined with SCIDX1 will also be used as

TABLE I. Lod Scores for Increasing Recombination Fraction (Θ) between SCIDX1 and Pericentromeric and Proximal Long-Arm Loci on the X Chromosome

Locus	Θ = 0	.001	.01	.05	.10	.15	.20	.25	.30	.35	.40	Θ̂[a]	Z[b]
DXS14	−∞	−0.61	3.50	5.06	**5.26**	5.05	4.65	4.12	3.49	2.76	1.94	**.092**	**5.27**
Centromere													
DXS1	−16.00	−4.78	−0.87	1.55	2.27	**2.46**	2.41	2.24	1.95	1.58	1.13	**.163**	**2.47**
DXS159[c]	−∞	3.16	7.95	10.48	**10.69**	10.19	9.33	8.24	6.94	5.45	3.77	**.083**	**10.73**
DXS153[d]	2.39	5.08	5.98	**6.24**	5.92	5.44	4.84	4.18	3.44	2.65	1.81	**.038**	**6.25**
DXS106	−∞	2.47	5.32	6.75	**6.79**	6.40	5.81	5.09	4.24	3.29	2.24	**.075**	**6.86**
DXS132	−∞	3.95	6.79	**8.19**	8.16	7.69	6.99	6.13	5.12	4.00	2.74	**.070**	**8.26**
DXS347[d]	**8.89**	8.87	8.75	8.19	7.45	6.67	5.85	4.99	4.08	3.10	2.08	**0**	**8.89**
DXS325[d]	**2.69**	2.69	2.65	2.49	2.28	2.06	1.83	1.58	1.31	1.02	0.71	**0**	**2.69**
DXS441[d]	**13.72**	13.70	13.50	12.63	11.49	10.29	9.04	7.71	6.31	4.83	3.26	**0**	**13.72**
PGK1	**8.12**	8.11	8.00	7.50	6.85	6.16	5.44	4.67	3.86	3.00	2.08	**0**	**8.12**
DXS447	4.18	6.87	7.73	**7.79**	7.27	6.59	5.82	4.98	4.07	3.11	2.09	**.027**	**7.89**
DXS72	**10.68**	10.66	10.51	9.83	8.94	8.01	7.03	6.00	4.90	3.74	2.52	**0**	**10.68**
DXYS1X	−3.58	4.02	6.85	**8.20**	8.13	7.64	6.94	6.09	5.11	4.02	2.82	**.067**	**8.26**
DSX3	−∞	−8.18	−1.33	2.92	4.18	**4.51**	4.43	4.10	3.57	2.88	2.05	**.162**	**4.52**

[a]Most likely recombination fraction.
[b]Z, Lod score at most likely recombination fraction. Lod scores at most likely recombination fraction shown in bold face.
[c]The relative order of DXS159, DXS153 has not been established.
[d]The relative order of DXS347, DXS325, DXS441 has not been established.

anchors to identify and order YACs in the Xq13 region. Transcribed sequences from these YACs will be evaluated as SCIDX1 gene candidates.

GENETIC MANAGEMENT

Determination of Recurrence Risks for X-Linked Immunodeficiencies

For families with pedigrees demonstrating multigenerational X-linked inheritance of all the diseases discussed above and shown in Fig. 5, straightforward linkage analysis can be used. Furthermore, autosomal phenocopies of agammaglobulinemia and Wiskott–Aldrich syndrome are thought to be extremely rare, with only isolated reports of similar clinical syndromes occurring in females. Therefore in families with only one male affected with XLA or WAS, X-linked inheritance can be assumed with reasonable probability. Much greater difficulty is encountered in families with a sporadic male case of the other immunodeficiencies because of the known occurrence of both autosomal and X-linked forms. The genetic management of SCID will be discussed in detail as a prototype for these immunodeficiencies.

Given the many autosomally inherited forms as well as the X-linked form of SCID, a combination of linkage analysis and X-inactivation determination provides the best genetic information, as illustrated in the flow diagram in Fig. 7. For each family with a possible case of SCID a complete pedigree should be obtained to ascertain all individuals affected and at risk for carrying the disease. All records should be reviewed in order to establish the diagnosis, recognizing that cell-surface phenotyping and currently accepted syndrome definitions were rarely used prior to 1980. In the United States, as many as 80% of SCID patients are male by some estimates, but fewer than 30% have a pedigree showing X-linked inheritance. In sporadic cases, specific testing for all the known autosomal forms of SCID should be sought, as discussed above. Females affected with SCID are likely to have autosomal forms, but karyotype analysis may be indicated to search for a chromosomal rearrangement or deletion involving Xq13.

In families with X-linked SCID by pedigree and living relatives available to furnish phase information, carrier testing and prenatal diagnosis can be accomplished by linkage analysis. There is no evidence for genetic heterogeneity in X-linked SCID. However, sporadic SCID in males may be X-linked or autosomal. In families of these cases, maternal T-cell X-inactivation studies can clarify the etiology. If only one X chromosome is found to function as the active X in a significant sample of maternal T cells, the X-linked form of SCID can be

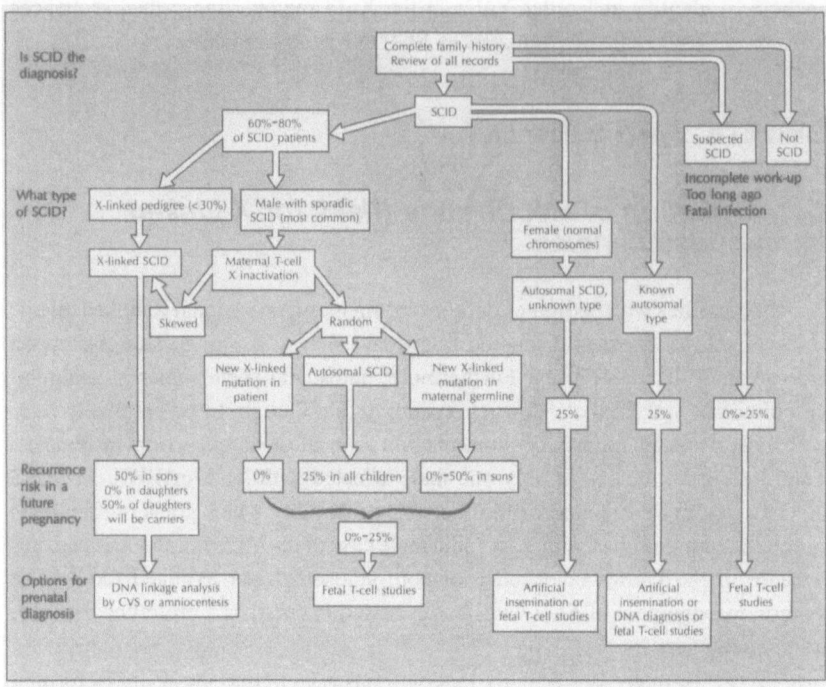

Fig. 7. Algorithm for evaluation of recurrence risks for severe combined immunodeficiency.

diagnosed; moreover, if the hybrid method is used, the non-SCID-bearing active X chromosome is isolated in the hybrid cell lines and can be used to assign marker alleles to the SCID and non-SCID maternal X chromosomes (phase determination). In a subsequent pregnancy, chorionic villus sampling at 9 weeks or amniocentesis at 14–16 weeks can be performed with cytogenetic ascertainment of the sex of the fetus. For a male fetus, who would be at 50% risk of being affected with SCID, linkage analysis can be carried out using DNA from the fetal cells and from the mother to show whether the fetus has inherited the SCIDX1 region in Xq13 from his mother's SCID-bearing X chromosome.

X-Inactivation studies in 22 mothers of sporadic males with SCID are shown in Fig. 4B. Ten, or 45%, of these women had skewed T-cell X inactivation and were therefore diagnosed as carriers of X-linked SCID, as shown in the lower left of the diagram in Fig. 7. The remainder had random T-cell X inactivation. As shown in the center of Fig. 7, the group with random X inactivation could include: (1) mothers of boys with new mutations at the X-chromosome SCID locus, for whom the recurrence risk for SCID would be 0%; (2) carriers of autosomal SCID, for whom there is a 25% recurrence risk for each subsequent pregnancy; and (3) women who have a new X-linked SCID mutation in their germ line. This latter group would be mosaics for X-linked SCID, with a SCID mutation on the X chromosome in some or all of their germ-line cells, but not in their T-cell progenitors. The recurrence risk for such women cannot be determined, and could vary from 0 to 50% for male pregnancies. Although not yet reported in SCID, such germ-line mosaics, first detected in hemophilia and muscular dystrophy, have been found using RFLP linkage studies in XLA (Hendriks et al., 1989) and WAS (Arveiler et al., 1990).

New mutations are expected to account for a substantial proportion of X-linked lethal conditions and may be the etiology of SCID in many of the sporadic cases seen in males. In the linkage studies of X-linked SCID performed in our laboratory the origin of five new SCID mutations has been determined in four pedigrees with sporadic cases, including Families Q and S in Fig. 6, and one X-linked pedigree, Family N. Four of these have been previously reported (Conley et al., 1990a; Puck et al., 1990b). Four of the five new mutations occurred in male germ lines, as illustrated in Families N and S, while one occurred in the grandmother in Generation I of Family Q. This woman had random T-cell X inactivation, but donated to her daughter an X chromosome which was never observed as the active X in 18 hybrids from her daughter's T cells and was subsequently inherited by the SCID-affected male in Generation III.

When the precise genetic etiology of SCID cannot be determined, sex determination is not helpful in assessing fetal risk. Periumbilical sampling of fetal blood at 18–20 weeks' gestation can be performed to obtain lymphocytes for surface marker staining and mitogenesis assays, which are diagnostic for SCID (Durandy et al., 1986; Blakemore et al., 1987; Lau and Levinski, 1988). After 15 weeks of pregnancy, lymphocytes are present in the peripheral blood and functional in normal fetuses. In contrast, fetuses affected with SCID due to either X-linked or autosomal mutations show absence of lymphocytes staining with monoclonal antibodies to CD3 and CD11 (pan-T-cell surface determinants) and lack of proliferation to phytohemagglutinin and other mitogens. These studies have been performed on small quantities of freshly obtained fetal blood by experienced research laboratories, but are not widely available.

Implementation of Diagnostic and Therapeutic Options

Reproductive Behavior of Families with Severe Genetic Diseases

Data on how genetic information is used are not available specifically for inherited immunodeficiencies. Genetic counseling services, guided by the formulation by Childs (1978), have recognized that the family's perceptions of both the risk of recurrence in a future child and the burden of the disorder to both child and family are critical to their decision making. Preliminary studies from several genetic diagnoses, combined and reviewed by Frets and Niermeijer (1990), have shown that when the recurrence risk of serious genetic diseases is at least 10%, a large majority of families desire and actually undergo carrier testing and prenatal diagnosis when these tests are made available. While abortion is chosen by some families upon diagnosis of an affected fetus, other families, while not contemplating termination of pregnancy, still find prenatal diagnosis beneficial. A normal result alleviates tremendous tension and worry, while knowledge that the fetus is affected gives the family and their physicians a chance to prepare for optimal neonatal management.

Improving Survival after Bone Marrow Transplantation for SCID

SCID was uniformly fatal in the first 1–2 years of life until the late 1960s, when bone marrow transplantation (BMT) became clinically feasible. In the review by Fischer *et al.* (1990) of 183 BMTs for SCID performed in European transplant centers between 1968 and 1989, overall survival rates were 76% and 52% for HLA identical and haploidentical BMTs, respectively. In the most recent years of this study period, various manipulations, including pretreatment of the patient with chemotherapy and depletion of mature T cells from the bone marrow graft, have produced survival rates above 95% for HLA-identical BMT, but only slight improvement for haploidentical BMT. Only about one quarter of patients have an HLA-identical sibling who can be a bone marrow donor.

Early diagnosis leading to transplantation before the onset of infectious complications of SCID is associated with faster and more successful immunoreconstitution. Thus the affected children born after the first one diagnosed in a family can be expected to have a better chance of being discovered in the neonatal period, and therefore a better likelihood of doing well. Experience with BMT for SCID in the United States has been similar to that in Europe, with an overall survival rate estimated at 75% (Conley, 1992). Unfortunately, the experience of Conley's review, echoing previous reviews (O'Reilly *et al.*, 1989; Hong, 1990; Lenarsky and Parkman, 1990; Parkman, 1991), was that none of eight patients with X-linked SCID tested greater than 2 years posttransplant had engrafted B cells; the patients' own B cells generally do not recover full function, so that the majority of transplanted SCID patients appear to require lifelong replacement therapy with gammaglobulin. It is unknown at present what will be the life expectancy of SCID patients post-BMT or whether X-linked SCID patients as a group fare better or worse than patients with other forms of SCID. There are males with X-linked SCID post-successful-BMT who have reached adulthood and had children. Late complications, including Epstein-Barr virus-associated lymphoma, graft failure, and infections related to incomplete immune reconstitution, have been noted, but no data on the frequency of these are available.

Preliminary Experience with Prenatal Diagnosis in X-Linked SCID

Each individual's perception of the burden of X-linked SCID may be different. This perception may be influenced by the clinical course of an affected relative, and it may change with the development of improved treatment prospects. Prenatal diagnosis has been performed in nine pregnancies at risk for X-linked SCID, and the outcome has been monitored (Table II). Karyotype analysis showed XX females in five, and the prediction of unaffected females was confirmed by immunological studies postnatally. Three of these female infants were found to be noncarriers of X-linked SCID by T-cell X inactivation, one was found to be a carrier, and one has not been tested by X inactivation and has a crossover between closely linked flanking DNA markers. Of four pregnancies in which karyotype analysis showed XY males, two inherited tightly linked flanking marker alleles from their mothers' non-SCID-bearing X chromosomes and were therefore pre-natally diagnosed as unaffected, with both diagnoses confirmed postnatally [the first reported by Puck *et al.* (1990b)]. In the remaining two pregnancies, linkage analysis showed at least a 99% likelihood that the male fetus had inherited the maternal SCID mutation. In both cases the families elected to continue pregnancy and plan for bone marrow transplantation in the neonatal period. The first male infant (Generation III, Family P), confirmed to have SCID at birth, received an HLA-identical BMT from his older sister. The second (Generation III, Family T), also confirmed postnatally to have SCID, was transplanted with cryopreserved maternal T-cell-depleted bone marrow. This marrow was left over from the BMT

TABLE II.   Outcome of Pregnancies at Risk for X-Linked SCID

Number monitored	Prenatal	Postnatal
5	XX Females	Unaffected girls
2	XY Males; flanking markers from mother's non-SCID-bearing X	Unaffected boys
2	XY Males; flanking markers from mother's SCID-bearing X	Affected boys

of his affected older brother, who had been diagnosed with SCID and successfully treated 2 years earlier (Himelstein *et al.*, 1992).

General Guidelines

Because diagnosis by linkage analysis for any genetic immunodeficiency depends on having DNA from family members to show which marker alleles travel with the defect, it is important to arrange for banking of frozen tissue or pretransplant cell lines. Such samples should be obtained from any patient with suspected genetic immunodeficiency and their normal siblings for the family's potential future needs. Even in cases which do not appear to be X-linked, the etiology may become clearer in the future, either through birth of further affected relatives, or through breakthroughs in research with the identification of specific gene defects for both X-linked and autosomal disease loci.

Finally, families of patients with genetic disorders often derive tremendous benefit from interacting with each other. The Immune Deficiency Foundation is an organization started by families with immunodeficiencies, which now has a growing national network of local chapters (Immune Deficiency Foundation, 1991).

REFERENCES

Anderson, W. F., 1990, September 14, 1990: The beginning, *Gene Ther.* **1**:371–372.

Arveiler, B., De Saint Basile, G., Fischer, A., Griscelli, C., and Mandel, J.-L., 1990, Germ-line mosaicism simulates genetic heterogeneity in Wiskott–Aldrich syndrome, *Am. J. Hum. Genet.* **46**:906–911.

Baehner, R. L., 1990, Chronic granulomatous disease of childhood: Clinical, pathological, biochemical, molecular, and genetic aspects of the disease, *Pediatr. Pathol.* **10**:143–154.

Blakemore, K., Scioscia, A., Grannum, P., Podell, D., Hobbins, J., and Mahoney, M., 1987, The value of fetal blood sampling in the prenatal diagnosis of severe combined immunodeficiency syndrome, *Am. J. Hum. Genet.* **41**(Suppl):A267.

Boyd, Y., and Fraser, N. J., 1990, Methylation patterns at the hypervariable X-chromosome locus DXS255 (M27b): Correlation with X-inactivation status, *Genomics* **7**:182–187.

Brahmi, Z., Lazarus, K. H., Hodes, M. E., and Baehner, R. L., 1983, Immunologic studies of three family members with the immunodeficiency with hyper-IgM syndrome, *J. Clin. Immunol.* **3**:127–134.

Brooks, E. G., Schmalsteig, F. C., Wirt, D. P., Rosenblatt, H. M., Adkins, L. T., Lookingbill, D. P., Rudloff, H. E., Rakusan, T. A., and Goldman, A. S., 1990, A novel X-linked combined immunodeficiency disease, *J. Clin. Invest.* **86**:1623–1631.

Bruton, O. C., 1952, Agammaglobulinemia, *Pediatrics* **9**:722–728.

Carrozzo, R., Ellison, J., Yen, P., Taillon-Miller, P., Brownstein, B. H., Persico, G., Ballabio, A., and Shapiro, L., 1992, Isolation and characterization of a yeast artificial chromosome (YAC) contig around the human steroid sulfatase gene, *Genomics* **12**:7–12.

Childs, B., 1978, A critical review of the published literature, in: *Genetic Issues in Public Health and*

Medicine (B. H. Cohen, A. Lilienfeld, and P. C. Huang, eds.), C. C. Thomas, Springfield, Illinois, pp. 339–357.

Cohen, D. I., Hedrick, S. M., Nielsen, E. A., D'Eustachio, P., Ruddle, F., Steinberg, A. D., Paul, W. E., and Davis, M. M., 1985, Isolation of a cDNA clone corresponding to an X-linked gene family (XLR) closely linked to the murine immunodeficiency disorder *xid*, *Nature* **314:**369–372.

Conley, M. E., 1992, X-linked severe combined immunodeficiency, *Clin. Immunol. Immunopathol.* **61**(Suppl.):S94–S99.

Conley, M. E., and Puck, J. M., 1988, Carrier detection in typical and atypical X-linked agammaglobulinemia, *J. Pediatr.* **112:**688–694.

Conley, M. E., Nowell, P. C., Henle, G., and Douglas, S. D., 1984, XX T cells and XY B cells in two patients with severe combined immunodeficiency, *Clin. Immunol. Immunopathol.* **31:**87–95.

Conley, M. E., Brown, P., Pickard, A. R., Buckley, R. H., Miller, D. S., Raskind, W. H., Singer, J. W., and Fialkow, P. J., 1986, Expression of the gene defect in X-linked agammaglobulinemia, *N. Engl. J. Med.* **315:**564–567.

Conley, M. E., Lavoie, A., Briggs, C., Brown, P., Guerra, C., and Puck, J. M., 1988, Nonrandom X chromosome inactivation in B cells from carriers of X chromosome-linked severe combined immunodeficiency, *Proc. Natl. Acad. Sci. USA* **85:**3090–3094.

Conley, M. E., Buckley, R. H., Hong, R., Guerra-Hanson, C., Roifman, C. M., Brockstein, J. A., Pahwa, S., and Puck, J. M., 1990a, X-linked severe combined immunodeficiency: Diagnosis in males with sporadic severe combined immunodeficiency and clarification of clinical findings, *J. Clin. Invest.* **85:**548–1554.

Conley, M. E., Sullivan, J. L., Neidich, J. A., and Puck, J. M., 1990b, X chromosome inactivation patterns in obligate carriers of X-linked lymphoproliferative syndrome, *Clin. Immunol. Immunopathol.* **55:**486–491.

Conley, M. E., Burks, A. W., Herrod, H. G., and Puck, J. M., 1991, Molecular analysis of X-linked agammaglobulinemia with growth hormone deficiency, *J. Pediatr.* **119:**392–397.

Davies, K. E., Mandel, J.-L., Monaco, A. P., Nussbaum, R. L., and Willard, H. F., 1990, Report of the committee on the genetic constitution of the X chromosome, HGM10.5, *Cytogenet. Cell. Genet.* **55:**254–313.

Davies, K. E., Mandel, J.-L., Monaco, A. P., Nussbaum, R. L., and Willard, H. F., 1991, Report of the committee on the genetic constitution of the X chromosome, HGM11, *Cytogenet. Cell. Genet.* **58:**853–966.

De Preval, C., Hadam, M. R., and Mach, B., 1988, Regulation of genes for HLA class II antigens in cell lines from patients with severe combined immunodeficiency, *N. Engl. J. Med.* **318:**1295–1300.

De Sainte Basile, G., Arveiler, B., Oberle, I., Malcolm, S., Levinsky, R. J., Lau, Y. L., Hofker, M., Debre, M., Fischer, A., Griscelli, C., and Mandel, J.-L., 1987, Close linkage of the locus for X chromosome-linked severe combined immunodeficiency to polymorphic DNA markers in Xq11–13, *Proc. Natl. Acad. Sci. USA* **84:**7576–7579.

Durandy, A., Dumez, Y., and Griscelli, C., 1986, Prenatal diagnosis of severe inherited immunodeficiencies: A five year experience, in: *Progress in Immunodeficiency Research and Therapy*, Volume 2 (J. Vossen and C. Griscelli, eds.), Elsevier, Amsterdam, pp. 323–327.

Edwards, A., Civitello, A., Hammond, H., and Caskey, C. T., 1991, DNA typing and genetic mapping with trimeric and tetrameric tandem repeats, *Am. J. Hum. Genet.* **49:**746–756.

Ezekowitz, A. B., Sieff, C. A., Dinauer, M. C., Nathan, D. G., Orkin, S. H., and Newburger, P. E., 1990, Restoration of phagocyte function by interferon-gamma in X-linked chronic granulomatous disease occurs at the level of a progenitor cell, *Blood* **76:**2443–2448.

Fearon, E. R., Winkelstein, J. A., Civin, C. I., Pardoll, D. M., and Vogelstein, B., 1987, Carrier detection in X-linked agammaglobulinemia by analysis of X-chromosome inactivation, *N. Engl. J. Med.* **316:**427–431.

Fearon, E. R., Ojhn, D. B., Winkelstein, J. A., Vogelstein, B., and Blaese, R. M., 1988, Carrier detection in the Wiskott–Aldrich syndrome, *Blood* **72:**1735–1739.

Fialkow, P. J., 1973, Primordial cell pool size and lineage relationships of five human cell types, *Ann. Hum. Genet.* **37**:39–48.

Fibison, W. J., Budarf, M., McDermid, H., Greenberg, F., and Emanual, B. S., 1990, Molecular studies of DiGeorge syndrome, *Am. J. Hum. Genet.* **46**:888–895.

Fischer, A., Landais, P., Friedrich, W., Morgan, G., Gerritsen, B., Fasth, A., Porta, F., Griscelli, C., Goldman, S. F., Levinsky, R., and Vossen, J., 1990, European experience of bone-marrow transplantation for severe combined immunodeficiency, *Lancet* **336**:850–854.

Fleisher, T. A., White, R. M., Broder, S., Nissley, S. P., Blaese, R. M., Mulvihill, J. J., Olive, G., and Waldmann, T. A., 1980, X-Linked hypogammaglobulinemia and isolated growth hormone deficiency, *N. Engl. J. Med.* **302**:1429–1433.

Francke, U., Ochs, H. D., Darras, B. T., and Swaroop, A., 1990, Origin of mutations in two families with X-linked chronic granulomatous disease, *Blood* **76**:602–606.

Frets, P. G., and Niermeijer, M. F., 1990, Reproductive planning after genetic counseling: A perspective from the last decade, *Clin. Genet.* **38**:295–306.

Garchon, H.-J., and Davis, M. M., 1989, The XLR gene product defines a novel set of proteins stabilized in the nucleus by zinc ions, *J. Cell Biol.* **108**:779–787.

Gealy, W. J., Dwyer, J. M., and Harley, J. B., 1980, Allelic exclusion of glucose-6-phosphate dehydrogenase in platelets and T-lymphocytes from a Wiskott–Aldrich syndrome carrier, *Lancet* **1**:63–65.

Geha, R., Hyslop, N., Alami, S., Farah, F., Schneeberger, E. E., and Rosen, F., 1979, Hyper immunoglobulin M immunodeficiency (dysgammaglobulinemia), *J. Clin. Invest.* **64**:385–391.

Gleich, G. J., Condemi, J. J., and Vaughn, J. H., 1965, Dysgammaglobulinemia in the presence of plasma cells, *N. Engl. J. Med.* **272**:331–340.

Goodship, J., Malcolm, S., Lau, Y. L., Pembrey, M. E., and Levinsky, R. J., 1988, Use of X-chromosome inactivation analysis to establish carrier status for X-linked severe combined immunodeficiency, *Lancet* **1**:729–732.

Goonewardena, P., Sjoholm, A., Nilsson, L.-A., and Patterson, U., 1988, Linkage analysis of the properdin deficiency gene: Suggestion of a locus in the proximal part of the short arm of the X chromosome, *Genomics* **2**:115–118.

Goundis, D., Holt, S. M., Boyd, Y., and Reid, K. B., 1989, Localization of the properdin structural locus to Xp11.23–Xp21.1, *Genomics* **5**:56–60.

Greer, W. L., Higgins, E., Sutherland, D. R., Novogrodsky, A., Brockhousen, I., Peacocke, M., Rubin, L. A., Baker, M., Dennis, J. W., and Siminovitch, K. A., 1989a, Altered expression of leucocyte sialoglycoprotein in Wiskott–Aldrich syndrome is associated with a specific deficit in *o*-glycosylation, *Biochem. Cell. Biol.* **67**:503–509.

Greer, W. L., Kwong, P. C., Peacocke, M., Ip, P., Rubin, L. A., and Siminovitch, K. A., 1989b, X-Chromosome inactivation in the Wiskott–Aldrich syndrome: A marker for detection of the carrier state and identification of cell lineages expressing the gene defect, *Genomics* **4**:60–67.

Grierson, H., and Purtilo, D. T., 1987, Epstein–Barr virus infections in males with the X-linked lymphoproliferative syndrome, *Ann. Intern. Med.* **106**:538–545.

Hendriks, R. W., Mensink, E. J. B., Kraakman, M. E. M., Thompson, A., and Schuurman, R. K. B., 1989, Evidence for male X chromosomal mosaicism in X-linked agammaglobulinemia, *Hum. Genet.* **83**:267–270.

Hendriks, R. W., Kraakman, M. E. M., Craig, I. W., Espanol, T., and Schuurman, R. K. B., 1990, Evidence that in X-linked immunodeficiency with hyperimmunoglobulinemia M the intrinsic immunoglobulin heavy chain class switch mechanism is intact, *Eur. J. Immunol.* **20**:2603–2608.

Himelstein, B. P., Puck, J. M., August, C., Pierson, G., and Bunin, N., 1992, T-cell depleted maternal bone marrow transplant for siblings with X-linked severe combined immunodeficiency. *J. Pediatr.* in press.

Hirschhorn, R., 1986, Inherited enzyme deficiencies and immunodeficiency: Adenosine deaminase

(ADA) and purine nucleoside phosphorylase (PNP) deficiencies, *Clin. Immunol. Immunopathol.* **40:**157–165.

Hirschhorn, R., Tzall, S., Ellenbogen, A., and Eng, F., 1990, "Hot-spot" mutations in adenosine deaminase (ADA) deficiency, *Proc. Natl. Acad. Sci. USA* **87:**6171–6175.

Hong, R., 1990, Update on the immunodeficiency diseases, *Am. J. Dis. Child.* **144:**983–991.

Immune Deficiency Foundation, 1991, Immune deficiency diseases: An overview [Available from the Immune Deficiency Foundation, P.O. Box 586, Columbia, Maryland 21045].

[The] International Chronic Granulomatous Disease Cooperative Study Group., 1991, A controlled trial of interferon gamma to prevent infection in chronic granulomatous disease, *N. Engl. J. Med.* **324:**509–516.

Jezyk, P. F., Felsburg, P. J., Haskins, M. E., and Patterson, D. F., 1989, X-Linked severe combined immunodeficiency in the dog, *Clin. Immunol. Immunopathol.* **52:**173–189.

Kara, C. J., and Glimcher, L. H., 1991, *In vivo* footprinting of MHC class II genes: Bare promoters in the bare lymphocyte syndrome, *Science* **252:**709–712.

Lafreniere, R. G., Brown, C. J., Powers, V. E., Carrel, L., Davies, K. E., Barker, D. F., and Willard, H. F., 1991, Physical mapping of 60 DNA markers in the p21.1 to q21.3 region of the human X chromosome, *Genomics* **11:**352–363.

Lathrop, G. M., Lalouel, J. M., Julier, C., and Ott, J., 1985, Multilocus linkage analysis in humans: Detection of linkage and estimation of recombination, *Am. J. Hum. Genet.* **37:**482–498.

Lau, Y. L., and Levinski, R. J., 1988, Prenatal diagnosis and carrier detection in primary immunodeficiency disorders, *Arch. Dis. Child.* **63:**758–764.

Lederman, H. M., and Winkelstein, J. A., 1985, X-Linked agammaglobulinemia: An analysis of 96 patients, *Medicine* **64:**145–156.

Lenarsky, C., and Parkman, R., 1990, Bone marrow transplantation for the treatment of immune deficiency states, *Bone Marrow Transplant.* **6:**361–369.

Levitt, D., Haber, P., Rich, K., and Cooper, M., 1983, Hyper IgM immunodeficiency, *J. Clin. Invest.* **72:**1650–1657.

Luty, J. A., Guo, Z., Willard, H. F., Ledbetter, D. H., Ledbetter, S., and Litt, M., 1990, Five polymorphic microsatellite VNTRs on the human X chromosome, *Am. J. Hum. Genet.* **46:**776–783.

Lyon, M. F., 1966, X-Chromosome inactivation in mammals, *Adv. Teratol.* **1:**25–54.

Lyon, M. F., Peters, J., Glenister, P. H., Ball, S., and Wright, E., 1990, The scurfy mouse mutant has previously unrecognized hematological abnormalities and resembles Wiskott–Aldrich syndrome, *Proc. Natl. Acad. Sci. USA* **87:**2433–2437.

Mahtani, M. M., Lafreniere, R. G., Kruse, T., and Willard, H. F., 1991, An 18-locus linkage map of the pericentromeric region of the human X chromosome: Genetic framework for mapping X-linked disorders, *Genomics* **10:**849–857.

Mandel, J.-L., Willard, H. F., Nussbaum, R. L., Romeo, G., Puck, J. M., and Davies, K. E., 1989, Report of the committee on the genetic constitution of the X chromosome, Human Gene Mapping 10, *Cytogenet. Cell Genet.* **51:**384–437.

Mayer, L., Kwan, S. P., Thompson, C., Ko, H. S., Chiorazzi, N., Waldmann, T., and Rosen, F., 1986, Evidence for a defect in "switch" T cells in patients with immunodeficiency and hyperimmunoglobulinemia M, *N. Engl. J. Med.* **314:**409–413.

Mensink, E. J. B. M., Thompson, A., Sandkuyl, L. A., Kraakman, M. E. M., Schot, J. D. L., Espanol, T., and Schuurman, R. K. B., 1987, X-Linked immunodeficiency with hyperimmunoglobulinemia M appears to be linked to the DXS42 restriction fragment length polymorphism locus, *Hum. Genet.* **76:**96–99.

Moreno, L. A., Gottrand, F., Turck, D., Manouvrier-Hanu, S., Mazingue, F., Morisot, C., Le Deist, F., Ricour, C., Nihoul-Fekete, C., Debeugny, P., Griscelli, C., and Farriaux, J.-P., 1990, Severe combined immunodeficiency syndrome associated with autosomal recessive familial multiple gastrointestinal atresias: Study of a family, *Am. J. Med. Genet.* **37:**143–146.

Mullins, L. J., Stephenson, D. A., Grant, S. G., and Chapman, V. M., 1990, Efficient linkage of 10 loci in the proximal region of the mouse X chromosome, *Genomics* 7:19–30.

Nahm, M. H., Paslay, J. W., and Davie, J. M., 1983, Unbalanced X chromosome mosaicism in B cells of mice with X-linked immunodeficiency, *J. Exp. Med.* 158:920–931.

Najiar, J. L., Phillips, J. A., Manness, K. J., Teague, D., Summar, M. L., and Lorenz, R. A., 1990, Some cases of nonclassical growth hormone deficiency may be due to an X-linked disorder [Abstract], Presented at the Seventy-second Annual Meeting of the Endocrine Society, Atlanta, Georgia, June 20–23, 1990.

Online Mendelian Inheritance in Man, OMIM (TM) [database online], 1988, Immunodeficiency, X-linked progressive combined variable [Lymphoproliferative disease, X-linked, LYP], MIM number 312060, The Johns Hopkins University, Baltimore, Maryland [last edited August 2, 1988].

Online Mendelian Inheritance in Man, OMIM (TM) [database online], 1991a, Agammaglobulinemia [Bruton type, AGMX1], MIM number 300300, The Johns Hopkins University, Baltimore, Maryland [last edited October 23, 1991].

Online Mendelian Inheritance in Man, OMIM (TM) [database online], 1991b, Agammaglobulinemia, Swiss type [X-linked severe combined immunodeficiency, SCIDX1], MIM number 300400, The Johns Hopkins University, Baltimore, Maryland [last edited December 12, 1991].

Online Mendelian Inheritance in Man, OMIM (TM) [database online], 1991c, Properdin deficiency, X-linked, MIM number 312060, The Johns Hopkins University, Baltimore, Maryland [last edited April 23, 1991].

Online Mendelian Inheritance in Man, OMIM (TM) [database online], 1991d, Severe combined immunodeficiency disease, X-linked, 2, MIM number 312863, The Johns Hopkins University, Baltimore, Maryland [last edited June 5, 1991].

Online Mendelian Inheritance in Man, OMIM (TM) [database online], 1991e, Hypogamma-globulinemia and isolated growth hormone deficiency, X-linked, MIM number 307200, The Johns Hopkins University, Baltimore, Maryland [last edited November 4, 1991].

Online Mendelian Inheritance in Man, OMIM (TM) [database online], 1992a, Aldrich syndrome [Wiskott–Aldrich syndrome], MIM number 301000, The Johns Hopkins University, Baltimore, Maryland [last edited January 21, 1992].

Online Mendelian Inheritance in Man, OMIM (TM) [database online], 1992b, Immunodeficiency with increased IgM [hyper-IgM syndrome, HIGM1], MIM number 308230, The Johns Hopkins University, Baltimore, Maryland [last edited January 24, 1992].

Online Mendelian Inheritance in Man, OMIM (TM) [database online], 1992c, Chronic granulomatous disease [CYBB], MIM number 306400, The Johns Hopkins University, Baltimore, Maryland [last edited February 11, 1992].

O'Reilly, R. J., Keever, C. A., Small, T. N., and Brockstein, J., 1989, The use of HLA-non-identical T-cell-depleted marrow transplants for correction of severe combined immunodeficiency disease, *Immunodeficiency Rev.* 1:273–309.

Pallant, A., Dskenazi, A., Mattei, M.-G., Fournier, R. E. K., Carlsson, S. R., Fukuda, M., and Frelinger, J. G., 1988, Characterization of cDNAs encoding human leukosialin and localization of the leukosialin gene to chromosome 16, *Proc. Natl. Acad. Sci. USA* 86:1328–1332.

Parkman, R., 1991, The biology of bone marrow transplantation for severe combined immune deficiency, *Adv. Immunol.* 49:381–410.

Patterson, D. F., Haskins, M. E., and Jezyk, P. F., 1982, Models of human genetic disease in domestic animals, *Adv. Hum. Genet.* 12:263–339.

Pearl, E. R., Vogler, L. B., Okos, A. J., Crist, W. M., Lawton, A. R., and Cooper, M. D., 1978, B Lymphocyte precursors in human bone marrow: An analysis of normal individuals and patients with antibody-deficiency states, *J. Immunol.* 120:1169–1175.

Pelham, A., O'Reilly, M.-A., Malcolm, S., Levinsky, R. J., and Kinnon, C., 1990, RFLP and deletion analysis for X-linked chronic granulomatous disease using the cDNA probe: Potential for improved prenatal diagnosis and carrier determination, *Blood* **76**:820–824.

Phillips, J. A., and Vnencak-Jones, C. L., 1989, Genetics of growth hormone and its disorders, *Adv. Hum. Genet.* **18**:305–363.

Prchal, J. T., Carroll, A. J., Prchal, J. F., Crist, W. M., Skalka, H. W., Gealy, W. J., Harley, J., and Malluh, A., 1980, Wiskott–Aldrich syndrome: Cellular impairments and their implication for carrier detection, *Blood* **56**:1048–1054.

Puck, J. M., 1991, The role of molecular genetics in carrier detection and prenatal diagnosis of immunodeficiencies, *Curr. Opinion Pediatr.* **3**:855–862.

Puck, J. M., Nussbaum, R. L., and Conley, M. E., 1987, Carrier detection in X-linked severe combined immunodeficiency based on patterns of X chromosome inactivation, *J. Clin. Invest.* **79**:1395–1400.

Puck, J. M., Nussbaum, R. L., Smead, D. L., and Conley, M. E., 1989, X-Linked severe combined immunodeficiency: Localization within the region Xq13.1–q21.1 by linkage and deletion analysis, *Am. J. Hum. Genet.* **44**:724–730.

Puck, J. M., Siminovitch, K. A., Poncz, M., Greenberg, C. R., Rotem, M., and Conley, M. E., 1990a, Atypical presentation of Wiskott–Aldrich syndrome: Diagnosis in two unrelated males based on studies of maternal T cell X chromosome inactivation, *Blood* **75**:2369–2374.

Puck, J. M., Krauss, C. M., Puck, S. M., Buckley, R. H., and Conley, M. E., 1990b, Prenatal test for X-linked severe combined immunodeficiency by analysis of maternal X-chromosome inactivation and linkage analysis, *N. Engl. J. Med.* **322**:1063–1066.

Puck, J. M., Stewart, C. C., and Nussbaum, R. L., 1992, Maximum likelihood analysis of human T-cell X chromosome inactivation patterns: Normal women vs. carriers of X-linked severe combined immunodeficiency, *Am. J. Hum. Genet.* **50**:742–748.

Purtilo, D. T., and Grierson, H. L., 1991, Methods of detection of new families with X-linked lymphoproliferative disease, *Cancer Genet. Cytogenet.* **51**:143–153.

Purtilo, D. T., Cassel, C. K., Yang, J. P. S., Harper, R., Stephenson, S. R., Landing, B. H., and Vewter, G. F., 1975, X-Linked recessive progressive combined variable immunodeficiency (Duncan's disease), *Lancet* **1**:935–941.

Purtilo, D. T., Sakamoto, K., Barnabei, V., Seeley, J., Bechtold, T., Rogers, G., Yetz, J., Harada, S., et al., 1982, Epstein–Barr virus-induced diseases in boys with the X-linked lymphoproliferative syndrome (XLP): Update on studies of the registry, *Am. J. Med.* **73**:49–56.

Ram, K. T., Barker, D. L., and Puck, J. M., 1992, Dinucleotide repeat polymorphism at the locus DXS441, *Nucleic Acids Res.* **20**:1428.

Reisinger, D., and Parkman, R., 1987, Molecular heterogeneity of a lymphocyte glycoprotein in immunodeficient patients, *J. Clin. Invest.* **79**:595–599.

Reith, W., Satola, S., Herrero-Sanchez, C., Amaldi, T., Lisowska-Grospierre, B., Griscelli, C., Hadam, M. R., and Mach, B., 1988, Congenital immunodeficiency with a regulatory defect in MHC class II gene expression lacks a specific HLA-DR promotor binding protein, RF-X, *Cell* **53**:897–906.

Remold-O'Donnell, E., Zimmerman, C., Kenney, D., and Rosen, F. S., 1987, Expression on blood cells of sialophorin, the surface glycoprotein that is defective in Wiskott–Aldrich syndrome, *Blood* **70**:104–109.

Sanger, W. G., Grierson, H. L., Skare, J., Wyandt, H., Pirruccello, S., Fordyce, R., and Purtilo, D. T., 1990, Partial Xq25 deletion in a family with the X-linked lymphoproliferative disease (XLP), *Cancer Genet. Cytogenet.* **47**:163–169.

Schlessinger, D., Little, R. D., Freije, D., Abidi, F., Zucchi, I., Porta, G., Pilia, G., Nagaraja, R., Johnson, S. K., Yoon, J.-Y., Srivastava, A., Kere, J., Palmieri, G., Ciccodicola, A., Montanaro, V., Romano, G., Casamassimi, A., and D'Urso, M., 1991, Yeast artificial chromosome-based genome mapping: Some lessons from Xq24–q28, *Genomics* **11**:783–793.

Siegel, J. N., Turner, C. A., Klinman, D. M., Wilkinson, M., Steinberg, A. D., MacLeod, C. L., Paul, W. E., Davis, M. M., and Cohen, D. I., 1987, Sequence analysis and expression of an X-linked lymphocyte-regulated gene family (XLR), *J. Exp. Med.* **166:**1702–1715.

Sitz, K. V., Burks, A. W., Williams, L. W., Kemp, S. F., and Steele, R. W., 1990, Confirmation of X-linked hypogammaglobulinemia with isolated growth hormone deficiency as a disease entity, *J. Pediatr.* **116:**292–294.

Skare, J. C., Grierson, H. L., Sullivan, J. L., Nussbaum, R. L., Purtilo, D. T., Sylla, B. S., Lenoir, G. M., Reilly, D. S., White, B. N., and Milunsky, A., 1989, Linkage analysis of seven kindreds with the X-linked lymphoproliferative syndrome (XLP) confirms that the XLP locus is near DXS42 and DXS37, *Hum. Genet.* **82:**354–358.

Sullivan, J. L., Byron, K. S., Brewster, F. E., Baker, S. M., and Ochs, H. D., 1983, X-Linked lymphoproliferative syndrome: Natural history of the immunodeficiency, *J. Clin. Invest.* **71:**1765–1778.

Verga, V., Hall, B. K., Wang, S., Johnson, S., Higgins, J. V., and Glover, T. W., 1991, Localization of the translocation breakpoint in a female with Menkes syndrome to Xq13.2–q13.3 proximal to PGK-1, *Am. J. Hum. Genet.* **48:**1133–1138.

Weinberg, K., 1991, Severe combined immunodeficiency disease, *Curr. Opinion Pediatr.* **3:**847–854.

Weinberg, K., and Parkman, R., 1990, Severe combined immune deficiency due to a specific defect in interleukin-2 production, *N. Engl. J. Med.* **322:**1718–1723.

WHO Scientific Group on Immunodeficiency., 1983, Meeting report: Primary immunodeficiency diseases, *Clin. Immunol. Immunopathol.* **28:**450–475.

Wicker, L. S., and Scher, I., 1986, X-Linked immune deficiency (*xid*) of CBA/N mice, *Curr. Top. Microbiol. Immunol.* **124:**87–101.

Wilson, D. K., Rudolph, F. B., and Quiocho, F. A., 1991, Atomic structure of adenosine deaminase complexed with a transition-state analog: Understanding catalysis and immunodeficiency mutations, *Science* **252:**1278–1284.

Winkelstein, J. A., and Fearon, E., 1990, Carrier detection of the X-linked primary immunodeficiency diseases using X-chromosome inactivation analysis, *J. Allergy Clin. Immunol.* **85:**1090–1097.

Chapter 3

Genetic Mutations Affecting Human Lipoproteins, Their Receptors, and Their Enzymes

Vassilis I. Zannis, Dimitris Kardassis,
and Eleni Economou Zanni

Section of Molecular Genetics, Cardiovascular Institute
Departments of Medicine and Biochemistry
Housman Medical Research Center
Boston University Medical Center
Boston, Massachusetts 02118

INTRODUCTION

Lipoproteins are macromolecular complexes of lipids and proteins which originate mainly from the liver and intestine, and are involved in the transport and redistribution of lipids in the body. The plasma lipoproteins are spherical particles with cores of nonpolar neutral lipid consisting of cholesteryl ester and triglycerides and coats of relatively polar materials consisting of phospholipid, free cholesterol, and proteins (Table I) (1–8). The protein components of lipoproteins are called apolipoproteins and have been designated apoA-I, apoA-II, apoA-IV, apoB, apoCI, apoCII, apoCIII, apoD, and apoE (9). Tables IIA and IIB summarize our knowledge on the structure–function relations, sites of synthesis, genetics, and physiological functions of the apolipoprotein moieties (3,10–22). The plasma lipoproteins have traditionally been grouped in four major lipoprotein classes: chylomicrons, very low-density lipoproteins (VLDL), low-density lipoproteins (LDL), and high-density lipoproteins (HDL) (Table I). A lipoprotein class

Advances in Human Genetics, Volume 21, edited by Henry Harris and Kurt Hirschhorn. Plenum Press, New York, 1993.

TABLE I. Properties and Composition of Human Plasma Lipoproteins[a]

Properties and composition	Chylomicrons	VLDL	LDL	HDL
Size (Å)	750–12,000	300–700	180–300	50–120
Density (g/ml)	0.94	0.94–1.006	1.019–1.063	1.063–1.21
Triglycerides (% wt)	80–95	45–65	4–8	2–7
Phospholipids (% wt)	3–6	15–20	18–24	26–32
Free cholesterol (% wt)	1–3	4–8	6–8	3–5
Esterified cholesterol (% wt)	2–4	16–22	45–50	15–20
Flotation rate	400[b]	20–400[b]	0–12[b]	0–9[c]
Electrophoretic mobility	Origin (cathode)	Pre-β	β	α
Proteins (% wt)	1–2	6–10	18–22	45–55
Major apoproteins	A-I, A-IV, B, CI, CIII, E	B, E, CI, CII, CIII	B	A-I, A-II, E
Minor apoproteins	A-II, CII	A-I, A-II, A-IV	CI, CII, CIII, E	CI, CII, CIII, D, E, J

[a]Modified from Herbert *et al.* (5).
[b]Corrected flotation rate at a density of 1.063 g/ml, S_f^o, expressed in svedbergs [10^{-13} cm/(sec dyne g)].
[c]Corrected flotation rate at a density of 1.20 g/ml, $F_{1.20}^o$, expressed in svedbergs [10^{-3} cm/(sec dyne g)].

TABLE IIA. Human Apolipoproteins: Structure and Biosynthesis

Apo-protein	Sites of synthesis	Plasma concentration (mg/ml)[a]	Molecular mass[b] (kD)	Amino acids in the mature protein	Signal peptide amino acids	Propeptide amino acids	Repeated units 22 amino acids[c]	Repeated units 11 amino acids[c]
A-I	Liver, intestine, minor sites[d]	1.0-1.2	28.1	243	18	6	8	2
A-II	Liver	0.3-0.5	8.6	77	18	5	1[i]	1
A-IV	Intestine, liver[e]	0.16	43	376	20	—	13[j]	1
B-100	Liver, intestine	0.7-1.0	513 (560)	4536	27 or 24	—	—	—
B-48	Liver, intestine	—	243	[h]	—	—	—	—
CI	Liver, adrenal[f]	0.04-0.06	6.6	57	26	—	—	1
CII	Liver, intestine[e]	0.03-0.05	8.8	79	22	—	1[i]	1
CIII	Liver, intestine[e]	0.12-0.14	8.8 (12)	79	20	—	—	—
D	Adrenal gland, kidney, pancreas, placenta, brain, intestine, testis, minor sites[g]	0.06-0.07	19.5 (35)	169	20	—	—	—
E	Liver, peripheral tissues	0.025-0.05	34.2 (38)	299	18	—	11[k]	1
J	Liver	—	49.0 (70)	427	—	—	—	—

[a]The plasma concentrations of apolipoproteins are based on ref. 10–15. Information on the major sites of synthesis was obtained from ref. 16–20.

[b]Numbers in parentheses indicate apparent Mr of the glycosylated protein.

[c]The information on the repeated units is based on ref. 21. All apolipoproteins contain a 33-amino acid unit in the amino-terminal domain with sequence homology to each other (21).

[d]Minor sites of apoA-I mRNA synthesis in fetal human tissues are the adrenal gland, kidney, gonads, ovaries, heart, and stomach, whereas apoE and apoJ mRNA are present in every tissue analyzed (17,20).

[e]Liver is a minor site of apoA-IV synthesis in primates and rats and the major site of synthesis in rabbits (18). Intestine is a minor site of synthesis of apoCII and apoCIII (17,18).

[f]V. I. Zannis, unpublished.

[g]Minor sites of apoD mRNA synthesis are the liver, spleen, lung, and heart (19,20).

[h]Similar to the amino-terminal portion (residues 1–2152) of apoB-100.

[i]The apoA-I and the apoCIII repeats contain 21 amino acids each.

[j]One of the apoA-IV repeats contains 19 amino acids.

[k]Two of the apoE repeats contain 20 and one contains 17 amino acids.

TABLE IIB. Human Apolipoproteins: Function, Genetics, and Association with Human Diseases

Apo-protein	Gene size (kb)	Exon	mRNA size (kb)	Chromosomal localization[a]	Affinity for lipid emulsions[b] K (μM)	Function	Association with disease
A-I	2	4	0.90	11 (11q13→qter)	0.16–4.24	Activates LCAT, binds to LCAT	Deletion of apoA-I CIII, A-IV loci and inversion of the apoA-I, CIII, A-IV loci are associated with atherosclerosis; some are associated with lipoprotein disorders
A-II	1.3	4	0.5	1 (1p21→1qter)	0.25–0.63	Activates LCAT	Gene deletion (see comments for apoA-I)
A-IV	2.8	3	1.3	11 (11q13-qter)	4.8–10	Activates LCAT	
B-100	43	29	14.1	2 (2p23→p24)	—	Receptor-mediated catabolism of LDL	Abetalipoproteinemia, hypobetalipoproteinemia, LDL-receptor binding defects
B-48	—	—	7.0	—	—		
CI	4.7	4	0.4	19	—	Activates LCAT (moderately), inhibits the catabolism of apoE-containing lipoproteins by the B/E and E/α2M receptor	—
CII	3.3	4	0.5	19	0.45–1.07	Activates lipoprotein lipase	Familial type I hyperlipoproteinemia
CIII	3.2	4	0.55	11 (11q13→qter)	0.53–1.07	Inhibits catabolism of apoE-containing lipoproteins by the LDL receptor	Deletion of apoA-I, CIII, A-IV loci and inversion of apoA-I, CIII loci (see comments for apoA-I)
D	—	—	0.65	3 (3q26.2→qter)	—		
E	3.6	4	1.2	19	—	Receptor-mediated catabolism of apoE-containing lipoproteins	Familial type III hyperlipoproteinemia
J	—	—	1.9	—	—		

[a] Information based on ref. 21.

[b] The affinity of apolipoproteins for lipid emulsions is reviewed in ref. 22.

[c] LCAT, lecithin:cholesterol acyltransferase. The isoprotein composition and the isoelectric points of human apolipoproteins are reviewed in ref. 3.

of density intermediate between VLDL and LDL, intermediate-density lipoproteins (IDL), and several subpopulations of VLDL, LDL, and HDL have been described. Finally, lipoprotein particles (Lp) with defined lipid and apolipoprotein composition have also been isolated from plasma and the media of cultured cells (23,24).

Lipoprotein metabolism can be viewed as a complex biological pathway which contains several steps, including (a) apolipoprotein synthesis, (b) intracellular modification of apolipoproteins, (c) lipoprotein assembly, (d) apolipoprotein and lipoprotein secretion, (e) extracellular modification of apolipoproteins and lipoproteins, (f) hydrolysis of triglycerides and phospholipids by lipoprotein lipase and hepatic lipase, (g) reverse transport of cholesterol from cells to lipoproteins, (h) esterification of lipoprotein cholesterol by lecithin:cholesterol acyltransferase, (i) enzyme-catalyzed exchange and/or transfer of cholesteryl esters and triglycerides and phospholipids, (j) exchange and/or transfer of apolipoproteins, and (k) receptor-mediated catabolism of lipoproteins. Figure 1 shows a schematic representation of the pathway of lipoprotein metabolism and indicates that lipid transport and homeostasis is achieved by the coordinate action of several proteins, including apolipoproteins, plasma proteins, and lipoprotein receptors. Mutations in any of these proteins may result in abnormal lipoprotein profiles and may contribute to the pathogenesis of atherosclerosis.

BIOSYNTHESIS AND MODIFICATIONS OF LIPOPROTEINS AND APOLIPOPROTEINS

Synthesis and Intracellular Modifications of Apolipoproteins

The sites of apolipoprotein synthesis have been determined using cell culture, organ culture, and organ perfusion systems (16,25–27). Additional information has been obtained by blotting analysis of messenger RNA (mRNA) isolated from different mammalian tissues (17,18,28). The combined results, presented in Table IIA, establish that the synthesis of individual apolipoproteins is tissue specific. The major sites of apolipoprotein synthesis in mammalian species are the liver and the intestine. A notable exception is the rabbit, where the liver is a minor site of apoA-I synthesis (18). Several minor sites of apolipoprotein synthesis have also been demonstrated (17,18,28). Following synthesis, apolipoproteins undergo a series of intracellular modifications, including N- and O-glycosylation, fatty acid acylation, phosphorylation, and rare modifications (29–43). The intracellular modifications of apoE, apoA-II, apoCII, and apoCIII were demonstrated

A

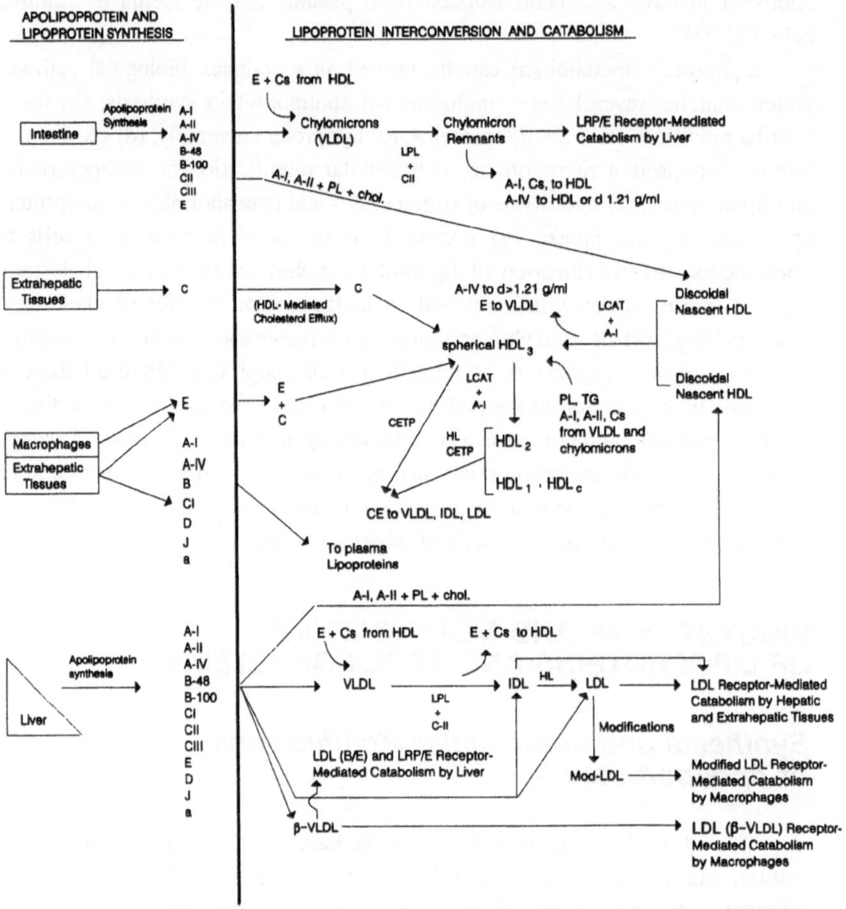

Fig. 1. (A) Schematic representation of the pathway of lipoprotein metabolism. The pathway is based on the information presented in the text. LCAT, Lecithin:cholesterol acyltransferase; CETP, cholesteryl ester transfer protein; LPL, lipoprotein lipase; HL, hepatic lipase; Ch, cholesterol; C's, apoCI, apoCII, apoCIII; TG, triglyceride; PL, phospholipid; A-1, E, . . ., apoA-I, apoE, . . .; DMPC, dimyristoylphosphatidyl choline; VLDL, very low-density lipoprotein; LDL, low-density lipoprotein. (B) Simplified representation of the pathway of lipoprotein metabolism showing biosynthesis, assembly, secretion, modifications, and catabolism of apolipoproteins and lipoproteins. (C) Schematic representation illustrating how a cell with the help of sets of lipoproteins and lipoprotein receptors can maintain cholesterol homeostasis. (D) Schematic representation showing cholesterol accumulation in macrophages caused by the recognition of the modified LDL by the modified LDL receptor.

B

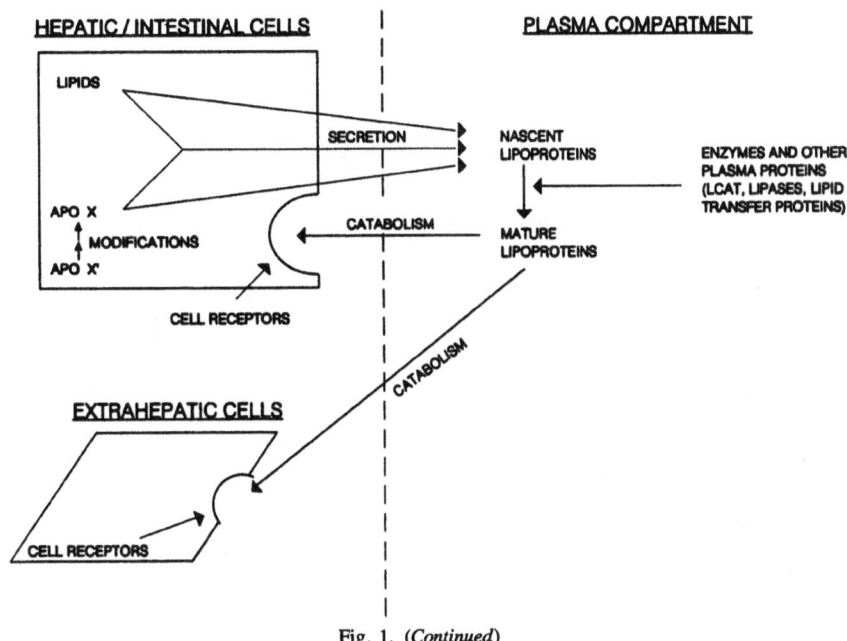

Fig. 1. (*Continued*)

by pulse chase experiments in HepG2 cells (33,43). All four apolipoproteins undergo intracellular O-linked glycosylation/sialylation. Ninety percent of apoE is secreted in the form of di-, tetra-, and hexasialylated isoproteins (33). ApoCIII is secreted exclusively in the disialylated form and apoCII in the disialylated and nonsialylated forms (43) (Fig. 2A–2C). The intra- and extracellular modifications of apoA-II are much more complex than those of apoE, apoCII, and apoCIII (Fig. 2D). Following synthesis, apoA-II undergoes an early modification in the endoplasmic reticulum, which results in the loss of two positive charges. Subsequent to this novel modification, apoA-II is glycosylated with O-linked carbohydrate chains containing two to four sialic acid residues. The sialylated apoA-II forms undergo intracellular proteolysis of the prosegment and cyclization of the N-terminal glutamine. Secreted apoA-II consists of a combination of the different apoA-II isoproteins (Fig. 2D). ApoA-I and apoE synthesized by HepG2 cells in the presence of ^{14}C palmitate are covalently modified by fatty acylation (32). It has been assumed that this modification occurs intracellularly since the plasma forms of apoA-I and apoE do not contain fatty acyl residues (32). ApoA-I has also been

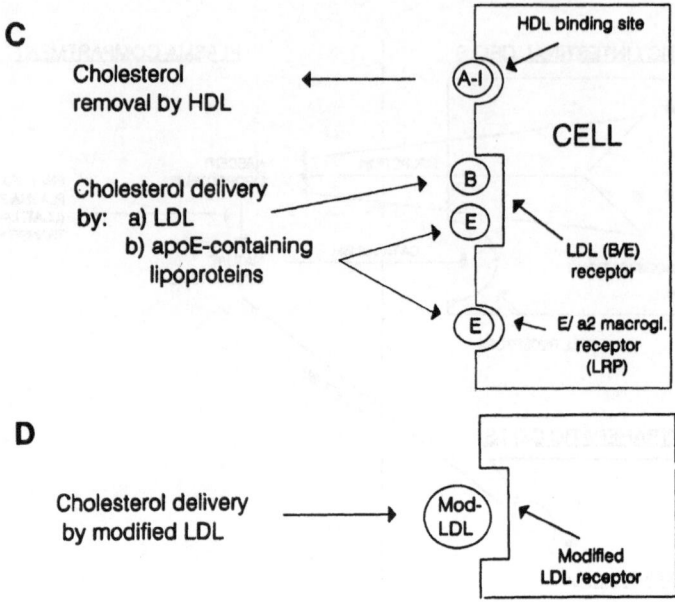

Fig. 1. (*Continued*)

reported to undergo an intracellular phosphorylation at Ser-201 by a calcium calmodulin-dependent kinase (41). The phosphorylated apoA-I form is not detectable in the plasma apoA-I. Different types of modifications have also been described for apoB. Early studies showed that apoB contains 8–10% (w/w) carbohydrate (29–31). Sequence analysis of the human apoB has shown that it contains 16 N-glycosylation sites that are occupied (37). Expression of segments of the apoB cDNA also demonstrated that apoB contains O-glycosylation sites (38). ApoB also undergoes phosphorylation at Ser and Tyr residues and is secreted as a phosphoprotein by rat hepatocytes (40). It has also been reported that human and rat plasma apoB contain both intermolecular thiolester bonds with palmitate and stearate, and intramolecular thiolester bonds involving cysteines 51 and 3734 and the COOH groups of glutamate 54 and aspartate 3737, respectively (34,35). The potential importance of fatty acylation of apoB for the intracellular assembly of LDL is considerable; however, the sites of these modifications on the apoB structure and the cellular compartment of this modification have not been established. The intra- as well as extracellular modifications of human apolipoproteins that will be described later are shown in Figs. 2A–2E.

A   **B**

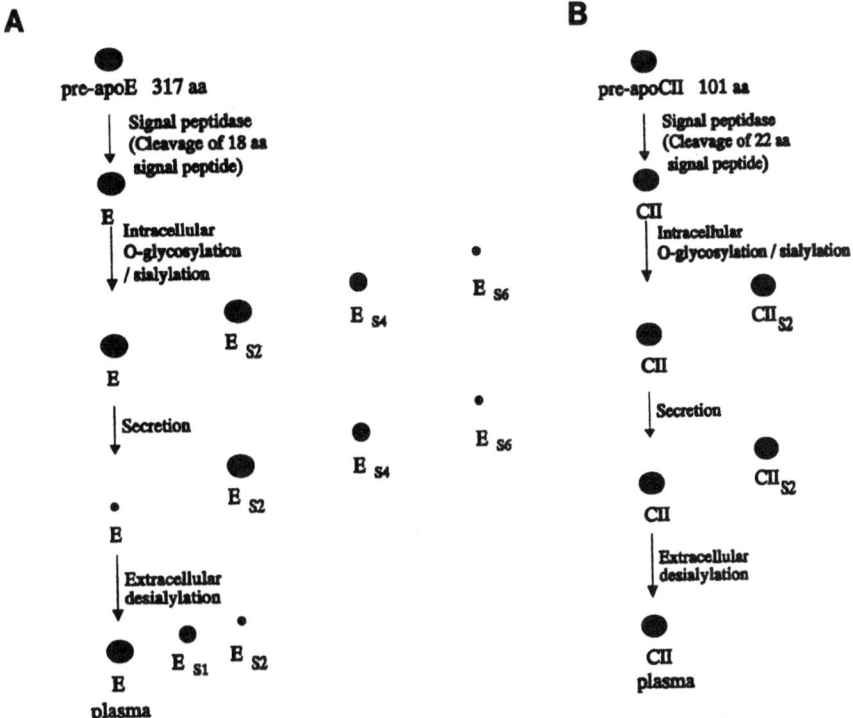

Fig. 2. Schematic representation of the intra- and extracellular modifications of (A) apoE, (B) apoCII, (C) apoCIII, (D) apoA-II, and (E) apoA-I.

Formation of Apolipoprotein B-Containing Lipoproteins by Intracellular Assembly

The intracellular assembly and secretion of LDL has been studied in cultures of chick hepatocytes (44) and HepG2 cells (45–47,47a). ApoB-100 and apoB-48 secreted by cells in culture are found entirely in the lipoprotein fraction. In HepG2 cells apoB-100 was found in VLDL, IDL, LDL, and HDL regions. In rat hepatocytes apoB-100 was distributed in VLDL plus IDL regions, whereas apoB-48 was found in all lipoprotein fractions (48,49). The kinetics of intracellular transport and distribution of apoB in the endoplasmic reticulum and the Golgi suggest that apoB-100 is integrated into lipoproteins cotranslationally (45–47,47a). The size of the lipoprotein particle formed increases as the length of the apoB sequence increases (47a,49a). The association of apoB-100 with neutral

C

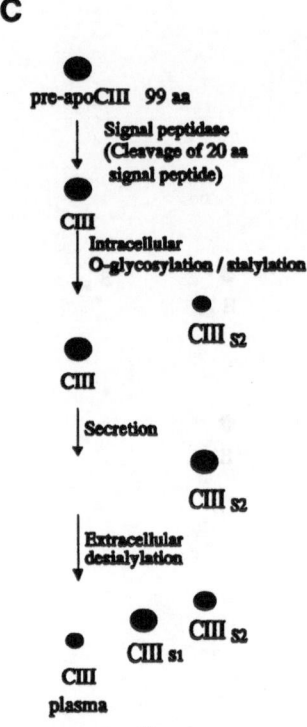

Fig. 2. (*Continued*)

lipids starts in the endoplasmic reticulum and leads to the formation of spherical lipoprotein particles of 200 Å diameter. The process continues during the intracellular transport of the membrane-bound apoB, leading to the enlargement of particles to 300 and 500 Å diameter. These particles are released into the lumen of the secretory pathway and are rapidly secreted. The major lipid component of nascent apoB-100-containing lipoprotein particles produced by HepG2 cells is triacylglycerol (70–85%), followed by cholesteryl esters (10–20%) (45–47). Davis *et al.* have shown that the apoB present in the endoplasmic reticulum exists in two pools (50). One pool is located in the lumen and participates in the assembly and secretion of VLDL, and the other is integrated into the cytoplasmic surface of the endoplasmic reticulum, is not translocated into the Golgi, and is destined for degradation (50). In another study, treatment of HepG2 cells with Brefeldin A resulted in degradation of 40–60% of apoB. The degradation was increased by

D

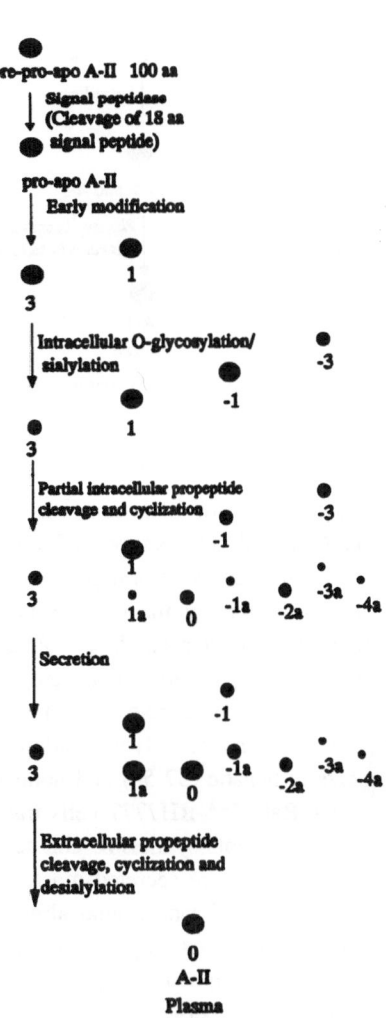

Fig. 2. (*Continued*)

E

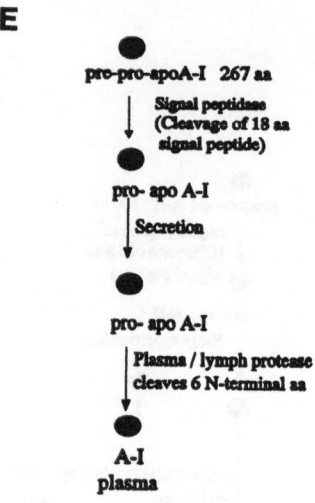

Fig. 2. (*Continued*)

preincubation of the cells with LDL and was not affected by inhibitors of lysosomal enzymes (51). Both studies support the concept that translocation across the endoplasmic reticulum represents a regulatory mechanism which determines the extent of apoB secretion by the cells. Data indicate that triacylglycerol incorporation into lipoproteins starts in the endoplasmic reticulum (52). However, newly synthesized phospholipid components and cholesterol and the majority of triacylglycerols associate with lipoproteins in the Golgi (53–56). The domains of apoB required for lipoprotein assembly were addressed recently by expression of apoB gene segments in hepatic (57,58) and nonhepatic cells (59) as well as in transgenic mice (60). Rat McA-RH7777 cells transfected with amino-terminal apoB cDNA or gene fragments synthesize and secrete the corresponding apoB form. ApoB-13 to apoB-37 were secreted mostly in the $d > 1.21$ g/ml fraction, whereas apoB-48 and apoB-53 were found almost exclusively associated with lipoproteins (57,58). Similar flotation properties are observed in truncated apoB forms that are found in patients with abetalipoproteinemia (61). These observations suggest that the region between apoB-37 and apoB-48 is important for lipoprotein assembly. Furthermore, apoB-17 produced by mouse C127 cells is secreted efficiently into the culture medium and exists 98% in the lipid-poor form. This lipid-poor form associates readily with dimyristoylphosphatidyl choline (DMPC) to form discoidal particles of 24 nm diameter and 6.1 nm thickness. The

DMPC-associated apoB-17 contains 39% α-helix and 36% β-sheet, suggesting that >70% of apoB-17 may be lipid bound (59). The findings suggest that the amino-terminal 17% of apoB-100 can bind lipid and may play an important role in the stabilization of triglyceride-rich VLDL and/or LDL particles. Expression of an internal apoB fragment 2878–3925 (region from 63–87% of apoB) in transgenic mice also produced a lipoprotein particle which floated in the LDL region (60), suggesting that this apoB region can also direct the assembly of lipoprotein particles.

Intracellular versus Extracellular Assembly of High-Density Lipoproteins

Ample evidence during the last 10 years has validated the concept of distinct lipoprotein families proposed by Alaupovic in 1971 (9). These particles, which are defined by their apolipoprotein and lipid composition and are designated LpB, LpB:E, LpB:C, LpB:C:E, LpAI, LpAII, LpAI:AII, LpAII:B, etc., have been isolated by immunoaffinity columns from plasma (23,62–72) and HepG2 cells (24,73). Characterization of nascent lipoproteins secreted by HepG2 has shown that these cells produce LDL- and HDL-like particles rich in phospholipids and free cholesterol. The HDL particles produced by hepatic cells and nonhepatic cells transfected with the apoA-I gene have both discoidal as well as spherical shape (74,75). The major population of the spherical particles isolated from media of hepatic cell cultures has 7.4 nm diameter and contains a cholesteryl ester core (76). A subpopulation of larger discoidal particles found in the HepG2 and Chinese hamster ovary (CHO) cell medium could be converted to 8- to 13-nm spheres by treatment with lecithin:cholesterol acyltransferase (LCAT) (75,77). The functional significance of the different Lp particles and their concentration in plasma in normal and pathological conditions is the subject of extensive investigation (69). Figure 3 shows a schematic representation of the pathway of biosynthesis and catabolism of VLDL, IDL, and LDL. Intracellular assembly of VLDL-like particles occurs in hepatic and intestinal cells (45–48). As will be discussed later, following secretion, VLDL is converted to IDL and LDL and is catabolized by the LDL receptor (78). Alternatively, modification of LDL in plasma makes it a substrate for the modified LDL receptor (79).

Numerous studies have also indicated that a large portion of the exchangeable nascent apolipoproteins produced by hepatic and nonhepatic cells and tissues is secreted in a lipid-poor form (49,74,75,80,81). C127 cells expressing the apoA-I, apoE, or apoCIII genes secrete 80–90% of these apolipoproteins in a lipid-free form. Addition of serum lipoproteins to the C127 cell culture medium resulted in

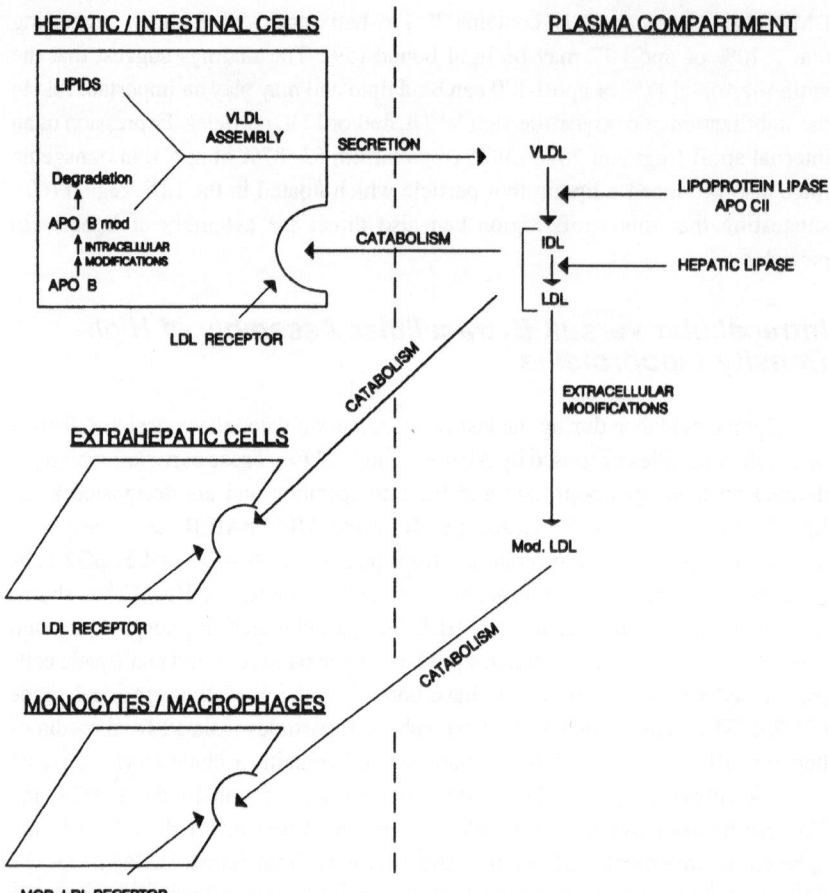

Fig. 3. Schematic representation of the pathway of biosynthesis and catabolism of LDL and modified LDL. VLDL/LDL biosynthesis occurs in hepatic and intestinal cells. Following assembly and secretion, VLDL is converted to IDL and LDL and is catabolized by the LDL receptor. Alternatively, modification of LDL in plasma makes it a substrate for the modified LDL receptor.

the association of the nascent human apolipoproteins with the exogenously added lipoprotein particles. Although apoA-I associates mostly with HDL, and apoE and apoCIII associate with both VLDL and HDL, some association of apoA-I with VLDL and LDL and of apoCIII and apoE with LDL was also observed in this system (81). It is possible that the increased flotation of apoA-I, apoCIII, and apoE in the presence of exogenous lipoproteins may indeed reflect extracellular formation of apoA-I- and apoB-containing Lp particles similar to those isolated from plasma or the culture medium of HepG2 cells (23,24,62,63,68–70,73). This observation is compatible with the findings that only a portion of the secreted nascent apoE may associate with lipoproteins and the remaining exists in a lipid-free form. A qualitatively similar pattern of partial intracellular assembly may occur for the other small apolipoproteins. As shown in Table IIA, peripheral tissues synthesize and secrete several (presumably lipid-poor) apolipoproteins, including apoA-I, apoA-IV, apoB, apoCI, apoD, and apoJ. Recent studies have shown that apoE synthesized by human epidermal grafts and by the kidney of transgenic mice reaches the systemic circulation and is subsequently incorporated into lipoproteins (82–84). Similar incorporation is expected to occur for the other small apolipoproteins that are secreted in a lipid-poor form. The question of whether apoA-I-containing particles, which are the HDL precursors, assemble intra- or extracellularly remains open. Intracellular assembly in the ER/Golgi compartment might involve initial association of apoA-I with phospholipid and cholesterol to form discoidal particles. These particles could be converted to spherical particles intracellularly by the action of acyl-CoA:cholesterol acyltransferase (ACAT) and then secreted. However, direct data showing the presence of discoidal or spherical HDL-like particles in the ER-Golgi system are not available (85). Alternatively, the discoidal particles could be secreted and then converted extracellularly to spherical particles by the action of LCAT. Extracellular assembly might involve secretion of lipid-poor apoA-I, extraction of phospholipid and cholesterol either from the plasma membrane or from plasma lipoproteins to form discoidal HDL, and esterification of the cholesterol by LCAT to form spherical HDL. Finally, lipid-free, newly secreted small apolipoproteins might equilibrate with preexisting lipoprotein VLDL, LDL, and HDL particles. It is also possible that all these processes occur simultaneously and that in plasma an equilibrium exists between free and lipid-bound small apolipoproteins. Isolation of lipoproteins following rupture of purified Golgi fractions showed the presence of nascent VLDL but could not demonstrate the presence of LDL and HDL or discoidal type particles (85). Golgi VLDL contained similar amounts of apoB-100 and apoB-48 but less apoC's as compared to the plasma VLDL (85). Immunogold labeling of hepatocytes also showed that apoE was localized both in compartments containing nascent VLDL as well as in Golgi vesicles and cisternae that lacked VLDL and on the matrix of peroxisomes (86). It is not known whether association of the small apolipoproteins apoE, apoC's, and apoAII with apoB-containing lipoproteins to generate LpB:E, LpB:C, LpB:C:E, LpAII:B, etc., type particles occurs intra- or extracellularly and whether this association involves lipid–protein or specific protein–protein interactions. The hypothetical alternate pathways of

HDL biosynthesis discussed above are shown in Fig. 4. The diagram also shows conversion of discoidal to spherical HDL particles, the interaction of HDL with cell surfaces, and the efflux of cholesterol from cells. All these events will be discussed in detail in later sections.

Extracellular Alterations of Apolipoproteins and Lipoproteins

Demodification of Apolipoproteins

ApoA-II and apoCII in plasma are nonsialylated. ApoCIII exists in nonsialylated (14%), monosialylated (59%), and disialylated (27%) forms (87,88), and apoE exists in nonsialylated (74%) and sialylated (26%) forms (89). This implies that the forms that are generated by intracellular modification are converted to the plasma form mainly by extracellular deglycosylation/desialylation (Figs. 2A, 2C). In the case of apoA-II conversion of the different isoforms to the plasma form, in addition to deglycosylation/desialylation, it also requires extracellular cleavage of the prosegment and N-terminal cyclization (Fig. 2D). Finally, apoA-I and apoE lose fatty acyl moieties and apoA-I a phosphorus moiety that is presumably added intracellularly (32,41).

Cleavage of the ApoA-I and ApoA-II Prosegments

Studies of human apoA-I synthesis and secretion in HepG2 cells (90,91) and DNA sequence analysis of the human apoA-I gene (92–94) have established that newly secreted apoA-I has a six-amino acid amino-terminal extension (prosegment) with the unusual sequence Arg–His–Phe–Trp–Gln–Gln. It was noted earlier that the two Gln residues might protect cleavage of apoA-I in the secretory vesicles. However, deletion or alteration of one or both Gln of the prosegment by Arg did not affect the intracellular transport, secretion, lipid-binding properties (95,96), or ability of proapoA-I to activate LCAT (97). Extracellular cleavage of the prosegment by a specific protease generates the plasma form of apoA-I (98–100). Several recent reports have described a proteolytic activity in human and rat plasma which converts proapoA-I to mature apoA-I; this conversion is relatively slow *in vivo* and even slower under the *in vitro* assay conditions employed (98–100). Similar to apoA-I, apoA-II contains a five-amino acid amino-terminal prosegment with the sequence Ala–Leu–Val–Arg–Arg. Pulse-chase experiments in HepG2 cells showed that the proapoA-II to apoA-II conversion occurs partially intracellularly (43). The same conclusion was originally reached by Gordon *et al.* (91), but was revised later to suggest almost exclusive extracellular conversion of

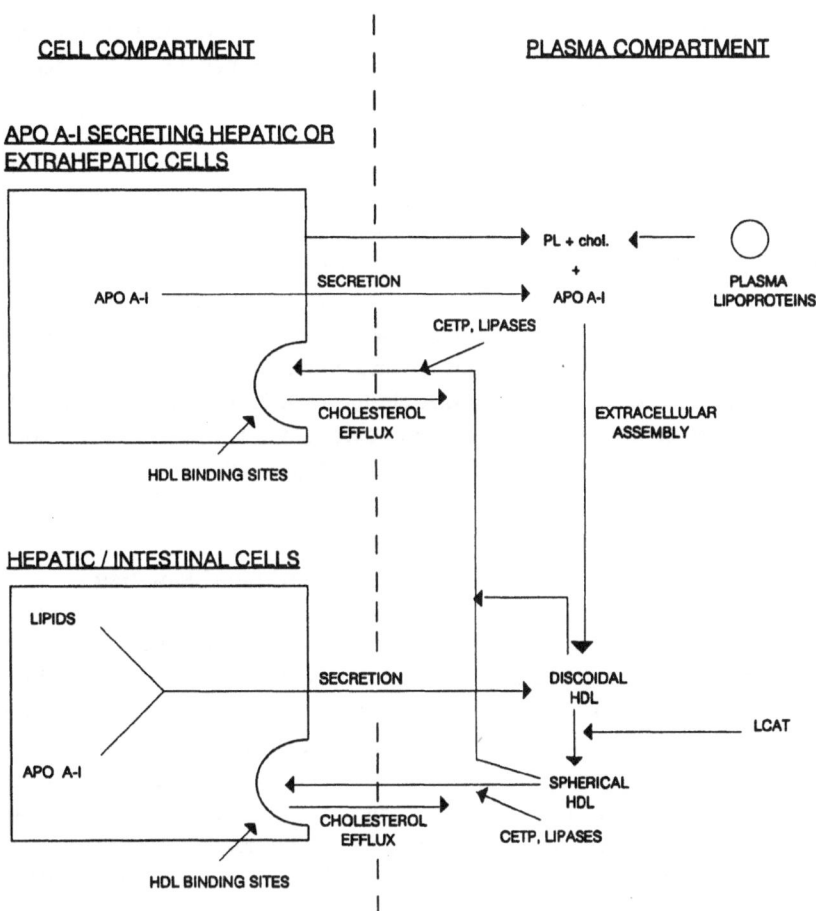

Fig. 4. Schematic representation showing alternate routes of biosynthesis of discoidal HDL by either intra- or extracellular assembly, the formation of spherical HDL by the action of LCAT, and the function of the HDL particles in the reverse transport of cholesterol.

proapoA-II to apoA-II by a cathepsin B-like proteolytic activity which is secreted by the HepG2 cells (101).

Extracellular Modifications of ApoB and Other Apolipoproteins

Nonenzymatic glycosylation of apolipoproteins (102), deamidation of apoA-I (103), and enzymatic modification of apolipoproteins in plasma by transglut-aminases have been reported (104). The nonenzymatic modification of apoB by malondialdehyde and derivatives of lipid oxidation (105) is of great importance. This modification renders the LDL a substrate for the modified LDL receptor present in monocyte-macrophages (106). The uptake of modified LDL leads to the generation of foam cells found in early atherosclerotic lesions (107). The formation of modified LDL particles is shown schematically in Figs. 1A, 1D, and 3. It has been shown that monoclonal antibodies against malondialdehyde-modified LDL, 4-hydroxynonal LDL, or Cu^{2+}-oxidized LDL recognize specific epitopes on apoB. These epitopes are generated by modification of the ϵ-amino groups of lysines by products of lipid oxidation (108). It is not known whether these adducts involve all the surface lysines of apoB or whether they are localized in specific epitopes in such a way as to make this region a binding site for the modified LDL receptor (108). It is very important that oleate-rich LDL particles (produced by diets enriched in oleic acid) are resistant to oxidation *in vitro* (109). The various intra- and extracellular modifications of apolipoproteins are summarized in Table III.

Exchange and Transfer of Apolipoprotein and Lipid Moieties

Several studies have established that, with the exception of apoB-100 and apoB-48, exchange and/or transfer of all small apoproteins occurs during various stages of lipoprotein metabolism. The major apoprotein components of lymph chylomicrons are apoB-48, apoA-I, and apoA-IV (110–112). When chylomicrons enter the plasma compartment they receive apoC and apoE from HDL. Following hydrolysis of chylomicrons by LPL, they transfer their apoA-I to HDL (113,114) and their apoA-IV either to HDL or to the $d>1.21$ g/ml fraction (13,115). ApoA-1 and apoA-II in plasma are found in HDL and exchange readily between the HDL_2

TABLE III.   Summary of Intra- and Extracellular
Posttranslational Modifications of Apolipoproteins[a]

Apolipoprotein	Intracellular	Extracellular
ApoA-I	Phosphorylation (41), fatty acid acylation (32)	Propeptide cleavage (90,91)
ApoA-II	Propeptide cleavage (43), sialylation, cyclization (43)	Propeptide cleavage (43,101), cyclization (43), desialylation (43)
ApoB	Phosphorylation (40), fatty acid acylation (34,36), N- and O-glycosylation, (37,38)	Malondialdehyde modification (105)
ApoE, ApoCIII, ApoCII	Sialylation (33,43)	Desialylation (43,87–89)

[a]All apolipoproteins undergo signal peptide cleavage.

and HDL_3 subclasses (116,117). Other studies have shown that hepatic Golgi VLDL contain mainly apoB and apoE but only small amounts of apoC's (85,118). This implies that when nascent VLDL enters the plasma compartment, it acquires apoC's from other plasma lipoproteins, possibly HDL_2 (113). In addition, nascent VLDL may acquire additional apoE from nascent HDL (119). When VLDL triglycerides are hydrolyzed by lipoprotein lipase, apoCII, apoCIII, and some apoE are transferred back to HDL (113,119–121). The enzyme-catalyzed lipolysis of chylomicron and VLDL triglycerides and esterification of HDL cholesterol and the ensuing exchange and transfer of lipid and protein moieties result in the conversion of one lipoprotein class into another. Plasma chylomicrons, when acted upon by lipoprotein lipase (122,123), lose the bulk of their triglycerides, transfer their phospholipids and apoproteins to HDL (114,124–126), and become chylomicron remnants. Chylomicron remnants are then cleared from plasma by the liver with a half-life of less than 10 min by a saturable high-affinity process (127,128). Similarly, sequential hydrolysis of VLDL by LPL and the ensuing loss of apoC's and apoE result in the generation of plasma IDL ($d = 1.006–1.019$ g/ml) and LDL (129,130), which are catabolized by extrahepatic as well as by hepatic tissues. The discoidal HDL found in patients with LCAT deficiency (131,132) can be converted into a spherical particle by the action of LCAT (119). This conversion is associated with transfer of the nascent HDL apoE to VLDL (119), transfer of VLDL apoC to the spherical plasma HDL (119), and possible altered affinity of the spherical HDL for apoA-I (133). As will be discussed later, apoE-containing HDL may be catabolized directly by tissues. In addition, the cholesteryl ester moiety of HDL may be transferred to VLDL and LDL for further catabolism (134–150). The VLDL and LDL_2 particles of unusual size and composition found in the plasma of patients with LCAT deficiency (119,132) could represent intermediates of lipoprotein metabolism that may be converted rapidly to other lipoproteins in normal plasma (151). The various events associated with the exchange and/or transfer of apolipoprotein and lipid moieties among lipoprotein classes are shown schematically in Fig. 1A.

The transfer of exchangeable small apolipoproteins, described above, is very important since apolipoproteins can act as modulators of enzymatic activities or as ligands for cell receptors (Table IIB). For instance, as discussed later, apoB and apoE are ligands for lipoprotein receptors (78,79). ApoCII and apoA-I activate the enzymes lipoprotein lipase and LCAT, respectively. Similarly, apoE promotes the binding of βVLDL to both the LDL receptor and the LDL receptor-related protein (LRP). In contrast, apoCI, and to a lesser extent apoCII, but not apoCIII, block the binding of βVLDL to the LDL receptor-related protein as well as to the LDL receptor (79).

LIPOPROTEIN RECEPTORS: STRUCTURE, FUNCTION, AND GENETIC VARIATION

The Low-Density-Lipoprotein Receptor

Introduction

Our current understanding of the molecular events involved in the receptor-mediated catabolism of lipoproteins has been shaped mainly by the pioneering work of Goldstein, Brown, and colleagues, who first demonstrated the presence of a specific receptor on cell surfaces which recognizes the low-density lipoprotein (LDL) (152,153). Since then, other distinct lipoprotein receptors have been described; these are the subject of outstanding recent reviews (78,79,154,155). The structures of these receptors are shown in Figs. 5A–5C and their properties are summarized in Table IV.

Functional Definition of the LDL Receptor

Early biochemical and genetic studies established that the cell surface of cultured human fibroblasts and other cell types contain high-affinity receptors for LDL, the major cholesterol transport protein in human plasma. Patients with familial hypercholesterolemia (FH) either totally lack or have defective receptors (152,153,156–158). It has been estimated that each normal fibroblast cell contains 10,000–100,000 receptors (157,159,160). Electron microscopy studies have shown that 50–80% of the receptors are clustered in regions of the plasma membrane called coated pits which have clathrin as their main protein (161–164). The binding of LDL to the LDL receptor is Ca^{2+} dependent and pronase sensitive (152,153,160).

Biosynthesis and Posttranslational Modifications of the LDL Receptor

Similar to other membrane and secretory proteins, the LDL receptor protein is modified posttranslationally in the endoplasmic reticulum (ER) and Golgi. The known modifications include cotranslational cleavage of the signal peptide and addition of the high-mannose, N-linked carbohydrate chains and posttranslational addition of the core O-linked carbohydrate chains in the ER or early Golgi (165). These modifications produce a precursor receptor form of molecular weight (Mr) = 120 kD (166). Subsequent processing of high mannose to the complex carbohydrate chains (endo H-insensitive form) and addition of galactose and sialic acid

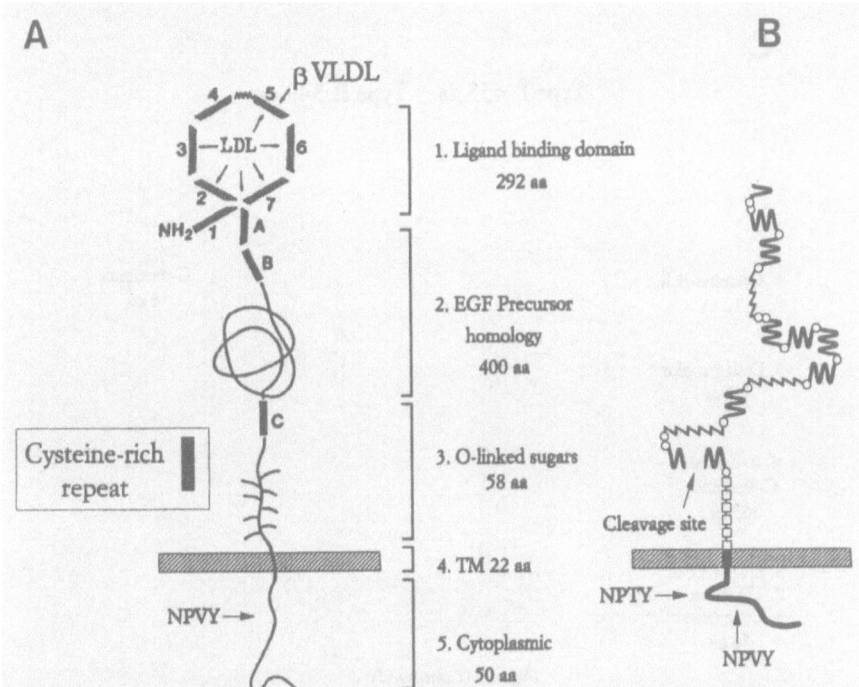

Fig. 5. (A) Schematic representation of the different domains of the LDL receptor. The regions designated 1, 2, 3, 4, 5, 6, 7, A, B, and C represent cysteine-rich regions. The regions involved in the binding of LDL or β-VLDL are indicated by arrows. The NPVY sequence directs the receptor to the coated pits. (Reproduced from reference 214 with permission). (B) Schematic representation of the domains of the LRP receptor. Jagged lines, cysteine-rich domains; wavy lines, EGF precursor homologous region, ∞, growth factor repeats; ○, the remaining spacer region; □, epidermal growth factor repeats, ■, transmembrane region. The NPVY and NPTY sequences in the cytoplasmic domain and the site of cleavage of the Mr = 600 kD precursor are indicated. (Reproduced from ref. 79 with permission.) (C) Schematic representation of the two types of modified LDL receptor. The different receptor domains are indicated. (Reproduced from ref. 289 with permission.)

to O-linked chains (165) increases the apparent Mr of the receptor to 160 kD. Failure of the receptor to be modified to the 160-kD form results in its entrapment and degradation in the ER.

Definition of the LDL Receptor Pathway and the Associated Feedback Regulatory Mechanisms

LDL binds to the LDL receptor at 4°C but is not internalized (152,153). The dissociation constant of the LDL–LDL receptor complex is 2.8 nM (160). When

C

Type I 453 aa   Type II 349 aa

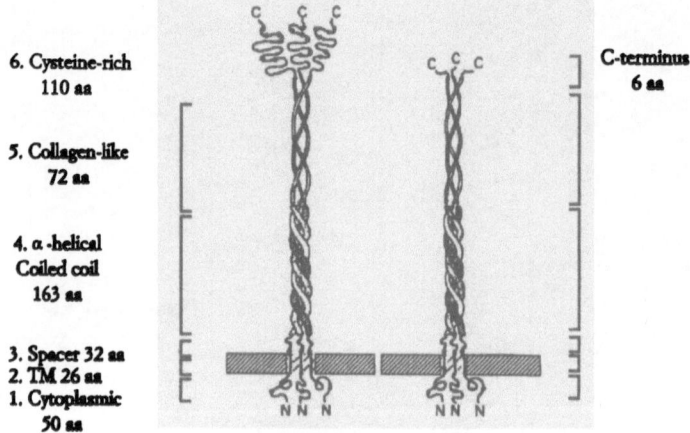

6. Cysteine-rich
110 aa

5. Collagen-like
72 aa

4. α-helical
Coiled coil
163 aa

3. Spacer 32 aa
2. TM 26 aa
1. Cytoplasmic
50 aa

C-terminus
6 aa

Fig. 5. (*Continued*)

cell cultures are incubated at 37°C, the coated pits with the LDL receptor complex invaginate into the cell and pinch off to form endocytic vesicles called endosomes that carry LDL to lysosomes (152,153,167). During this time, the endosome develops an acidic pH that is believed to mediate the dissociation of the lipoprotein receptor complex (168). The free receptor recycles to the cell surface prior to the fusion of endosomes with primary lysosomes (168,169). Recycling of the receptors is inhibited by the carboxylionophore monensin (169). Fusion of endosomes with primary lysosomes results in the hydrolytic degradation of apoB to amino acids and hydrolysis of cholesteryl esters by the lysosomal enzyme acid lipase (156,159). Liberated cholesterol is used by cells for membrane synthesis (170). It also triggers three regulatory responses that assure cellular cholesterol homeostasis: (a) suppression of 3-hydroxy-3-methylglutaryl (HMG) CoA reductase, the rate-limiting enzyme of cholesterol biosynthesis, turning off cellular cholesterol synthesis (171), (b) activation of ACAT which reesterifies excess cholesterol preferentially with oleic acid, resulting in cytoplasmic storage of cholesteryl ester droplets (172), and (c) decrease in the number of surface LDL receptors, which prevents additional cholesterol influx into the cell (152,153,157). The decrease of the LDL receptor number is regulated at the level of transcription (173). In addition to fibroblasts, all other animal and human cells tested, including lymphoid cells (174,175), arterial smooth muscle cells (176,177), endothelial cells (170), and adrenocortical cells (178,179), have LDL receptors. The LDL receptors are particularly important in adrenocortical cells, where the cholesterol derived from LDL may provide one source of substrate for steroid hormone synthesis (178,179).

Ligands for the LDL Receptor (Also Designated B/E Receptor)

The ligands for the LDL receptor are the LDL, which contain apoB as the sole protein, and apoE-containing lipoproteins, such as IDL, βVLDL, and HDL with apoE (HDLwE) (180–185), as well as apoE-containing proteoliposomes (186,187). For this reason, the LDL receptor was also designated the B/E receptor. LDL binds to the LDL receptor by interacting with apoB. Earlier studies showed that modification of Arg or Lys residues abolished the binding of the LDL or HDLwE to the LDL receptor (188–190), indicating that positively charged residues are important for receptor binding. The binding and the intracellular regulatory events following the binding of apoE-containing lipoproteins are similar to those described for LDL (180). The affinity of HDLwE for the LDL receptor was more than 20 times greater than that for LDL itself. The dissociation constant K_d of the lipoprotein–receptor complex was 0.12 nM for HDLwE and 2.8 nM for LDL (160). Furthermore, four times as many LDL particles as HDLwE particles were required for saturation of the receptors at maximal binding (160). Since HDLwE and LDL bind the same receptor, these data indicate that one HDLwE particle may bind to four LDL receptor sites (186,187). In other studies, it has been shown that, besides HDLwE, other apoE-containing lipoproteins, such as chylomicron remnants, βVLDL, and HDL, show apoE-mediated binding to cultured fibroblasts (182–185). Liver membranes obtained from normal rabbits have been shown to have high-affinity binding sites ($K_d = 0.5$ μg/ml) for βVLDL. The receptor that binds βVLDL resembles the extrahepatic LDL receptor in having high affinity for apoB- and/or apoE-containing lipoproteins, requiring Ca^{2+}, and being sensitive to pronase. This receptor appears to mediate the rapid hepatic clearance of βVLDL in normal rabbits (185). After cholesterol feeding of the rabbits, saturation of these receptors by the high concentration of endogenous βVLDL and a downregulation of the number of hepatic receptors appeared to be responsible for the accumulation of βVLDL in plasma and profound hypercholesterolemia (185). The βVLDL particles rich in cholesterol and apoE are found in the plasma of other animals (191,192) and in patients with type III hyperlipoproteinemia (193). The βVLDL obtained from various animals caused uptake and degradation of cholesteryl ester

TABLE IV.   Lipoprotein Receptors[a]

Name of receptor	Cell or tissue origin	Ligand	Affinity K_d, (nM)	Binding characteristics
LDL (B/E) receptor	Skin fibroblasts (present in all cell types)	LDL	2.8	Binds one LDL moleculer/ receptor
LDL (B/E) receptor	Skin fibroblasts	HDL with apoE	0.12	Binds one HDLwE molecule/four binding sites
LDL (B/E) receptor	Skin fibroblasts	DMPC–apoE	0.1	Binds one HDLwE molecule/four binding sites
LDL (B/E) receptor	Liver	LDL	11.0	—
βVLDL (LDL) receptor	Monocyte-derived macrophages	βVLDL	—	—
E/α2M receptor (LRP) (also referred to as chylomicron remnant receptor)	Liver (present in other cell types)	HDL with apoE, DMPC–apoE	0.23	—
HDL binding sites	Liver	HDL (apoE-free)	82.0	Nonspecific
HDL binding sites	Skin fibroblasts	HDL (apoE-free)	~200	Nonspecific
HDL binding sites	Adrenal, ovary, gonads	HDL	—	—
Modified LDL receptor	Monocyte-macrophage	Modified LDL	—	—

in cultured mouse peritoneal macrophages (194,195). The uptake of βVLDL by macrophages may lead to foam cell formation (195–197). Other studies have also shown that the uptake of βVLDL (197,198) and chylomicron remnants (198) by mouse peritoneal macrophages is mediated by the classical LDL receptor. These studies include ligand blotting and [125I]βVLDL and [125I]LDL, immunoblotting with antireceptor immunoglobulin G (IgG), and downregulation of receptor activity and numbers with either βVLDL or LDL (197,198). The binding affinity

TABLE IV. *(Continued)[a]*

Other features	Receptor–ligand internalization	Receptor regulation[b]	Function
Ca^{2+} dependent, pronase sensitive	Yes	Yes	Regulation of body and cellular cholesterol homeostasis (steriod hormone synthesis by adrenal gland, ovaries, and gonads)
Ca^{2+} dependent, pronase sensitive	Yes	Yes	—
Ca^{2+} dependent, pronase sensitive	—	—	—
Ca^{2+} dependent, pronase sensitive, age dependent, inducible by hypocholesterolemic agents (cholestyramine)	Yes	Yes	Clearance of βVLDL, LDL, and HDLwE cholesterol by the liver for excretion, regulation of body and cellular cholesterol homeostasis
Ca^{2+} dependent	Yes	Yes[c]	Clearance of diet-induced lipoprotein particles; role in foam cell formation
Ca^{2+} dependent, pronase sensitive, age independent, noninducible	Yes	Yes	Clearance of chylomicron remnant, βVLDL, IDL, and HDLwE cholesterol by the liver for excretion, regulation of body and cellular cholesterol homeostasis
Ca^{2+} independent, pronase insensitive	No	—	Reverse transport of cholesterol
Ca^{2+} independent, pronase insensitive	No	Yes	Reverse transport of cholesterol
—	—	—	Provision of cholesterol for steroid hormone synthesis
Ca^{2+} independent, pronase sensitive	Yes	No	Scavenging particles; role in foam cell formation

[a]The references on which this table is based are given in the text.

[b]With the exception of HDL binding sites, "receptor regulation" is manifested by (a) decrease of intracellular cholesterol synthesis, (b) increase in intracellular cholesterol esterification, and (c) decrease in number of surface receptors. Regulation of HDL binding sites following binding of HDL (apoE-free) is manifested by (a) increase in number of LDL receptors, (b) increase in cellular sterol synthesis, and (c) decrease in cellular cholesterol esterification.

[c]Regulation occurs at higher ligand concentrations.

of the mouse macrophage LDL receptor is $\frac{1}{18}$ that of human fibroblasts (197). In addition, macrophage cultures display defective regulation of the ACAT activity (199) and defective downregulation in response to cholesterol uptake via VLDL or LDL (194,195,197). Similar studies with human monocyte-macrophage cultures showed that the putative human βVLDL receptor present in these cultures was structurally and functionally indistinguishable from the classical LDL receptor of cultured human fibroblasts (200). Furthermore, monocyte-macrophage cultures from several patients with homozygous familial hypercholesterolemia were unable to metabolize either βVLDL or LDL, thus providing genetic evidence supporting the identity of the two receptors (200). The finding that monoclonal antibodies to apoE inhibit the binding of βVLDL and chylomicron remnants to cultured skin fibroblasts and hepatic membrane preparations strongly suggests that apoE is the protein determinant for the recognition and catabolism of these particles by the LDL receptor (201). Previous studies also showed that apoE is responsible for the uptake of hypertriglyceridemic VLDL by the LDL receptor (202). The combined data indicate that apoE mediates the catabolism of a variety of different lipoproteins found in the VLDL, IDL, or HDL density range by the classical LDL (B/E) receptor.

Purification of the LDL Receptor; Cloning cDNAs and the Gene Encoding for the LDL Receptor

The development of an assay for LDL receptor activity in cell membrane preparations (203) has allowed the purification of the LDL receptor from bovine adrenal cortex (204,205). The mature receptor is an acidic glycoprotein of apparent molecular weight 160,000 and isoelectric point 4.3 (166,206). Sequencing of peptides obtained from the purified bovine LDL receptor allowed synthesis of degenerate oligonucleotides which were used as probes for cloning of cDNAs encoding for the bovine LDL receptor (207) and subsequently for the cDNA and the gene encoding the human LDL receptor (173,208).

The LDL receptor mRNA is 5.3 kilobases (kb) long, including a 2.6-kb-long 3′ untranslated region (173). The LDL receptor gene is 45 kb long and contains 17 exons and 18 introns (208). It maps on the short arm of chromosome 19 (p13.1–p13.3) (209,210). As shown later, the LDL receptor gene is closely linked to the human apoE, apoCI, and apoCII genes (211,212).

The human LDL receptor cDNA and gene encode a protein of 860 amino acids, including a 21-residue signal peptide (207,208). Computer analysis of the receptor sequence showed that it consists of five domains (Fig. 5A). Domain 1 at the amino terminus has 292 residues, is rich in Cys, and consists of seven homologous sequences containing an average of 40 amino acids each. These sequences are encoded mostly by separate exons and have homology with a sequence found in the component C8 of complement. Mutations in this domain affect the binding of LDL to the LDL receptor. Domain 2 consists of 400 amino acids and has 35% amino acid sequence homology to a segment of the precursor of the epidermal growth factor (EGF). Domain 3 consists of 58 amino acids, 18 of which are Ser that are the sites of O-linked glycosylation. Domain 4 is the membrane-spanning region and consists of 22 hydrophobic amino acids. Domain 5 is the cytoplasmic tail and consists of 50 amino acids. The cytoplasmic domain contains a conserved NPVY sequence between residues 804 and 807 which is required for movement of the receptor to the coated pits. Similar sequences have been found in the cytoplasmic domain of the LDL receptor-related protein (LRP) (79,213). Computer analysis of the structure of the LDL receptor showed the presence of seven imperfect amino-terminal repeats, designated 1–7 (Fig. 5A). Two adjacent repeats, designated A and B, are located in the EGF precursor homology region (208). Deletion or oligonucleotide-directed mutagenesis within each of the repeated sequences and functional analysis of the mutant receptors following expression of COS cells confirmed the importance of these regions for ligand binding. This analysis also showed that repeats 2, 3, 6, 7, and A are required for maximum binding of LDL (via apoB) but not βVLDL (via apoE), whereas repeat 5 is required for maximum binding of both LDL and βVLDL (214,215).

Transcriptional Regulation of the LDL Receptor Gene

The promoter of the LDL receptor contains three direct repeated sequences between nucleotides −110 and −40 (216). Repeated units 1 and 3 bind the previously described transcription factor Sp1 (217,218). It has been proposed that the middle regulatory element, designated the sterol regulatory element, is recognized by another factor which in the absence of sterols acts synergistically with the transcription factor Sp1 to promote the transcription of the LDL receptor gene (219).

Animal Models with LDL Receptor Deficiency or Abundance

The Watanabe heritable hyperlipidemic (WHHL) rabbit has been characterized by an absence of functional LDL receptors and has been used as an animal model for familial hypercholesterolemia (220). This rabbit strain has a mutation which prevents the transport of the receptor to the cell surface (221,222). As a result of the LDL receptor defect (223,241), the WHHL rabbit accumulates in its plasma apoB-containing lipoproteins (VLDL, IDL, and LDL) but not apoB-48-

containing chylomicron remnants (222,224,225). The metabolic block in the hepatic clearance of IDL that exists in the WHHL rabbit causes an increase in the conversion of IDL to LDL and results in elevated plasma LDL levels in this animal model as well as in patients with familial hypercholesterolemia (222,224). Molecular analysis showed that the WHHL defect results from the deletion of residues 117–120 of the rabbit LDL receptor (226). In addition to the WHHL rabbit, several members of a Rhesus monkey colony have been identified with an LDL receptor deficiency. Affected members are heterozygous for a nonsense mutation at residue Trp-284 (227). In contrast to the WHHL rabbit and the LDL receptor-deficient monkeys, transgenic mice have been generated which overexpress the human LDL receptor. The transgenic mice overcatabolize LDL, leading to very low levels of plasma LDL (228). Feeding of the transgenic mice with atherogenic diets with high cholesterol and saturated fat content failed to increase the lipid and lipoprotein levels, suggesting that overexpression of the LDL receptor is sufficient to protect from hypercholesterolemia and atherosclerosis (229).

Mutations in the LDL Receptor Gene Are Associated with Familial Hypercholesterolemia

The functional, biochemical, and genetic analysis of the LDL receptor facilitated enormously the delineation of the molecular defects which underlie familial hypercholesterolemia. Based on new molecular information, a variety of defects in the LDL receptor identified previously with binding studies could be assigned to one of four classes. Class I mutants are characterized by the absence of receptor protein due to either gross alterations (insertions, deletions) in the receptor gene or nonsense mutations leading to premature chain termination (154,230–235). Class 2 mutants are characterized by defective modification of the precursor N- and O-linked carbohydrate (221,236). These mutations consist mostly of amino acid deletions or substitutions, have been localized in exons 2, 4, 6, 11, and 14 (154,237–244), and cause the entrapment of a precursor form of $Mr = 120$ kD in the ER (236). The WHHL rabbit mutation also belongs to this category (226). Class 3 mutants are characterized by receptors of either normal or aberrant apparent Mr which are modified, normally reach the cell surface, but have reduced affinity for LDL. Such mutants result from deletions/insertions or amino acid substitutions in exons 2–8 (154,244–250) in the cysteine-rich domain of the receptor. Finally, class 4 mutants are characterized by normal synthesis, modification, and transport to the cell surface, but inability to cluster into the coated pits. As a result, these mutant receptors bind LDL normally, but do not internalize the complex (154,235,241,251–256). All these mutants have alterations within the first 21 residues of the cytoplasmic tail. Functional analysis of either naturally occurring mutants or mutants generated by *in vitro* mutagenesis showed that substitution of Tyr to Cys or other nonaromatic amino acids was sufficient to create an internalization defect (257). The mutations described for the LDL receptor have been reviewed extensively (154) and are shown in Fig. 6. It was reported recently that fibroblasts from patients with familial hypercholesterolemia contain a high-affinity binding site (K_d 22 nM) for LDL, modified LDL, and other apoB-containing lipoproteins which is stimulated by long-chain fatty acids. The physiological importance of this binding site is not known (258).

Chylomicron Remnants [Apolipoprotein E/ α2 Macroglobulin (E/α2M) Receptor]

Functional Definition of the Chylomicron Remnant ApoE Receptor

Early studies by Carrela and Cooper identified a specific high-affinity chylomicron remnant receptor on rat liver plasma membranes from animals that were not treated with estradiol and therefore were not expected to exhibit LDL (B/E) receptor activity (259). Subsequent studies using hepatic membranes from adult dogs, which presumably lack the LDL (B/E) receptor, showed that the

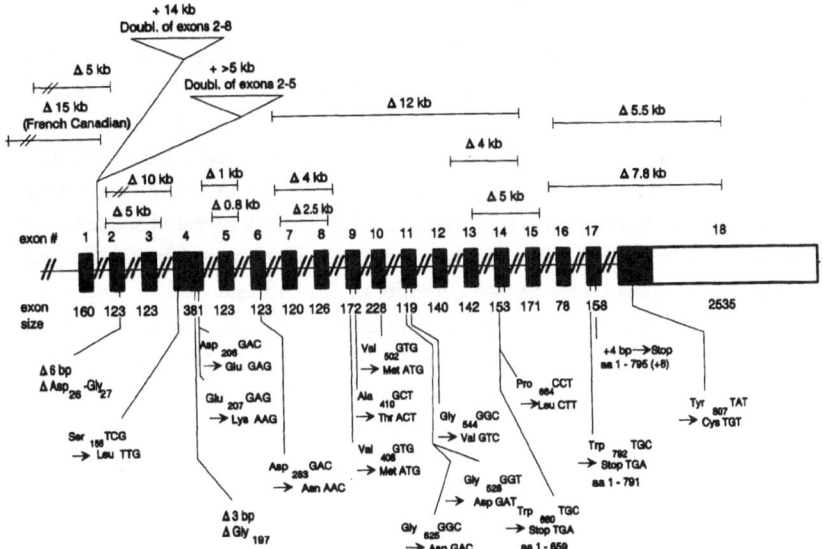

Fig. 6. Schematic representation of the human LDL receptor gene and the various mutations associated with familial hypercholesterolemia. The mutations are described in refs. 154, 221, 230–256. Δ, Deletion.

hepatic membranes contained high-affinity putative receptors specific for apoE-containing lipoproteins. These putative receptors were initially designated apoE receptors (260). Scatchard analysis of the binding of [^{125}I]HDLwE to adult canine hepatic membranes suggested the presence of two binding sites: a higher-affinity ($K_d = 0.23$ nM), calcium-dependent pronase-sensitive site and a lower-affinity ($K_d = 20$ nM), calcium-independent, pronase-resistant site. Arginine or lysine modification of HDLwE abolished binding to the higher- but not to the lower-affinity site (260). The specificity of the high-affinity receptor for apoE was established by competition experiments. [^{125}I]HDLwE, which can bind to either the LDL or the putative apoE receptor, was only partially displaced by excess unlabeled LDL, but was totally displaced by excess unlabeled HDLwE (260). The difference between the LDL receptor and the chylomicron remnant receptor was further suggested by the observation of normal hepatic chylomicron remnant catabolism in patients with homozygous familial hypercholesterolemia, who are LDL receptor negative (78,153,261,262), as well as by the WHHL rabbit (224,225).

Structure and Functions of the ApoE Receptor

In 1988, overlapping cDNA clones were isolated from a human lymphocyte cDNA library which encode for a protein containing 4525 amino acids that has structural and biochemical similarities to the human LDL (B/E) receptor (213). This protein has been named lipoprotein receptor-related protein (LRP) and is the topic of a recent review (79). The structural similarities between LRP and the LDL receptor encompass the cysteine-rich ligand-binding domain, the EGF precursor homologous region, and a 100-residue, carboxy-terminal cytoplasmic tail. This cytoplasmic sequence contains two copies of the NPXY sequence which directs the LDL receptor to the clathrin-coated pits (263). The O-linked sugar domain present in the LDL receptor is replaced in the LRP by six repeated sequences that are found in the EGF. LRP is synthesized as a precursor of Mr-600 kD and is cleaved in the trans Golgi to two subunits of Mr-515 and 85 kD, respectively, which associate noncovalently on the cell surface (264). A schematic representation of the different domains of the LRP is shown in Fig. 5B. LRP mRNA and protein are present in various tissues, including liver, brain, and lung (213). Studies with LDL receptor-negative skin fibroblasts showed that the LRP-mediated binding and the uptake of βVLDL by the cells are stimulated by apoE and are inhibited by apoCI and to a lesser extent by apoCII (265,266).

The binding of apoE-enriched βVLDL is calcium dependent (265). Indirect evidence using monoclonal antibodies directed to the 515-kD and 85-kD subunits suggests that LRP may follow a similar pathway as the LDL receptor, which includes internalization, recycling of the receptor, and degradation of the ligand in the lysosomes (267). Parallel studies have established that the receptor for protease-activated α2 macroglobulin has identical structure and amino acid sequence with LRP (268–271). LRP also binds with high affinity and mediates the uptake of activated α_2 macroglobulin-protease complexes (272), plasminogen activator-inhibitor complexes (272a), lactopherin (272a), and lipoprotein lipase (272b). Thus, LRP may perform multiple physiological functions which involve the hepatic clearance of lipoprotein remnants containing apoE or lipoprotein lipase as well as the clearance of protease-inhibitor complexes (272,272a,272b). At the present time, however, native lipoprotein ligands that bind to the LRP have not been identified. There are no reported mutations in the LRP in humans or other animal species.

The Modified Low-Density-Lipoprotein Receptor

Functional Definition of the Modified LDL Receptor

It has been estimated that in normal humans two-thirds of LDL is catabolized through the LDL receptor pathway and one-third by a receptor-independent pathway which may reside in scavenger cells (152,153,273). The molecular basis of LDL degradation by scavenger cells has been investigated using mouse peritoneal macrophages and cultures of human monocyte-derived macrophages. Mouse peritoneal macrophages contain a receptor that binds specifically and with high affinity to acetylated LDL, but not to native LDL (274,275). The human monocyte-derived macrophages contain specific, high-affinity receptors for both LDL and modified LDL (276). The bound, modified LDL is internalized and the protein moiety degraded to amino acids (274,275). The regulation of cellular cholesterol homeostasis following degradation of the modified LDL is fundamentally different from that following LDL degradation by reticuloendothelial cells (153). The cholesteryl ester moiety of modified LDL is hydrolyzed, presumably by a nonlysosomal cholesteryl esterase (277). When the cells are grown in medium containing serum, half of the free cholesterol is secreted and the remainder is reesterified by ACAT and stored in the cytoplasm as cholesteryl ester droplets (275). In the continuous presence of modified LDL in the culture medium, the macrophages apparently fail to downregulate their receptor activity and this results in a dramatic increase in cellular cholesteryl ester content (275). Such cholesteryl ester accumulation in the monocyte-macrophages may lead to the formation of the foam cells that are found in atherosclerotic lesions (196,276,278–280). This hypothesis is further supported by the observation that aortic foam cells have receptors for both βVLDL as well as modified LDL (196,281).

Other experiments have shown that, similar to the LDL receptor, the modified

LDL receptor recycles from the cytoplasm to the cell surface every 18 min and this process is inhibited by monensin (282). The physiological importance of this receptor and the manner in which the modified LDL may be actively produced *in vivo* are the subject of extensive investigation. It was initially proposed by Fogelman *et al.* (273) that one form of modified LDL may be produced *in vivo* by modification of the lysines of apoB malonyldialdehyde (MDA). MDA and other short-chain aldehydes are released by platelets or produced during lipoid peroxidation (283). This hypothesis has been supported by observations showing that monoclonal antibodies developed against MDA, modified LDL, and copper-oxidized LDL recognize the MDA–lysine adducts (108). LDL can also be modified and become a ligand of the modified LDL receptor, by incubation with cultured vascular endothelial cells, or by chemical oxidation with Cu^{2+} (106,284,285).

Structure and Functions of the Modified LDL Receptor

The modified LDL receptor was initially purified from tumors induced by injection of the murine macrophage cell line p388D into syngeneic mice (286). The dissociation constant K_d of the modified LDL–modified LDL receptor complex is 30 nM (286). The purified receptor had an apparent Mr of 260 kD and isoelectric point of 6.0 and consisted of three subunits (286). Krieger and colleagues purified the modified LDL receptor to near homogeneity from bovine lung (287). This protein was localized on the surfaces of sinusoidal liver cells and cultures of bovine alveolar macrophages (287). Partial protein sequence of the purified receptor allowed the synthesis of degenerate oligonucleotides that were used as probes to isolate cDNAs encoding the bovine modified LDL receptor (288,289). cDNAs encoding the human modified LDL receptor have also been isolated from a monocyte-derived cell line (THP1) (209). This analysis showed two forms of the human and the bovine receptor, type I (453 amino acids) and type II (349 amino acids) (288,289). Computer analysis of the predicted primary protein sequence showed that the type I receptor contains a 50-amino acid N-terminal cytoplasmic domain, a 260-amino acid membrane-spanning domain, a 32-amino acid membrane-spacer region with two potential N-glycosylation sites, a rodlike structure which contains 163 residues which form an α-helical coiled coil and contains five potential N-glycosylation sites, 72 residues which form a collagenlike domain, and a 110-amino acid, cysteine-rich carboxy-terminal domain. The type II receptor is identical with the type I receptor except that it contains only a six-residue amino-terminal domain. A schematic representation of the different domains of the LDL receptor is shown in Fig. 5C. The rodlike structure is generated in the trimeric receptor by the merger of a triple-stranded, left-handed superhelix formed by the α-helical coiled coil and a right-handed collagenlike triple helix (288). The binding of modified LDL to both types of the receptor is similar (288), suggesting that the carboxy-terminal cysteine-rich domain is not involved in receptor binding. *In vitro* mutagenesis and expression of the mutated receptor indicated that binding of both acetyl LDL and oxidized LDL is diminished by deletion of residues -320 to -342 of the collagenlike domain (155,291). Purified modified LDL receptors and those generated by expression of the receptor cDNA in cells bind both acetyl LDL and oxidized LDL, although the maximum binding of oxidized LDL is 30–50% of that of the acetyl LDL (155,291–293). The binding of acetyl LDL is competed for poorly by oxidized LDL. Furthermore, the binding of oxidized LDL to purified or expressed receptors is competed for completely by acetyl LDL (287). The findings suggest that the same receptor recognizes with different specificities both acetyl LDL and oxidized LDL (155,291). There are no reported mutations in the modified LDL receptor in humans or other animal species.

High-Density-Lipoprotein Binding Sites on Cell Surfaces

Earlier studies showed that apoA-I, as a component of HDL, promotes cholesterol efflux from cells, and through this mechanism might be important in maintaining cellular cholesterol homeostasis (294). Several subsequent studies demonstrated binding of HDL to primary cultures of hepatocytes (295–300), HepG2 cells (300), nonparenchymal cells (296,301), cultured human fibroblasts (302–306), adrenal cells (178,179,298), and adipose cells (62,63) as well as to membrane preparations of rat liver (260,307–309), kidney (309), testis (310), and adipose tissue (311). The number of HDL binding sites per hepatocyte and nonparenchymal cell has been estimated to be 10^5–10^6 (297,301,307,308). The dissociation constant of the HDL–receptor complex is in the range of 10^{-6}–10^{-7} M (295,297,301). The binding is Ca^{2+} independent (260,297,307) and is not affected by acetylation or reductive methylation of HDL (260,297,301) or pronase treatment of the membrane preparations or of cell cultures (260,297,305,307). Various lipoproteins, including HDL_3, HDL_2, HDL with apoE (260), LDL, and VLDL (295,301), as well as liposomes containing apoA-I (312–314), apoA-II (313,34), apoC's (313), apoA-IV (315), and phospholipid vesicles alone (316), compete to varying degrees for the HDL-binding site. Human fibroblasts and arterial smooth muscle cells contain specific, high-affinity sites for HDL free of apoE (304). HDL bound to these sites was not internalized or degraded by these cells; instead it caused an increase in LDL receptor activity, sterol synthesis, and cholesterol efflux. These processes were saturated at HDL concentrations of 20 $\mu g/ml$, as was the specific, high-affinity HDL binding (304). Subsequent studies by the same group of investigators suggested that the binding of HDL to cells promotes the translocation of cholesterol from intracellular membranes to cell surfaces (317,318). The translocated cholesterol can then be removed from the plasma membrane by cholesterol acceptors. Other studies showed that HDL binds to cell surface receptors of human monocytes. It was suggested that, following the binding of HDL to normal monocytes, apoA-I was internalized, transported through cytoplasmic organelles, and finally exocytosed undegraded (319–321). The hypothesis advanced by Oram, Bierman, and colleagues (317) was that regardless of the subsequent events, binding of HDL to cells via apoA-I promotes the efflux of cholesterol from cells, which is subsequently delivered to the liver for excretion or reutilization (322). This process has been termed reverse cholesterol transport (94,323). Other studies suggested that the binding of dimyristoyl-phosphatidyl choline (DMPC) and apoA-I or apoA-IV, but not apoA-II, particles promotes cholesterol efflux from cultured mouse adipose cells (63,315), suggesting different functions for these and possibly for other types of lipoprotein particles in promoting cholesterol efflux. In addition to the role of HDL in promoting cholesterol efflux from peripheral tissues through binding to cells, it has been suggested that HDL may directly or indirectly deliver cholesterol to steroid-hormone-producing cells (178,179,310). Other studies showed that, in addition to HDL, a variety of cholesterol acceptors which do not bind to cells, such as nitrosylated HDL (324,325), albumin phospholipid complexes, phospholipid vesicles alone, and intact erythrocytes, efficiently promote cholesterol efflux from cells, thus suggesting bidirectional equilibration of membrane and HDL cholesterol (325–328). Numerous efforts to clone and sequence the putative HDL receptor have been unsuccessful. Ligand blotting indicated that the molecular mass of an HDL-binding protein varies from 70 to 100 kD (309,318,329). Radiation inactivation studies showed that the subunit molecular mass of the HDL-binding site is 10–16 kD (330,331). Despite the remaining ambiguities, it is generally accepted that HDL directly (through binding to cells) or indirectly (by serving as a cholesterol acceptor) plays an important role in reverse cholesterol transport.

GENETIC VARIATION OF PLASMA ENZYMES AND PROTEINS INVOLVED IN LIPOPROTEIN METABOLISM

Introduction

Following secretion, lipoproteins undergo a series of alterations in plasma, prior to their uptake and catabolism by lipoprotein receptors. These alterations include (a) hydrolysis of chylomicron and VLDL triglycerides by lipoprotein lipase, (b) hydrolysis of lipoprotein triglycerides and phospholipids by hepatic lipase, (c) reverse transport of cholesterol from cells to lipoproteins, (d) esterification of HDL cholesterol by lecithin:cholesterol acyltransferase (LCAT), (e) enzyme-catalyzed exchange or transfer of cholesteryl esters, triglycerides, and phospholipids, and (f) exchange and transfer of apolipoproteins. In this section we review the various enzymes and transfer proteins which are responsible for these alterations.

Lipoprotein Lipase (LPL)

Function and Biosynthesis

LPL hydrolyzes preferentially the 1,3 ester bonds of chylomicron and VLDL triglyercides, generating free fatty acids and mainly 2-monoglycerides (122,123). The enzyme is activated by apoprotein CII (apoCII) (123,332–334) and inhibited by 1 M NaCl (123,335). LPL activity has been demonstrated in postheparin plasma, skeletal muscle, heart, lung, mammary gland, adipose tissue, and brain (123,336–338). LPL synthesis and secretion have also been observed in the murine macrophage cell line J774 (339). High levels of lipoprotein lipase mRNA have been observed in adipose tissue, heart, muscle, adrenal, and brain and low levels of mRNA are detectable in most other tissues with the exception of liver, spleen, and white blood cells (340–343). Figures 7A–7C show the distribution of LPL mRNA in human, cynomolgous monkey, and rabbit tissues. It is interesting that the concentration of the LPL mRNA is high in the fetal human brain but very low in the adult rabbit brain (Figs. 7A and 7C). The LPL enzymatic activity and mRNA levels are regulated by nutrients (344–347) and hormones (348–358). Following synthesis, LPL is modified with N-linked carbohydrate chains. Following secretion, LPL associates with the luminal surface of the vascular endothelium, where it is attached to a membrane-bound glycosaminoglycan and can be released into plasma by injection of heparin (336,359,360). Two forms of rat LPL with high (heart enzyme) and low (adipose tissue enzyme) substrate affinity have been described (335). The mature LPL has a dimeric structure (360a). The reported apparent subunit molecular mass values of LPL were in the range of 60–70 kD (123) and the carbohydrate content was 3–10% (334,336,361,362). Inhibition of LPL by antibodies (338) or genetic defects of LPL (361–363) and/or apoCII (364,365) cause excessive accumulation of VLDL and chylomicrons in plasma (363–365). This indicates that the initial hydrolysis of VLDL and/or phospholipids by lipoprotein lipase is a necessary step for the subsequent catabolism of these particles.

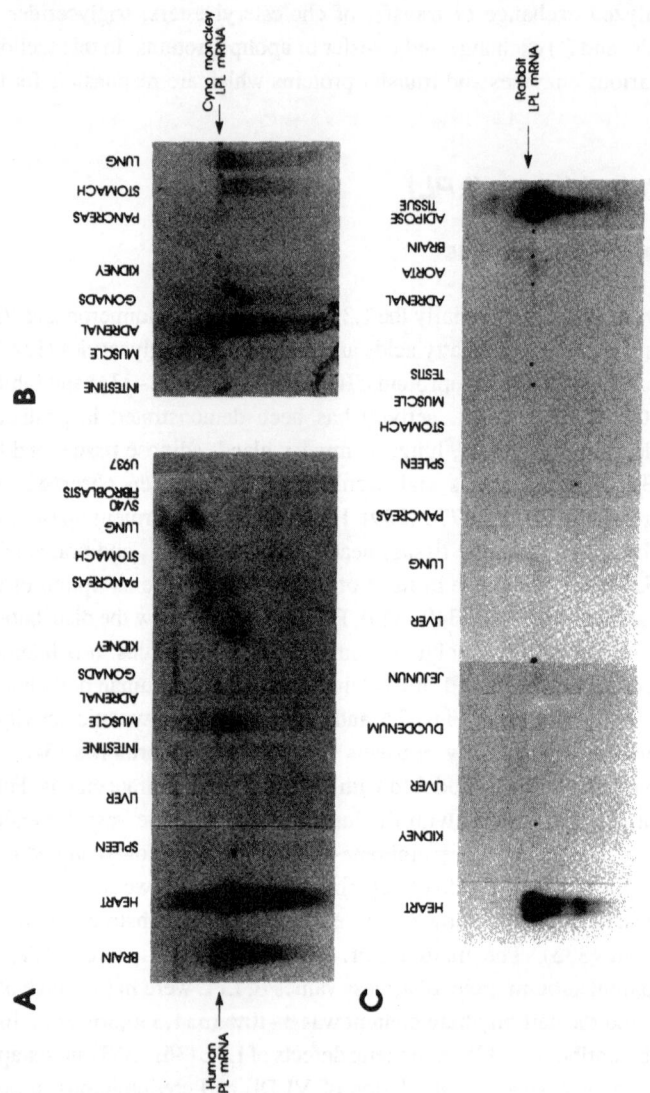

Fig. 7. Distribution of LPL mRNA in (A) fetal human, (B) adult cynomolgous monkey, and (C) adult rabbit.

Structure and Genetics

The structure of the human LPL has been deduced from the cDNA sequence of clones encoding this protein (343). The human lipoprotein lipase gene contains ten exons and nine introns and encodes a protein of 475 amino acids, including a 27-residue signal peptide (366,367). The human LPL maps on chromosome 8, region 8p22 (366,367). The human lipoprotein lipase belongs to a larger lipase gene family which encompasses hepatic lipase (HL) and pancreatic lipase (368). All three proteins have a triad of catalytic residues, Asp–His–Ser. These residues were previously described in the active site of pancreatic lipase as well as in serine esterases such as chymotrypsin and were thought to participate in the catalytic mechanism through a putative charge relay system. The Ser residue assumed to participate in the catalytic mechanism occurs in a consensus sequence Gly–X–Ser–X–Gly previously found in several proteases (362). Mutagenesis of this residue in the hepatic lipase abolished the enzymatic activity (369). Following synthesis, LPL is N-glycosylated at Asn residues 43 and 359. *In vitro* mutagenesis at residue 43, which prevented glycosylation at this site, inhibited the enzymatic activity and abolished the secretion of LPL (370). Furthermore, the cld/cld mouse (371,372) produces an inactive form of lipoprotein lipase and hepatic lipase. The N-linked carbohydrate chains of this protein cannot be processed from high-mannose to complex carbohydrate and as a result the protein is not secreted (371,372). Mutations in LPL are associated with type I hyperlipoproteinemia, an autosomal recessive disorder which is characterized by the presence of chylo-microns and triglycerides in fasting plasma, pancreatitis, eruptive xanthomas, abdominal pain, and lipemia retinalis (122,123,373–375). The frequency of the disease is $1/10^6$ for homozygotes and 1/500 for heterozygotes (362). Subjects heterozygous for LPL defects may also have reduced levels of adipose tissue LPL activity (376). Depending on other genetic or environmental factors, these hetero-zygous subjects may present a lipid profile resembling that of familial combined hyperlipidemia (377,378). The activity of LPL is inversely related to the HDL cholesterol levels. Thus, low HDL levels observed in LPL-deficient subjects may represent a risk for atherosclerotic disease (377). The mutations in lipoprotein lipase have been classified in three broad classes. Class I patients lack any immunoreactive LPL mass in their postheparin plasma. This condition results from nonsense mutations early in the protein sequence or deletions/insertions in the LPL gene. Class II mutations contain immunoreactive LPL mass in the post- but not in the preheparin plasma. These mutations are clustered in exons 4–6, which contain residues involved in the binding of lipids as well as the catalytic triad consisting of Ser-132, Asp-156, and His-241 (379). Finally, class III mutations have immunoreactive LPL mass both in the pre- and postheparin plasma. These mutations may cause conformation changes which alter the heparin-binding site of LPL, which is encoded by exon 2 (376). The mutations in the lipoprotein lipase gene have been reviewed recently (361,362) and are shown in Fig. 8A (380–392,392a–g). The functional significance of some of these mutations has been determined by *in vitro* mutagenesis and expression of the LPL cDNA and functional analysis of the mutant LPL forms (376,387,388,392,392c,392d,392g).

Hepatic Lipase (HL)

Function and Biosynthesis

Early biochemical and genetic evidence suggested that HL is a lipolytic activity distinct from LPL (393–397). The purified enzyme has been characterized as a glycoprotein with subunit molecular mass of 62.5 kD (120,398) or 53 kD (399,400). HL does not require apoCII for activation (123), is not affected by protamine and high salt (336,393,395), and is inhibited by sodium dodecyl sulfate (SDS) (396). Protein and mRNA analysis has shown that HL is synthesized in the hepatocytes (361,401). Following synthesis, HL is modified with N-linked carbohydrate chains. This modification is required for the secretion of HL (402). Immunofluorescence and immunoelectron microscopy studies have shown that HL is localized on the surface of sinusoidal endothelial cells (403) and can be released from there by treatment with heparin (404).

Structure and Genetics

The structure of the human hepatic lipase gene has been deduced from the sequence of cDNA and genomic clones encoding for this protein (405,406). The human hepatic lipase gene contains nine exons and eight introns and encodes a protein of 499 amino acids, including a 23-residue signal peptide. The human HL gene maps on chromosome 15, region 15q21 (367,405), and has similar intron–exon organization as the LPL gene (368). The Ser involved in catalysis is localized at amino acid 145. Deficiency of HL in three kindreds is associated with increased plasma cholesterol and triglycerides, abnormal lipoprotein profiles, and premature atherosclerosis (407–409). In contrast with LPL-deficient patients, HL-deficient patients have elevated plasma HDL and βVLDL and increased concentration of triglycerides and phospholipids in the LDL and HDL (397). The unusual lipoprotein particles that accumulate in the plasma of these patients may constitute the

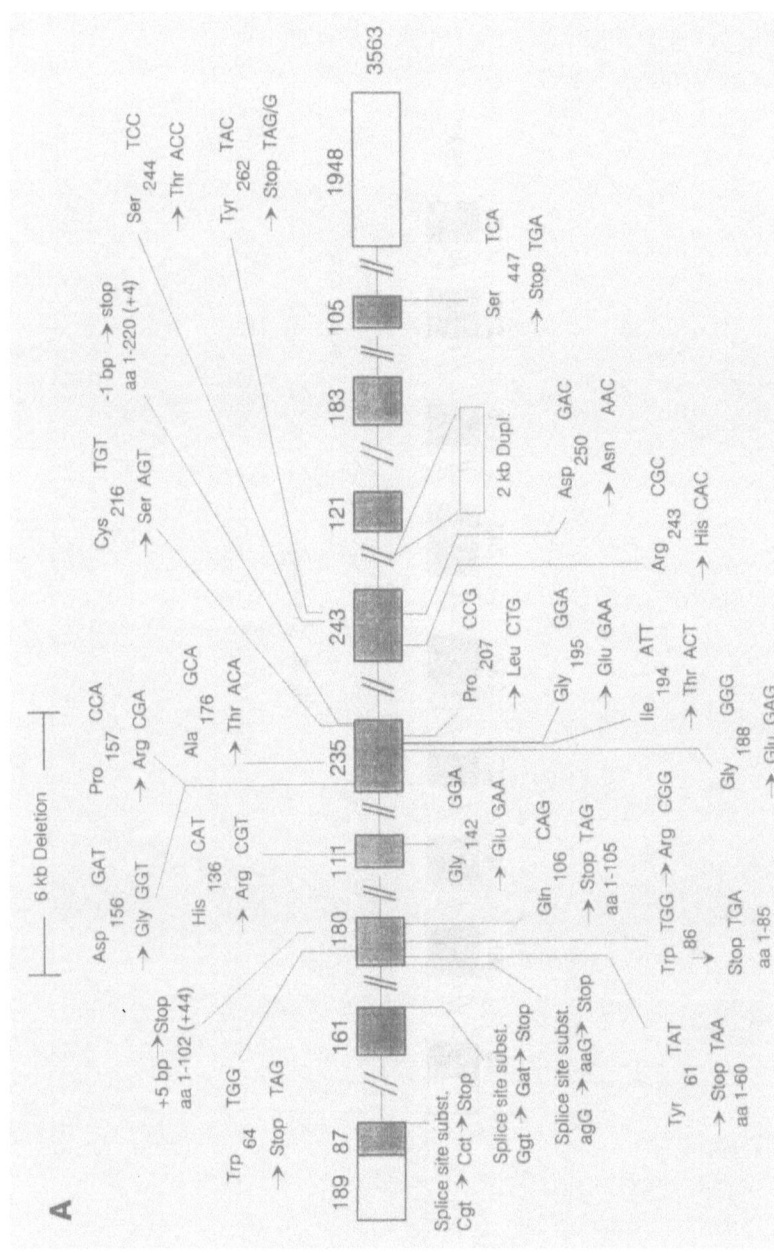

Fig. 8. (A) Schematic representation of the human lipoprotein lipase gene and the various mutations associated with familial type I hyperlipoproteinemia. The mutations are described in refs. 361, 362, 376, 380–392, 392a–g. (B) Schematic representation of the human hepatic lipase gene showing the position of a heterozygous mutation associated with HL deficiency (410).

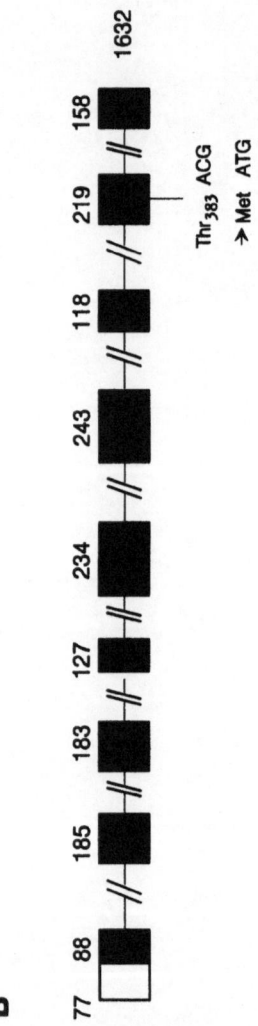

Fig. 8. (*Continued*)

natural substrates of normal HL. A substitution of Met-383 for Thr was observed in two of the families with complete deficiency of HL (410) (Fig. 8B). Since the subjects were heterozygous for the Met-383 mutation, it has been postulated that a second mutation may also exist in the other allele which causes a complete absence of enzymatic activity.

Lecithin:Cholesterol Acyltransferase (LCAT)

Function and Biosynthesis

LCAT is the enzyme responsible for the formation of cholesteryl esters in plasma (411). It has been proposed that LCAT forms a complex in plasma with components of HDL (412) and catalyzes the esterification of HDL cholesterol using HDL lecithin as the acyl donor. The C-2 fatty acyl group of lecithin is preferentially transferred and the specific fatty acyl groups at the C-1 and C-2 positions of lecithin affect the rate of the reaction (413,414). The products of the reaction are 2-lysolecithin and cholesteryl ester. LCAT also catalyzes the reverse reaction (esterification of lysolecithin), and this process is activated by LDL (415,416), the hydrolysis of lecithin to lysolecithin and free fatty acids (416), and the transfer of fatty acyl groups from esterified to free cholesterol molecules (417). LCAT is activated by apoA-I and to a lesser extent by apoCI (418–420), apoA-IV (421,422), and apoE (423). LCAT is inhibited by sulfhydryl reagents (411). It was proposed recently that the cysteine residues 31 and 184 may participate in the catalytic mechanism (424). However, *in vitro* mutagenesis of both residues did not affect the enzymatic activity of the mutant enzyme (425). The preferred substrates of LCAT are nascent HDL cholesterol (77,426–430), HDL_3 cholesterol, and to a lesser extent HDL_2 cholesterol (431) and LDL cholesterol (432). VLDL cholesterol is a poor substrate. The rate of HDL cholesterol esterification depends on the cholesteryl ester content of acceptor VLDL and LDL particles (433), suggesting that the synthesis and transfer of cholesteryl esters in plasma may be functionally linked. It has been proposed that the nascent discoidal βHDL particles containing phospholipid and apoA-I or apoA-I and apoA-II (23,434) may extract cholesterol from cells (63) as well as the LDL (430,432,435). The complex may further acquire apoA-IV (436,437), which facilitates the esterification reaction, and cholesteryl ester transfer protein (C/ETP) and possibly apoD (434,437,438). The sequential action of LCAT and C/ETP in this complex results in the conversion of the precursor HDL particles to spherical HDL_3 particles, and the transfer of newly formed cholesteryl esters from HDL_3 to VLDL and LDL (439). Discoidal HDL particles detected in the plasma of LCAT-deficient patients, HepG2 culture medium, and peripheral lymph are most likely lipid bilayer in structure, and are converted to spherical plasmalike HDL upon incubation with LCAT (77,428,429). It has been shown that "core-poor" lipid bilayer discs formed *in vitro* by mixing apoA-I with phospholipid-cholesterol vesicles are also transformed into cholesteryl-ester-containing spheres upon incubation with LCAT (440). The abnormal lipoprotein particles found in the plasma of LCAT-deficient patients may indeed represent intermediates of lipoprotein metabolism that do not reach significant concentrations in normal plasma.

LCAT is synthesized mainly by the liver, is modified with N-linked carbohydrate chains, and is subsequently secreted into plasma (441,442). LCAT mRNA has been detected in liver HepG2 cells, brain, and testis (443). Purification of the plasma enzyme suggested that LCAT is a glycoprotein which contains 24% carbohydrate and has a molecular mass of 59–70 kD.

Structure and Genetics

The structure of the human LCAT has been deduced from the sequence of cDNA and genomic clones encoding this protein (444,445). The LCAT gene contains six exons and four introns and encodes a protein of 440 amino acids, including a 24-residue signal peptide, with Mr = 53.2 kD (445). The LCAT gene maps on the long arm of chromosome 16, region 16q22 (443). Familial LCAT deficiency is a rare autosomal recessive disorder characterized by corneal opacities, xanthomas, and lipoproteins of abnormal shape and composition (132,428). The concentrations of plasma unesterified cholesterol, phosphatidyl choline, and triglycerides are highly elevated, whereas the plasma concentrations of lysolecithin, HDL and LDL cholesterol, apoB, and apoA-I are decreased. The lipoprotein abnormalities that result from LCAT deficiency include βVLDL enriched in unesterified cholesterol, large- and intermediate-size LDL_2 particles depleted of apoB, and the discoidal HDL particles (131,132,428,446) with abnormal lipid and apoprotein composition. The patients' VLDL has increased apoE and apoCI and decreased apoCII and apoCIII concentrations (132,447). The large- and intermediate-size LDL_2 have an increased concentration of cholesterol and lecithin and decreased concentration of cholesteryl esters (132). The discoidal HDL particles contain either apoE or apoA-I and apoA-II and have increased concentration of unesterified cholesterol and lecithin compared to normal HDL (132). These particles are most likely lipid bilayers in structure and resemble the discoidal HDL particles found in the peripheral lymph (429) and those produced by the perfused liver (426,448), the intestine (427), and cultures of HepG2 cells (76). There are numerous tissue and organ abnormalities (428) which probably result from excessive amounts of unesterified cholesterol and lecithin in plasma

(132,224,449,450). The most serious complication of the disease is renal failure (132,451–453). Initial quantitation of plasma LCAT mass in deficient patients by radioimmunoassay revealed 0–20% of normal amounts of LCAT (454–457). The LCAT activity in heterozygote subjects is 50% of the normal values (456,457). The reduction of the enzymatic activity in affected individuals was consistent with structural mutations in the LCAT gene that inactivate the enzyme and affect to various degrees the synthesis and/or the stability of the mutant LCAT protein (454,458). The molecular defects underlying some cases of LCAT deficiency have been determined by amplification of genomic DNA obtained from LCAT-deficient patients. These include eight homozygous substitutions: one bp insertion following Pro-10, Tyr-83→Stop, Ala-93→Thr, Tyr-156→Asn, Arg-158→Cys, Lys-209→ Pro, Asn-228→Lys, Met-293→Ile, and Thr-321→Ile; a compound heterozygote with a substitution of Arg-135→Trp in one allele and a frameshift after Ile-375 in the other allele; an insertion of a Gly residue between amino acids 140 and 141; and a compound heterozygote with an Arg-147→Trp substitution in one allele plus an unidentified mutation in the other allele (459–461,461a–d,463,463a). These mutations are shown in Fig. 9A. *In vitro* mutagenesis and expression studies showed that in one of the cases a substitution of Arg-147 to Trp produced an unstable LCAT form that was degraded intracellularly (459). A special case of LCAT deficiency is the fish-eye disease (428,462,463). This condition is characterized by massive corneal opacities, greatly reduced HDL cholesterol and apoA-I levels, mild hypertriglyceridemia, and small-size HDL with increased amounts of unesterified cholesterol. Fish-eye disease results from two structural mutations in the LCAT gene which caused either a homozygous substitution of the Thr-123 to Ile (463) or substitution of Thr-123→Ile in one allele and Thr-347→Met in the other (363a). (Fig. 9A). This mutation eliminates the esterification of cholesterol on the HDL particles (αLCAT activity), but does not affect the ability of the enzyme to esterify cholesterol on the LDL particle (βLCAT activity). This interesting mutation demonstrates that the LCAT substrates HDL and LDL may bind to different domains on the catalytic site of the enzyme.

FACILITATED EXCHANGE AND TRANSFER OF LIPID AMONG LIPOPROTEIN CLASSES

Introduction

Several investigators have observed that the lipid and protein moieties are not statically fixed on the lipoprotein molecules, but exchange readily among lipoprotein classes. The specific enzymatic activities and the other factors that may

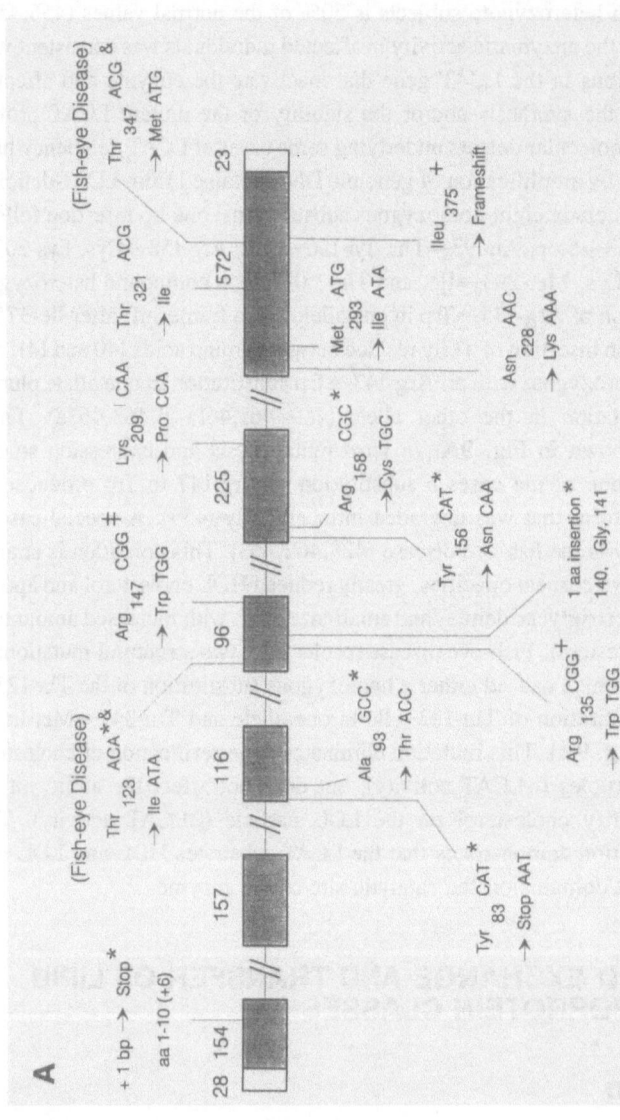

Fig. 9. (A) Schematic representation of the human LCAT gene and the various mutations associated with LCAT deficiency. The mutations are described in refs. 459–461, 461a–d, 463, 463a. +, Compound heterozygote carrying an Arg-135→Trp and a frameshift after Ile-375. ‡, Compound heterozygote carrying an Arg-147 substitution and an unknown mutation. &, Fish-eye disease resulting from a heterozygote substitution of Thr-123→Ile and Thr-347→Met. Asterisk indicates homozygous substitution. (B) Schematic representation of the human C/ETP gene and the mutation which causes C/ETP deficiency (485).

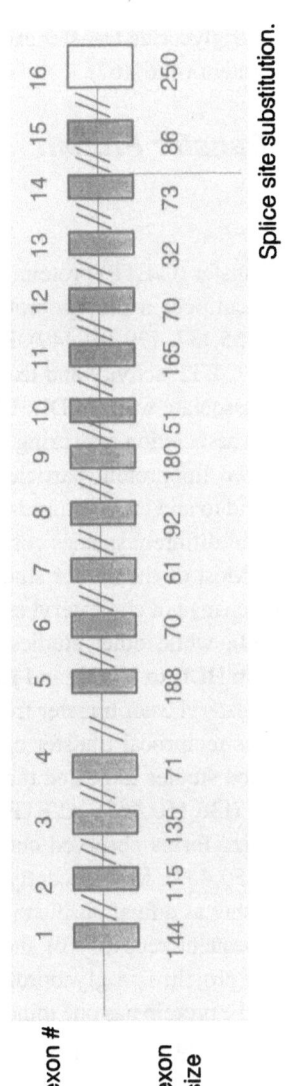

Fig. 9. (*Continued*)

regulate the change of composition of the lipoprotein molecules are the subject of extensive investigations. Two activities have been identified in plasma that catalyze the exchange and/or transfer of lipid moieties between lipoproteins. These include the cholesteryl ester/triglyceride transfer exchange protein (464,465) and the phospholipid transfer protein (466,467).

Cholesterol Ester Transfer Protein

Functions

The cholesteryl ester transfer (C/ETP) protein, also designated lipid transfer protein I (467), has been identified in the lipoprotein-free plasma of all mammalian species studied (134,135,137–139,141,149). Rats, mice, and pigs (141,142, 149) have very low levels of C/ETP activity and tissue mRNA (468). It has been shown that C/ETP can also associate with VLDL, LDL, and HDL and phospholipid vesicles (469,470). This association may bring about cholesteryl ester and/or triglyceride transfer when two lipoprotein particles form a transient complex. Alternatively, C/ETP may bind to and transfer lipids from one lipoprotein class to another. The C/ETP activity of different species correlates with the concentration of VLDL cholesterol (142). Most of the earlier studies indicate that this protein promotes the bidirectional exchange of cholesteryl ester of triglycerides among all lipoprotein classes (136–139), while other studies indicate that there is a net cholesteryl ester transfer from HDL to VLDL and LDL (143–146). Several early studies also suggest that cholesteryl ester transfer from one lipoprotein particle to another is associated with the reciprocal transfer of triglycerides (143,145,147). The earlier protein purification studies indicated that C/ETP is a glycoprotein of molecular mass 62–78 kD (136,150,467). C/ETP is susceptible to oxidative degradation. The different size forms observed could be the result of oxidative degradation of the enzyme (150,467). More recently the C/ETP has been purified to homogeneity (464,465) using as a final purification step the binding of C/ETP to lipid emulsions and subsequent recovery of the lipid-bound C/ETP by gel filtration (465). The purified protein is a glycoprotein. The apparent Mr of the purified C/ETP is 74 kD and the protein has one minor and one major isoform with apparent isoelectric points 5.6 and 5.3, respectively. The purified C/ETP displayed five times greater cholesteryl ester transfer activity than triglyceride transfer activity from the triolein emulsions to HDL and catalyzes the transfer of phospholipids from phospholipid vesicles to HDL (465). The protein contains binding sites for cholesteryl ester, triglycerides, and phospholipids. The lipids bound to C/ETP could be transferred to LDL (471). Monoclonal and polyclonal antibodies to C/ETP neutralize all the cholesteryl ester transfer protein activities, possibly by blocking the binding of these lipids to C/ETP. The same antibodies neutralized only half of the phospholipid transfer activity (472,473). Injection of monoclonal antibodies which inhibit the C/ETP activity into rabbits reduced catabolism of HDL cholesteryl esters and the overall clearance of cholesteryl ester from plasma (474–476).

Structure, Biosynthesis, and Genetics

The primary sequence of C/ETP has been deduced from the sequence of cDNA and genomic clones (477,478). The protein is highly hydrophobic and consists of 493 amino acids, including a 17-residue signal peptide. The calculated Mr of the unmodified protein is 53.1 kD. The human C/ETP gene is 25 kb long and contains 16 exons and 15 introns and maps on chromosome 16 (478). C/ETP mRNA synthesis occurs primarily in the liver, small intestine, spleen, adipose tissue, muscle, and the adrenal and to a lesser extent in the kidney (477,479). The C/ETP mass and mRNA increase 30-fold in response to atherogenic diets (480). C/ETP synthesis and secretion has also been demonstrated in HepG2, Caco-2 cells, and human monocyte-derived macrophage cultures (480–483). Studies with HepG2 cells have shown that C/ETP undergoes intracellular glycosylation with N-linked carbohydrate chains (482). These modifications may account for the increased Mr of the secreted protein as compared to the calculated Mr of 53.1 kD based on the amino acid sequence. Our understanding of the physiological significance of the C/ETP has increased recently following the description of a subject with familial hyperalphalipoproteinemia associated with C/ETP deficiency (484). The plasma cholesteryl ester and triglyceride transfer activity in the proband is zero, while the phospholipid transfer activity is 47% of the control value (474,485). This observation provides conclusive genetic evidence that additional plasma proteins are involved in phospholipid transfer. The C/ETP deficiency is characterized by increased HDL cholesterol and plasma apoA-I, apoA-IV, and apoE levels, decreased LDL cholesterol and plasma apoB levels, and the presence of HDL_c-type particles in plasma enriched in apoE (474,486,487). The increase in HDL and apoA-I levels results from decreased apoA-I and HDL catabolic rates (488). Heterozygote subjects have 50% normal cholesteryl ester and triglyceride transfer activity and near normal phospholipid transfer activity (486). The C/ETP-deficient subject has no clinical signs of atherosclerosis and this is consistent with the overall lipid profile. It has been well established that high HDL cholesterol and apoA-I levels combined with low LDL cholesterol and apoB levels is protective against atherosclerosis and confers longevity (489–493).

Analysis by radioimmunoassay and Western blotting showed complete absence of plasma C/ETP protein in the C/ETP proband (485). Molecular analysis of the gene of the homozygous proband and heterozygote relatives showed a G-to-A transition at the splice site of intron 14 (Fig. 9B), which apparently accounts for the absence of C/ETP activity in the plasma of the homozygous C/ETP-deficient subject (485). In contrast to the C/ETP deficiency syndrome, overexpression of the human C/ETP gene in transgenic mice reduced significantly the HDL and apoA-I levels, reduced the size of HDL, and stimulated cholesterol esterification in all lipoprotein fractions (494–496).

Lipid Transfer Protein II (LTP-II)

As discussed earlier, monoclonal and polyclonal antibodies directed against C/ETP abolish the cholesteryl ester and triglyceride transfer activities, but suppress the phospholipid transfer activity to approximately 50% (472), suggesting that other proteins may be involved in lipid transfer. Prior work by Albers *et al.* led to the characterization of two distinct lipid transfer proteins, LTP-I and LTP-II (467). LTP-I is identical to C/ETP and is heat stable following heat treatment at 58°C for 1 hr. LTP-II was purified from human plasma. The protein has an apparent molecular mass of 69 kD, apparent isoelectric point of 5, and is labile following heat treatment at 58°C for 1 hr (497). The LTP-II facilitates the exchange and net transfer of phospholipids from VLDL to HDL, lacks cholesteryl ester and triglyceride transfer activities, and is not recognized by anti-C/ETP antibodies (497). Conclusive genetic evidence for the existence of a distinct phospholipid transfer protein has been provided by the C/ETP deficiency syndrome. As discussed earlier, the plasma of the homozygous C/ETP-deficient subject who totally lacks any immunoreactive C/ETP contains 47% phospholipid transfer activity (474).

Lipid Transfer Inhibitory Protein

Morton and Zilversmit initially purified from plasma a protein of molecular weight 35,000 and isoelectric point ≥4 which inhibited the transfer or exchange of cholesteryl ester and triglycerides between lipoproteins (498,499). More recently, a protein of Mr = 29 kD and apparent isoelectric point of 4.6 was isolated from HDL. This protein, which was designated lipid transfer inhibitor protein (LTIP), inhibited the cholesteryl ester, triglyceride, and phospholipid transfer activities of C/ETP as well as the phospholipid transfer activity of LTP-II (500). Removal of the

LTIP by immunoaffinity column from human, rat, and pig plasma increased the C/ETP by 17, 125, and 200%, respectively (500). This observation may account partially for the low levels of C/ETP activity observed in rat and pig plasma (137,141).

The Role of Plasma Enzymes and Apolipoproteins in the Reverse Transport of Cholesterol

Glomset and colleagues have proposed that HDL may be involved in the reverse transport of cholesterol from the peripheral tissues to the liver and other organs (411,449). According to this hypothesis, HDL removes the excess of cholesterol from plasma membranes of tissues or red blood cells. The free cholesterol is esterified by LCAT (411) and this esterification reaction is activated by apoA-I and to a lesser extent by apoA-IV, apoCI, and apoE (418,419,421–423). The esterified cholesterol transferred by the C/ETP to VLDL and LDL (134–141, 143–150,475), with reciprocal transfer of triglycerides, is subsequently removed from the circulation by LDL and/or the apoE/α2M receptors (78,79). Additional triglycerides and phospholipids may be transferred to HDL following hydrolysis of VLDL as well as by the action of LTP-II (497). A portion of the cholesteryl ester delivered to the liver is secreted into the bile either as free cholesterol or in the form of bile acids. Direct uptake of apoE-containing HDL by the LDL and apoE/α2M receptors is possible. Overall the concerted action of LCAT, apoA-I, and C/ETP on the HDL particle plays a pivotal role in the reverse transport of cholesterol. In addition, other apolipoproteins (apoCI, apoE, apoA-IV) and plasma enzymes, such as the lipases, LTP-II, and LTIP, may modulate the overall reverse transport of cholesterol.

APOLIPOPROTEINS: STRUCTURE, FUNCTION, AND GENETIC VARIATION

Apolipoprotein A-I

Structure

ApoA-I is the major protein component of high-density lipoprotein (HDL). Plasma apoA-I consists of a single polypeptide composed of 243 amino acids of known sequence (501,502). Both the amino acid (503–505) and the nucleotide (21,89,93,94) sequence analyses of apoA-I showed that the protein and the gene sequence contain repeated units (21). The protein repeat consists of 22 amino acids and contains two symmetrical 11-residue segments with similar arrangements of charged and hydrophobic residues (503). ApoA-I contains eight 22-amino acid repeats and two 11-amino acid repeats (Table V) (21).

Similar repeated units have been found in the other exchangeable apolipoproteins (21). The 22-mer units have amphipathic α-helical arrangements (506–508) which have been suggested to constitute the lipid-binding domains of the apolipoproteins (506). According to the model proposed by Segrest, the amphipathic α-helices have two phases, one hydrophobic, which can associate with the hydrophobic fatty acyl chains of the phospholipids, and another hydrophilic, which can associate with the phospholipid head groups and the aqueous phase (506). More efficient binding to lipids has been achieved with a 44-mer composed of two repeated units having a proline residue in the middle (509,510). The concept of amphipathic α-helices has been verified recently with the derivation of the three-dimensional structure of migratory locust apolipophorin III (apoLp-III) (511) and the amino-terminal portion of human apolipoprotein E (apoE) (512). Both molecules fall into the category of up-and-down α-helical bundles. ApoLp-III forms five helix bundles. The amino-terminal portion of apoE also forms an elongated four-helix bundle. It is reasonable to speculate that apoA-I may consist of similar sets of up-and-down amphipathic α-helical bundles to those observed in apoLp-III and apoE. Analysis of the apoA-I structure by monoclonal antibodies showed that the epitopes of the lipid-free protein in the region of the repeated units 4 and 5 recognize one but not both of these repeats. This suggests that these helices

TABLE V. ApoA-I Repeated Units[a]

Repeat number	Residue		Residue
1	44	L K L L D N W D S V T S T F S K L R E Q L G	65
2	66	P V T Q E F W D N L E K E T E G L R Q E M S	8
3	88	K D*L E E V K A K V Q	98
4	99	P Y L D D*F Q K K*W Q E*E M E L Y R Q K V E	20
5	121	P L R A E L Q E G A R Q K L H E*L Q E*K L S	142
6	143	P L G E E*M R D R A R A H V D A*L R T H L A	164
7	165	P*Y S D E*L R Q R*L A A R*L E A L K E N G G	186
8	187	A R L A E Y H A K A T E*H L S T L↑S E K A K	208
9	209	P A L E D*L R Q G L L	219
10	220	P V L E S F K V S F L S A L E E Y T K K L N	

[a]Asterisk indicates amino acid substitution (see Fig. 11A). Arrow indicates frameshift causing incorporation of 27 random carboxy-terminal residues.

may be continuous or at an angle (513). In contrast, antibodies recognizing repeated units 6 and 7 have overlapping epitopes, suggesting that these helices may have antiparallel configuration, possibly stabilized by leucine zippers described previously for the eukaryotic transcription factor C/EBP (514). Other monoclonal antibodies have discontinuous epitopes which recognize residues 45–51, 83–92, 119–126, and 136–143, suggesting that these apoA-I regions are in close proximity and that this region has a compact tertiary structure. There were no monoclonal antibodies recognizing the carboxy-terminal repeats 9 and 10 (513).

ApoA-I Biosynthesis

ApoA-I synthesis has been demonstrated predominantly in the liver and the intestine in most (16–18,25,26) but not all (18) mammalian species. Tissue apoA-I mRNA levels are controlled by hormonal (515–519) and nutritional (520,521) factors and are developmentally regulated (515,522). It has also been reported that the steady-state apoA-I mRNA levels also increase during liver regeneration and cirrhosis by posttranscriptional control mechanisms (523). Newly synthesized apoA-I contains a six-residue amino-terminal prosegment (90,91) which is subsequently removed extracellularly by proteolysis (98–100) (Fig. 2D). In plasma and media of HepG2 cells apoA-I is found predominantly in two populations of lipoprotein particles containing either only apoA-I (LpAI) or both apoA-I and apoA-II (LpAI:AII) (23,64,66–69,71,73,524). The site of assembly and the functional significance of these particles are not known. LpAI particles of Mr-269 kD isolated from plasma bind saturably and are taken up by surface receptors present in HepG2 cells. The binding and uptake of LpAI particles was significantly higher than the binding of LpAI:AII particles (68). Previous studies suggested that the binding of LpAI but not of LpAI:AII particles promotes cholesterol efflux from cultured mouse adipose cells (62,63), suggesting different functions for these two and possibly for other types of Lp particles.

ApoA-I Functions

Binding of ApoA-I to Lipids, Formation of HDL. The association of apoA-I with phospholipids leads to the formation of discoidal particles and increases the helical content of apoA-I from 55 to 75% (525,526). The diameter of the particles decreases from 150 to 75 Å as the ratio of phospholipid to protein decreases (22). X-ray scattering analysis showed that at a 2.5:10 DMPC to apoA-I ratio the discs have diameter 110 Å with a bilayer thickness of 55 Å. The apoA-I covers the hydrocarbon chains of the phospholipids on the outside perimeter of the disc (527). The interaction of apoA-I with liposomes increases in the presence of small amounts of cholesterol (528). The dissociation constant K_d of apoA-I from proteoliposomes (PL) of egg yolk lecithin vesicles and the saturation number N (moles apoA-I/1000 moles PL) are 0.9 μM and 1.74, respectively (529). For proteoliposomes containing 20% cholesterol or small emulsions containing egg yolk lecithin and triolein the K_d is 0.3 and 0.2 μM, respectively, and the N is 3.7 and 2.6, respectively (529,530). Finally, the dissociation constant of apoA-I from model lipid emulsions (microemulsions and large emulsions containing egg yolk lecithin and triolein) range from 0.16 to 4.24 μM and the saturation number N ranges from 0.6 to 3.9, depending on the size and the composition of the lipid emulsions (22,529).

Activation of the Enzyme Lecithin:Cholesterol Acyltransferase (LCAT). HDL is the site of cholesterol esterification by lecithin:cholesterol acyltransferase (438). ApoA-I is an effective cofactor for LCAT activity (418), particularly with physiological lecithins, which typically are long-chain molecules with 1-saturated and 2-unsaturated acyl groups. ApoA-I also stimulates the efflux of cholesterol from peripheral cells, providing substrate for the LCAT reaction (531). As a result of these activities apoA-I may play an important role in regulating the cholesterol content of peripheral tissues through the putative reverse cholesterol transport pathway (438,531). The lipid bilayer discs formed by binding of apoA-I to phospholipid vesicles containing cholesterol are efficient substrates for LCAT (440,532). Discoidal HDL particles detected in peripheral lymph and the culture medium of HepG2 cells are also converted to spherical plasmalike HDL upon incubation with LCAT (429). Following the introduction of the amphipathic α-helix hypothesis in 1974 (506), several studies have examined the ability of peptide fragments of apoA-I and a variety of synthetic peptides to associate with phospholipid vesicles. This association requires a minimum length of approximately 20 amino acids (5.5 helical turns) and sequence organization into distinct polar and nonpolar faces (533). Several laboratories have shown that synthetic peptides with these characteristics bind phospholipid and activate LCAT, although enzyme activation appears to have been in all cases significantly less effective than that obtained with native apoA-I in reactions with physiological lecithins (440). Specifically, LCAT activation, approximately 25–30% of native apoA-I, has been attained with the amino (1–85)- and carboxyl (149–243)-terminal cyanogen bromide fragments derived from native apoA-I (419) and synthetic peptides corresponding to the 145–185 and 121–164 apoA-I sequences (420,534). Studies by Sparrow and Gotto using synthetic peptides showed that the 148–185 region can both bind phospholipids and activate LCAT, whereas the carboxy-terminal region 200–243 can bind to phospholipids but cannot activate LCAT (534). Activation of LCAT was also achieved with synthetic peptides corresponding to the 121–164 region (420). Finally, substitution of Pro-143 or Pro-165 for arginine results in LCAT activation which is 40–70% of normal (535,536). Preliminary results in our laboratory with mutant apoA-I forms also suggest that the carboxy-terminal 149–243 apoA-I region is important for LCAT activation. Other studies suggest the importance of a more upstream region of the apoA-I molecule for LCAT activation. Thus, deletion of Lys-107 results in LCAT activation which is approximately 40–60% of normal (537). In addition, monoclonal antibodies which bind to the 95–121 apoA-I sequence inhibit the activation of LCAT by apoA-I (538). Based on experiments with 44-residue synthetic peptides consisting of a dimer resembling the consensus apoA-I repeated unit, Anantharamaiah et al. (510) have proposed that residues 66–121, particularly Glu-78 and Glu-113, are important for LCAT activation. This region contains two 22-amino acid repeats which can form amphipathic α-helices separated by one 11-amino acid repeat (21). Furthermore, the sequence between amino acids 66 and 98 is one of only two well-conserved regions of apoA-I (21,539). Screening for amino acid substitutions which change the charge of the protein showed that substitutions of basic or acidic amino acids and prolines occurred with high frequency in the 103–198 region (539). The various apoA-I mutations observed are marked with asterisks in Table V. Analysis of several mutations shows that, with the exception of the Ala-158 to Glu substitution, none of the other mutations represented alterations of the hydrophobic residues of the amphipathic α-helices (535–537,540–545). Preliminary data in our laboratory showed that amino acid substitution within this region has no effect and an internal deletion caused a 50% reduction in LCAT activation (97). The analysis based on synthetic peptides, natural mutations, and large protein deletions can only provide suggestive evidence on the domains of apoA-I which are important for the activation of LCAT and possibly other functions.

Binding of HDL and ApoA-I Proteoliposomes to Cell Surfaces; Reverse Cholesterol Transport. It was shown earlier that apoA-I, as a component of HDL, promotes cholesterol efflux from cells, and through this mechanism might be important in maintaining cellular cholesterol homeostasis (294,531). Glomset and colleagues have proposed that HDL may be involved in the reverse transport of cholesterol from the peripheral tissues to the liver and other organs (411,449). According to this hypothesis, HDL attracts the excess of cholesterol from plasma membranes of tissues or red blood cells and the free cholesterol is esterified by LCAT (411). HDL either acquires apoE and is cata-bolized directly by some tissues (178,179,310) or transfers its cholesteryl esters to VLDL and LDL (310,464,474), which are subsequently removed from the circulation by the LDL receptor. A portion of the cholesteryl esters delivered to the liver is secreted into the bile either as free cholesterol or in the form of bile acids (546). Several subsequent studies have attempted to identify a cell surface receptor which recognizes apoA-I (309,312–315,318,319,329). Although ligand blotting has identified a protein of 70–110 kD in different cell types (309,318,329,547), efforts to clone and sequence a putative HDL receptor have been unsuccessful.

Importance of ApoA-I

High HDL levels are protective against atherosclerosis. Epidemiological and genetic data have shown convincingly that low (489,490) or high (491–493) levels of HDL or apoA-I are associated with increased or decreased risk of developing atherosclerosis, respectively, indicating the importance of apoA-I in cellular cholesterol homeostasis. The protective role of apoA-I against atherosclerosis has been suggested recently with transgenic mouse experiments. Overexpression of the human apoA-I gene in transgenic mice generated HDL_3- and HDL_2-type particles and increases in HDL cholesterol and apoA-I to 60–100% of control (548,549). The transgenic mice overexpressing the human apoA-I gene were fed with atherogenic diets containing 15% (w/w) fat (cocoa fat or dairy butter) and 1–1.25% (w/w) cholesterol for 14–18 weeks. These atherogenic diets did not produce significant fatty streak lesions in the transgenic animals. In contrast, control animals developed fatty streak lesions (548–550).

Genetics

ApoA-I was the first apolipoprotein for which a cDNA sequence was obtained, in 1982 (551). Several other cDNA and gene sequences were reported in the 1982–1984 (89,92–94). The apoA-I gene is 2 kb long and contains four exons and three introns (89).

The apoA-I gene is closely linked to the human apoCIII and apoA-IV genes (552,553). The apoA-I gene is localized 2.5 kb upstream of the apoCIII gene and 7.5 kb upstream of the apoA-IV gene. The direction of transcription of the apoCIII gene is opposite to that of apoA-I and apoA-IV (Fig. 10A). The cluster of the three genes is mapped on the long arm of chromosome 11, region 11q13→qter (554). Figure 10B shows the cluster of apoE, apoCI, apoCII, and the LDL receptor genes, which will be discussed later.

The first structural mutation of apoA-I described was found in Italy and has been called apoA-I Milano (540,555). Subjects with apoA-I Milano have low HDL

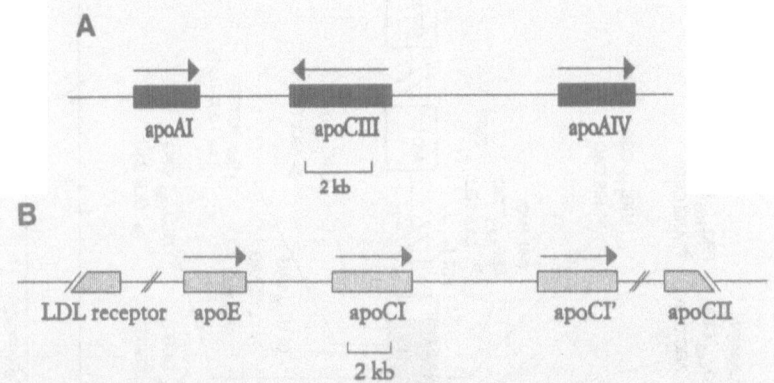

Fig. 10. (A) Schematic representation of human apoA-I, apoCIII, apoA-IV gene cluster (552,553). (B) Schematic representation of human apoE, apoCI, apoCII, and LDL receptor gene cluster (209–212).

cholesterol and apoA-I levels, but do not appear to suffer from premature atherosclerotic disease (555,556). Further analysis indicated that the mutant apoA-I Milano allele results from a structural mutation that causes a 173-Arg→Cys substitution (557). Affected individuals are heterozygotes and carry one normal and one variant apoA-I allele (557). Subsequently, other apoA-I variants were described and named apoA-I Giessen and apoA-I Marburg or Munster-2 (535, 537,558). The subjects with apoA-I Marburg had hypertriglyceridemia and reduced HDL cholesterol levels (558). Subjects with apoA-I Marburg and apoA-I Giessen also represent heterozygosity for one normal and one variant allele. The apoA-I Giessen results from a Pro-143→Arg substitution (535) and the apoA-I Marburg from a deletion of Lys-107 (537). Both mutant apoA-I forms have 40–70% ability to activate LCAT as compared to normal apoA-I (535,537). A third apoA-I mutation which is associated with HDL deficiency, partial LCAT deficiency, and corneal opacities results from a frameshift starting at residue 202 which causes incorporation of 27 random carboxy-terminal residues, including three cysteines (545). Several other mostly heterozygote variants of apoA-I have been described and are shown in Fig. 11A. The majority of the mutations are the result of amino acid substitutions (559–566). Most of the subjects with apoA-I mutations are asymptomatic and have normal lipids and lipoprotein profiles (543). As shown in Table VI, in addition to apoA-I Milano and Giessen, the Pro-165→Arg was associated with low apoA-I and HDL levels (567) and displayed 50–60% of normal ability to activate LCAT (536). The substitutions of Pro-3 with Arg or His were associated with increased plasma concentration of proapoA-I, suggesting a

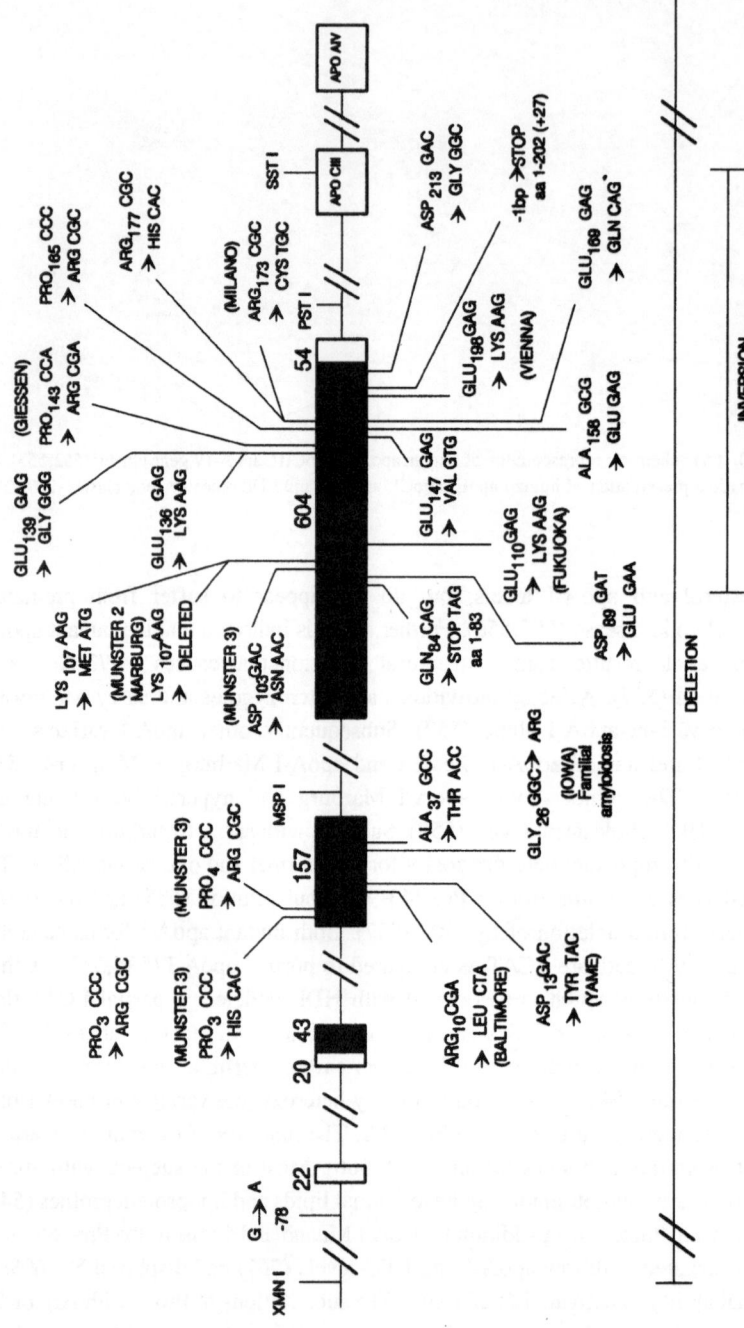

Fig. 11. (A) Schematic representation of the human apoA-I gene and various mutations (535,536,539,541–545,557–567,569,570). (B) Schematic representation of the human apoA-II gene and the mutation associated with apoA-II deficiency (611).

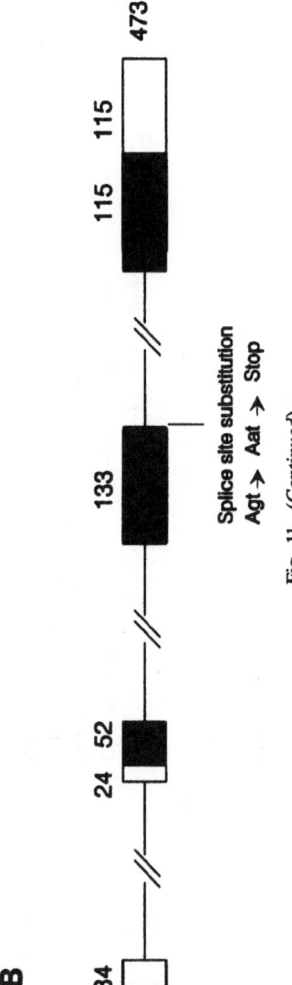

Fig. 11. (*Continued*)

TABLE VI. ApoA-I Variants with Abnormal Lipid Profiles or ApoA-I Properties

Variant form	ApoA-I and HDL levels	ApoA-I properties	Atherosclerosis	Other condition
Pro-3±Arg (Munster-3), Pro-3→His	—	Reduced proapoA-I to apoA-I conversion	No	—
Gly-26→Arg (Iowa)	Low	—	—	Familial amyloidotic neuropathy
Gln-84→Stop	Deficiency	—	Mild	—
Lys-107 deletion (Marburg, Munster-2)	—	40–60% LCAT activation	—	—
Gln-110→Lys (Fukuoka)	Low	—	No	—
Pro-143→Arg (Giessen)	Low	60–70% LCAT activation	No	—
Pro-165→Arg	Low (55–60%)	50–60% LCAT activation	—	—
Arg-173→Cys (Milano)	Low (33–60%) abundance HDL$_{3b}$-type particles	Low HDL and apoA-I levels	No	Reduced LCAT mass and activity
Thr 202→frameshift addition 27 amino acids	Deficiency	—	—	Partial LCAT deficiency, corneal opacities
ApoAI, CIII deficiency	Deficiency of HDL, apoA-I, CIII	—	Severe	Partial LCAT deficiency
ApoAI, CIII, A-IV deficiency	Deficiency of HDL, apoA-I, CIII, A-IV	—	Severe	Partial LCAT deficiency

low proapoA-I to apoA-I conversion (567). The Gly-26 to Arg substitution is associated with familial amyloidotic polyneuropathy III and is inherited as an autosomal dominant trait (562,563). The amyloid fibrils formed in this condition consist of the 83 amino-terminal residues of apoA-I. Two human conditions characterized by severe deficiencies of plasma apoCIII and HDL are associated with premature atherosclerosis. One condition has resulted from deletion of the apoA-I, apoCIII, apoA-IV loci (568,569) and the other from an inversion of the apoA-I, apoCIII loci. The breakpoints of this inversion are in the fourth exon of the apoA-I gene and the first intron of the apoCIII gene (570) (Fig. 11A). Development of premature atherosclerosis has been manifested in a patient deficient in apoA-I synthesis. This condition results from a homozygous nonsense mutation at the codon specifying residue 84 of apoA-I (565). The rather moderate severity of this condition and the absence of atherosclerosis in the HDL-deficient patient with the partial LCAT deficiency (545) suggest that some of the functions of apoA-I may be assumed by other apolipoproteins, perhaps apoA-IV.

The apoA-I, apo-CIII, apoA-IV gene loci contain several polymorphic sites (7,571), some of which show considerable linkage disequilibrium (7,566,571). Some of these polymorphic sites are shown in Fig. 11A. The *Pst*I polymorphism at the 3' end of the apoA-I gene provides restriction fragments of 2.2 and 3.3 kb. The *Xmn*I polymorphism at the 5' end of the gene provides restriction fragments of 8.3 and 6.6 kb and the *Sst*I polymorphism provides restriction fragments of 4.5 and 3.2 kb. Attempts to associate these polymorphisms with lipid abnormalities and coronary heart disease gave variant results in different populations (544).

Tangier Disease Is Not Associated with Structural or Regulatory ApoA-I Gene Mutations

Tangier disease, first described by Fredrickson *et al.* in 1961 (572), is a rare autosomal recessive disorder of lipoprotein metabolism characterized by extremely low levels of plasma HDL cholesterol and extensive tissue cholesteryl ester storage (572–575). It was shown that Tangier plasma contains very low levels of apoA-I, of which over 90% is found in the $d>1.21$ g/ml fraction (574,576,577). In contrast, Tangier intestinal epithelial cells contain normal amounts of apoA-I, as demonstrated by immunofluorescent antibody techniques (578–580). These observations suggested that the defect in Tangier disease might not be in apoA-I synthesis, but rather in other aspects of apoA-I or HDL structure which caused enhanced apoA-I catabolism. Analysis of the isoprotein composition of the normal and Tangier plasma apoA-I showed that in normal plasma proapoA-I represents 1–2% of the total apoA-I, whereas in Tangier plasma the pro form is 40–50% of the total. In addition, the Tangier apoE was enriched in the sialylated apoE forms (89,581). Thus, both Tangier plasma apoA-I and apoE closely resembled the nascent apoA-I and apoE forms which are synthesized by the HepG2 cells and hepatic organ cultures (Figs. 2A and 2E) (33,582). The increased concentration of the proapoA-I in Tangier disease could be the result of either a structural apoA-I gene mutation that retards conversion of proapoA-I to apoA-I or a mutation in the converting protease.

Several recent reports have described a proteolytic activity in human and rat plasma which converts proapoA-I to mature apoA-I; this conversion is relatively slow *in vivo* and even slower under the *in vitro* assay conditions employed (98–100). However, Tangier plasma as well as normal plasma convert the pro form to the plasma form at the same rate (100). These observations suggested that Tangier disease may not be caused by a mutation in the converting protease (100). The possibility of structural or regulatory apoA-I gene mutations was investigated by sequencing the cDNA (583) and the gene from two different patients with Tangier disease. This analysis showed normal apoA-I mRNA and gene sequence and normal expression of the Tangier apoA-I gene (89). These findings established that Tangier disease is not related to apoA-I structure or regulation of expression, but rather to other factors pertinent to apoA-I and high-density lipoprotein metabolism. Tangier disease remains an enigma that, when solved, may shed light on new aspects in the biosynthesis and/or catabolism of HDL.

Apolipoprotein A-II

Structure and Biosynthesis

ApoA-II is a major component of HDL. Plasma apoA-II contains two identical polypeptide chains of 77 amino acid residues of known sequence (584) which are linked by disulfide bonds at residue 6. Complexes of apoA-II and apoE arising from mixed disulfide bridges have been observed in plasma (585) and may hinder the apoE-mediated catabolism of lipoproteins (575). Protein (16) and mRNA analysis (43) showed that apoA-II is synthesized almost exclusively by the liver, whereas the intestine may contribute less than 1% of the total plasma apoA-II pool (43). The apoA-II mRNA encodes a protein of 101 amino acids, including an 18-residue signal peptide. The newly synthesized apoA-II contains a five-residue N-terminal prosegment which is cleaved both intra- and extracellularly (43, 586,587). ApoA-II is also subject to other intra- and extracellular modifications, which include a new type of modification, O-glycosylation, and cyclization of the

N-terminal glutamine (43) (Fig. 2D). Following synthesis, apoA-II is mostly incorporated into lipoprotein particles containing apoA-I and apoA-II (LpAI:AII) (23,66–69,71,73,524,588). Particles containing only apoA-II (Lp:AII) (65,589–591) as well as particles containing apoA-II and apoB (LpAII:B) and apoA-II, apoB, apoC, apoE, and apoD (LpAII:B:C:D:E) have also been described in normal subjects and patients with lipoprotein disorders (72). The functional significance of these particles is not known. Comparison of the nucleotide and amino acid sequences of the apolipoproteins showed that they contain similar units, with sequences that are homologous to each other. ApoA-II contains a 33-, an 11-, and a 21-residue unit at nucleotides 7–39, 40–50, and 51–71, respectively (21). Analysis of the secondary structure of apoA-II according to the Chou and Fasman procedure predicted α-helices at residues 24–30, 42–47, and 51–58, β-sheet at residues 10–21 and 60–71, and β-turns at residues 4–7, 20–23, and 31–34. Association of apoA-II with egg yolk phosphatidyl choline leads to the formation of discoidal particles and increases the α-helical content of the monomeric protein from 37 to 56% and of the dimeric form from 46 to 61% (592). The phospholipid-binding domains of apoA-II have been localized at residues 12–31, 39–47, and 50–61 (593,594). Electron microscopy studies showed that the dimensions of the discoidal particles depend on the lipid to protein ratio. At a ratio of lipid to protein of 45:1, the dimensions of the particles are 17.5 nm $\times$ 6.2 nm, and at a ratio of 240:1, the dimensions are 50 nm $\times$ 5.5 nm (595,596). The dissociation constant K_d of apoA-II from model lipid emulsions (microemulsions and large emulsions containing egg yolk lecithin and triolein) ranges from 0.25 to 0.63 μM and the saturation number N (number of apolipoprotein molecules/1000 phospholipid molecules) ranges from 4.7 to 13.0 (529).

Functions

ApoA-II can displace apoA-I from HDL particles, indicating that both proteins may occupy overlapping domains on the HDL surface (597). Earlier studies showed that apoA-II-containing lipoproteins and proteoliposomes compete for the specific binding of HDL cell-surface binding sites (313,314,598,599). As discussed, HDL consists mainly of LpAI or LpAI:AII particles. Compared to LpAI, LpAI:AII particles bind less efficiently to HepG2 cells and cultured adipose cells (62,63,68). In addition, LpAI:AII particles fail to promote the efflux of cholesterol from mouse adipose cells (62,63). The proposed roles of apoA-II in the activation of hepatic lipase (600–602) and the inhibition of LCAT (6,418,419,603) are uncertain. It has also been shown that apoA-II interferes with the association of factor III with factor VIIα *in vitro*, thus causing inhibition of the activation of factor X (604).

Genetics

cDNAs (605–607) and the gene (608) encoding the human apoA-II have been isolated and sequenced. The apoA-II gene contains four exons and three introns (608,609) and maps on chromosome 1, region 1p21→1qter (605,606,610). A patient has been described in Japan with deficiency of plasma apoA-II. This condition is caused by a donor splice mutation of the third intron of the apoA-II gene (611) (Fig. 11B). The condition is associated with normal lipids and lipoprotein profiles and the absence of atherosclerosis (611). A Pro-5→Gln substitution in a strain of mice is associated with an amyloidosis and senescence (612). It has been reported that inbred strains of mice with increased apoA-II levels are associated with increased HDL size, suggesting that the plasma concentration of apoA-II may affect HDL structure and function(s) (613,614). Overexpression of the human apoA-II gene in transgenic mice affected the size and the distribution of HDL particles, but not the plasma HDL concentration (615). Thus, the overall physiological significance of apoA-II remains unknown.

Apolipoprotein A-IV

Structure and Biosynthesis

ApoA-IV is a major component of rat HDL and chylomicrons (616). Human apoA-IV (Mr = 46 kD) has been found mainly in the $d > 1.21$ g/ml plasma fraction. Small quantities of apoA-IV are also found in lymph, chylomicrons, VLDL, and HDL (114,617–621). The cDNA-derived protein structure of the human and rat apoA-IV showed that the protein and the gene contain repeated units (21,622,623). ApoA-IV contains twelve 22-amino acid repeats, one 11-amino acid repeat, and one 19-residue truncated repeat which lacks the first three amino-terminal residues (21). Similar to apoA-I and the other apolipoproteins, the 22-mer repeated units have α-helical arrangement (506–508) and may contribute to the lipid-binding properties of apoA-IV (506,508,624–627). Protein (114) and mRNA analyses (623) indicated that in humans and nonhuman primates and rats apoA-IV synthesis occurs predominantly in the intestine, whereas in rabbits both the liver and the intestine are major sites of apoA-IV mRNA synthesis (16,18). ApoA-IV mRNA and protein synthesis in mammalian species is controlled by hormonal (519,628) and nutritional factors (521,629–631) and is developmentally regulated (515,628).

Physicochemical Properties

Human apoA-IV binds to triglyceride emulsions with a K_d of 2.3 μM and can be displaced from these particles by HDL_2-associated apoC's (624). The K_d of apoA-IV from intralipid triglyceride-rich particles ranges from 4.8 to 10 μM and N (moles apoA-IV/1000 moles PL) ranges from 0.96 to 0.37 in the same pH range of 6–8.5 (625).

Functions

ApoA-IV has been reported to potentiate the activation of lipoprotein lipase by apoCII (632). ApoA-IV has also been reported to activate efficiently the enzyme LCAT when DMPC was used as a substrate (421). With physiological lecithins, however, the activation was 30% of that of apoA-I. Similarly to apoA-I, in tissue culture experiments apoA-IV complexed to liposomes promotes cholesterol efflux from cells (633). Proteoliposomes containing apoA-IV bind saturably to cell surface sites (315,634) as well as to hepatic cell membranes (635). The binding of apoA-IV to mouse adipose cells promotes the efflux of cholesterol from mouse adipose cell cultures and it has been suggested that apoA-IV occupies the same binding site that is occupied by LpAI:AII particles (315). These observations suggest a potential role of apoA-IV in the reverse cholesterol transport similar to that described for apoA-I.

Genetics

The sequence of human apoA-IV has been determined from the sequences of cDNA and genomic clones encoding for this protein (553,622,623,636). The apoA-IV gene is 2.8 kb long and contains three exons and two introns and encodes a protein of 396 amino acids, including a 20-amino acid signal peptide (622,623,636). The intron–exon arrangement of the apoA-IV gene is different from that of the other small apolipoprotein genes, all of which contain four exons and three introns. The human apoA-IV gene is closely linked with the human apoA-I and apoCIII genes (552,553) (Fig. 10A). The gene cluster maps on the long arm of chromosome 11, region 11q13→qter (554). Analysis of plasma apoA-IV by isoelectric focusing and two-dimensional polyacrylamide gel electrophoresis showed the presence of different phenotypes (3,558,637–644). Family studies showed that the phenotypes follow Mendelian inheritance and result from the presence of two common and at least three rare alleles (510,638,640–643). The different apoA-IV alleles and the expected phenotypes are shown schematically in

Fig. 12. Following the nomenclature proposed by Lohse and Brewer (644), the homozygous phenotypes have been designated apoA-IV-0, apoA-IV-1, apoA-IV-2, apoA-IV-3, and apoA-IV-4. Phenotypes apoA-IV-1 and apoA-IV-2 and the heterozygote form apoA-I-2/1 are the most common. ApoA-IV-0 is the most acidic protein and apoA-IV-4 the most basic. The positive charges of apoA-IV-1, apoA-IV-2, apoA-IV-3, and apoA-IV-4 differ by one, two, three, and four units from that of apoA-IV-0. The difference in charge allowed the separation of the proteins by isoelectric focusing techniques. Similar to apoE phenotypes that will be discussed later, the apoA-IV phenotypes are the result of structural mutations in the apoA-IV gene. The apoA-IV allele frequency differs in different populations (644–648). In Caucasians the apoA-IV-1 allele frequency ranges from 2 to 6% and that of apoA-IV-2 ranges from 94 to 98% (3,646,647). The frequency of the apoA-IV-2 allele is

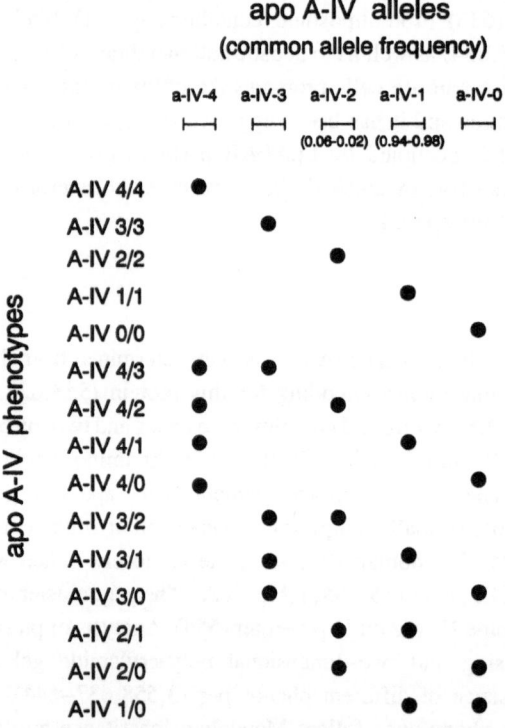

Fig. 12. Schematic representation of the currently described apoA-IV alleles and the expected phenotypes. The diagram also shows the expected isoelectric focusing patterns of the different phenotypes.

low in certain populations (639,645,648,649). Since the apoA-IV proteins are separated on the basis of their charge difference, it is expected that each homozygous phenotype may originate from more than one genotype. The molecular basis of apoA-IV phenotypes was elucidated by amplification of genomic DNA obtained from subjects with different phenotypes using the polymerase chain reaction (644,650,651). This analysis identified amino acid substitution at residues 165 (Glu→Lys), 167 (Lys→Gln), 230 (Glu→Lys), and 347 (Thr→Ser), and an insertion of four residues between amino acids 361 and 362 which were responsible for the different apoA-IV phenotypes. These mutations are shown in Fig. 13. The inversion of the apoA-I, apoCIII cluster (570) and the deletion of the human apoA-I, apoCIII, apoA-IV clusters (569) have been discussed; these are shown in Fig. 11A. Several studies indicated that the apoA-IV-2/1 phenotype is associated with increased HDL cholesterol and decreased plasma triglyceride levels (641,647,652). Physicochemical studies have shown an increase in the helical content of apoA-IV-2 (75%) as compared to the apoA-IV-1 (56%) (653). Furthermore, functional studies have suggested increased activation of LCAT (653) and of lipoprotein lipase (652) by apoA-IV-2 as compared to apoA-IV-1. These properties of the apoA-IV-2 may account for the lipid and lipoprotein profile observed in subjects with this apoA-IV-2/1 phenotype (641,647,652).

Apolipoprotein B

Structure

Apolipoprotein B is the main protein component of LDL and makes up 23.8% of the weight of the LDL particle (153,654). The primary sequence of human apoB-100 was originally deduced by four independent research groups from the corresponding sequence of overlapping cDNA clones (655–658), and was verified by direct protein (37) and gene (659,660) sequences. The mature apoB-100 protein contains 4536 amino acids (37,655–660). Based on an Mr of 513 kD, it can be calculated that there is one apoB-100 molecule per LDL particle (655).

Cysteines and N-Glycosylation Sites in the ApoB-100 Sequence

ApoB-100 contains a total of 25 cysteine residues (655–658) and 16 N-glycosylated Asn residues (37,657) (Figs. 14A and 14B). Most of the cysteines form disulfide bridges (661). Eighteen of the cysteine residues are clustered in the amino-terminal portion of the apoB-100 molecule between residues 1 and 1635. It

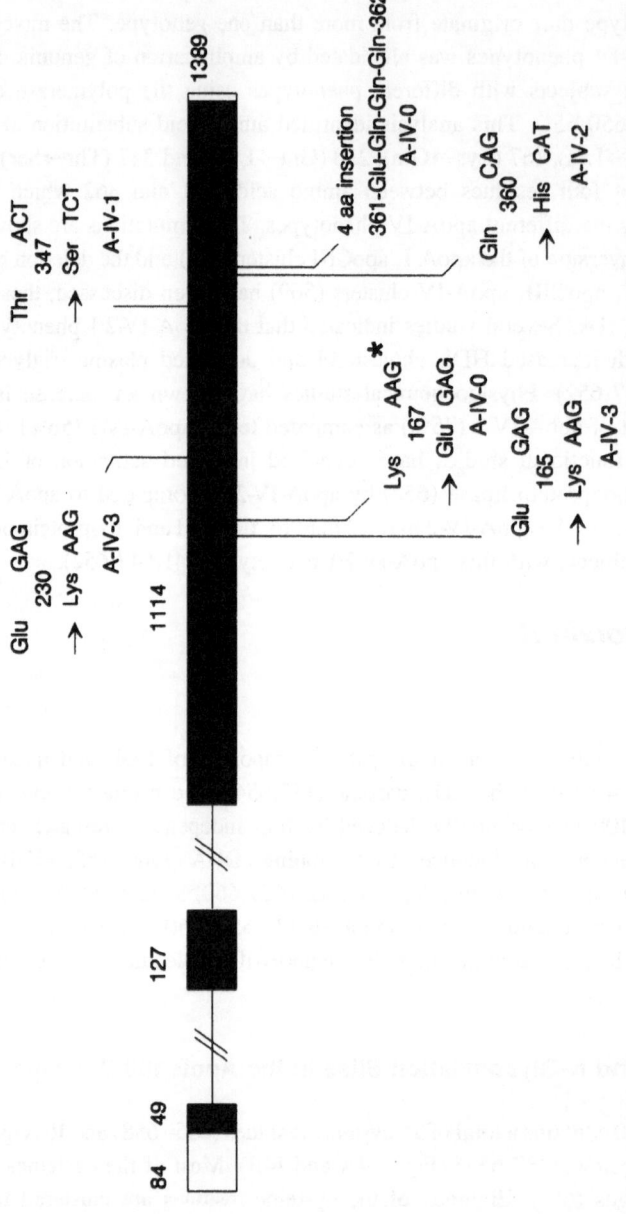

Fig. 13. Schematic representation of the human apoA-IV gene and various mutations (644,650,651) which account for the different apoA-IV phenotypes shown in Fig. 12. Asterisk indicates that this phenotype results from a double mutation causing a Lys-167→Glu and a Gln-360→His substitution.

A

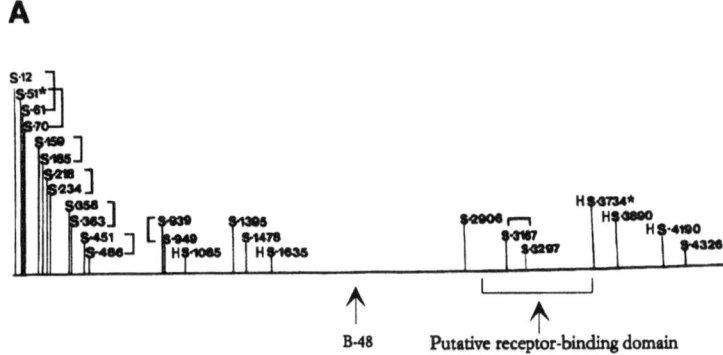

AMINO ACID RESIDUE

B

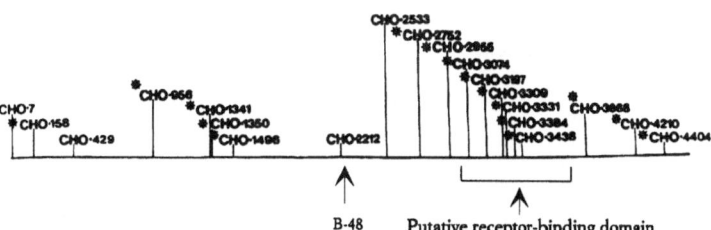

AMINO ACID RESIDUE

Fig. 14. Distribution of (A) cysteines and (B) potential N-glycosylation sites in the apoB-100 sequence. The stars in panel B indicate actual glycosylation sites (7,9,37,657). In panel A, the HS and brackets indicate free cysteines, and disulfide bridges, respectively (20,662), and the asterisk indicates amino acids which may participate in the formation of intramolecular thiolesters (23,24,34,35). (C) The position of fatty acid acylated cysteines has not been identified. Schematic representation of the putative location of the receptor-binding domains of apoB-100. The supportive evidence which implicates the apoB region between residues 3000 and 3700 in receptor binding is provided in refs. 173, 188, 189, 655, 657, 693–695. CHO indicates N-linked carbohydrate chains. 1, 2, 3, 4: Regions with strong polar character, strong positive charge, and high hydrophobic moments (655). Asterisks indicate heparin-binding sites (696,697). (D) The top two lines show the homology between the positively charged apoE 142–150 and apoB 3359–3367 regions presumed to participate in receptor binding and the bottom line shows the consensus ligand-binding domain of the LDL receptor, which is negatively charged. (E) Schematic representation of the putative lipid-binding domain of apoB-100 obtained from ref. 7. (Reproduced from ref. 37 with permission.)

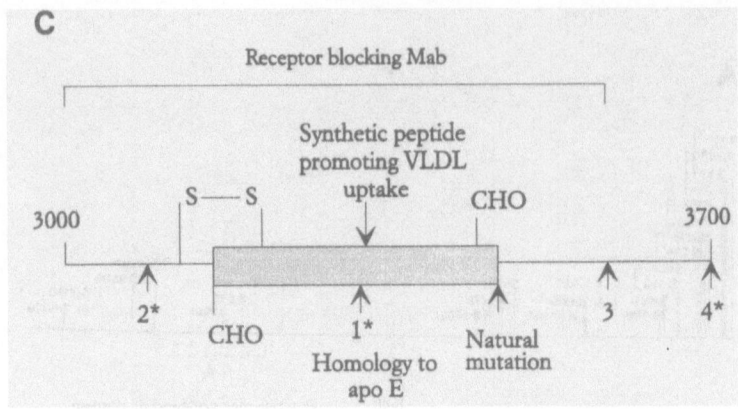

D

Apo E aa 141-150	Leu-Arg-Lys-Leu-Arg-Lys-Arg-Leu-Leu-Arg
Apo B-100 aa 3358-3367	Thr-Arg-Leu-Thr-Arg-Lys-Arg-Gly-Leu-Lys
LDL receptor binding domain (consensus sequence)	Asp-X-X-X-Asp-Cys-X-Asp-Gly-Ser-Asp-Glu

Fig. 14. (*Continued*)

was reported recently that 14 of the 16 bridges identified so far link adjacent cysteine residues of the amino-terminal domain (662) (Fig. 14A). Previous studies suggested that apoB contains covalently bound fatty acids (36,633). It has also been reported that apoB contains both intermolecular thiolester bonds with palmitate and stearate, and intramolecular thiolester bonds involving Cys-51 and Cys-3734 and the COOH groups of Glu-54 and Asp-3737, respectively (34,35). It must be noted, however, that Cys-51 and Cys-3734 are also reported to participate in the formation of disulfide bridges (662). The importance of these putative modifications of cysteine remains unclear.

Biosynthesis of ApoB-100 and ApoB-48 Forms

ApoB is synthesized by the liver and the intestine and has two forms of protein and mRNA, designated apoB-100 and apoB-48 (655,664). Minor sites of apoB-100 mRNA synthesis are placenta, kidney, colon, and stomach (18,665, 666). ApoB secretion from hepatic cell cultures increases in response to oleic acid

E

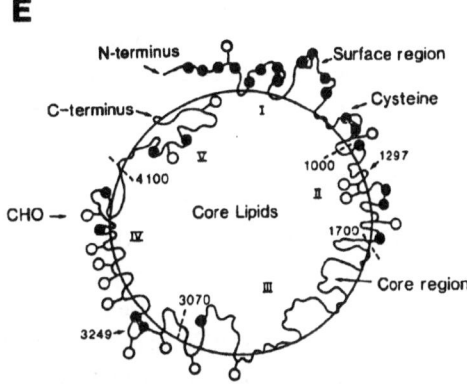

Fig. 14. (*Continued*)

(667,668) and decreases in response to ω'-3 fatty acids, insulin, glucagon, albumin, and calcium channel blockers (667–672) without change to apoB mRNA synthesis (668–670,673). In CaCo-2 cells, apoB secretion increases in response to calcium ionophores and is unaffected by oleic acid (674). It was shown recently that the apoB-48 mRNA is generated by a posttranscriptional mRNA modification which results in the change of the codon CAA specifying Gln-2153 to UAA (675–677). The editing activity is a specific cytidine deaminase with a K_m of 0.2–2 nM (678,679). The editing activity requires a nucleotide motif GAUCAGUAU at residues 6672–6680 in the mRNA sequence. Nucleotide substitutions within this motif abolished or reduced mRNA editing (680). The editing activity is expressed in intestine, liver, and a variety of extrahepatic tissues, including colon, kidney, and stomach (665). The activity is developmentally regulated in fetal intestine (681,682). It is also regulated by thyroid hormone and growth hormone in liver (683–685) and by fasting following a feeding of high-carbohydrate diets (686).

Functions

ApoB is the ligand which mediates the recognition of LDL by the LDL receptor. The LDL receptor–apoB interaction mediates the clearance of LDL from plasma and regulates cellular cholesterol biosynthesis (152,153). ApoB is also required for the assembly and secretion of LDL. Inhibition of apoB synthesis by cycloheximide prevents the secretion of VLDL by hepatic cell cultures (687,688). Assembly and secretion of chylomicrons and VLDL are prevented in abetalipoproteinemia and some forms of homozygous hypobetalipoproteinemia (5,689). It has also been shown recently that apoB obtained by delipidation of LDL contains phospholipase A2 activity which is inhibited by reagents which modify histidine residues (690). The physiological role of this activity is not clear.

The Receptor-Binding Domain of ApoB-100

The receptor-binding domain of human apoB-100 resides within the carboxy-terminal half of the molecule, but has not been precisely localized. Several independent studies mentioned below have suggested that the carboxy-terminal domain of apoB-100 between residues 3000 and 3700 might be involved in receptor binding. The data supporting such an assignment include the following: (a) Early studies had shown that modification of Arg or Lys residues abolishes the binding of apoB or apoE to the LDL receptor (188,189). It has been proposed that positively charged amino acids of apoB or apoE interact with negatively charged residues of the ligand-binding site of the LDL receptor (173). (b) The epitopes of a variety of receptor-blocking monoclonal antibodies have been mapped in the region of amino acids 3000–3700 (691–693). (c) A peptide corresponding to residues 3345–3381 promotes uptake of apoE-depleted hypertriglyceridemic VLDL (657). (d) A region of homology exists between residues 142–150 of apoE (694) and 3359–3367 of apoB (655–657). (e) A naturally occurring mutation at residue 3500 is responsible for decreased receptor binding of the variant LDL (695). Utilization of hydrophobicity and hydrophobic moment calculations for analysis of the secondary structure of apoB identified several potential helical domains with strong polar character with a strong net positive charge and high hydrophobic moment. Four of these regions, defined by residues 3144–3162, 3352–3369, 3595–3613, and 3666–3681, have five, five, four, and three overall positive charges, respectively (655). Three domains (1, 2, and 4 in Fig. 14C), defined by residues 3150–3157, 3361–3368, and 3670–3677, overlap with three heparin-binding domains of apoB (696–697). Four other heparin-binding peptides are defined by residues 5–99, 205–279, 875–932, and 2016–2151 (696,697). Domain 1 has the highest hydrophobic moment and occurs between residues 3352 and 3369 (655). As shown in Fig. 14D, the sequence between residues 3359 and 3367 (domain 1) has 63% homology to the apoE sequence between residues 142 and 150 (694), which has been implicated in receptor binding (698,699). Finally, truncated apoB forms extending to apoB-75 bind LDL efficiently, whereas those expressing shorter apoB forms do not (700,701). Despite the above evidence, recent genetic data showed that subjects with defective LDL that has reduced affinity for the LDL receptor do not have sequence alteration in the apoB gene region encoding residues 3130–3630, suggesting that alterations in other domains of apoB must be responsible for the defective LDL particles (702). Thermal unfolding of apoB has suggested the presence of interacting domains (703). It is plausible that the interaction of these domains confers the proper conformation to apoB which allows its association with the LDL receptor. Other experiments using *in vitro* mutagenesis of the LDL receptor (214,215) indicated that the ligand-binding domain of the LDL receptor is complex. One region (domain 5) interacts with apoE in βVLDL (residues 140–160), and six others (2, 3, 5, 6, 7, and A) interact with apoB in LDL (Fig. 5A). This indicates that the two ligands of the LDL receptor, apoB and apoE, interact with different regions of the receptor-binding site.

The Lipid-Binding Domains and Assembly of Human ApoB-100

The lipid-binding domains of apoB-100 have been assessed by direct protein sequencing by Yang *et al.* (37,657). This analysis suggested that the amino-terminal end (residues 1–1000) and a domain near the carboxy terminal (residues 3017–4100) are greatly enriched in lipid-free peptides. The carboxy-terminal end (residues 4101–4536) and a domain in the middle of the molecule (residues 1701–3070) are greatly enriched in lipid-bound peptides. Finally, the domains between residues 1001–1700 contain both lipid-bound and lipid-free peptides (37,657). The model proposed by Yang *et al.* is shown in Fig. 14E. Amphipathic helical structures have been strongly implicated in the lipid binding of other water-soluble, exchangeable apolipoproteins; however, many water-soluble proteins of known tertiary structure contain segments of amphipathic helices which are frequently associated with each other (704). The unique physicochemical properties of apoB-100, its insolubility and nonexchangeability, suggest that structures other than amphipathic α-helices may be involved in the binding of lipids to the lipoprotein surface. For instance, strongly apolar segments of secondary structure may associate to form a hydrophobic anchor domain, perhaps similar to that found in membrane proteins, which are then associated with the lipid interface. ApoB-100 secreted by HepG2 and apoB-100 and apoB-48 secreted by rat hepatocytes associate intracellularly with lipids and are secreted in the form of lipoprotein particles (49). The kinetics of intracellular assembly of apoB and the regulation of translocation of the assembled lipoprotein across the endoplasmic reticulum leading to apoB secretion have been discussed earlier. Labeling of newly synthesized lipoproteins by deuterated leucine showed that newly synthesized apoB can be assembled and secreted in the form of either VLDL or LDL particles (705). Studies with naturally occurring truncated apoB forms or those expressed in cell culture suggest that fragments larger than apoB-41 as well as 63–87% of the carboxy-terminal region of apoB can assemble intracellularly to form lipoproteins (57,58,60). In addition, the shorter amino-terminal fragments such as apoB-17 can bind to lipids and may play an important role in the stabilization of triglyceride-rich VLDL and/or LDL (59). Utilization of immunoelectron microscopy to visualize pairs of monoclonal antibodies bound to LDL allowed the mapping of different apoB epitopes on the LDL sphere. The epitopes mapped extend over half a hemisphere on the LDL surface (706). Additional work in this direction will provide information on the arrangement of apoB on the LDL surface.

Genetics

Polymorphisms in the Human ApoB-100 Gene.   The apoB-100 gene is 43 kb long and contains 29 exons and 28 introns (659,660) and maps on the distal end of the short arm of chromosome 2 (707,708). The initial sequencing of the apoB-100 cDNA indicated that this protein is highly polymorphic (Fig. 15A). Some of the polymorphic sites could be detected by restriction fragment length polymorphisms such as the *Xba*I site (566,709), the *Eco*RI polymorphism at codon 4154 (499), and the *Msp*I site (710,711). These and other polymorphic sites were utilized as markers in population studies in an attempt to correlate individual alleles or haplotypes with lipid levels or risk for myocardial infarction (709,712). The conclusions reached in these studies vary in different populations (712). Another substitution, of Arg-4019→Tyr (apoB$_{Hopkins}$), was reported in a family with familial high plasma cholesterol (713). Two additional interesting polymorphisms in apoB have been described. The first consists of variation in the length of the signal peptide to either 24, 27 or 29 amino acids thus producing the 5′βSP-24, or 27 or 29 variants (655,713a). The second consists of a hypervariable region localized 0.5 kb 3′ of the stop codon of the apoB gene. This polymorphism results from a variable number of tandemly repeated A+T rich DNA sequences (VNTR). At least 12 VNTR alleles differing in the number of the 3′ repeats have been identified and have been designated 3′β35, 3′β36, etc. (713b,713c). Earlier work had established immunological polymorphisms in apoB that were determined by the presence of five different antigenic sites, designated Ag (c/g), Ag (al/d), Ag (h/i), Ag (t/z), and Ag (x/y). The immunological polymorphism of apoB is the result of amino acid substitutions at residues 71 Ag (c/g), 591 Ag (al/d), 2212 Ag (x/y), 3611 Ag (h/i), and 4154 Ag (t/z) (710,711,714–718) (Fig. 15B).

Abetalipoproteinemia.   Abetalipoproteinemia is a rare autosomal recessive disorder characterized by fat malabsorption, failure to thrive, ataxic neuropathy, retinitis pigmentosa, and acanthocytosis (5,689). The disorder is characterized by low levels of plasma cholesterol and triglycerides and complete absence of plasma apoB, VLDL, LDL, and chylomicrons (689,719). In abetalipoproteinemia all of the other apolipoproteins are found in the HDL fraction (720). Liver

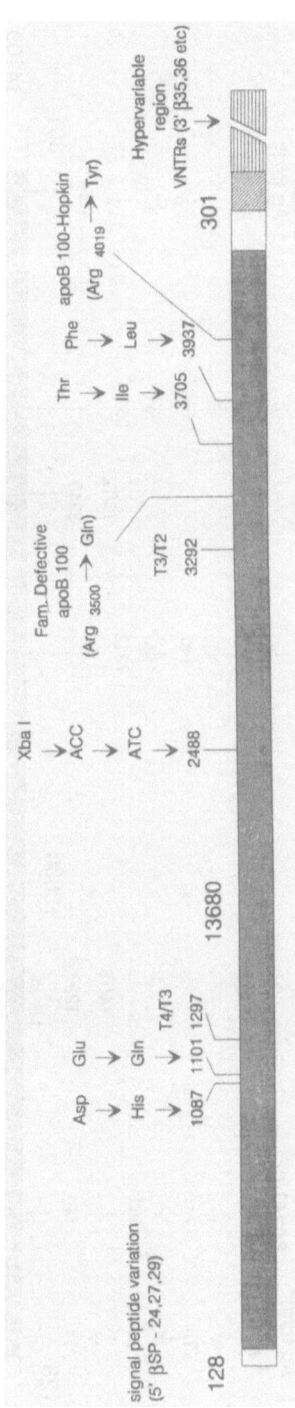

Fig. 15. (A) Schematic representation of human apoB cDNA and the position of several early detected mutations which suggested more extensive genetic variation. The mutations are described in refs. *566,655–658,660,709–711,713,713a–c. 5'βsp* etc. indicates 35 bp tandem A + T rich repeats.). (B) Schematic representation of the mutations responsible for the immunochemical apoB polymorphisms. The mutations are described in refs. 710, 711, 714–717. (C) Schematic representation of the different apoB mutations associated with hypobetalipoproteinemia. The mutations are described in refs. 738–750.

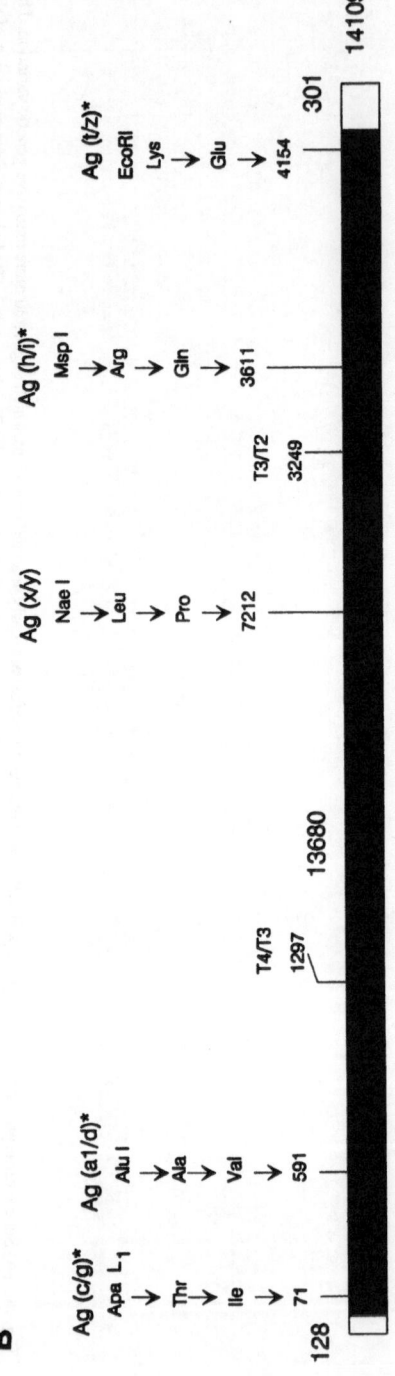

Fig. 15. (*Continued*)

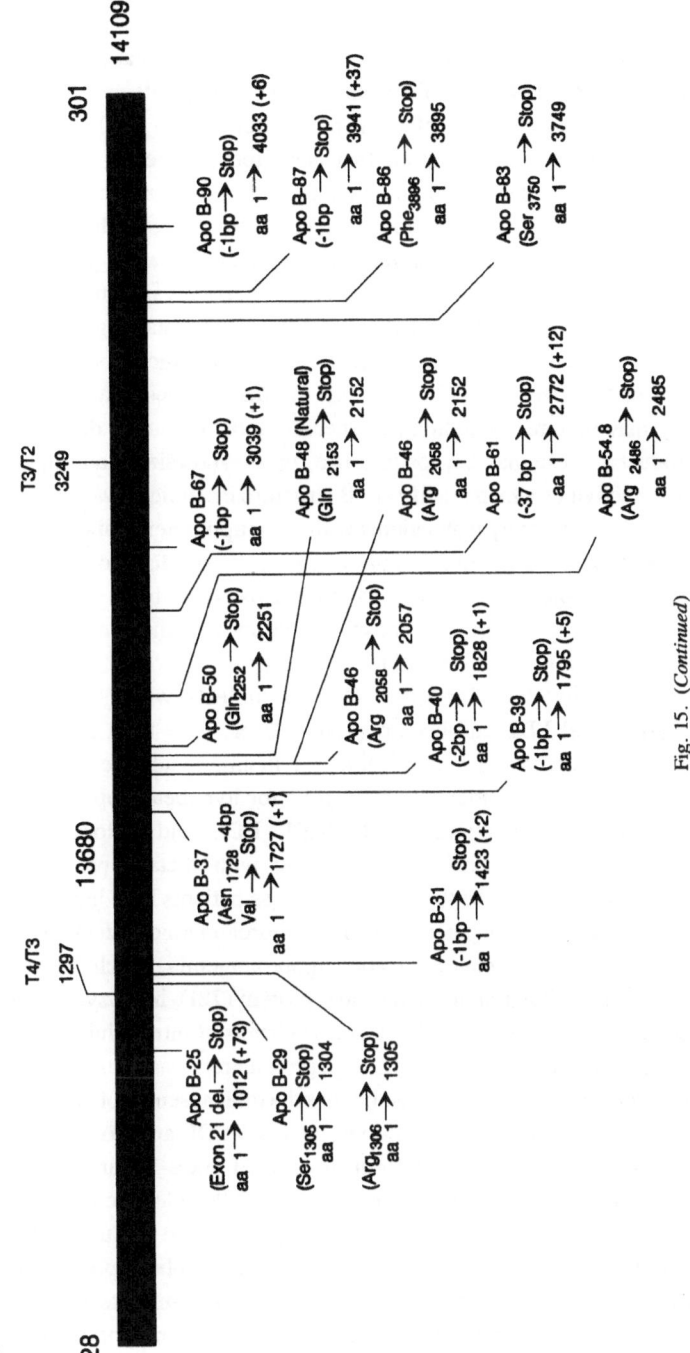

Fig. 15. (*Continued*)

biopsies obtained from abetalipoproteinemic patients contain near normal or increased levels of apoB-100 mRNA and intracellular apoB protein, suggesting that this condition represents a defect in apoB secretion (721,722). Intestinal organ cultures of biopsies obtained from two abetalipoproteinemic patients showed normal apoB-48 synthesis and modification, but an absence of intracellular lipoprotein particles (722). Genetic evidence also suggests that abetalipoproteinemia results from mutations in loci different from the apoB gene (723–725). Finally, small quantities of lipid-poor apoB–apoa particles are found in the $d > 1.21$ g/ml plasma fraction of patients with abetalipoproteinemia, suggesting that this condition is caused by a defect in LDL assembly which also precludes the formation of Lpa particles (726). Overall the data suggest that the lipoprotein assembly may require, in addition to specific apoB domains, different cellular gene products. These may be, for instance, lipid-synthesizing and lipid-transfer activities or activities which modify apoB structure in a unique way. In this regard, it has been shown recently that patients with abetalipoproteinemia are deficient in the 88-kDa subunit of the microsomal triglyceride transfer protein (726a). The findings suggest that this activity, which normally mediates the transport of cholesteryl ester triglycerides and phospholipids between membranes (726b), may indeed participate in VLDL and LDL assembly.

Hypobetalipoproteinemia. Human subjects with genetically determined half-normal LDL and betalipoprotein levels have been described (727–732). These people have a hypobetalipoproteinemia syndrome and are usually phenotypically normal. Patients homozygous for hypobetalipoproteinemia have been described who lack plasma apoB, VLDL, LDL, and chylomicrons and are phenotypically indistinguishable from patients with abetalipoproteinemia. The only difference appears to be that parents of these patients have half-normal LDL levels (727–732), whereas parents of abetalipoproteinemic patients have normal LDL levels (733–735). In contrast to abetalipoproteinemia, which displays normal or increased intracellular apoB mRNA and protein (721), homozygous hypobetalipoproteinemia is characterized by decreased levels of intracellular apoB mRNA and protein (725). A case of hypobetalipoproteinemia has been reported which was characterized by very low levels, but not the total absence, of LDL cholesterol (4–8 mg/dl) (732). Analysis of chylomicron and VLDL apoB of this patient and his kin revealed a distinct apoB form designated apoB-37, in addition to the normal apoB-100 and apoB-48 forms (701,736–738). Genetic analysis showed that the apoB-100 and apoB-37 forms of these patients are products of two alleles (701,736–738). Since the initial description of the hypobetalipoproteinemic patient with the apoB-37 form of deficiency, several other amino-terminal-truncated forms of apoB extending from apoB-25 to apoB-90 have been reported (61,738–750) in subjects with familial hypobetalipoproteinemia. The molecular defects which cause premature apoB protein termination are shown in Fig. 15C. In general, the shorter forms have reduced capacity to assemble within lipoprotein particles as compared to those that are larger than apoB-48 (61). It is also interesting that apoB-75 and apoB-90 have increased affinity for the LDL receptor (742,743,751), whereas the shorter forms are not recognized by the LDL receptor (701).

Familial Defective ApoB-100. In 1986, patients with moderate forms of hypercholesterolemia were described who have normal LDL receptor function but display decreased fractional catabolic rates for autologous LDL (752). LDL obtained from these patients had reduced affinity for the LDL receptor of normal cultured human skin fibroblasts (753). Genetic analysis indicated autosomal codominant transmission of the defect, which has been named familial defective apoB-100. The affected members were heterozygous for this disorder (753). DNA sequence analysis of the defective allele has shown a CGG–CAG transition at the codon specifying amino acid 3500, resulting in a Gln–Arg substitution (695). Screening of different populations showed that heterozygosity for the mutation at 3500 occurs with a frequency of 0.002% (754). This mutation affects the LDL–LDL receptor interactions and is reminiscent of familial type III hyperlipoproteinemia (HLP), which is caused by structural mutations in the apoE gene (755).

Other Potential Defects in ApoB-100. Numerous population studies indicate that elevated LDL cholesterol levels are associated with an increased risk of coronary heart disease (490,756). In contrast, it has been suggested that moderately decreased LDL cholesterol levels are associated with longevity (492). Finally, subjects with a condition characterized as hyperapobetalipoproteinemia have been described (757). These people have elevated LDL apoB but normal LDL cholesterol levels and are at increased risk for developing coronary atherosclerosis (758). Some of the variations in LDL cholesterol and apoB levels observed in these subjects may be the result of as yet unidentified structural or regulatory apoB gene abnormalities or alterations in the LDL assembly pathway.

Human Apolipoprotein CI

Structure and Biosynthesis

ApoCI is a major component of very low-density lipoprotein (VLDL) and a minor component of high-density lipoprotein (HDL) and consists of 57 amino acids of known sequence (759,760). The major site of apoCI mRNA and protein synthesis is the liver (16,18). A major site of apoCI mRNA synthesis in the cynomolgous monkey is also the adrenal (V. I. Zannis, unpublished). The apoCI mRNA encodes a protein of 83 amino acids, including a 26-amino acid signal peptide which is cleaved cotranslationally in the rough endoplasmic reticulum

(761,762). In plasma, lipoprotein particles have been described which contain apoB, apoCI, apoCII, and apoCIII (LpB:C) or apoB, apoCI, apoCII, apoCIII, and apoE (LpB:C:E) (64,69). The physiological significance as well as the sites of assembly of these particles are not known.

Comparison of the nucleotide and amino acid sequences of the apolipoproteins showed that they contain similar repeated units with sequence homologies. ApoCI contains a 33- and an 11-residue unit at nucleotides 7–39 and 40–50, respectively (21). Analysis of the secondary structure of apoCI according to the Chou and Fasman procedure predicted α-helices at residues 7–14, 18–20, and 33–53 (763). Association of apoCI with egg yolk phosphatidyl choline increases its α-helical content from 56 to 73% (763,764). The phospholipid-binding domain of apoCI is between residues 1 and 38 (763,764). The association of apoCI with egg yolk lecithin forms discoidal particles with a minor axis of 4 nm and a major axis of 20 nm (765). The initial binding studies of apoCI to egg yolk lecithin led to the hypothesis of the amphipathic α-helix (506). According to the model proposed by Segrest, the amphipathic α-helices have two phases, one hydrophobic, which can associate with the hydrophobic fatty acyl chains of the phospholipids, and another hydrophilic, which can associate with the phospholipid head groups and the aqueous phase.

Functions

ApoCI moderately activates LCAT. This activation is 25% of that achieved with apoA-I (419). The region at residues 17–57 activates LCAT as effectively as the intact apoCI (766). There are conflicting reports on the role of apoCI in the regulation of lipoprotein lipase (334). It was shown previously that apoC peptides inhibit the catabolism of apoE-containing lipoproteins by perfused liver (767–770). Recently, it has been shown that apoCI and to a lesser extent apoCII (but not apoCIII) inhibit the binding of apoE-containing lipoproteins such as βVLDL to the LDL receptor-related protein (LRP) as well as to the LDL receptor (265,266). Transgenic animals expressing apoCI have increased triglyceride levels, suggesting a physiological role for apoCI in the catabolism of triglyceride-rich lipoproteins (771). Recently it has been shown that apoCI binds to phospholipids and cell membranes and inhibits cellular phospholipase A2, possibly by preventing the access of the enzyme to its phospholipid substrate (722).

Genetics

The cDNA and gene sequences of the human apoCI gene have been reported (761,762) (Fig. 16A). The human apoCI gene is 4.7 kb long, contains four exons and three introns (762), and is closely linked to the human apoE, apoCII, and LDL

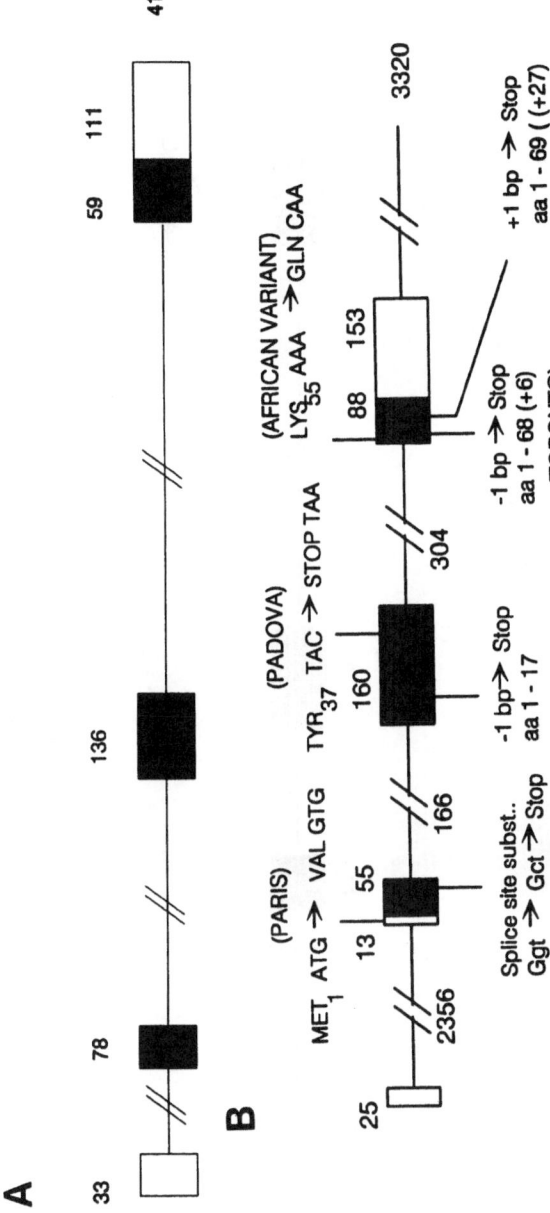

Fig. 16. (A) Schematic representation of the human apoCI gene. (B) Schematic representation of the human apoCII gene and the various mutations that are associated with apoCII deficiency and familial type I hyperlipidemia (782–784, 792–795). (C) Schematic representation of the human apoCIII gene and two point mutations (809, 810).

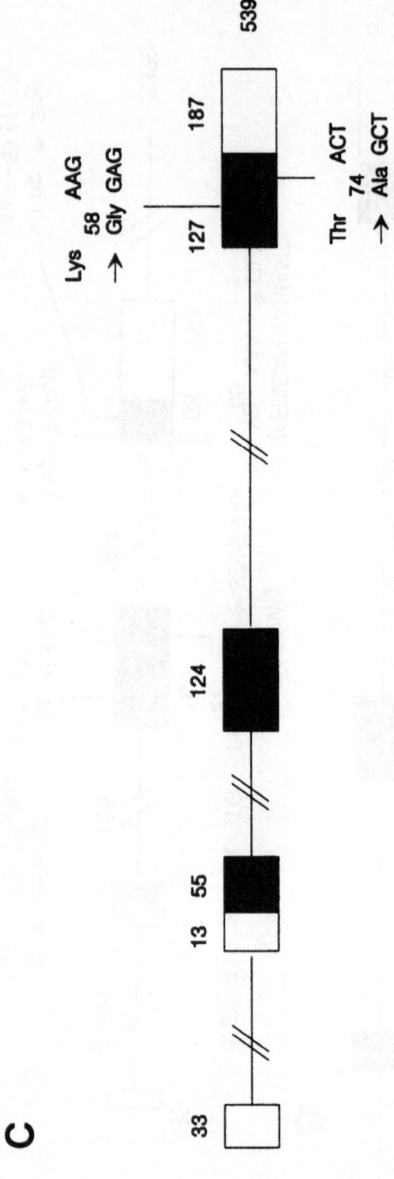

Fig. 16. (*Continued*)

receptor genes (211) (Fig. 10B). The apoCI gene is located 5.5 kb downstream of the apoE gene, and the apoCI pseudogene is located 7.5 kb 3' to the apoCI gene (762). The human apoCII gene is found 20 kb downstream of the apoCI' pseudogene (212), and the LDL receptor genes are closely linked with the apoE, apoCI cluster (209–211) (Fig. 10B). The cluster of the four genes maps on the long arm of chromosome 19 (211). There are no reports of genetic alterations involving the human apoCI gene.

Human Apolipoprotein CII

Biosynthesis and Structure

Plasma apoCII is a major component of chylomicrons and very low-density lipoprotein (VLDL) and a minor component of high-density lipoprotein (HDL) and consists of 79 amino acids of known sequence (773). The major site of apoCII mRNA and protein synthesis is the liver, and a minor site is the intestine (16–18). The apoCII mRNA encodes a 101-amino acid protein which contains a 22-residue signal peptide. Following cotranslational cleavage of the signal peptide, the 79-residue protein is partially modified in the ER/Golgi by O-glycosylation and sialylation (43). The glycosylation is inefficient and the protein is secreted in the disialylated (CII_{s2}) and nonsialylated (CII) forms at a 1:1 ratio. ApoCII$_{s2}$ is subsequently desialylated in plasma to generate the asialo apoCII form (Fig. 2B). The physiological importance of the modification of apoCII is not known. In plasma lipoprotein particles, lipoproteins have been described which contain apoB, apoCI, apoCII, and apoCIII (LpB:C) or apoB, apoCI, apoCII, apoCIII, and apoE (LpB:C:E) (64,69). The sites of assembly and the physiological significance of these particles are not known.

Comparison of the nucleotide and amino acid sequences of the apolipoproteins showed that they contain similar repeated units with sequence homologies. ApoCII contains a 33- and a 22-residue unit at nucleotides 18–50 and 51–72, respectively (21). Analysis of the secondary structure of apoCII according to the Chou and Fasman procedure predicted amphipathic α-helices at residues 13–22, 28–39, and 42–51, β-turns at residues 9–12, 23–26, and 52–55, and a β-sheet at residues 60–74 (774). Association of apoCII with phospholipids increases its α-helical content from 35 to 59% (774,775). Association of apoCII with egg lecithin yields discoidal particles with a minor axis of 4 nm and a major axis of 20 nm (765). The dissociation constant K_d of apoCII from model lipid emulsions (microemulsions and large emulsions containing egg yolk lecithin and triolein and cholesterol) ranges from 0.45 to 1.07 μM and the saturation number N (number of apolipoprotein molecules/1000 phospholipid molecules) ranges from 8.3 to 11.8 (529).

Functions

ApoCII is a potent activator of lipoprotein lipase but not hepatic lipase (122,123,332–334,776). Maximal stimulation of LPL is achieved at a molar ratio of apoCII to enzyme of 1:1. The reported dissociation constant of the complex is 10^{-8}–10^{-10} M (334,777). The complex dissociates at high NaCl concentrations (778). The region 43–79, which can bind to VLDL and phospholipid vesicles, is sufficient for maximum activation of the lipoprotein lipase, and the carboxy-terminal region 55–79 retains 90% of the activation capacity (779–781). The physiological importance of apoCII in activating LPL has been established by the finding of individuals with inherited apoCII deficiency who have great difficulty in clearing triglyceride-rich lipoprotein particles from their plasma (109,364). Naturally occurring mutations in the carboxy-terminal apoCII region, such as apo-CII$_{Toronto}$ or apoCII$_{St.\ Michael}$ (782–784), or mutations generated by *in vitro* mutagenesis (785) abolish the ability of the mutant protein to activate lipoprotein lipase. Recently it has been shown that apoCI and to a lesser extent apoCII (but not apoCIII) inhibit the binding of apoE-containing lipoproteins such as βVLDL to the LDL receptor-related protein (LRP) as well as to the LDL receptor (265,266).

Genetics

The cDNA and gene sequences encoding the human apoCII have been reported (607,786–789). The human apoCII gene is 3.3 kb long and contains four exons and three introns (789) and is closely linked to the genes for apoE, apoCI, and the LDL receptor. The gene cluster maps on the long arm of chromosome 19 (209–212) (Fig. 10B). Initial protein analysis of patients with apoCII deficiency by two-dimensional polyacrylamide gel electrophoresis showed the presence of small quantities of variant protein forms (790,791). Subsequent nucleotide sequence analysis of the patients' genes showed a variety of mutations which are associated with apoCII deficiencies and type I hyperlipoproteinemia (782–784,792–795, 795a). This condition is characterized by markedly increased plasma triglycerides, chylomicrons, and VLDL and is associated with eruptive xanthomas, pancreatitis, and premature vascular disease (375,791). The molecular changes in the apoCIII gene that are associated with type I hyperlipoproteinemia are summarized in Fig. 16B.

Human Apolipoprotein CIII

Structure and Biosynthesis

Plasma apoCIII is a major component of chylomicrons and very low-density lipoprotein (VLDL) and a minor component of high-density lipoprotein (HDL) and consists of 79 amino acids of known sequence (796). Small quantities of apoCIII are also found in intermediate-density lipoprotein (IDL) and low-density lipoprotein (LDL) (5,87). ApoCIII protein and mRNA synthesis occurs predominantly in the liver and to a lesser extent in the intestine (16–18), and it is developmentally regulated (515). The apoCIII mRNA encodes a 99-amino acid protein which contains a 20-residue signal peptide. Following cotranslational cleavage of the signal peptide, the 79-residue protein is modified in the ER/Golgi by O-glycosylation and sialylation and is secreted in the disialylated form (CIII$_{s2}$) (43). The protein is subsequently desialylated in plasma to generate a mixture consisting of 14% of the asialo form (CIII-0), 59% of the monosialo form (CIII$_{s1}$), and 27% of the disialo form (CIII$_{s2}$) (87,88) (Fig. 2C). The physiological importance of the modifications of apoCIII is not known. In tissue culture experiments, 90% of secreted apoCIII is found in the lipoprotein-free fraction, and it can associate with exogenously added VLDL and HDL and to a lesser extent with LDL (81). Secreted apoCIII is found as a component of particles with defined lipid and apolipoprotein composition, designated LpB:CIII, LpB:E:CIII, LpB:AII:CIII, and LpAI:AII:CIII, as well as complexes containing apoCI and apoCII. Similar particles containing LpB:C or LpB:C:E have been isolated from plasma (64,69). The physiological significance as well as the sites of assembly of these particles are not known.

Comparison of the nucleotide and amino acid sequences of the apolipoproteins showed that they contain similar repeated units with sequence homologies. ApoCIII contains a 33-, an 11-, and a 21-residue unit at nucleotides 8–40, 41–51, and 52–72, respectively (21). Analysis of the secondary structure of apoCIII according to the Chou and Fasman procedure predicted α-helices at residues 1–39 and 54–69, β-turns at residues 39–42 and 72–75 (763), and an amphipathic α-helix at residues 40–67 (506). Association of apoCIII with phospholipids increases its α-helical content from 26 to 70% (797–799). Based on phospholipid-binding properties of apoCIII peptides, it has been proposed that the phospholipid-binding domain of apoCIII is localized in the 39 carboxy-terminal amino acid residues 41–79 (798). This region contains the 11- and the 22-residue repeated units. Association of apoCII with egg lecithin yields discoidal particles with a minor axis of 4 nm and a major axis of 20 nm (765). Small-angle X-ray scattering studies showed that the particles of DMPC–apoCIII complexes have a 17 × 5 nm ellipsoidal structure and a 1-nm-thick shell representing the apolipoprotein on the surface of the hydrophobic chains of the phospholipids (800). The dissociation constant K_d of apoCIII from model lipid emulsions (microemulsions and large emulsions containing egg yolk lecithin and triolein) ranges from 0.53 to 1.07 and the saturation number N (number of apolipoprotein molecules/1000 phospholipid molecules) ranges from 8.2 to 13.2 (22,529).

Functions

ApoCIII inhibits *in vitro* the function of lipoprotein lipase, the enzyme that hydrolyzes the triglyceride moieties of chylomicrons and VLDL (334,801,802). ApoCIII also inhibits the binding of apoE-containing lipoproteins to the LDL receptor, but not to the LDL receptor-related protein (LRP) (265,266). The receptor-mediated binding and catabolism of lipoproteins is enhanced by apoE and is inhibited by apoCIII (767–770). These observations suggested a role of apoCIII in the catabolism of triglyceride-rich lipoprotein. Recent experiments showed that overexpression of the human apoCIII gene in transgenic mice results in severe hypertriglyceridemia with fasting triglyceride levels exceeding 950 mg/dl (803). These findings reinforce a potential role of apoCIII in the catabolism of triglyceride-rich lipoproteins *in vivo*.

Genetics

The cDNA and gene sequences for human apoCIII have been reported (607, 609,804,805). The apoCIII gene is 3.2 kb long and contains four exons and three introns, and is closely linked to the human apoA-I and apoA-IV genes (552,553). The apoCIII gene is localized 2.5 kb downstream of the apoA-I gene and 5 kb upstream of the apoA-IV gene (552,553,636) (Fig. 10A). The direction of transcription of the apoCIII gene is opposite to that of the apoA-I and apoA-IV genes. The cluster of the three genes is mapped on the long arm of chromosome 11, in the region 11q13→qter (554). Two human conditions characterized by severe deficiencies in plasma apoCIII and HDL, associated with premature atherosclerosis, have been described and are shown in Fig. 11A. One condition has resulted from deletion of the apoA-I, apoCIII, apoA-IV loci (569) and the other from an inversion of the apoA-I, apoCIII loci (570). Patients with type III, type IV, and type V hyperlipoproteinemia have elevated plasma apoCIII levels (88,806). Variations in the relative concentration of apoCIII isoproteins have also been observed which deviate from normal values (88,807). For instance, patients with type V hyperlipoproteinemia are reported to have statistically lower relative apoCIII-0 concentrations (807). Similarly, a familial preponderance of apoCIII-0 and absence of sialylated apoCIII forms ($CIII_{s1}$ and $CIII_{s2}$) has been reported (808). This condition is caused by substitution of Thr-74, which is the site of O-glycosylation, by an Ala (809) (Fig. 16C). Recently, a subject with hyperalpha-lipoproteinemia has been described who is heterozygous for an apoCIII gene mutation which results in substitution of Lys-58→Glu (810) (Fig. 16C). This condition was associated with increased HDL and apoA-I levels, enlarged HDL_{2b} particles enriched in apoE, and relatively low apoCIII and triglyceride levels. In general, hyperalphalipoproteinemia has been associated with longevity and it is possible that reduced expression of the apoCIII gene might be beneficial (810).

Apolipoprotein E

Structure

Apolipoprotein E is a component of VLDL, IDL, HDL, chylomicrons, and chylomicron remnants. The primary sequence of human apoE was obtained by conventional protein sequencing methods (694) and was verified in 1983–1985 by cloning cDNAs and the gene encoding for this protein (811–814). Similar to the other apolipoproteins, both the amino acid (694) and the nucleotide sequence of apoE showed that the protein, as well as the gene, contains repeated units which can form amphipathic α-helices. ApoE contains seven 22-residue repeats, one 11-residue repeat, two 20-residue repeats, and one 17-residue repeat. The 17- and 20-residue repeats have deletions of amino-terminal, carboxy-terminal, or internal residues as compared to the 22-residue repeats (21). As discussed earlier, the single 22-residue repeat or a 44-residue sequence containing two tandem repeats with a proline in the middle may provide the necessary secondary and tertiary structure which allows the binding of lipids (509,510). As will be discussed later, in the context of apoE mutations and type III hyperlipoproteinemia, such amphipathic α-helices have been observed in the X-ray-derived three-dimensional structure of the amino-terminal portion of human apoE (512). Biochemical studies combined with secondary structure predictions indicate that apoE contains two independently folded domains which are separated by a region of random coil (815,816). Digestion with thrombin produces a 22-kD amino-terminal fragment (residues 1–191) and a 10-kD carboxy-terminal fragment (residues 216–299). The receptor-binding activity is localized in the amino-terminal fragment (698,699), whereas the lipid binding is mediated by the carboxy-terminal fragment (202,817). The association of apoE with phospholipids leads to the formation of discoidal particles of 16–21 nm major axis by 5–7 nm minor axis (186) and increases the helical content of apoE from 45 to 66% (818). The structure of the DMPC–apoE particles is very similar to that of the apoA-I particles (186,818). The dissociation constant K_d of apoE from model lipid emulsions (containing egg yolk, lecithin, and triolein and 6% cholesterol) and the saturation number N (molecules apoE/1000 molecules phospholipid) are 1.12 μM and 3.8, respectively (22,819). Increase of the cholesterol content of the emulsion to 30% decreased the N to 0.7 with little effect on the dissociation constant ($K_d = 1.05$ μM) (819).

Biosynthesis

The early work indicated that the major site of apoE synthesis was the liver (16). Detailed information on the sites of apoE synthesis has been obtained by blotting analysis of RNA isolated from human and other mammalian tissues (17, 18,28,48,820). Quantitative studies in nonhuman primates have also shown that the peripheral tissues may contribute 20–40% of total plasma apoE pool (28). ApoE protein and mRNA synthesis is regulated by hormonal nutritional factors (821–825) and is developmentally regulated (820,826,827). ApoE synthesis in rat ovarian granulosa cells is stimulated through transcriptional and posttranscriptional mechanisms by cAMP and phorbol esters, two agents which activate protein kinases A and C, whereas mevinolin, an inhibitor of hydroxymethyl glutaryl-CoA reductase, inhibits apoE synthesis (828). ApoE mRNA and protein synthesis in macrophages is stimulated by the cholesterol content of the cells (829–831) and is inhibited by endotoxin treatment (832). Induction of apoE mRNA and protein synthesis occurs during the differentiation of mouse 3T3-LI preadipocytes to adipocytes, and this process is regulated by the lipid content of the cells (827). In contrast to the apoE that is synthesized by other cell types, most of the apoE synthesized by 3T3-LI adipocytes is not secreted (827). Similarly, the induction of apoE mRNA and protein synthesis in human monocyte-macrophage cultures correlates with monocyte-macrophage differentiation (17). In contrast, apoE mRNA synthesis decreases during liver regeneration and cirrhosis, presumably by transcriptional control mechanisms (523). ApoE synthesis has also been reported in epithelial cells (28,833), smooth muscle cells (833,834), and lipocytes (835). ApoE synthesis has also been demonstrated in rat astrocyte cultures of the central nervous system, the glial-like cells of the peripheral nervous system, the non-myelinated Schwann cells near or around the sensory and motor neurons (836,837), and human glial neoplasms (836). ApoE synthesis was not observed in the glial cells and neurons of the central nervous system (838,839). In addition, astrocytes and Schwann cells express the LDL B/E receptor (837,839). Injury of the central or peripheral nervous system is associated with increased apoE synthesis (840,841). Furthermore, following injury, macrophages and monocytes recruited to the site of injury synthesize and secrete apoE. The cholesterol produced by the degenerating myelin is absorbed by the macrophages and Schwann cells and is reutilized later by the regenerating cell (839). ApoE-containing lipoproteins produced by the regenerating nerve are taken up by B/E receptors present in the tips of the growing axons of regenerating nerves (839). In addition to apoE, the regenerating peripheral nerve increases the synthesis of apoD and accumulates intracellularly apoA-I and apoA-IV (842). The system of the injured and regenerating nerve provides an example of the role of apoE and possibly of other apolipoproteins in lipid transport and in the redistribution of cholesterol. In other cases, nascent apoE synthesized by the peripheral tissues may participate in the reverse transport of cholesterol. Thus, following cholesterol efflux from cells by the HDL and subsequent cholesterol esterification by LCAT, nascent apoE is incorporated into HDL and mediates its catabolism by both hepatic as well as extrahepatic tissues (843). A large body of evidence indicates that a large portion of apoE synthesized by cells (49,81,844,845) and liver perfusion systems (80) is secreted in a lipid-poor form and is subsequently incorporated into preexisting lipoprotein particles such as IDL, HDL, and chylomicron remnants. In plasma, apoE can be isolated as a component of lipoprotein particles (LP) with defined lipid and apoprotein composition, designated LpB:E, LpB:C:E, LpAII:B: C:D:E (64,69,72). Similar particles have been isolated from the culture medium of HepG2 cells (24,73). The physiological significance as well as the sites of assembly of these particles are not known.

Modification and Flotation Properties of Nascent ApoE

The apoE mRNA encodes a 317-amino acid protein containing an 18-residue signal peptide which is cleaved cotranslationally. The protein is then modified in the ER/Golgi at the threonine residue 194 (39,42) with carbohydrate chains containing sialic acid. The degree of glycosylation differs in different cells. Thus, in hepatic cells 92% of secreted apoE is glycosylated with carbohydrate chains containing either two, four, or six residues of sialic acid; these forms of apoE are denoted E_{s2}, E_{s4}, E_{s6}. The relative abundance of these forms is E_{s2} 60%, E_{s4} 33%, E_{s6} 7% (33). Similarly, apoE secreted by human monocyte-derived macrophages is entirely glycosylated (846). In contrast to the nascent apoE, the plasma apoE consists of only 24% of modified forms of apoE, which contain either one or two molecules of sialic acid per molecule of apoE and are denoted E_{s1} and E_{s2} (89). The

remaining (76%) of the plasma apoE is not modified and, depending on the phenotype, which will be explained later, is denoted E4, E3, or E2 (Fig. 2A). These
observations imply that, following secretion, apoE is desialylated in the plasma
compartment. The effects of apoE glycosylation on the synthesis and secretion of
apoE was studied with mutant Chinese hamster ovary cells with a reversible defect
in protein O-glycosylation (ldlD cells). Permanent ldlD cells transfected with the
apoE gene synthesize and secrete only the unmodified form of apoE. The
synthesis, rate, and extent of apoE secretion were unaffected by O-glycosylation
in this cell system, suggesting that the modification is not required for the
intracellular transport and secretion of apoE (39).

ApoE Functions

Receptor Binding. ApoE, as discussed previously, is the ligand which
promotes the recognition and catabolism of apoE-containing lipoproteins such as
chylomicron remnants, βVLDL, hypertriglyceridemic VLDL, IDL, and HDLwE
by the LDL receptor (160,182,187) and the E/α2M receptor (79). The receptor-
binding domain of apoE was identified by binding studies of normal variant and
truncated apoE forms as well as by the use of receptor-blocking monoclonal
antibodies (698,699,847,848). Binding studies of apoE–DMPC complexes
showed that the receptor-binding domain of apoE was located in an amino-
terminal thrombolytic peptide (residues 1–191) as well as in a CNBr fragment
between residues 126 and 218 (698). The binding of these peptides was inhibited
by modification of Lys or Arg residues (698). ApoE fragments of residues 1–183
and 1–170 generated by *in vitro* mutagenesis of the apoE gene contains 85 and 1%
of the normal receptor binding activity, respectively (848). Utilization of monoclonal antibodies showed that antibodies recognizing the carboxy-terminal thrombolytic fragment (residues 216–299) of apoE had no effect on receptor binding. A
monoclonal antibody (ID7) recognizing the amino-terminal thrombolytic fragment (residues 1–191) inhibited receptor binding by 90% (699). Utilization of
synthetic peptides indicated that the epitope recognized by ID7 is in the vicinity of
residues 139–146 (699). Site-directed *in vitro* mutagenesis also verified that
residues in the region 140–160 are important for receptor binding (847). A
receptor-binding domain in this region was also predicted by hydrophobicity and
hydrophobic moment calculations (849). As mentioned before, the sequence of
apoE between residues 142 and 149 has 63% homology with the region between
residues 3359 and 3366 of apoB-100 (Fig. 14D). The heparin-binding domains of
apoE were localized at amino acids 142–147, 243–272, and 211–218 (850,851).

Other Functions. As discussed above, apoE synthesized by cells can be
incorporated into lipoprotein particles and redistribute cholesterol to other cell

types in the vicinity which may require cholesterol. Such redistribution of cholesterol, for instance, occurs during nerve regeneration following injury (837, 839–841,852). In a similar fashion apoE synthesized by peripheral cells can be incorporated into HDL and redistribute cholesterol directly to the liver, thus contributing to reverse cholesterol transport or to other peripheral cells (178,179, 310,843). The ability of apoE to direct the catabolism of apoE-containing lipoproteins by the LDL and the LDL receptor-related protein (LRP) or E/α2M receptor is inhibited by apoCI and to a lesser extent by apoCII and apoCIII proteins (265,266). Introduction by transfection of a functional apoE gene in mouse Y1 adrenal cells dramatically decreases steroidogenesis, suggesting a potential regulatory role of apoE in adrenal steroidogenesis (853). Immunogold labeling of hepatocytes showed that apoE could be localized in the Golgi as well as on the matrix of the peroxisomes, suggesting a potential role of apoE in interorganellar cholesterol transport (86). Finally, *in vitro* experiments have shown that apoE promotes the conversion of IDL to LDL (854).

Genetics

The structures of the human apoE mRNA and the apoE gene have been derived from sequences of cDNA and genomic clones (811,813,814). ApoE mRNA consists of 1163 nucleotides, including 5' and 3' untranslated regions of 67 and 142 nucleotides, respectively (812,814). The apoE gene is 3.6 kb long, contains four exons and three introns (811,812), and maps on chromosome 19 (811,855). The apoE gene is closely linked with apoCII and the LDL receptor genes. The apoCI gene is located 5.5 kb downstream of the apoE gene and the apoCI pseudogene is located 7.5 kb 3' to the apoCI gene (762). The human apoCII gene is located 20 kb downstream of the apoCI pseudogene (212), and the LDL receptor gene is located 5' to the apoE, apoCI, apoCII gene cluster (209–211) (see Fig. 10B).

Common Genetic Polymorphisms in Human ApoE. The first evidence of genetic polymorphism in human apoE was obtained by Utermann and colleagues in 1975 (856). Extensive studies of the human plasma apoE by two-dimensional polyacrylamide gel electrophoresis showed the presence of six common apoE phenotypes with different isoelectric points (857–860). The phenotypes could be classified in two categories with three phenotypes each. In the first category the phenotypes consisted of one major nonmodified isoprotein which differed by either one or two charges from the others along with O-glycosylated/sialylated forms described in Fig. 2E. In the second category the phenotypes contained two major nonmodified isoproteins along with the sialylated forms (3) (Fig. 17A). Initial studies of the inheritance of the apoE in families showed that the phenotypes are determined by a single apoE locus with three common alleles

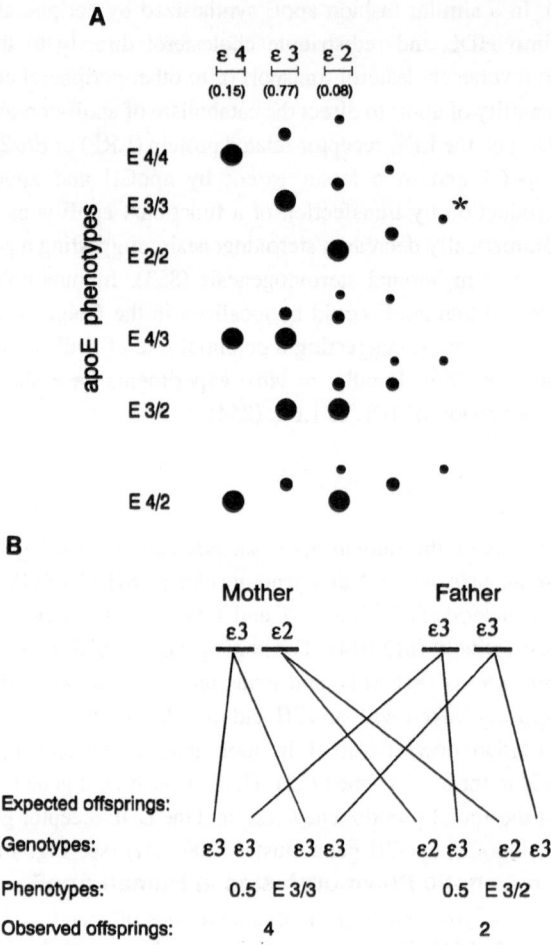

Fig. 17. (A) Schematic representation of the common apoE alleles and apoE phenotypes. Large and small spheres represent the unmodified and the O-glycosylated sialylated forms of apoE, respectively. Asterisk indicates phenotype associated with type III hyperlipoproteinemia. (B) Schematic representation of the expected and observed pattern of inheritance of the apoE phenotypes. (C) Displacement of 125I-LDL from monolayers of cultured fibroblasts by apoE3 and apoE2 phospholipid vesicles. The phenotype of the apoE used is indicated. (Reproduced from ref. 893 with permission.) (D) Schematic representation of the antiparallel amphipathic α-helical bundles observed in human apoE. The zigzags and arcs indicate α-helical structure and turns of the polypeptide chain, respectively. (Reproduced from ref. 512 with permission.)

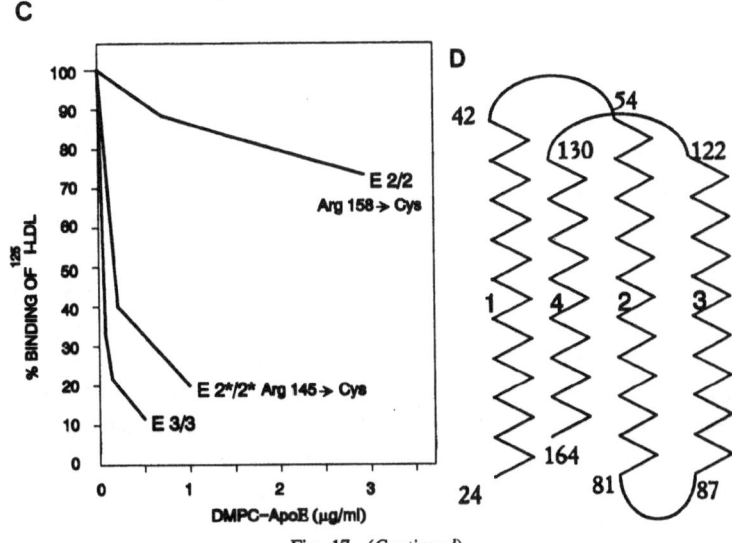

Fig. 17. (*Continued*)

designated ε4, ε3, and ε2. We proposed that these alleles were the result of structural mutations in the apoE gene (857–860). Individuals homozygous for alleles ε4, ε3, and ε2 have the phenotypes E4/4, E3/3, and E2/2, respectively. An example of inheritance of the apoE phenotypes is shown in Fig. 17B. The genetic model we proposed was subsequently verified by a large number of family studies (reviewed in 1985 and 1988) (860,861). The frequency of the apoE alleles differs in different populations. In Caucasians, the average allele frequencies are ε4 = 0.15, ε3 = 0.77, and ε2 = 0.08. This corresponds to calculated phenotype frequencies of E4/4 = 3%, E3/3 = 60%, E2/2 = 1%, E4/3 = 22%, E3/2 = 12%, and E4/2 = 2% (861). The isolation of cDNAs and the gene encoding human apoE provided the molecular explanation for the common polymorphism of apoE (811–814). The three different apoE isoproteins, E4, E3, and E2, observed in humans could be explained by mutations in amino acids 112 and 158. The E4 isoprotein contains Arg-112 and Arg-158, the E3 contains Cys-112 and Arg-158, and the E2 contains Cys-112 and Cys-158. The substitution of one or two molecules of Arg by Cys explains the observed charge differences in apoE isoproteins E4, E3, and E2 (857–860).

Common ApoE Alleles Affect Plasma Lipid and Lipoprotein Levels in the General Population. In addition to the contribution of the apoE phenotype to the expression of type III hyperlipoproteinemia, Utermann and colleagues originally observed that carriers of the $\epsilon2$ allele had lower levels of plasma and LDL cholesterol than did those with homozygosity for the $\epsilon3$ allele (862). These findings suggested that the apoE phenotypes may be involved in determining the plasma lipid and lipoprotein levels in the general population (862). A variety of subsequent studies have shown uniformly that the effect of apoE genotypes on plasma and LDL cholesterol follows the order $\epsilon4/\epsilon4 = \epsilon4/\epsilon3 = \epsilon4/\epsilon2 > \epsilon3/\epsilon3 > \epsilon3/\epsilon2 > \epsilon2/\epsilon2$ (861,863–870). Homozygosity or heterozygosity for the $\epsilon2$ allele is also associated with increases in VLDL cholesterol and triglycerides (869) compared to $\epsilon2/\epsilon3$ genotypes. Homozygosity or heterozygosity for $\epsilon2$ or $\epsilon4$ alleles is also associated with increased and decreased plasma apoE, respectively (871,872). Furthermore, the $\epsilon2/\epsilon2$ genotype is associated with decreased plasma apoB levels as compared to $\epsilon4/\epsilon3$ and $\epsilon4/\epsilon4$ genotypes (865,871). No association has been found between the apoE genotypes and plasma triglyceride and HDL cholesterol levels (861,863,865–868,870,873). A variety of recent studies with hyperlipidemic populations have shown that homozygosity (873) and/or heterozygosity (873–876) for the $\epsilon2$ allele is associated with hyperlipidemia and hypertriglyceridemia. In addition, homozygosity (873) and/or heterozygosity (877,878) for the $\epsilon4$ allele is associated in some studies with hypercholesterolemia. Attempts by several groups to correlate apoE phenotypes with myocardial infarction (864,879–881) or coronary artery disease (864) gave either negative or inconclusive results.

Certain ApoE Phenotypes and Genotypes Are Associated with Type III Hyperlipoproteinemia (Type III HLP).   Familial type III HLP, also called familial dysbetalipoproteinemia or broad β, or floating β disease, is characterized by premature atherosclerosis, xanthomas, elevated plasma cholesterol and triglyceride levels, cholesterol-enriched βVLDL and IDL particles, increased plasma apoE levels, and premature coronary and peripheral atherosclerosis (882–884). The most reliable criterion used in the past for diagnosis of this disease was an increase in the ratio of VLDL cholesterol to total triglycerides ($r > 0.30$) and a plasma triglyceride concentration between 150 and 1000 mg/dl (882,883,885). The frequency of the disease was estimated to be 0.1–0.01% in the population (883,884,886). Initial studies of the phenotypes of patients who were diagnosed with the above criteria to have type III HLP showed that 31 of the 34 patients had the E2/2 phenotype, 2 the E4/2 phenotype, and 1 the E3/2 phenotype (887,888). Thus, the apoE phenotype E2/2 could serve as a molecular marker in 91% of the patients with type III HLP.

Reduced Binding to Lipoprotein Receptors of ApoE Derived from Individuals with the E2/2 Phenotype.   Turnover studies showed that the catabolism of ^{125}I-labeled apoE derived from an individual with type III HLP and the E2/2 phenotype was slower than that of the apoE3 (889,890) or E4 (891) isoproteins. In addition to these *in vivo* studies, extensive experiments involving fibroblast cultures and membrane preparations have shown that DMPC–apoE complexes generated with apoE of different phenotypes display variable degrees of competition for the LDL (B/E) receptor (892–894). ApoE from individuals with the E3/3 and E4/4 phenotypes displays the same competition for the LDL (B/E) receptor as described previously (894). However, apoE derived from individuals with the E2/2 phenotype does not compete as well for the LDL (B/E) receptor (893). This analysis also showed functional heterogeneity within the E2/2 pheno-type (893). ApoE that has arisen from an Arg-158→Cys substitution competes inefficiently for the binding to the LDL (B/E) receptor (893), whereas apoE that has arisen from an Arg-145→Cys substitution competes more effectively for binding to the LDL receptor (Fig. 17C). In other studies apoE of the E2/2 phenotype with a severe receptor-binding defect was functionally restored by treatment with cysteamine (894,895). This treatment restores the positive charge at amino acid 158 and increases the receptor-binding activity 20-fold (895). It has been proposed that the cysteamine reaction changes the conformation of the protein in such a way as to form the interaction of residues 140–150 with the receptor-binding site and that residue 158 may not be involved directly in receptor binding (895). These binding and catabolic experiments are consistent with earlier observations showing the accumulation of remnant lipoproteins in the plasma of patients with type III HLP which are enriched in cholesteryl esters and apoE (11, 14,193,203,896–900). These apoE-rich lipoprotein remnants are apparently the result of slow clearance *in vivo* of apoE-containing lipoproteins due to the described structural defect in apoE.

Dominant and Recessive Forms of Type III Hyperlipoproteinemia. The majority of apoE2/2 phenotypes result from the Arg-158→Cys substitution. The Arg-158→Cys mutation in the homozygous state, depending on other genetic or environmental factors, may result in type III hyperlipoproteinemia. This form of type III HLP is inherited in an autosomal recessive mode (901). A variety of rare apoE mutations have also been described which are associated with a dominant mode of inheritance of type III HLP which is expressed at an early age (902,903). These include the substitutions Arg-136→Glu, Arg-142→Cys, Arg-145→Cys, Lys-146→Gln, and Lys-146→Glu and an insertion of seven amino acids (duplica-tion of residues 121–127) (904–909).

With the exception of apoE$_{Leiden}$ with a tandem repeat of residues 121–127 (907,908), the apoE mutations which are associated with dominant forms of type III HLP are between residues 136 and 152. It has been proposed that these mutations affect directly the binding of apoE-containing lipoproteins to the LDL receptor (512,901). A mutation with a substitution of Arg-228→Cys has normal receptor-binding activity (910). It has been shown that, in contrast to the full-length mutant apoE, which has 25% normal receptor-binding activity (698), the 1–191 apoE fragment of the mutant protein, which contains the 121–127 tandem repeat, has normal receptor binding (907). This suggests that the defective apoE binding which is observed in this patient results from an altered conformation of the protein which may indirectly affect receptor binding. The importance of the 136–152 region of apoE for receptor binding was also assessed by *in vitro* mutagenesis. Mutations within this region reduced the receptor binding activity to 10–50% of control (847). New insights into the structure–function relationship of human apoE have been obtained with X-ray crystallography of the 22-kD amino-terminal thrombolytic fragment of apoE (512). This analysis showed that the amino-terminal portion of apoE forms a four-helix bundle. The four defined antiparallel helical bundles extend between residues 24 and 42, 54 and 81, 87 and 122, and 130 and 164 (Fig. 17D). Adjacent helices 1 and 4, and 2 and 3, are stabilized by hydrophobic interactions and salt bridges. It is interesting that residues 130–150, which have been implicated in receptor binding, are within the fourth helical bundle. It has been proposed that the residues 130–150 may participate directly in receptor binding, whereas the adjacent residues, including Arg-158, may affect the conformation and the lipid-binding properties of apoE (512). The X-ray crystallographic data show that Arg-158 is within the fourth α-helical bundle at a distance greater than 15 Å from residue 150 and has the potential to form salt bridges with the adjacent Asp-154 of bundle 4 or Glu-96 of bundle 3 (512) (Fig. 17D). Following the elucidation of the three-dimensional structure of the amino-terminal domain of apoE, it has been suggested that the Arg-158→Cys mutation affects the conformation of the protein and this indirectly affects receptor binding (182,512).

Additional information on the mechanisms which may be responsible for the expression of type III hyperlipoproteinemia came from the work of Innerarity and colleagues (292). These investigators have shown that the receptor-binding properties of apoE-containing lipoproteins obtained from type III patients with the Cys-158 mutation depend on the lipid composition of the particles. Particles obtained from untreated patients with high values of cholesterol and triglycerides bind poorly to cultured skin fibroblasts and their binding could be improved by treatment of the particles with cysteamine. In contrast, particles obtained following weight loss could bind efficiently and the binding was not affected by cysteamine treatment of the lipoprotein particles (292,901).

Some of the apoE variants are the result of double mutations in the apoE gene. These include the Ala-99→Thr and Ala-152→Pro variant (911), the Gly-127→Asp and Arg-158→Cys variant (912,913), the Arg-145→Cys and Glu-13→Lys variant (913), the Pro-84→Arg and Cys-112→Arg variant (914), and the Glu-244→Lys and Glu-245→Lys variant (915,916). A variant has been described which has the

E4/4 phenotype, carries a Gln-13→Lys and an Arg-145→Cys substitution, and is associated with a severe dominant form of type III HLP (917). A single mutation, Arg-145→Cys, was shown previously to be connected to 45% normal receptor binding and less severe clinical symptoms (901), whereas a subject with a Glu-13→Cys substitution has a normal lipid profile (918). Thus, it is possible that changes in the receptor-binding domain of apoE combined with distal changes may more severely affect apoE functions. Consistent with this notion is that an E5 variant with Glu-3→Lys substitution (914,919,920) displayed 190% of normal receptor binding and this mutation was associated with a dyslipoproteinemia (914,921). A rare form of type III HLP associated with familial apoE deficiency was first described in blood relatives in 1981 (922). Plasma apoE was undetectable in the affected individuals (923). The absence of apoE was associated with atherosclerosis and accumulation of lipoprotein remnants in the plasma of the patients, suggesting that the formation of remnants does not require the participation of apoE (923). DNA sequencing and expression studies showed that this condition results from an A→G substitution in the 3′ splice junction (donor site) of the third intron of the apoE gene (924). This single base substitution abolishes the correct 3′ splice site, thus creating two abnormally spliced mRNA forms. The smaller form contains 53 nucleotides, and the larger form contains the entire third intron of the apoE gene. Both of these mRNA species contain chain-termination codons within the intronic sequence and code for short apoE peptides that are not detectable by gel electrophoretic techniques. The A→G substitution also generated a new polymorphic site which can be detected by DNA blotting analysis. This condition is inherited in an autosomal recessive mode and illustrates the importance of basal levels of apoE for the clearance of the lipoprotein remnants. Another case of apoE deficiency results from a nonsense mutation at codon TGG specifying Trp-210 (925). It is interesting that apoE-deficient mice generated by inactivation of the apoE gene develop severe hypercholesterolemia and atherosclerosis (925a,b). The different apoE mutations described are shown in Fig. 18.

Factors Affecting Phenotypic Expression of Type III HLP. Numerous studies suggest that the E2/2 phenotype alone which results from an Arg→Cys substitution is not sufficient to cause type III HLP and that other genetic factors may be required for the phenotypic expression of the disease (926–931). In some cases, type III HLP is expressed in families with a tendency toward hypertriglyceridemia based on environmental or other genetic factors (871,930, 931). The concept that other genetic and/or environmental factors may be required for the expression of type III HLP has received direct biochemical support from studies by Rall *et al.* (932). These investigators have shown that apoE from subjects with the apoE2/2 phenotype who are normolipidemic and even hypolipidemic behaves in the same way in competition experiments and presumably

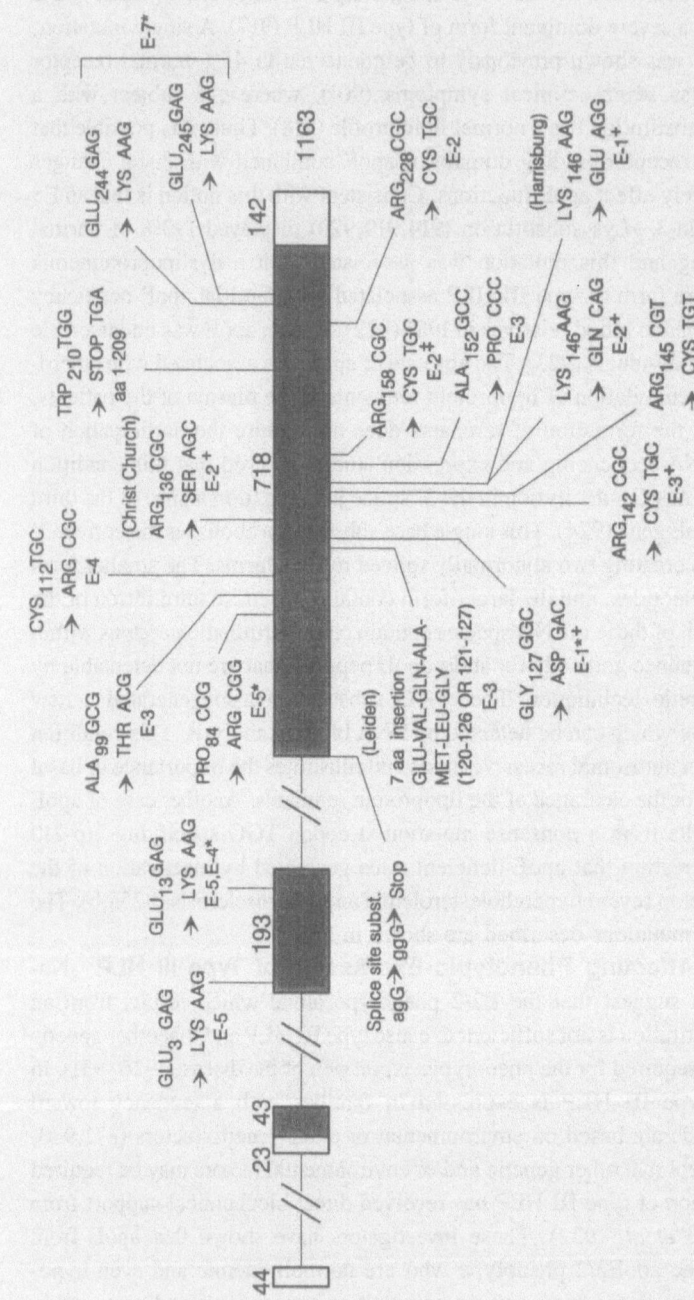

Fig. 18. Schematic representation of the human apoE gene, the various apoE mutations, and the phenotypes they generate. The Arg-158→Cys substitution is associated with the recessive form of type III HLP. ApoE forms associated with dominant forms of type III HLP are indicated by +; recessive inheritance is indicated by ‡. The symbols E1*, E4*, E5*, and E7* indicate double mutations as follows: E1* = Gly-127→Asp and Arg-158→Cys; E4* = Glu-13→Lys and Arg-145→Cys; E5* = Pro-84→Arg and Cys-112→Arg; E7* = Glu-244→Lys and Glu-245→Lys. The apoE mutations are described in refs. 904–920, 924.

contains the same molecular defect as does apoE from patients with the E2/2 phenotype (Cys-158) who express type III HLP (932).

PROTEINS ASSOCIATED WITH LIPOPROTEIN PARTICLES

ApoD

ApoD, originally designated apoA-III (933), is a glycoprotein found in HDL and the $d>1.21$ g/ml fraction (934) and contains 18% carbohydrate by weight (935). The initially reported Mr of apoD was 20 kD (933,936). Subsequent purification of apoD showed that apoD consists of three isoproteins with apparent isoelectric points of 5, 5.08, and 5.2 and has an apparent molecular weight of 33–35 kD (10). The apoD sequence has been determined from the sequence of cDNA and genomic clones encoding this protein (19,937). The apoD gene is 12 kb long and contains six exons and five introns. The apoD gene encodes a 189-amino acid protein, including a 20-amino acid signal peptide (937). The apoD gene maps to the distal long arm of chromosome 3 (region 3p14.2→qter) (937). Protein and mRNA analysis showed that apoD is a ubiquitous protein which is synthesized by a variety of tissues, including placenta, brain, adrenal gland, kidney, pancreas, spleen, liver, intestine, testis, heart, skin, and urinary bladder (19,938).

The distribution of apoD in plasma resembles that of LCAT, C/ETP, and apoA-I, and while it has been suggested that a complex of apoD with these proteins may participate in the esterification and transfer of cholesterol from HDL, VLDL, and LDL (146,434), the role of apoD in cholesterol esterification and the transfer of cholesteryl esters requires further clarification (146,939,940). Despite its association with HDL, apoD is not structurally related to the other apolipoproteins. Rather, it belongs to the α2 microglobulin superfamily (941). Molecular modeling has suggested that, similar to the other members of the α2 microglobulin superfamily which bind hydrophobic ligands, apoD may be involved in the transport of heme-related compounds (941).

Lpa

Structure and Biosynthesis

Lpa represents a distinct class of lipoprotein particles found mostly in the LDL density region but in other lipoprotein classes as well, including triglyceride-rich fractions (942–944). The lipid moieties of these particles are similar to that of

LDL (942,945,946). The protein moiety of Lpa consists of a complex of molecules of apoB-100 linked by an intradisulfide bridge with a molecule of apoa (947,948). The cysteine residues which participate in this linkage have not been identified. Apoa is a glycoprotein containing approximately 30% of carbohydrate moieties. Apoa mRNA and protein synthesis occurs mainly in the liver (949,950). Minor sites of apoa mRNA synthesis are the brain and testis. Plasma Lpa levels decrease by 50% in males with estrogen treatment (951). The site of assembly of the Lpa particle has not been established, but it is expected to be similar to that of LDL. A mutation which prevents LDL assembly in a patient with abetalipoproteinemia results in the production of a lipid-poor apoB−apoa complex (726).

Functions

Recent data indicate that apoa displays proteolytic activity but has different specificity than that of plasmin (952). Unmodified Lpa binds to the modified LDL receptor but fails to cause accumulation of cholesteryl esters in macrophages (953). The longer residence time of Lpa in plasma, however, may enhance its modification and may lead to uptake and cholesterol accumulation in macrophages. The association of Lpa particles with triglyceride-rich particles may also enhance their uptake by macrophages (954,955). It has been suggested that Lpa may contribute to the pathogenesis of atherosclerosis and possibly thrombosis by interfering with the metabolism of LDL and plasminogen. Transgenic mice expressing human apoa develop fatty streak lesions in response to atherogenic diets as compared to control littermates (955a). Evidence suggests that Lpa may inhibit fibrinolysis by interfering with the binding of plasminogen to the plasminogen receptor, a process which leads to the activation of the zymogen and the generation of plasmin (946,956–961). It has been reported that Lpa inhibits the streptokinase-mediated activation of plasminogen (962), competes for the binding of plasminogen to fibrin (963), and affects the synthesis of plasminogen activator inhibitor-1 (964). Despite these possibilities, the relationship between plasma Lpa levels and thrombogenicity has not been established (957–959,965). Evidence suggests that Lpa binds and is catabolized by the LDL receptor and that such catabolism is accelerated by the action of lipoprotein lipase (943,966–968). Turnover studies have shown that the catabolism of Lpa *in vitro* is lower than that of LDL (969).

Genetics

The structure of apoa has been derived from the cDNA sequence of overlapping cDNA clones (949). Human apoa has homology to plasminogen. It consists of a protease domain which has 94% homology to that of plasminogen, one domain which has homology to Kringle V of plasminogen, and a variable number of domains (ranging from 15 to 40) with homology to Kringle IV of plasminogen (949). The apoa gene maps on chromosome 6 (region q26→q27) and is closely linked to the plasminogen gene locus (970–972). Extensive polymorphism for apoa has been observed in humans. The different apoa isoforms differ in molecular mass from 400 to 800 kD and their transmission follows Mendelian inheritance (942). The difference in Lpa size arises from the different number of Kringle IV repeats, which ranges from 15 to 40 (973–975). Epidemiological studies have shown that the Lpa size is inversely related to the plasma Lpa concentration (973,976–978). Differences in the Lpa concentration in subjects with the same phenotype are determined primarily by differences in the production rates (979). In patients with familial hypercholesterolemia (980,981) or with high LDL levels (982), the Lpa plasma levels are correlated with the risk of coronary artery disease (978,983,984). The plasma Lpa levels are controlled by the Lpa locus as well as by other genetic factors such as LDL receptor defects and by environmental factors such as drugs and hormones (942,966).

Minor Apolipoproteins

ApoF and apoG are minor proteins present in HDL. The former has a molecular weight of 26,000–32,000 and an isoelectric point of 3.7 (985), and the latter has a molecular weight of 72,000 (986). ApoH, or β2-glycoprotein-I (987–989), is a glycoprotein of molecular weight 40,000–43,000 containing 18% carbohydrate by weight; it is present in HDL and the $d>1.21$ g/ml fraction (987). ApoH binds with high affinity to chylomicrons and artificial triglyceride emulsions (990) and acts synergistically with apoCII in the activation of lipoprotein lipase (988). Proline-rich protein (PRP) has been isolated from chylomicrons and has a molecular weight of 74,000 (991). A new apolipoprotein designated apoJ has been cloned and characterized (20). RNA blotting analysis has shown that, like apoE, the protein is ubiquitous, and is relatively abundant in brain, liver, testis, and ovary. A single apoJ gene encodes a protein of 427 amino acids which is then cleaved posttranslationally between Arg-205 and Ser-206 to generate two sub-units, Ja (amino acids 1–205) and Jb (amino acids 206–427), which form a disulfide-linked dimer. Secondary structure predictions indicate the presence of amphipathic α-helices (20). The plasma apoJa and apoJb subunits are glycosylated and have apparent Mr of 34–36 and 36–39 kD, respectively, and isoelectric points ranging from 4.9 to 5.4 (992). Plasma apoJ is distributed in HDL_3 and $d=1.21$– 1.25 g/ml fractions and to a lesser extent in HDL_2 (992). The function of this apolipoprotein remains unknown. A group of two major and four minor proteins designated as serum amyloid A proteins (SAA) have been found as components of

HDL and intestinal chylomicrons (3,993–997). The proteins are synthesized by the liver (996) and intestine (3). The SAA isoforms observed in plasma have approximately the same molecular weight (11,000) but different isoelectric points (3). The two major human amyloid proteins, apoSAA$_1$ and apoSAA$_2$, have very similar amino acid compositions and N-terminal amino acid sequences. However, apoSAA$_2$ lacks the N-terminal arginine residue that is found in apoSAA$_1$ (997). Human apoSAA$_1$ is a polypeptide composed of 104 amino acids of known sequence. Cleavage of the COOH-terminal hexapeptide 99–104 of SAA$_1$ generates the amyloid protein that accumulates in tissues in certain inflammatory conditions (998). Evidence indicates that there might be at least three different genes encoding the human SAA (998–1003). Similar to the human proteins, SAA forms of different mammalian species are the products of different genes (1004,1005). The physiological significance of all these proteins for lipoprotein metabolism and their firm classification as apolipoproteins require further investigation.

REGULATION OF APOLIPOPROTEIN SYNTHESIS

Importance and Background

As discussed in the introduction (Fig. 1A), lipoprotein metabolism is a complex pathway in which a variety of proteins, including apolipoproteins, plasma proteins, and lipoprotein receptors, participate. It is apparent from the preceding presentation that genetic changes which alter either the function or the concentration of any of these proteins will perturb this pathway and may directly or indirectly contribute to the pathogenesis of atherosclerosis. Similar perturbations in lipoprotein metabolism may be caused by increased concentration of these proteins. Transgenic animal experiments, for instance, have shown that increase in the plasma concentration of apoA-I or increase in the number of LDL receptors is beneficial (228,229,548–550). In contrast, increase in the plasma concentration of apoCI, apoCIII, or C/ETP may be detrimental (474,485,486,771,803,810). Numerous studies, summarized and cited in Table VII and discussed under the individual apolipoprotein sections, have provided evidence that the synthesis and/ or secretion of apolipoproteins may be influenced by hormonal and nutritional factors, and are developmentally regulated. The plasma concentration of apolipoproteins may be influenced either at the level of gene transcription or during posttranscriptional events. Thus, it is important to understand the molecular mechanisms which allow the switching on and off and the modulation of expression of the apolipoprotein and the other genes described in the preceding sections.

TABLE VII.   Regulation of Apolipoprotein Synthesis[a]

Apolipo-protein	Species differences on sites of synthesis	Developmental regulation	Regulation by hormones and drugs	Nutritional regulation
ApoA-I	Yes	Yes	Insulin, dexamethasone, estrogen, testosterone, thyroid hormones	Fats
ApoA-II	—	—	—	Fats
ApoA-IV	Yes	Yes	Insulin, dexamethasone	
ApoB	Yes	—	Insulin,[b] oleic acid,[b] ω-3 fatty acid,[b] Ca[2+] channel blockers[b]	
ApoB-48	Yes	Yes	Thyroid hormones	Carbohydrates
ApoCI	—	—	—	
AppCII	—	—	—	
ApoCIII	—	Yes	—	
ApoE	Yes	Yes	Insulin, glucagon	Fats, cholesterol

[a]Table based on ref. 17, 18, 515–523, 628–631, 665–674, 681, 683–686, 820–832.
[b]Secretion affected with no changes in mRNA levels.

The transcription of eukaryotic genes is controlled by the interaction of regulatory gene sequences (regulatory elements) with specific nuclear proteins (transcription factors). This interaction controls tissue-specific gene expression, gene expression during differentiation and development, and gene expression in response to intracellular and extracellular stimuli such as hormones and metabolites. Promoter regulatory elements essential for transcription have always been found in front of the transcription initiation site at distances ranging from a few hundred bases to a few kilobases. In several cases, however, important regulatory elements which control tissue-specific or developmental expression of genes have been localized several kilobases either 5' or 3' of the gene or within the transcribed sequences. All these elements are the binding sites of transcription factors.

New Methodologies for the Study of Gene Regulation

Several experimental advances (reviewed in ref. 1006) have facilitated the study of eukaryotic promoters and have led to the identification and characterization of several eukaryotic transcription factors. These include: (a) Definition of the long-range regulatory elements which confer tissue specificity or developmentally regulated expression. This analysis utilizes transgenic mouse technologies (for review see refs. 1007,1008). (b) Definition of the promoter region a few kilobases upstream of the transcription initiation site which is necessary for gene transcription. For this analysis a promoterless gene, usually chloramphenicol acetyl transferase (CAT), is placed under the control of the promoter to be studied and inserted in a plasmid. Following transfection of cells with this plasmid, cell extracts are prepared and analyzed for their ability to convert ^{14}C-chloramphenicol to the mono- and diacetylated forms. (c) Identification of the different factors which bind to a specific promoter region and definition of their binding domains. For this purpose a variety of techniques are used, including DNase I footprinting. DNA-binding gel electrophoretic assays, methylation interference assays, and *in vitro* mutagenesis. The relationship of a factor which binds to a specific regulatory element to previously described factors can be assessed by competition DNA binding and competition footprinting assays and by the direct comparison with purified factors and by utilization of antifactor antibodies in DNA-binding assays. (d) *In vitro* mutagenesis of the promoter region and analysis of the promoter activity in order to assess the importance of specific residues for transcription *in vivo* (using CAT assays) and *in vitro* (using *in vitro* transcription assays). This information can then be correlated with the ability of the mutated sequence to bind nuclear factors. (e) Purification and characterization of transcription factors. (f) Cloning of cDNAs and genes encoding transcription factors and study of their function in the regulation of the target genes. Examples of these types of analyses are provided in refs. 83, 771, 1008–1032.

Distal Regulatory Elements of Human ApoB, ApoCIII, ApoA-I, ApoA-II, ApoCI, and ApoE Genes

The human apoA-I gene is closely linked to the apoCIII and apoA-IV genes (552,553) (Fig. 10A). Expression of segments of the apoA-I, apoCIII, apoA-IV gene cluster in transgenic mice indicated that the hepatic transcription requires only 5' regulatory elements in the apoA-I and apoCIII gene, whereas the intestinal transcription of the apoA-I, apoCIII, and apoA-IV genes requires elements localized in the intergenic sequence between apoCIII and apoA-IV genes, which extend to −2.4 kb upstream of the apoCIII gene (548,803,1033,1033a) (Fig. 19A). Similarly, expression of segments of the apoAI, apoCI, apoCII gene cluster in transgenic mice have identified the distal regulatory elements which dictate the transcription of the apoE gene in different tissues (83,771,1008). This analysis showed that expression of apoE in the kidney requires only 650 nucleotides of the 5' region and 72 nucleotides of the 3' regions flanking the apoE gene. Various regions in the intragenic sequence between apoE and apoCI are required for the expression of the apoE gene in brain, skin, and testis, and of the apoCI gene in skin. This sequence also contains a silencer which inhibits the expression of the

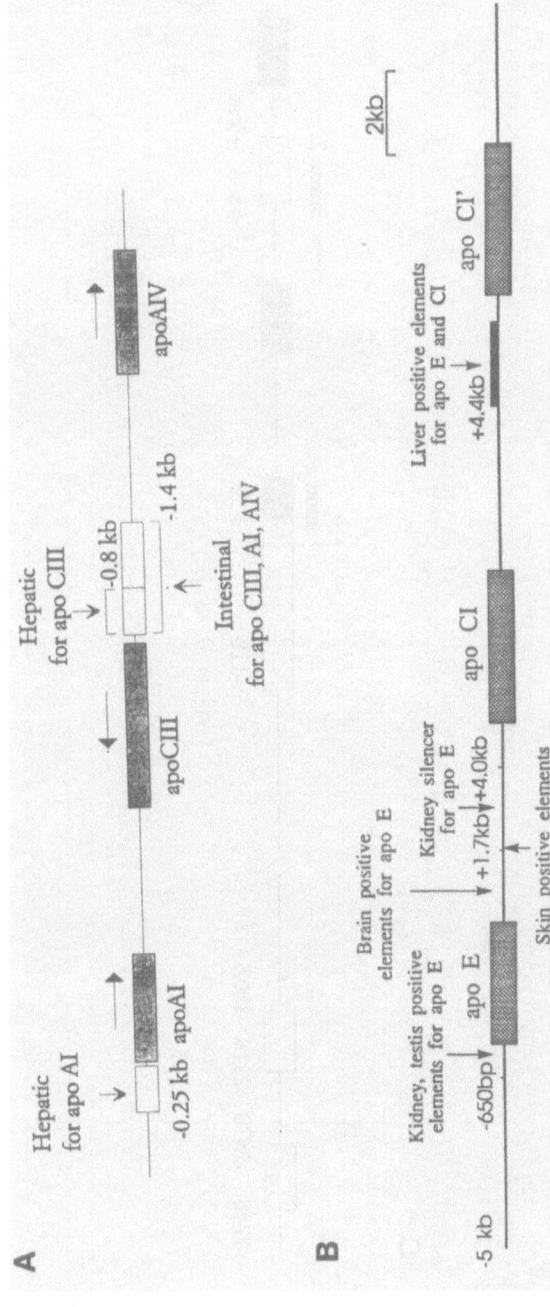

Fig. 19. Schematic representation of the long-distance regulatory elements which may participate in the regulation of (A) the human apoA-I, apoCIII, apoA-IV complex, (B) the human apoE, apoCI complex, and (C) the human apoB gene. The diagram is based on refs. 83, 548, 771, 803, 1008, 1014, 1015, 1033, 1033a.

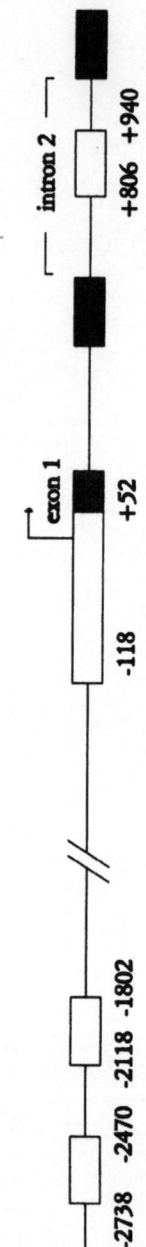

Fig. 19. (*Continued*)

apoE gene in the kidney in a construct containing 650 bp of the 5′ region and 4 kb of the 3′ region flanking the apoE gene. The intragenic sequence between the apoCI gene and the apoCI pseudogene contains a 2-kb element originating 4.4 kb 3′ of the apoCI gene which is required for the hepatic expression of the human apoE and apoCI genes (Fig. 19B). The factors which bind to the distal regulatory elements of the apoA-I, apoCIII, apoA-IV and of the apoE, apoCI gene complexes have not been identified. Finally, CAT assays have suggested that the regulation of the apoB gene is controlled by negative elements upstream of nucleotide -128 (1010,1012,1013) and in the -2.7 to -1.8 kb region upstream of the initiation of transcription (1015). The factors responsible for the negative regulation have not been identified (Fig. 19C).

Promoter Elements and Factors Involved in the Regulation of Transcription of the Human ApoB Gene. Isolation and Characterization of the Transcription Factor NF-BA1

The transcription of the apoB gene in hepatic and intestinal cells is mainly controlled by the interaction of factors which bind to three elements (II, III, and IV) present in the -120 to -33 proximal promoter region. Element II (or BCB) contains the motif AAAAGCAAACAG and is recognized by three nuclear activities designated BCB1,2,3. Element IV (or BA2,3) binds a family of heat-stable proteins designated NF-BA2,3 as well as the previously described transcription factor C/EBP (514). Element III (or BA1) contains the motif GCGCCCTTTGGACCTTT between nucleotides -79 and -63 and is recognized by a nuclear factor of Mr-60 kD which was purified to homogeneity and was designated NF-BA1 (1011) (Fig. 20A). Factors NF-BA1 and NF-BA2,3 also recognize the promoters of the human apoA-I, apoA-II, and apoCIII genes.

The importance of the hepatic factors identified by DNA-binding assays in crude nuclear extracts for transcription was assessed by *in vitro* mutagenesis of the promoter region. Mutations in elements II (BCB1), III (BA1), and IV (BA2,3) which reduced dramatically the binding of BCB1,2,3, NF-BA1, C/EBP, and NF-BA2,3 to these elements reduced hepatic and intestinal transcription to 9, 2, and 10–13% of control, respectively, whereas other mutations in elements I and E had small effects on transcription (1009,1010). This analysis clearly demonstrated that the transcription of the apoB gene in hepatic and intestinal cells is controlled by the synergistic action of factors which interact with elements II (BCB1), III (BA1), and IV (BA2,3). Subsequent work also demonstrated that the regulatory element III (BA1) is recognized by members of the steroid hormone superfamily Ear-2,

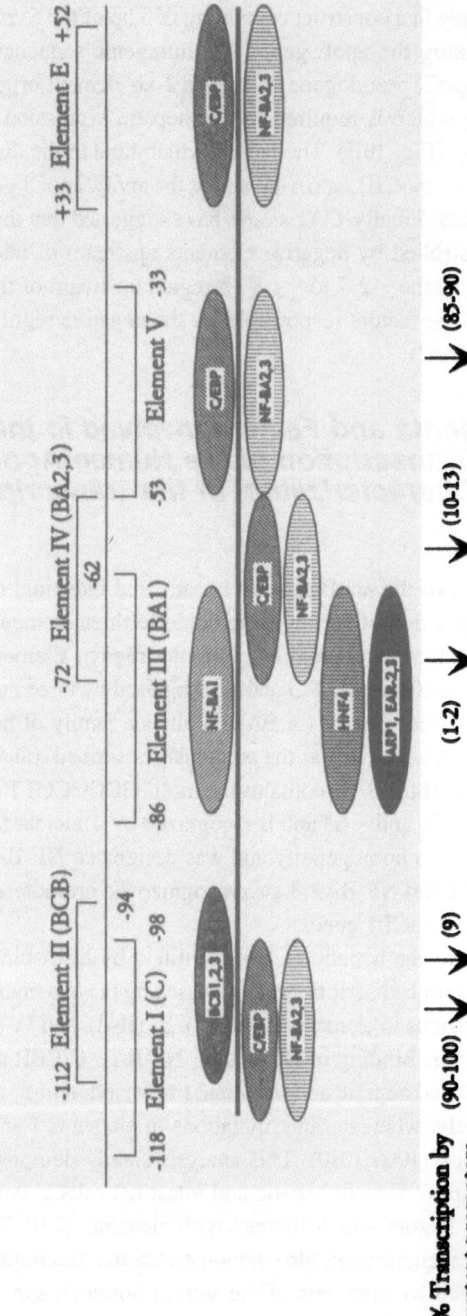

Fig. 20. Schematic representation of the regulatory elements and the transcription factors which may participate in the regulation of transcription of (A) the human apoB, (B) apoCIII, (C) apoA-I, (D) apoA-II, and (E) apoE genes. The locations of the regulatory elements on the sequence promoters are indicated by negative numbers. The factors are symbolized by ellipsoids carrying their names. The diagrams are based on refs. 1009–1031, 1034.

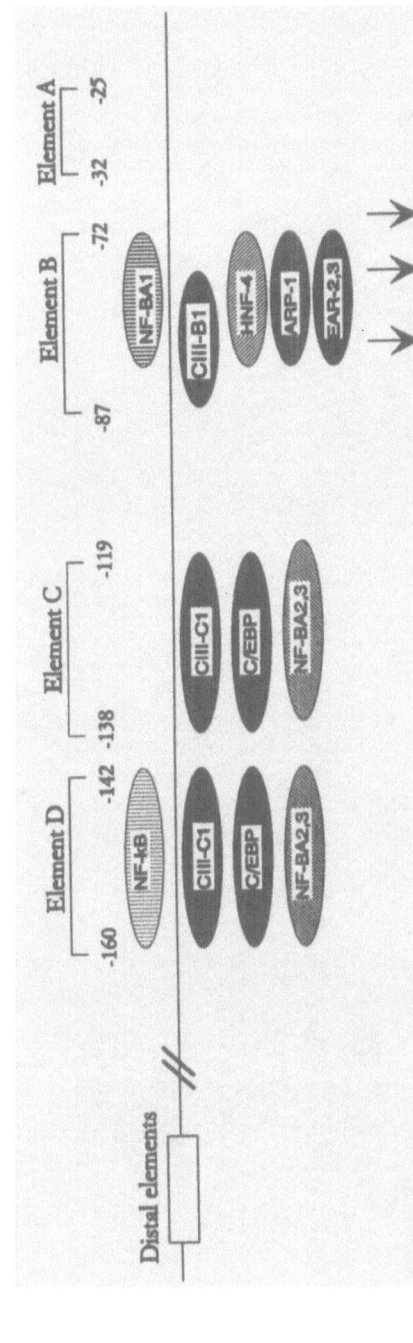

Fig. 20. (*Continued*)

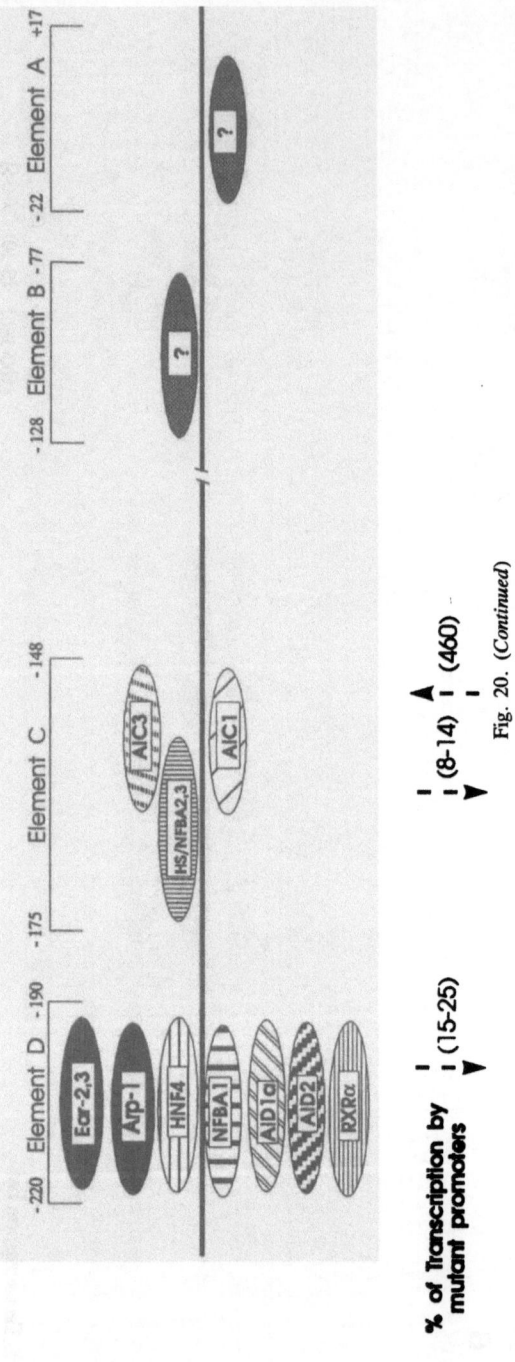

Fig. 20. (Continued)

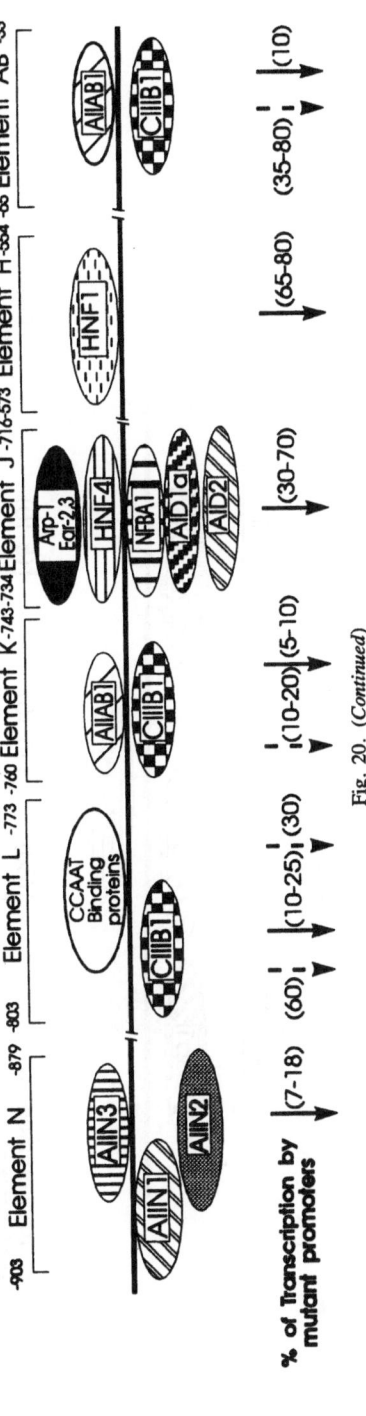

Fig. 20. (*Continued*)

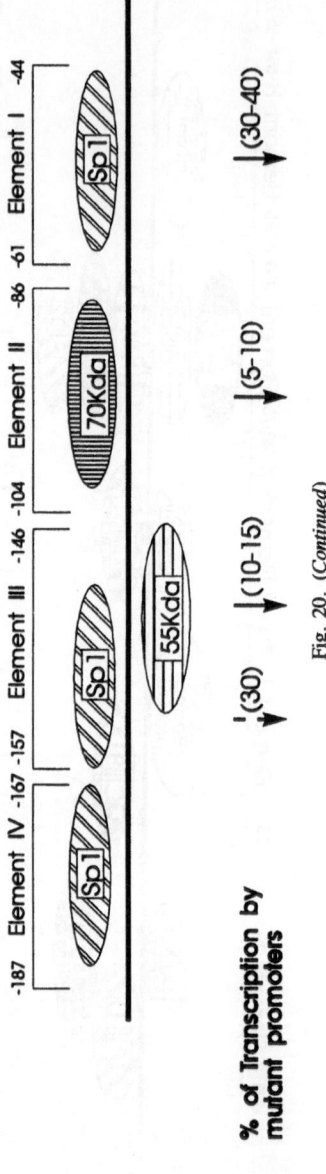

Fig. 20. (*Continued*)

Ear-3 (1016), Arp-1 (1021), and HNF4 (1017). Cotransfection experiments in HepG2 cells showed that Ear-2, Ear-3, and Apr-1 are strong repressors of the element III-dependent transcription. In contrast, HNF4 is a positive activator (1031).

Promoter Elements and Factors Involved in the Transcriptional Regulation of the Human ApoCIII and ApoA-I Genes. Isolation and Characterization of the Transcription Factor CIIIB1

The transcription of the apoCIII gene in hepatic and intestinal cells is controlled by ten regulatory elements (A–J) which are present between nucleotides −792 and −25 of the apoCIII promoter. Element B binds two activities in overlapping binding motifs. One of the factors may be similar or identical to the NF-BA1 and recognizes the human apoB promoter region −79 to −63. The other factor is a new protein of Mr-41 kD which has been purified to homogeneity and has been designated CIIIB1 (1022) (Fig. 20B). The binding site of factor CIIIB1 contains the octameric motif CAGGTGAC. Nucleotide substitutions within this sequence abolished the binding of the purified factor. CIIIB1 also binds to the apoA-II element B (−65 to −48), which contains an identical octameric CAGGTGAC motif in the antisense strand, as well as to the regulatory elements K and L of the apoA-II promoter. Mutations in element B which eliminated the binding of NF-BA1 abolished transcription, whereas mutations which eliminated the binding of factor CIIIB1 reduced hepatic transcription to 36% of control (1018) (Fig. 20B). These data indicated that both factors NF-BA1 and CIIIB1 are positive regulators, although the former has greater activation potential than the latter. The mutagenesis analysis of the apoCIII promoter also demonstrated that the factors which bind to the distal regulatory elements are not by themselves sufficient to drive transcription and require the synergistic action of the factors which bind to the proximal element B. The regulatory region B also binds the previously characterized factor HNF4 (1017) which is a positive regulator, and Arp-1 (1021), Ear-2, and Ear-3 (1031), which are negative regulators. Elements C and D bind C/EBP as well as the nuclear activities NF-BA2 and NF-BA3, which bind in several locations in the human apoB promoter. The CD region contains two binding sites for a new activity designated CIIIC1. Mutations in element C which prevented the binding of CIIC1 and of heat-stable activities NF-BA2 and NF-BA3 in this region reduced hepatic and intestinal transcription to 60–70% of control. Element D also binds purified NF-kB, suggesting a potential role of this region in the acute-phase response (1034). The organization of the different activities on the proximal apoCIII promoter is shown in Fig. 20B. The findings indicate that factors NF-BA1,

CIIIB1, and CIIC1 and the heat-stable factors NF-BA2 and NF-BA3 are positive activators which act synergistically with factors binding to the distal elements to modulate the transcription of the human apoCIII gene.

Positive and Negative Regulation of ApoA-I Gene Transcription

The hepatic transcription of the apoA-I gene is controlled by four regulatory elements between nucleotides -220 and $+17$. The identity of the factors which bind to these elements was established by DNA-binding and competition assays. Element D binds three activities which are present in rat liver nuclear extracts. One of the activities is similar or identical to NF-BA1 and recognizes the -79 to -63 apoB region and the other two represent new types of activities that have been designated AID2 and AID1a (Fig. 20C). Nucleotide substitution within this region which affected the binding of these activities reduced hepatic transcription to 15–25% of control (I. Tzameli, C. Cladaras, and V. I. Zannis, unpublished), indicating that these proteins are positive activators of the apoA-I gene (Fig. 20C). The same region also binds the protein HNF4, which is a positive activator of the apoCIII gene, members of the steroid receptor superfamily Arp-1 (1021), Ear-2 and Ear-3 (1016), which act as negative regulators, and members of the retinoic acid receptor, RXRa and RARa (1025). The physiological function of Arp-1, Ear-2, Ear-3, and RXRa in the hepatic transcription of the human apoA-I gene is not known and will require careful investigation. The regulatory element C is recognized by both positive and negative regulators, which bind in overlapping domains. The region -148 to -168 is recognized by two activities, designated AIC1 and AIC3. Mutations which affected the binding of AIC1 increased transcription 4.6-fold, indicating that this protein acts as a negative regulator. The region -175 to -155 is recognized by the heat-stable activities NF-BA2 and NF-BA3, which bind to several elements of the apoB promoter. This region also binds C/EBP. Mutations which affected the binding of these activities reduced transcription to 8–14% of control (1024) (Fig. 10C). The finding suggests that the factors which bind to regulatory element C can either upregulate or downregulate the transcription of the apoAI gene.

Transcriptional Regulation of the ApoA-II Gene by CIIIB1 and Other Factors

The transcription of the apoA-II gene in hepatic and intestinal cells is controlled by 14 regulatory elements (A–N) between nucleotides -911 and $+29$. Elements N to I between nucleotides -903 and -680 display enhancer-type activity in hepatic cells when placed in front of heterologous promoters (1026). Elements H to D between nucleotides −537 and −255 have a slight effect on the transcription of the apoA-II gene in both cell types (1027). The proximal region contains elements C to A between nucleotides −126 and −33 (Fig. 20D).

DNA-binding and competition assays showed that elements AB, K, and L bind to the heat-stable factor CIIIB1, which binds to the −93 to −76 apoCIII promoter region and is a transcription activator of the human apoCIII gene. Elements AB and K also bind to another heat-labile activity, designated AIIAB1, and element L binds to proteins that recognize the CCAAT motif. Mutations in domain L which prevented the binding of the CCAAT-box binding activities reduced both the hepatic and intestinal transcription to 30% of control, indicating the importance of these factors in transcription. Simultaneous nucleotide substitutions which prevented the binding of CIIIBI activity in elements AB, K, and L reduced hepatic and intestinal transcription to 7 and 6% of control, respectively (1028). The findings suggest that the level of transcription of the apoAII gene can be modulated by the synergistic interaction of the CIIB1, which binds to the proximal and distal regulatory elements, with CCAAT box proteins which bind to element L (1028). Element N binds two minor (AIIN1, AIIN2) and one major (AIIN3) activities. Deletion of this element reduces hepatic and intestinal transcription to 7 and 18% of control, respectively (1027). Element H binds the previously described factor HNF1/LFB1 (1027,1032). Finally, element J binds the same activities which bind to the apoA-I D region (−220 to −190), NF-BA1, AID1a, and AID2. Deletions in element H or J reduced hepatic transcription to 80 and 70% of control, respectively. Region J is also recognized by HNF4, which is a positive regulator of the human apoCIII and apoB genes, and by Arp-1, Ear-2, and Ear-3, which are negative regulators of the human apoCIII, apoB, and apoA-I genes (1031). The role of these factors in apoA-II gene transcription requires further study.

Transcriptional Regulation of the Human ApoE Gene

Deletion and footprinting analysis of the apoE promoter identified six regulatory elements (I–VI) in the −336 to −44 region and two elements (DI and DII) within the +69 to +191 region (1029) (Fig. 20E). DNA-binding and competition assays using HeLa nuclear extracts showed that elements I, III, and IV are recognized by the previously described transcription factor SP1 (217), and element I is also recognized by a protein of Mr = 70 kD (1029,1030) (Fig. 20E). Element III also binds a protein of Mr = 55 kD. Deletion analysis of the −246 to −81 region inserted in front of the herpes thymidine kinase promoter indicated that element III is important for transcription of the reporter CAT gene in CHO cells.

Elements II and III were also essential for transcription of the apoE gene in hepatic cells (1029). Similar information was obtained by *in vitro* transcription using the −383 to +73 region as a template and HeLa nuclear extracts (1030) (Fig. 20E). The *in vitro* transcription analysis also showed that element I contributes to the optimum transcription in HeLa cells (1030). The role of these regulatory elements and factors in hepatic apoE gene transcription requires further study.

FUTURE DIRECTIONS

The advances in molecular and cell biology have revolutionized the field of lipoprotein research. The derivation of cDNA and gene probes during the 1982–1990 period facilitated the study of the genetic variability of apolipoproteins, plasma enzymes, transfer proteins, and lipoprotein receptors. This allowed the description of several disorders of lipoprotein metabolism at the molecular level. Simultaneously, the naturally occurring mutations provided important clues to the function and physiological significance of the proteins involved. In some instances, the functions of the proteins were further probed with *in vitro* mutagenesis and structure–function analysis of the mutant protein forms. The physiological role of several proteins was also assessed by overexpression of the corresponding genes in transgenic mice. Finally, methodologies have been utilized to probe the molecular mechanisms responsible for gene regulation.

What we have learned during the past decade has reinforced the concept that lipoprotein metabolism is a complex and very interactive pathway. Any perturbations in this pathway may alter the lipid and lipoprotein profile in humans or experimental animals. Some of the changes, such as the increase in HDL cholesterol caused, for instance, by C/ETP deficiency or the overexpression of the apoA-I gene, may be beneficial (548–550), whereas others, such as overexpression of the apoCI, apoCIII, or apoB genes (771,803), may be detrimental. Thus, there is a need to better understand the *in vivo* and *in vitro* functions of several proteins of the lipid transport system. This direction is being pursued by *in vitro* mutagenesis of the genes and analysis of the functions of the mutant proteins and is expected to provide comprehensive answers for all the proteins involved, particularly for those proteins for which some of the functions are not fully understood (transfer proteins, LRP, and several apolipoproteins). Impressive progress in this direction has been achieved for apoE and the LDL receptor (214,215,847). Understanding the functions of the different proteins may allow the design of specific inhibitors to control the function of certain proteins such as hepatic lipase and C/ETP.

In the case of apoE the effect of the mutations on structure can be partially explained by alterations in the three-dimensional structure of the protein (512).

Examining the three-dimensional structure is clearly a future goal for several other proteins of the lipid transport system. The function of the different proteins *in vivo* can be explored by a combination of transgenic animal methodologies and *in vitro* gene inactivation techniques (1035–1039). These approaches can test the overall importance of a protein of the lipid transport system for overall lipoprotein metabolism. Introduction of homologous normal or mutated genes into animals in which the host genes have been inactivated by homologous recombination can generate animal models for specific disorders of lipoprotein metabolism. This research direction is in progress in different laboratories and is expected to intensify and improve in the near future. Work in progress in different laboratories when this chapter was submitted has already generated mice deficient in apoA-I (1040), apoE (1041), as well as mice defective in one apoB allele leading to hypo-betalipoproteinemia (1042). ApoE-deficient mice develop severe hypercholes-terolemia and atherosclerosis on normal diets. The apoA-I-deficient mice did not develop spontaneous atherosclerosis in a period of a few months following birth (1040). The plasma levels of some of the proteins of the lipid transport system can be controlled at the level of gene transcription. The identification and characteriza-tion of the transcription factors which participate in the regulation of expression of apolipoproteins (1011,1016,1017,1021,1022,1024,1028), lipoprotein receptors (218), and plasma proteins is in progress in several laboratories and is also expected to intensify in the future. This approach promises to provide rational pharmacological approaches to control plasma lipid, lipoprotein, and apolipopro-tein levels. Overall, these exciting directions open new chapters in lipoprotein research which may permit the description of most of the diseases of lipoprotein metabolism at the molecular level and possibly suggest rational approaches for their correction or treatment.

ACKNOWLEDGMENTS. The review is dedicated to the memory of our parents Yiannis and Markella Zannis and Kostas Economou, to the memory of our sister Liana Economou, and to the memory of Dr. Allan Wilson, an inspired scientist and friend at the University of California, Berkeley. This work was supported by grants from the National Institutes of Health (HL33952) and the March of Dimes Birth Defects Foundation (1-0658) and Kos Pharmaceuticals, Inc. Eleni Zanni is a trainee of the Cardiovascular Institute, training grant HL07224. We would like to thank Elizabeth Walsh, Alex Mitsialis, and Philippe Cardot for excellent technical assistance and Drs. Christopher Fielding, Alan Tall, Tom Innerarity, Laurence Chan, and Michael Schotz for helpful discussion. Experimental work included in this review was performed in the Housman Research Center, Boston, Massa-chusetts. The cost of reprints has been defrayed by a grant from Kos Pharmaceuti-cals, Inc. of Miami, Florida.

REFERENCES

1. Atkinson, D., Davis, M. A. F., and Leslie, R. B., 1974, The structure of a high density lipoprotein (HDL) from porcine plasma, *Proc. R. Soc. Lond. Biol.* **b186**:165–180.
2. Laggner, P., Kostner, G. M., Rakusch, U., and Worcester, D., 1981, Neutron small angle scattering on selectively deuterated human plasma low density lipoproteins, *J. Biol. Chem.* **255**: 11832–11838.
3. Zannis, V. I., and Breslow, J. L., 1985, Genetic mutation affecting human lipoprotein metabolism, in *Advances in Human Genetics*, Vol. 14 (H. Harris and K. Hirschhorn, eds.), Plenum Press, New York, pp. 125–215, 383–386.
4. Smith. L. C., Pownall, H. J., and Gotto, A. M., 1978, The plasma lipoproteins: Structure and metabolism, *Annu. Rev. Biochem.* **47**:751–778.
5. Herbert, P. N., Assmann, G., Gotto, Jr., A. M., and Fredrickson, D. S., 1982, Familial lipoprotein deficiency: Abetalipoproteinemia, hypobetalipoproteinemia, and Tangier disease, in: *The Metabolic Basis of Inherited Disease* (J.B. Stanbury, J. B. Wyngaarden, D. S. Fredrickson, J. L. Goldstein, and M. D. Brown, eds.), McGraw-Hill, New York, pp. 589–651.
6. Scanu, A. M., Byrne, R. E., and Mihovilovic, M., 1982, Functional roles of plasma high density lipoproteins, *CRC Crit. Rev. Biochem.* **13**:109–140.
7. Breslow, J. L., 1988, Apolipoprotein genetic variation and human disease, *Physiol. Rev.* **68**:85–132.
8. Havel, R. J., and Kane, J. P., 1982, Introduction: Structure and metabolism of plasma lipoproteins, in: *The Metabolic Basis of Inherited Disease* (C. R. Scriver, A. I. Beaudet, W. S. Sly, and D. Valle, eds.), McGraw-Hill, New York, pp. 1129–1138.
9. Alaupovic, P., 1971, Conceptual development of the classification systems of plasma lipoproteins, in: *Protides of the Biological Fluids* (H. Peeters, ed.), Pergamon Press, Oxford, pp. 9–20.
10. Albers, J. J., Cheung, M. C., Ewens, S. L., and Tollefson, J. H., 1981, Characterization and immunoassay of apolipoprotein D, *Atherosclerosis* **39**:395–409.
11. Blum, C. B., Aron, L., and Sciacca, R., 1980, Radioimmunoassay studies of human apolipoprotein E, *J. Clin. Invest.* **66**:1240–1250.
12. Carlson, L. A., and Holmquist, L., 1982, Concentrations of apolipoproteins, B, C-I, C-II, C-III and E in sera from normal men and their relation to serum lipoprotein levels, *Clin. Chim. Acta* **124**:163–178.
13. Green, P. H. R., Glickman, R. M., Riley, J. W., and Quinet, E., 1980, Human apolipoprotein A-IV. Intestinal origin and distribution in plasma, *J. Clin. Invest.* **65**:911–919.
14. Havel, R. J., Kotite, L., Vigne, J. L., Kane, J. P., Tun, P., Phillips, N., and Chen, G. C., 1980, Radioimmunoassay of human arginine-rich apolipoprotein, apoprotein E. Concentration in blood plasma and lipoproteins as affected by apoprotein E-3 deficiency, *J. Clin. Invest.* **66**: 1351–1362.
15. Schonfeld, G., Patsch, W., Rudel, L. L., Nelson, C., Epstein, M., and Olson, R. E., 1982, Effects of dietary cholesterol and fatty acids on plasma lipoproteins, *J. Clin. Invest.* **69**:1072–1080.
16. Wu, A. I., and Windmueller, H. G., 1979, Relative contribution of liver and intestine to individual plasma apolipoproteins in the rat, *J. Biol. Chem.* **254**:7316–7322.
17. Zannis, V. I., Cole, F. S., Jackson, C. L., Kurnit, C. L., and Karathanasis, S. K., 1985, Distribution of apoA-I, apoCII, apoCIII, and apoE mRNA in human tissues, time dependent induction of apoE mRNA by cultures of human monocyte-macrophages, *Biochemistry* **24**: 4450–4455.
18. Lenich, C., Brecher, P., Makrides, S., Chobanian, A., and Zannis, V. I., 1988, Apolipoprotein gene expression in the rabbit: Abundance, size and distribution of apolipoprotein mRNA species in different tissues, *J. Lipid Res.* **29**:755–764.
19. Drayna, D., Fielding, C., McLean, J., Baer, B. Castro, G., Chen, E., Comstock, L., Henzel,

W., Kohr, W., Rhee, L., *et al.*, 1986, Cloning and expression of human apolipoprotein D cDNA, *J. Biol. Chem.* **261**:16535–16539.

20. De Silva, H. V., Harmonty, J. A. K., Stuart, W. D., Gil, C. M., and Robbins, J., 1990, Apolipoprotein J: Structure and tissue distribution, *Biochemistry* **29**:5380–5389.

21. Li, W. H., Tanimura, M., Luo, C. C., Datta, S., and Chan, L., 1988, The apolipoprotein multigene family: Biosynthesis, structure, structure–function relationships, and evolution, *J. Lipid Res.* **29**:245–271.

22. Atkinson, D., and Small, D. M., 1986, Recombinant lipoproteins: Implications for structure and assembly of native lipoproteins, *Annu. Rev. Biophys. Biophys. Chem.* **15**:403–456.

23. Cheung, M. C., and Albers, J. J., 1984, Characterization of lipoprotein particles isolated by immunoaffinity chromatography, *J. Biol. Chem.* **259**:12201–12209.

24. Dashti, N., Alaupovic, P., Knight-Gibson, C., and Koren, E., 1987, Identification and partial characterization of discrete apolipoprotein B containing lipoprotein particles produced by human hepatoma cell line HepG2, *Biochemistry* **26**:4837–4846.

25. Zannis, V. I., Breslow, J. L., SanGiacomo, T. R., Aden, D. P., and Knowles, B. B., 1981, Characterization of the major apolipoproteins secreted by two hepatoma cell lines, *Biochemistry* **20**:7089–7096.

26. Zannis, V. I., Kurnit, D., and Breslow, J. L., 1982, Hepatic apoA-I and apoE and intestinal apoA-I are synthesized in precursor isoprotein forms by organ cultures of human fetal tissues, *J. Biol. Chem.* **275**:536–544.

27. Hamilton, R. I., Williams, M. C., Fielding, C. J., and Havel, R. J., 1976, Discoidal bilayer structure of nascent high density lipoproteins from perfused rat liver, *J. Clin. Invest.* **58**:667–680.

28. Newman, T. C., Dawson, P. A., Rudel, L. L., and Williams, D. L., 1985, Quantitation of apolipoprotein E mRNA in the liver and peripheral tissues of nonhuman primates, *J. Biol. Chem.* **260**:2452–2457.

29. Lee, P., and Breckenridge, W. C., 1976, Isolation and carbohydrate composition of glycopeptides of human apo low-density lipoprotein from normal and type II hyperlipoproteinemic subjects, *Can. J. Biochem.* **54**:829–833.

30. Swaminathan, N., and Aladjem, F., 1976, The monosaccharide composition and sequence of the carbohydrate moiety of human serum low density lipoproteins, *Biochemistry* **15**:1515–1522.

31. Vauhkonen, M., Viitala, J., Parkkinen, J., and Rauvala, H., 1985, High-mannose structure of apolipoprotein-B from low-density lipoproteins of human plasma, *Eur. J. Biochem.* **152**:43–50.

32. Hoeg, J. M., Meng, M. S., Ronan, R., Fairwell, T., and Brewer, Jr., H. B., 1986, Human apolipoprotein A-I. Post-translational modification by fatty acid acylation, *J. Biol. Chem.* **261**:3911–3914.

33. Zannis, V. I., vanderSpek, J., and Silverman, D., 1986, Intracellular modifications of human apolipoprotein E, *J. Biol. Chem.* **261**:13415–13421.

34. Huang, G., Lee, D. M., and Singh, S., 1988, Identification of the thiol ester linked lipids in apolipoprotein B, *Biochemistry* **27**:1395–1400.

35. Lee, D. M., and Singh, S., 1988, Presence and localization of two intramolecular thiolester linkages in apolipoprotein 3, *Circulation* **78**:II-286.

36. Hoeg, J. M., Meng, M. S., Ronan, R., Demosky, Jr., S. J., Fairwell, T., and Brewer, Jr., H. B., 1988, Apolipoprotein B synthesized by HepG2 cells undergoes fatty acid acylation, *J. Lipid Res.* **29**:1215–1220.

37. Yang, C. Y., Gu, Z. W., Weng, S. A., Kim, T. W., Chen, S. H., Pownall, H. J., Sharp, P. M., Liu, S. W., Li, W. H., Gotto, Jr., A. M., and Chan, L., 1989, Structure of apolipoprotein B-100 of human low density lipoproteins, *Arteriosclerosis* **9**:96–108.

38. Hadzopoulou-Cladaras, M., Kritis, A., Zanni, E. E., Cladaras, C., and Zannis, V. I., 1989, Expression of segments of the apoB-100 cDNA, *Circulation* **79**:434.

39. Zanni, E. E., Kouvatsi, A., Hadzopoulou-Cladaras, M., Krieger, M., and Zannis, V. I., 1989, Expression of apoE gene in Chinese hamster cells with a reversible defect in O-glycosylation.

Glycosylation is not required for apoE secretion, *J. Biol. Chem.* **264**:9137–9140.

40. Sparks, J. D., Sparks, C. E., Roncone, A. M., and Amatruda, J. M., 1988, Secretion of high and low molecular weight phosphorylated apolipoprotein B by hepatocytes from control and diabetic rats. Phosphorylation of apo BH and apo BL, *J. Biol. Chem.* **263**:5001–5004.

41. Beg, Z. H., Stonik, J. A., Hoeg, J. M., Demosky, S. J., Fairwell, T., and Brewer, H. B., 1989, Human apolipoprotein A-I. Post-translational modification by covalent phosphorylation, *J. Biol. Chem.* **264**:6913–6921.

42. Wernette-Hammond, M. E., Lauer, S. J., Corsini, A., Walker, D., Taylor, J. M., and Rall, Jr., S. C., 1989, Glycosylation of human apolipoprotein E. The carbohydrate attachment site is threonine 194, *J. Biol. Chem.* **264**:9094–9101.

43. Hussain, M. M., and Zannis, V. I., 1990, Intracellular modification of human apolipoprotein AII (apoAII) and sites of apoAII mRNA synthesis: Comparison of apoAII with apoCII and apoCIII isoproteins, *Biochemistry* **29**:209–217.

44. Bambeger, M. J., and Lane, M. D., 1988, Assembly of very low density lipoprotein in the hepatocyte, *J. Biol. Chem.* **263**:11868–11878.

45. Bostrom, K., Wettesten, M., Boren, J., Bondjers, G., Wiklund, O., and Olofsson, S. O., 1986, Pulse-chase studies of the synthesis and intracellular transport of apolipoprotein B-100 in HepG2 cells, *J. Biol. Chem.* **261**:13800–13806.

46. Bostrom, K., Boren, J., Wettesten, M., Sjoberg, A., Bondjers, G., Wiklund, O., Carlsson, P., and Olofsson, S. O., 1988, Studies on the assembly of apoB-100-containing lipoproteins in HepG2 cells, *J. Biol. Chem.* **263**:4434–4442.

47. Boren, J., Wettesten, M., Sjoberg, A., Thorlin, T., Bondjers, G., Wiklund, O., and Olofsson, S. O., 1990, The assembly and secretion of apoB-100 containing lipoproteins in HepG2 cells, *J. Biol. Chem.* **265**:10556–10564.

47a. Boren, J., Graham, L., Wettesten, M., Scott, J., White, A., and Olofsson, S., 1992, The assembly and secretion of apoB100-containing lipoproteins in HepG2 cells, *J. Biol. Chem.* **267**:9858–9867.

48. Zannis, V. I., 1989, Molecular biology of human apolipoproteins B and E and associated diseases of lipoprotein metabolism, *Adv. Lipid Res.* **23**:1–64.

49. Hussain, M. M., Zanni, E. E., Kelly, M., and Zannis, V. I., 1989, Synthesis, modification and flotation properties of rat hepatocyte apolipoproteins, *Biochim. Biophys. Acta* **1001**:90–101.

49a. Spring, D. J., Chen-Liu, W., Chatterton, E., Elovson, J., and Schumaker, V. N., 1992, Lipoprotein Assembly. Apolipoprotein B size determines lipoprotein core circumference, *J. Biol. Chem.* **267**:14839–14845.

50. Davis, R. A., Thrift, R. N., Wu, C. C., and Howell, K. E., 1990, Apolipoprotein B is both integrated into and translocated across the endoplasmic reticulum membrane. Evidence for two functionally distinct pools, *J. Biol. Chem.* **265**:10005–10011.

51. Sato, R., Imanaka, T., Takatsuki, A., and Takano, T., 1990, Degradation of newly synthesized apolipoprotein B-100 in a pre-Golgi compartment, *J. Biol. Chem.* **265**:11880–11884.

52. Higgins, J. A., and Hutson, J. L., 1984, The roles of Golgi and endoplasmic reticulum in the synthesis and assembly of lipoprotein lipids in rat hepatocytes, *J. Lipid Res.* **25**:1295–1305.

53. Vance, J. E., and Vance, D. E., 1986, Specific pools of phospholipids are used for lipoprotein secretion by cultured rat hepatocytes, *J. Biol. Chem.* **261**:4486–4491.

54. Higgins, J. A., 1988, Evidence that during very low density lipoprotein assembly in rat hepatocytes most of the triacylglycerol and phospholipid are packaged with apolipoprotein B in the Golgi complex, *FEBS Lett.* **232**:405–408.

55. Vance, J. E., 1989, The use of newly synthesized phospholipids for assembly into secreted hepatic lipoproteins, *Biochim. Biophys. Acta* **1006**:59–69.

56. Vance, J. E., and Vance, D. E., 1990, Lipoprotein assembly and secretion by hepatocytes, *Annu. Rev. Nutr.* **10**:337–356.

57. Yao, Z., Blackhart, B. D., Linton, M. F., Taylor, S. M., Young, S. G., and McCarthy, B. J., 1991, Expression of carboxy-terminal truncated forms of human apolipoprotein B in rat hepatoma cells, *J. Biol. Chem.* **266**:3300–3308.

58. Graham, D. L., Knott, T. J., Jones, T. C., Pease, R. J., Pullinger, C. R., and Scott, J., 1991, Carboxyl-terminal truncation of apolipoprotein B results in gradual loss of the ability to form buoyant lipoproteins in cultured human and rat liver cell lines, *Biochemistry* **30**:5616–5621.

59. Herscovitz, H., Hadzopoulou-Cladaras, M., Walsh, M. T., Cladaras, C., Zannis, V. I., and Small, D. M., 1991, Expression, secretion, and lipid-binding characterization of the N-terminal 17% of apolipoprotein B, *Proc. Natl. Acad. Sci. USA* **88**:7313–7317.

60. Xiong, W. J., Sigmond, E., Gotto, Jr., A. M., Lei, K. Y., and Chan, L., 1991, Locating a low density lipoprotein-targeting domain of human apolipoprotein B-100 by expressing a minigene construct in transgenic mice, *J. Biol. Chem.* **266**:20893–20898.

61. Innerarity, T. L., 1990, Familial hypobetalipoproteinemia and familial defective apolipoprotein B100: Genetic disorders associated with apolipoprotein B, *Curr. Opin. Lipidol.* **2**:104–109.

62. Barbaras, R., Grimaldi, P., Negrel, R., and Ailhaud, G., 1986, Characterization of high density lipoprotein binding and cholesterol efflux in cultured mouse adipose cells, *Biochim. Biophys. Acta* **888**:143–156.

63. Barbaras, R., Puchois, P., Fruchart, J. C., and Ailhaud, G., 1987, Cholesterol efflux from cultured adipose cells is mediated by LpAI particles but not by LpAI:AII particles, *Biochem. Biophys. Res. Commun.* **142**:63–69.

64. Alaupovic, P., Tavella, M., Bard, J. M., Wang, C. S., Attman, P. O., Koren, E., Corder, C., Knight-Gibson, C., and Downs, D., 1988, Lipoprotein particles in hypertriglyceridemic states, *Adv. Exp. Med. Biol.* **243**:289–297.

65. Marz, W., and Gross, W., 1988, Immunochemical evidence for the presence in human plasma of lipoproteins with apolipoprotein A-II as the major protein constituent, *Biochim. Biophys. Acta* **962**:155–158.

66. Ohta, T., Hattori, S., Nishiyama, S., and Matsuda, I., 1988, Studies on the lipid and apolipoprotein compositions of two species of apoA-I-containing lipoproteins in normolipidemic males and females, *J. Lipid Res.* **29**:721–728.

67. James, R. W., Proudfoot, A., and Pometta, D., 1989, Immunoaffinity fractionation of high density lipoprotein subclasses 2 and 3 using anti-apolipoprotein A-I and A-II immunoabsorbent gels, *Biochim. Biophys. Acta* **1002**:292–301.

68. Kilsdonk, E. P. C., VanGent, T., and VanTol, A., 1990, Characterization of human high-density lipoprotein subclasses LpA-I and LpA-II/A-II and binding to HepG2 cells, *Biochim. Biophys. Acta* **1045**:205–212.

69. Fruchart, J. C., and Bard, J. M., 1991, Lipoprotein particle measurement: An alternative approach to classification of lipid disorders, *Curr. Opin. Lipidol.* **2**:362–366.

70. Bard, J. M., Candelier, L., Agnani, G., Clavey, V., Torpier, G., Steinmetz, A., and Fruchart, J. C., 1991, Isolation and characterization of human Lp-B lipoprotein containing apolipoprotein B as the sole apolipoprotein, *Biochim. Biophys. Acta* **1082**:170–176.

71. Bekaert, E. D., Alaupovic, P., Knight-Gibson, C., Blackett, P., and Ayrault-Jarrier, M., 1991, Composition of plasma ApoA-I-containing lipoprotein particles in children and adults, *Pediatr. Res.* **29**:315–321.

72. Alaupovic, P., Knight-Gibson, C., Wang, C. S., Downs, D., Koren, E., Brewer, Jr., H. B., and Gregg, R. E., 1991, Isolation and characterization of an apoA-II-containing lipoprotein (LP-A-II:B complex) from plasma very low density lipoproteins of patients with Tangier disease and type V hyperlipoproteinemia, *J. Lipid Res.* **32**:9–19.

73. Dashti, N., Koren, E., and Alaupovic, P., 1989, Identification and partial characterization of discrete apolipoprotein A-containing lipoprotein particles secreted by human hepatoma cell line HepG2, *Biochem. Biophys. Res. Commun.* **163**:574–580.

74. Thrift, R. N., Forte, T. M., Cahoon, B. E., and Shore, V. G., 1986, Characterization of lipoproteins produced by human liver cell line, HepG2, under defined conditions, *J. Lipid Res.* **27**:236–250.

75. Forte, T. M., McCall, M. R., Amacher, S., Nordhausen, R. W., Vigne, J. L., and Mallory, J. B., 1990, Physical and chemical characteristics of apolipoprotein A-I–lipid complexes produced by Chinese hamster ovary cells transfected with the human apolipoprotein A-I gene, *Biochim. Biophys. Acta* **1047**:11–18.

76. McCall, M. R., Forte, T. M., and Shore, T. G., 1988, Heterogeneity of nascent high density lipoproteins secreted by the hepatoma derived cell lines, HepG2, *J. Lipid Res.* **29**:1127–1137.

77. McCall, M. R., Nichols, A. V., Blanche, P. J., Shore, V. G., and Forte, T. M., 1989, Lecithin:cholesterol acyltransferase-induced transformation on HepG2 lipoproteins, *J. Lipid Res.* **30**:1579–1589.

78. Goldstein, J. L., and Brown, M. S., 1989, Familial hypercholesterolemia, in: *The Metabolic Basis of Inherited Disease* (C. R. Scriver, A. I. Beaudet, W. S. Sly, and D. Valle, eds.), McGraw-Hill, New York, pp. 1215–1250.

79. Brown, M. S., Herz, J., Kowal, R. C., and Goldstein, J. L., 1991, The low-density lipoprotein receptor-related protein: Double agent or decoy, *Curr. Opin. Lipidol.* **2**:65–72.

80. Jones, L. A., Teramoto, T., Juhn, D. J., Goldberg, R. B., Rubenstein, A. H., and Getz, G. S., 1984, Characterization of lipoproteins produced by the perfused monkey liver, *J. Lipid Res.* **25**:319–335.

81. Hussain, M. M., Roghani, A., Cladaras, C., Zanni, E. E., and Zannis, V. I., 1991, Secretion of lipid-poor nascent human apolipoprotein apoAI, apoCIII, and apoE by cell clones expressing the corresponding genes, *Electrophoresis* **12**:273–283.

82. Fenjves, E. S., Gordon, D. A., Pershing, L. K., Williams, D. L., and Taichman, L. B., 1989, Systemic distribution of apolipoprotein E secreted by grafts of epidermal keratinocytes: Implications for epidermal function and gene therapy, *Proc. Natl. Acad. Sci. USA* **86**:8803–8807.

83. Smith, J. D., Plump, A. S., Hayek, T., Walsh, A., and Breslow, J. L., 1990, Accumulation of human apolipoprotein E in the plasma of transgenic mice, *J. Biol. Chem.* **265**:14709–14712.

84. Simonet, W. S., Bucay, N., Lauer, S. J., Wirak, D. O., Stevens, M. E., Weisgraber, K. H., Pitas, R. E., and Taylor, J. M., 1990, In the absence of a downstream element, the apolipoprotein E gene is expressed at high levels in kidneys of transgenic mice, *J. Biol. Chem.* **265**:10809–10812.

85. Hamilton, R. L., Moorehouse, A., and Havel, R. J., 1991, Isolation and properties of nascent lipoproteins from highly purified rat hepatocytic Golgi fractions, *J. Lipid Res.* **32**:529–543.

86. Hamilton, R. L., Wong, J. S., Guo, L. S., Krisans, S., and Havel, R. J., 1990, Apolipoprotein E localization in rat hepatocytes by immunogold labeling of cryothin sections, *J. Lipid Res.* **31**:1589–1603.

87. Zannis, V. I., and Breslow, J. L., 1980, Two dimensional maps of human apolipoproteins in normal and diseased states, in: *Electrophoresis* (B. J. Radola, ed.), Walter de Gruyter, Berlin, pp. 437–473.

88. Kashyap, M. L., Srivastava, L. S., Hynd, B. A., Gartside, P. S., and Perisutti, G., 1981, Quantitation of human apolipoprotein C-III and its subspecies by radioimmunoassay and analytical isoelectric focusing: Abnormal plasma triglyceride-rich lipoprotein apolipoprotein C-III subspecie concentrations in hypertriglyceridemia, *J. Lipid Res.* **22**:800–810.

89. Makrides, S. C., Ruiz-Opazo, N., Hayden, M., Nussbaum, A. L., Breslow, J. L., and Zannis, V. I., 1988, Sequence and expression of Tangier apoA-I gene, *Eur. J. Biochem.* **173**:465–471.

90. Zannis, V. I., Karathanasis, S. K., Keutmann, H. T., Goldberger, G., and Breslow, J. L., 1983, Intracellular and extracellular processing of human apoA-I. Secreted apoA-I isoprotein 2 is a propeptide, *Proc. Natl. Acad. Sci. USA* **80**:2574–2578.

91. Gordon, J. I., Sims, H. F., Lentz, S. R., Edelstein, C., Scanu, A. M., and Strauss, A. W., 1983,

Proteolytic processing of human preproapolipoprotein A-I. A proposed defect in the conversion of pro A-I to A-I in Tangier's disease, *J. Biol. Chem.* **258**:4037–4044.

92. Shoulders, C. C., Kornblihtt, A. R., Munro, B. S., and Baralle, F. E., 1983, Gene structure of human apolipoprotein A1, *Nucleic. Acids Res.* **11**:2827–2837.

93. Karathanasis, S. K., Zannis, V. I., and Breslow, J. L., 1983, Isolation and characterization of the human apolipoprotein A-I gene, *Proc. Natl. Acad. Sci. USA* **80**:6147–6151.

94. Cheung, P., and Chan, L., 1983, Nucleotide sequence of cloned cDNA of human apolipoprotein A-I, *Nucleic Acids Res.* **11**:3703–3715.

95. Roghani, A., and Zannis, V. I., 1988, Mutagenesis of the glycosylation site of human ApoCIII. O-Linked glycosylation is not required for ApoCIII secretion and lipid binding, *J. Biol. Chem.* **263**:17925–17932.

96. Fennewald, S. M., Hamilton, Jr., R. L., and Gordon, J. I., 1988, Expression of human preproapo AI and pre(delta pro)apoAI in a murine pituitary cell line (AtT-20, A comparison of their intracellular compartmentalization and lipid affiliation, *J. Biol. Chem.* **263**:15568–15577.

97. Minnich, A. M., Collet, X., Hamilton, R. L., Fielding, C. J., and Zannis, V. I., 1991, Relationship between apoA-I gene mutations and effects on LCAT activation, lipid binding and disc formation, *Circulation* **84**:55.

98. Edelstein, C., Gordon, J. I., Toscas, K., Sims, H. F., Strauss, A. W., and Scanu, A. M., 1983, *In vitro* conversion of proapoprotein A-I to apoprotein A-I. Partial characterization of an extracellular enzyme activity, *J. Biol. Chem.* **258**:11430–1433.

99. Sliwkowski, M. B., and Windmueller, H. G., 1984, Rat liver and small intestine produce proapolipoprotein A-I which is slowly processed to apolipoprotein A-I in the circulation, *J. Biol. Chem.* **259**:64259–64265.

100. Bojanovski, D., Gregg, R. E., and Brewer, Jr., H. B., 1984, Tangier disease. *In vitro* conversion of proapo-A-I Tangier to mature apo-A-I Tangier, *J. Biol. Chem.* **259**:6049–6051.

101. Gordon, J. I., Sims, H. F., Edelstein, C., Scanu, A. M., and Strauss, A. W., 1984, Human proapolipoprotein A-II is cleaved following secretion from HepG2 cells by a thiol protease, *J. Biol. Chem.* **259**:15556–15563.

102. Kim, H., and Kwrup, I., 1982, Nonenzymatic glycosylation of human plasma low density lipoprotein. Evidence for *in vitro* and *in vivo* glycosylation, *Metabolism: Clinical and Experimental* **31**:348–353.

103. Ghiselli, G., Rohde, M. F., Tanenbaum, S., Krishnan, S., and Gotto, Jr., A. M., 1985, Origin of apolipoprotein A-I polymorphism in plasma, *J. Biol. Chem.* **260**:15662–15668.

104. Cocuzzi, E., Piacentini, M., Beninat, S., and Chung, S. I., 1990, Posttranslational modification of apolipoprotein B by transglutaminases, *Biochem. J.* **265**:707–713.

105. Palinski, W., Rosenfeld, M. E., Yia-Herttuala, S., Gurtner, G. C., Socher, S. S., Butler, S. W., Parthasarathy, S., Carew, T. E., Steinberg, D., and Witztum, J. L., 1989, Low density lipoprotein undergoes oxidative modification *in vitro*, *Proc. Natl. Acad. Sci. USA* **86**:1372–1376.

106. Parthasarathy, S., Fong, L. G., Otero, D., and Steinberg, D., 1987, Recognition of solubilized apoproteins from delipidated, oxidized low density lipoprotein (LDL) by the acetyl-LDL receptor, *Proc. Natl. Acad. Sci. USA* **84**:537–540.

107. Steinberg, D., 1988, Metabolism of lipoproteins and their role in the pathogenesis of atherosclerosis, *Atherosclerosis Rev.* **18**:1–23.

108. Palinski, W., Yia-Herttuala, S., Rosenfeld, M. E., Butler, S. W., Socher, S. S., Parthasarathy, S., Curtiss, L. K., and Witztum, J. L., 1990, Antisera and monoclonal antibodies specific for epitopes generated during oxidative modification of low density lipoprotein, *Arteriosclerosis* **10**:325–335.

109. Parthasarathy, S., Khoo, J. C., Miller, E., Barneti, J., Witztum, J. L., and Steinberg, D., 1990, Low density lipoprotein rich in oleic acid is protected against oxidative modification: Implications for dietary prevention of atherosclerosis, *Proc. Natl. Acad. Sci. USA* **87**:3894–3898.

110. Glickman, R. M., and Green, P. H. R., 1977, The intestine as a source of apolipoprotein A-I, *Proc. Natl. Acad. Sci. USA* **74:**2569–2573.

111. Schonfeld, G., Bell, E., and Alpers, D. H., 1978, Intestinal apoproteins during fat absorption, *J. Clin. Invest.* **61:**1539–1550.

112. Green, P. H. R., Lefkowitch, J. H., Glickman, R. M., Riley, J. W., Quinet, E., and Blum, C. B., 1982, Apolipoprotein localization and quantitation in the human intestine, *Gastroenterology* **83:**1223–1230.

113. Havel, R. J., Kane, J. P., and Kashyap, M. L., 1973, Interchange of apolipoproteins between chylomicrons and high density lipoproteins during alimentary lipemia in man, *J. Clin. Invest.* **52:**32–38.

114. Green, P. H. R., Glickman, R. M., Sauder, C. D., Blum, C. B., and Tall, A. R., 1979, Human intestinal lipoproteins. Studies in chyluric subjects, *J. Clin. Invest.* **64:**233–242.

115. Fidge, N. H., 1980, The redistribution and metabolism of iodinated apolipoprotein A-IV in rats, *Biochim. Biophys. Acta* **619:**129–141.

116. Shepherd, J., Patsch, J. R., Packard, C. J., Gotto, Jr., A. M., and Taunton, O. D., 1978, Dynamic properties of human high density lipoprotein apoproteins, *J. Lipid Res.* **19:**383–389.

117. Grow, T. E., and Fried, M., 1978, Interchange of apoprotein components between the human plasma high density lipoprotein subclasses HDL2 and HDL3 *in vitro*, *J. Biol. Chem.* **253:**8034–8041.

118. Swift, L. L., Manowitz, N. R., Dunn, G. D., and LeQuire, V. S., 1980, Isolation and characterization of hepatic Golgi lipoproteins from hypercholesterolemic rats, *J. Clin. Invest.* **66:**415–425.

119. Glomset, J. A., Mitchell, C. D., King, W. C., Applegate, K. A., Forte, T., Norum, K. R., and Gjone, E., 1980, *In vitro* effects of lecithin:cholesterol acyltransferase on apolipoprotein distribution in familial lecithin:cholesterol acyltransferase deficiency, *Ann. N. Y. Acad. Sci.* **348:**224–243.

120. Kushwaha, R. S., and Hazzard, W. R., 1978, Catabolism of very low density lipoproteins in the rabbit. Effect of changing composition and pool size, *Biochim. Biophys. Acta* **528:**176–189.

121. Eisenberg, S., 1978, Effects of temperature and plasma on the exchange of apolipoproteins and phospholipids between rat plasma very low and high density lipoproteins, *J. Lipid Res.* **19:**229–236.

122. Fielding, C. J., and Havel, R. J., 1977, Lipoprotein lipase, *Arch. Pathol.* **101:**225–229.

123. Nilsson-Ehle, P., Garfinkel, A. S., and Schotz, M. C., 1980, Lipolytic enzymes and plasma lipoprotein metabolism, *Annu. Rev. Biochem.* **49:**667–693.

124. Patsch, J. R., Gotto, Jr., A. M., Olivecrona, T., and Eisenberg, S., 1978, Formation of high density lipoprotein2-like particles during lipolysis of very low density lipoproteins *in vitro*, *Proc. Natl. Acad. Sci. USA* **75:**4519–4523.

125. Redgrave, T. G., and Small, D. M., 1979, Quantitation of the transfer of surface phospholipid of chylomicrons to the high density lipoprotein fraction during the catabolism of chylomicrons in the rat, *J. Clin. Invest.* **64:**162–171.

126. Blum, C. B., 1982, Dynamics of apolipoprotein E metabolism in humans, *J. Lipid Res.* **23:**1308–1316.

127. Cooper, A. D., and Yu, P. Y. S., 1978, Rates of removal and degradation of chylomicron remnants by isolated perfused rat liver, *J. Lipid Res.* **19:**635–643.

128. Sherrill, B. C., and Dietschy, J. M., 1978, Characterization of the sinusoidal transport process responsible for uptake of chylomicrons by the liver, *J. Biol. Chem.* **253:**1859–1867.

129. Berman, M., Hall, M., Levy, R. I., Eisenberg, S., Bilheimer, D. W., Phair, R. D., and Goebel, R. H., 1978, Metabolism of apoB and apoC lipoproteins in man: Kinetic studies in normal and hyperlipoproteinemic subjects, *J. Lipid Res.* **19:**38–56.

130. Sigurdsson, G., Nicoll, A., and Lewis, B., 1975, Conversion of very low density lipoprotein to

low density lipoprotein. A metabolic study of apolipoprotein B kinetics in human subjects, *J. Clin. Invest.* **56**:1481–1490.

131. Mitchell, C. D., King, W. C., Applegate, K. R., Forte, T., Glomset, J. A., Norum, K. R., and Gjone, E., 1980, Characterization of apolipoprotein E-rich high density lipoproteins in familial lecithin:cholesterol acyltransferase deficiency, *J. Lipid Res.* **21**:625–634.

132. Glomset, J. A., Norum, K. R., and Gjone, E., 1982, Familial lecithin:cholesterol acyltransferase deficiency, in: *The Metabolic Basis of Inherited Disease* (J. B. Stanbury, J. B. Wyngaarden, D. S. Fredrickson, J. L. Goldstein, and M. S. Brown, eds.), McGraw-Hill, New York, pp. 643–654.

133. Lux, S. E., Kirz, R., Shrager, R. L., and Gotto, A. M., 1972, The influence of lipid on the conformation of human plasma high density apolipoproteins, *J. Biol. Chem.* **247**:2598–2606.

134. Barter, P. J., Hopkins, G. J., and Calvert, G. D., 1982, Transfers and exchanges of esterified cholesterol between plasma lipoproteins, *Biochemistry* **208**:1–7.

135. Zilversmit, D. B., Hughes, L. B., and Balmer, J., 1975, Stimulation of cholesterol ester exchange of lipoprotein-free rabbit plasma, *Biochim. Biophys. Acta* **409**:393–398.

136. Pattnaik, N. M., Montes, A., Hughes, L. B., and Zilversmit, D. B., 1978, Cholesteryl ester exchange protein in human plasma: Isolation and characterization, *Biochim. Biophys. Acta* **530**:428–438.

137. Barter, P. J., and Lally, J. I., 1978, The activity of an esterified cholesterol transferring factor in human and rat serum, *Biochim. Biophys. Acta* **531**:233–236.

138. Barter, P. J., and Jones, M. E., 1979, Rate of exchange of esterified cholesterol between human plasma low and high density lipoproteins, *Atherosclerosis* **34**:67–74.

139. Barter, P. J., and Jones, M. E., 1980, Kinetic studies of the transfer of esterified cholesterol between human plasma low and high density lipoproteins, *J. Lipid Res.* **21**:238–249.

140. Barter, P. J., Ha, Y. C., and Calvert, G. I., 1981, Studies of esterified cholesterol in subfractions of plasma high density lipoproteins, *Atherosclerosis* **38**:165–175.

141. Ha, Y. C., Calvert, G. D., McIntosh, G. H., Barter, P. J., 1981, A physiologic role for the esterified cholesterol transfer protein: *In vivo* studies in rabbits and pigs, *Metabolism: Clinical and Experimental* **30**:380–383.

142. Ha, Y. C., and Barter, P. J., 1982, Differences in plasma cholesteryl ester transfer activity in sixteen vertebrate species, *Comp. Biochem. Physiol. B* **71**:265–269.

143. Chajek, T., and Fielding, C. J., 1978, Isolation and characterization of a human serum cholesteryl ester transfer protein, *Proc. Natl. Acad. Sci. USA* **75**:3445–3449.

144. Nestel, P. J., Reardon, M., and Billington, T., 1979, *In vivo* transfer of cholesteryl esters from high density lipoproteins to very low density lipoproteins in man, *Biochim. Biophys. Acta* **573**:403–407.

145. Hopkins, G. J., and Barter, P. J., 1980, Transfers of esterified cholesterol and triglyceride between high density and very low density lipoproteins: *In vitro* studies of rabbits and humans, *Metabolism: Clinical and Experimental* **29**:546–550.

146. Morton, R. E., and Zilversmit, D. B., 1981, The separation of apolipoprotein D from cholesteryl ester transfer protein, *Biochim. Biophys. Acta* **663**:350–355.

147. Barter, P. J., Gorjatschko, L., and Calvert, G. D., 1980, Net mass transfer of esterified cholesterol from human low density lipoproteins to very low density lipoproteins incubated *in vitro*, *Biochim. Biophys. Acta* **619**:436–439.

148. Hopkins, G. J., and Barter, P. J., 1982, Dissociation of the *in vitro* transfers on esterified cholesterol and triglyceride between human lipoproteins, *Metabolism: Clinical and Experimental* **31**:78–81.

149. Barter, P. J., Gooden, J. M., and Rajaram, O. V., 1979, Species differences in the activity of a serum triglyceride transferring factor, *Atherosclerosis* **33**:165–169.

150. Morton, R. E., and Zilversmit, D. B., 1982, Purification and characterization of lipid transfer protein(s) from human lipoprotein-deficient plasma, *J. Lipid. Res.* **23**:1058–1067.

151. Norum, K. R., Glomset, J. A., Nichols, A. V., Forte, T., Albers, J. J., King, W. C., Mitchell, C. D., Applegate, K. R., Gong, E. L., Cabana, V., and Gjone, E., 1975, Plasma lipoproteins in familial lecithin:cholesterol acyltransferase deficiency: Effects of incubation with lecithin: cholesterol acyltransferase *in vitro*, *Scand. J. Clin. Lab. Invest.* **35**(Suppl. 142):31–55.
152. Goldstein, J. L., and Brown, M. S., 1977, The low density lipoprotein pathway and its relation to atherosclerosis, *Annu. Rev. Biochem.* **46**:897–930.
153. Goldstein, J. L., and Brown, M. S., 1982, Familial hypercholesterolemia, in: *The Metabolic Basis of Inherited Disease* (J. B. Stanbury, J. B. Wyngaarden, D. S. Fredrickson, J. L. Goldstein, and M. S. Brown, eds.), McGraw-Hill, New York, pp. 672–712.
154. Hobbs, H. H., Russell, D. W., Brown, M. S., and Goldstein, J. L., 1990, The LDL receptor locus in familial hypercholesterolemia: Mutational analysis of a membrane protein, *Annu. Rev. Genet.* **24**:133–170.
155. Kurihara, Y., Matsumoto, A., Itakura, H., and Kodama, T., 1991, Macrophage scavenger receptors, *Curr. Opin. Lipodol.* **2**:295–300.
156. Goldstein, J. L., and Brown, M. S., 1974, Binding and degradation of low density lipoproteins by cultured human fibroblasts: Comparison of cells from a normal subject and from a patient with homozygous familial hypercholesterolemia, *J. Biol. Chem.* **249**:5153–5162.
157. Brown, M. S., and Goldstein, J. L., 1975, Regulation of the activity of the low density lipoprotein receptor in human fibroblasts, *Cell* **6**:307–316.
158. Brown, M. S., Kovanen, P. T., and Goldstein, J. L., 1981, Regulation of plasma cholesterol by lipoprotein receptors, *Science* **212**:628–635.
159. Goldstein, J. L., Basu, S. K., Brunschede, G. Y., and Brown, M. S., 1976, Release of low density lipoprotein from its cell surface receptor by sulfated glycosamino glycans, *Cell* **7**:85–95.
160. Pitas, R. E., Innerarity, T. L., Arnold, K. S., and Mahley, R. W., 1981, Rate equilibrium constants for binding of apoE HDL (a cholesterol-induced lipoprotein) and low density lipoproteins to human fibroblasts. Evidence for multiple receptor binding of apoE HDL, *Proc. Natl. Acad. Sci. USA* **76**:2311–2315.
161. Anderson, R. G. W., Goldstein, J. L., and Brown, M. S., 1976, Localization of low density lipoprotein receptors on plasma membrane of normal human fibroblasts and their absence in cells from a familial hypercholesterolemia homozygote, *Proc. Natl. Acad. Sci. USA* **73**:2434–2438.
162. Anderson, R. G. W., Brown, M. S., and Goldstein, J. L., 1977, Role of the coated endocytic vesicle in the uptake of receptor-bound low density lipoprotein in human fibroblasts, *Cell* **10**:351–364.
163. Orci, L., Carpenter, J. L., Perrelet, A., Anderson, R. G. W., Goldstein, J. L., and Brown, M. S., 1978, Occurrence of low density lipoprotein receptors within large pits on the surface of human fibroblasts as demonstrated by freeze-etching, *Exp. Cell. Res.* **113**:1–13.
164. Anderson, R. G. W., Brown, M. S., Beisiegel, U., and Goldstein, J. L., 1982, Surface distribution and recycling of the low density lipoprotein receptor as visualized with anti-receptor antibodies, *J. Cell. Biol.* **93**:523–531.
165. Cummings, R. D., Kornfeld, S., Schneider, W. J., Hobgood, K. K., Tolleshaug, H., Brown, M. S., and Goldstein, J. L., 1983, Biosynthesis of the N- and O-linked oligosaccharides of the low density lipoprotein receptor, *J. Biol. Chem.* **258**:15261–15273.
166. Tolleshaug, H., Goldstein, J. L., Schneider, W. J., and Brown, M. S., 1982, Post-translational processing of the LDL receptor and its genetic disruption in familial hypercholesterolemia, *Cell* **30**:715–724.
167. Goldstein, J. L., Anderson, R. G. W., and Brown, M. S., 1979, Coated pits, coated vesicles, and receptor-mediated endocytosis, *Nature* **279**:679–685.
168. Brown, M. S., Anderson, R. G. W., and Goldstein, J. L., 1983, Recycling receptors: The round-trip itinerary and migrant membrane proteins, *Cell* **32**:663–667.

169. Basu, S. K., Goldstein, J. L., Anderson, R. G. W., and Brown, M. S., 1981, Monensin interrupts the recycling of low density lipoprotein receptors in human fibroblasts, *Cell* **24**: 493–502.

170. Stein, O., and Stein, Y., 1976, High density lipoproteins reduce the uptake of low density lipoproteins by human endothelial cells in culture, *Biochim. Biophys. Acta* **431**:363–368.

171. Brown, M. S., Dana, S. E., and Goldstein, J. L., 1974, Regulation of 3-hydroxy-3-methylglutaryl coenzyme A reductase activity in cultured human fibroblasts: Comparison of cells from a normal subject and from a patient with homozygous familial hypercholesterolemia *J. Biol. Chem.* **249**:789–796.

172. Goldstein, J. L., Dana, S. E., and Brown, M. S., 1974, Esterification of low density lipoprotein human fibroblasts and its absence in homozygous familial hypercholesterolemia, *Proc. Natl. Acad. Sci. USA* **71**:4288–4294.

173. Yamamoto, T., Davis, C. G., Brown, M. S., Russell, D. W., and Schneider, W. J., 1984, The human LDL receptor: A cysteine-rich protein with multiple Alu sequences in its mRNA, *Cell* **39**:27–38.

174. Ho, Y. K., Brown, M. S., Kayden, H. J., and Goldstein, J. L., 1976, Binding, internalization, and hydrolysis of low density lipoproteins in long term lymphoid cell lines from a normal subject and a patient with homozygous familial hypercholesterolemia, *J. Exp. Med.* **144**: 444–455.

175. Ho, Y. K., Brown, M. S., Bilheimer, D. W., and Goldstein, J. L., 1976, Regulation of low density lipoprotein receptor activity in freshly isolated human lymphocytes, *J. Clin. Invest.* **58**: 1465–1474.

176. Goldstein, J. L., and Brown, M. S., 1975, Lipoprotein receptors, cholesterol metabolism and atherosclerosis, *Arch. Pathol.* **99**:181–184.

177. Albers, J. J., and Bierman, E. L., 1976, The effect of hypoxia on uptake and degradation of low density lipoproteins by cultured human arterial smooth muscle cells, *Biochim. Biophys. Acta* **454**:422–429.

178. Kovanen, P. T., Schneider, W. J., Hillman, G. M., Goldstein, J. L., and Brown, M. S., 1979, Separate mechanism for the uptake of high and low density lipoprotein by mouse adrenal gland *in vivo*, *J. Biol. Chem.* **254**:5498–5505.

179. Gwynne, J. T., and Hess, B., 1980, The role of high density lipoproteins in rat adrenal cholesterol metabolism and steroidogenesis, *J. Biol. Chem.* **255**:10875–10883.

180. Bersot, T. P., Mahley, R. W., Brown, M. S., and Goldstein, J. L., 1976, Interaction of swine lipoproteins with the low density lipoprotein receptor in human fibroblasts, *J. Biol. Chem.* **251**: 2395–2398.

181. Mahley, R. W., Innerarity, T. L., Weisgraber, K. H., and Fry, D. L., 1977, Accumulation of lipid by aortiomedial cells *in vivo* and *in vitro*, *Am. J. Pathol.* **87**:205–226.

182. Gianturco, S. H., Gotto, Jr., A. M., Jackson, R. L., Patsch, J. R., Sybers, H. D., Taunton, O. D., Yeshurin, D. L., and Smith, L. C., 1978, Control of 3-hydroxy-3-methylglutaryl-CoA reductase activity in cultured human fibroblasts by very low density lipoproteins of subjects with hypertriglyceridemia, *J. Clin. Invest.* **61**:320–328.

183. Innerarity, T. L., Pitas, R. E., and Mahley, R. W., 1980, Disparities in the interaction of rat and human lipoproteins with cultured rat fibroblasts and smooth muscle cells, *J. Biol. Chem.* **255**: 11163–11172.

184. Floren, C. H., Albers, J. J., Kudchodkar, B., and Bierman, E. L., 1981, Receptor-dependent uptake of human chylomicron remnants by cultured skin fibroblasts, *J. Biol. Chem.* **256**: 425–433.

185. Kovanen, P. T., Brown, M. S., Basu, S. K., Bilheimer, D. W., and Goldstein, J. L., 1981, Saturation and suppression of hepatic lipoprotein receptors: A mechanism for the hypercholesterolemia of cholesterol-fed rabbits, *Proc. Natl. Acad. Sci. USA* **78**:1396–1400.

186. Innerarity, T. L., Pitas, R. E., and Mahley, R. W., 1979, Binding of arginine-rich (E) apoprotein after recombination with phospholipid vesicles to the low density lipoprotein receptors of fibroblasts, *J. Biol. Chem.* **254**:4186–4190.

187. Pitas, R. E., Innerarity, T. L., and Mahley, R. W., 1980, Cell surface receptor binding of phospholipid–protein complexes containing different ratios of receptor-active and inactive E apoprotein, *J. Biol. Chem.* **255**:5454–5460.

188. Mahley, R. W., Innerarity, T. L., Pitas, R. E., Weisgraber, K. H., Brown, J. H., and Gross, E., 1977, Inhibition of lipoprotein binding to cell surface receptors of fibroblasts following selective modification of arginyl residues in arginine rich and B-apoproteins, *J. Biol. Chem.* **252**:7279–7287.

189. Weisgraber, K. H., Innerarity, T. L., and Mahley, R. W., 1978, Role of the lysine residues of plasma lipoproteins in high affinity binding to cell surface receptors, *J. Biol. Chem.* **253**:9053–9062.

190. Mahley, R. W., Weisgraber, K. H., and Innerarity, T. L., 1979, Interaction of plasma lipoprotein containing apolipoproteins B and E with heparin and cell surface receptors, *Biochim. Biophys. Acta* **575**:81–91.

191. Mahley, R. W., Weisgraber, K. H., Innerarity, T. L., Brewer, Jr., H. B., and Assmann, G., 1975, Swine lipoproteins and atherosclerosis. Changes in the plasma lipoproteins and apoproteins induced by cholesterol feeding, *Biochemistry* **14**:2817–2823.

192. Guo, L. S., Hamilton, R. L., Kane, J. P., Fielding, C. J., and Chen, G. C., 1982, Characterization and quantitation of apolipoproteins A-I and E of normal and cholesterol-fed guinea pigs, *J. Lipid Res.* **23**:531–542.

193. Fainaru, M., Mahley, R. W., Hamilton, R. L., and Innerarity, T. L., 1982, Structural and metabolic heterogeneity of b-very low density lipoproteins from cholesterol-fed dogs and from humans with type III hyperlipoproteinemia, *J. Lipid Res.* **23**:702–714.

194. Goldstein, J. L., Ho, Y. K., Brown, M. S., Innerarity, T. L., and Mahley, R. W., 1980, Cholesteryl ester accumulation in macrophages resulting from receptor mediated uptake and degradation of hypercholesterolemic canine beta-very low density lipoproteins, *J. Biol. Chem.* **255**:1839–1848.

195. Mahley, R. W., Innerarity, T. L., Brown, M. S., Ho, Y. K., and Goldstein, J. L., 1980, Cholesteryl ester synthesis in macrophages: Stimulation of beta-very low density lipoproteins from cholesterol-fed animals of several species, *J. Lipid Res.* **21**:970–980.

196. Pitas, R. E., Innerarity, T. L., and Mahley, R. W., 1983, Foam cells in explants of atherosclerotic rabbit aortas have receptors for beta-very low density lipoproteins and modified low density lipoproteins, *Arteriosclerosis* **3**:2–12.

197. Koo, C., Wernette-Hammond, M. E., and Innerarity, T. L., 1986, Uptake of canine beta-very low density lipoproteins by mouse peritoneal macrophages is mediated by a low density lipoprotein receptor, *J. Biol. Chem.* **261**:11194–11201.

198. Ellsworth, J. L., Kraemer, F. B., and Cooper, A. D., 1987, Transport of beta-very low density lipoproteins and chylomicron remnants by macrophages is mediated by the low density lipoprotein receptor pathway, *J. Biol. Chem.* **262**:2316–2325.

199. Tabas, I., Boykow, G. C., and Tall, A. R., 1987, Foam cell-forming J774 macrophages have markedly elevated acyl coenzyme A:cholesterol acyl transferase activity compared with mouse peritoneal macrophages in the presence of low density lipoprotein (LDL) despite similar LDL receptor activity, *J. Clin. Invest.* **79**:418–426.

200. Koo, C., Wernette-Hammond, M. E., Garcia, Z., Malloy, M. J., Uauy, R., East, C., Bilheimer, D. W., Mahley, R. W., and Innerarity, T. L., 1988, Uptake of cholesterol-rich remnant lipoproteins by human monocyte-derived macrophages is mediated by low density lipoprotein receptors, *J. Clin. Invest.* **81**:1332–1340.

201. Hui, D. Y., Innerarity, T. L., Milne, R. W., Marcel, Y. L., and Mahley, R. W., 1984, Binding of

chylomicron remnants and beta-very low density lipoproteins to hepatic and extrahepatic lipoprotein receptors. A process independent of apolipoprotein B48, *J. Biol. Chem.* **259**:15060–15068.

202. Gianturco, S. H., Gotto, Jr., A. M., Hwang, S. L., Karlin, J. B., Lin, A. H., Prasad, S. C., and Bradley, W. A., 1983, Apolipoprotein E mediates uptake of Sf 100–400 hypertriglyceridemic very low density lipoproteins by the low density lipoprotein receptor pathway in normal human fibroblasts, *J. Biol. Chem.* **258**:4526–4533.

203. Basu, S. K., Goldstein, J. L., and Brown, M. S., 1978, Characterization of the low density lipoprotein receptor in membranes prepared from human fibroblasts, *J. Biol. Chem.* **253**:3852–3856.

204. Schneider, W. J., Goldstein, J. L., and Brown, M. S., 1980, Partial purification and characterization of the low density lipoprotein receptor from bovine adrenal cortex, *J. Biol. Chem.* **255**:11442–11447.

205. Schneider, W. J., Beisiegel, U., Goldstein, J. L., and Brown, M. S., 1982, Purification of the low density lipoprotein receptor, an acidic glycoprotein of 164,000 molecular weight, *J. Biol. Chem.* **257**:2664–2673.

206. Beisiegel, U., Schneider, W. J., Brown, M. S., and Goldstein, J. L., 1982, Immunoblot analysis of low density lipoprotein receptors in fibroblasts from subjects with familial hyper-cholesterolemia, *J. Biol. Chem.* **257**:13150–13156.

207. Russell, D. W., Yamamoto, T., Schneider, W. J., Slaughter, C. J., Brown, M. S., and Goldstein, J. L., 1983, cDNA cloning of the bovine low density lipoprotein receptor: Feedback regulation of a receptor mRNA, *Proc. Natl. Acad. Sci. USA* **80**:7501–7505.

208. Sudhof, T. C., Goldstein, J. L., Brown, M. S., and Russell, D. W., 1985, The LDL receptor gene: A mosaic of exons shared with different proteins, *Science* **228**:815–822.

209. Francke, U., Brown, M. S., and Goldstein, J. L., 1984, Assignment of the human gene for the low density lipoprotein receptor to chromosome 19: Synteny of a receptor, a ligand and a genetic disease, *Proc. Natl. Acad. Sci. USA* **81**:2826–2830.

210. Lindgren, V., Luskey, K. L., Russell, D. W., and Francke, U., 1985, Human genes involved in cholesterol metabolism: Chromosomal mapping of the loci for the low density lipoprotein receptor and 3-hydroxy-3-methylglutaryl-coenzyme A reductase with cDNA probes, *Proc. Natl. Acad. Sci. USA* **82**:8567–8571.

211. Lusis, A. J., Heinzmann, C., Sparkes, R. S., Scott, J., Knott, T. J., Geller, R., Sparkes, M. C., and Mohandas, T., 1986, Regional mapping of human chromosome 19: Organization of genes for plasma lipid transport (APOC1, -C2, and -E and LDLR) and the genes C3, PEPD, and GPI, *Proc. Natl. Acad. Sci. USA* **83**:3929–3933.

212. Smit, M., van der Kooij-Meijs, E., Frants, R. R., Havekes, L., and Klasen, E. C., 1988, Apolipoprotein gene cluster on chromosome 19, *Hum. Genet.* **78**:90–93.

213. Herz, J., Hamann, U., Rogne, S., Myklebost, O., Gausepohl, H., and Stanley, K. K., 1988, Surface location and high affinity for calcium of a 500 Kd liver membrane protein closely related to the LDL-receptor suggest a physiological role as lipoprotein receptor, *EMBO J.* **7**:4119–4127.

214. Esser, V., Limbird, L. E., Brown, M. S., Goldstein, J. L., and Russell, D. W., 1988, Mutational analysis of the ligand binding domain of the low density lipoprotein receptor, *J. Biol. Chem.* **263**:13282–13290.

215. Russell, D. W., Brown, M. S., and Goldstein, J. L., 1989, Different combinations of cysteine-rich repeats mediate binding of low density lipoprotein receptor to two different proteins, *J. Biol. Chem.* **264**:21682–21688.

216. Sudhof, T. C., 1987, Three direct repeats and a TATA like sequence are required for regulated expression of the human low density lipoprotein receptor gene, *J. Biol. Chem.* **262**:10173–10179.

217. Kadonaga, J. T., and Tjian, R., 1986, Affinity purification of sequence-specific DNA binding proteins, *Proc. Natl. Acad. Sci. USA* **83**:5889–5893.

218. Dawson, P. A., Hofmann, S. L., van der Westhuyzen, D. R., Sudhof, T. C., Brown, M. S., and Goldstein, J. L., 1988, Sterol-dependent repression of low density lipoprotein receptor promoter mediated by 16-base pair sequence adjacent to binding site for transcription factor Sp1, *J. Biol. Chem.* **263**:3372–3379.

219. Smith, J. R., Osborne, T. F., Goldstein, J. L., and Brown, M. S., 1990, Identification of nucleotides responsible for enhancer activity of sterol regulatory element in low density lipoprotein receptor gene, *J. Biol. Chem.* **265**:2306–2310.

220. Watanabe, Y., 1980, Serial inbreeding of rabbits with hereditary hyperlipidemia (WHHL-rabbit). Incidence and development of atherosclerosis and xanthoma, *Atherosclerosis* **36**: 261–268.

221. Schneider, W. J., Brown, M. S., and Goldstein, J. L., 1983, Kinetic defects in the processing of the low density lipoprotein receptor in fibroblasts from WHHL rabbits and a family with familial hypercholesterolemia, *Mol. Biol. Med.* **1**:353–367.

222. Goldstein, J. L., Kita, T., and Brown, M. S., 1983, Defective lipoprotein receptors and atherosclerosis: Lessons from an animal counterpart of familial hypercholesterolemia, *N. Engl. J. Med.* **309**:288–295.

223. Tanzawa, K., Shinada, Y., Kuroda, M., Tsujiota, Y., Arai, M., and Watanabe, Y., 1980, WHHL-rabbit: A low density lipoprotein receptor-deficient animal model for familial hyper-cholesterolemia, *FEBS Lett.* **118**:81–84.

224. Kita, T., Brown, M. S., Bilheimer, D. W., and Goldstein, J. L., 1982, Delayed clearance of very low density and intermediate density lipoproteins with enhanced conversion to low density lipoprotein in WHHL rabbits, *Proc. Natl. Acad. Sci. USA* **79**:5693–5697.

225. Kita, T., Goldstein, J. L., Brown, M. S., Watanabe, Y., Hornick, C. A., and Havel, R. J., 1982, Hepatic uptake of chylomicron remnants in WHHL rabbits: A mechanism genetically distinct from the low density lipoprotein receptor, *Proc. Natl. Acad. Sci. USA* **79**:3623–3627.

226. Yamamoto, T., Bishop, R. W., Brown, M. S., Goldstein, J. L., and Russell, D. W., 1986, Deletion in cysteine-rich region of LDL receptor impedes transport to cell surface in WHHL rabbit, *Science* **232**:1230–1237.

227. Hummel, M., Li, Z., Pfaffinger, D., Neven, L., and Scanu, A. M., 1990, Familial hyper-cholesterolemia in a rhesus monkey pedigree: Molecular basis of low density lipoprotein receptor deficiency, *Proc. Natl. Acad. Sci. USA* **87**:3122–3126.

228. Hofmann, S. I., Russell, D. W., Brown, M. S., Goldstein, J. L., and Hammer, R. E., 1988, Overexpression of low density lipoprotein (LDL) receptor eliminates LDL from plasma in transgenic mice, *Science* **239**:1277–1281.

229. Yokode, M., Hammer, R. E., Ishibashi, S., Brown, M. S., and Goldstein, J. L., 1990, Diet-induced hypercholesterolemia in mice: Prevention by overexpression of LDL receptors, *Science* **250**:1273–1275.

230. Hobbs, H. H., Lehrman, M. A., Yamamoto, T., and Russell, D. W., 1985, Polymorphism and evolution of Alu sequences in the human low density lipoprotein receptor gene, *Proc. Natl. Acad. Sci. USA* **82**:7651–7655; Erratum, *Proc. Natl. Acad. Sci. USA* **83**(6):1964, 1986.

231. Lehrman, M. A., Russell, D. W., Goldstein, J. L., and Brown, M. S., 1986, Exon–Alu recombination deletes 5 kilobases from the low density lipoprotein receptor gene, producing a null phenotype in familial hypercholesterolemia, *Proc. Natl. Acad. Sci. USA* **83**:3679–3683.

232. Horsthemke, B., Beisiegel, U., Dunning, A., Havinga, J. R., Williamson, R., and Humphries, S., 1987, Unequal crossing-over between two Alu-repetitive DNA sequences in the low-density-lipoprotein-receptor gene. A possible mechanism for the defect in a patient with familial hypercholesterolemia, *Eur. J. Biochem.* **164**:77–81.

233. Hobbs, H. H., Brown, M. S., Russell, D. W., Davignon, J., and Goldstein, J. L., 1987, Deletion

in the gene for the low-density-lipoprotein receptor in a majority of French Canadians with familial hypercholesterolemia, *N. Engl. J. Med.* **317**:734–737.

234. Hobbs, H. H., Leitersdorf, E., Goldstein, J. L., Brown, M. S., and Russell, D. W., 1988, Multiple mutations that prevent synthesis of LDL receptors in FH: Evidence for 13 alleles, including 5 deletions, *J. Clin. Invest.* **81**:909–917.

235. Rudiger, N. S., Heinsvig, E. M., Hansen, F. A., Faergeman, O., Bolund, L., and Gregersen, N., 1991, DNA deletions in the low density lipoprotein (LDL) receptor gene in Danish families with familial hypercholesterolemia, *Clin. Genet.* **39**:451–462.

236. Tolleshaug, H., Hobgood, K. K., Brown, M. S., and Goldstein, J. L., 1983, The LDL receptor locus in familial hypercholesterolemia: Multiple mutations disrupt transport and processing of a membrane receptor, *Cell* **32**:941–951.

237. Lehrman, M. A., Schneider, W. J., Brown, M. S., Davis, C. G., Elhammer, A., Russell, D. W., and Goldstein, J. L., 1987, The Lebanese allele at the low density lipoprotein receptor locus. Nonsense mutation produces truncated receptor that is retained in endoplasmic reticulum, *J. Biol. Chem.* **262**:401–410.

238. Esser, V., and Russell, D. W., 1988, Transport-deficient mutations in the low density lipoprotein receptor. Alterations in the cysteine-rich and cysteine-poor regions of the protein block intracellular transport, *J. Biol. Chem.* **263**:13276–13281.

239. Leitersdorf, E., Hobbs, H. H., Fourie, A. M., Jacobs, M., van der Westhuyzen, D. R., and Coetzee, G. A., 1988, Deletion in the first cysteine-rich repeat of low density lipoprotein receptor impairs its transport but not lipoprotein binding in fibroblasts from a subject with familial hypercholesterolemia, *Proc. Natl. Acad. Sci. USA* **85**:7912–7916.

240. Hobbs, H. H., Leitersdorf, E., Leffert, C. C., Cryer, D. R., Brown, M. S., and Goldstein, J. L., 1989, Evidence for a dominant gene that suppresses hypercholesterolemia in a family with defective low density lipoprotein receptors, *J. Clin. Invest.* **84**:656–664.

241. Leitersdorf, E., van der Westhuyzen, D. R., Coetzee, G. A., and Hobbs, H. H., 1989, Two common low density lipoprotein receptor gene mutations cause familial hypercholesterolemia in Afrikaners, *J. Clin. Invest.* **84**:954–961.

242. Knight, B. L., Gavigan, S. J., Soutar, A. K., and Patel, D. D., 1989, Defective processing and binding of low-density lipoprotein receptors in fibroblasts from a familial hypercholesterolemic subject, *Eur. J. Biochem.* **179**:693–698.

243. Soutar, A. K., Knight, B. L., and Patel, D. D., 1989, Identification of a point mutation in growth factor repeat C of the low density lipoprotein-receptor gene in a patient with homozygous familial hypercholesterolemia that affects ligand binding and intracellular movement of receptors, *Proc. Natl. Acad. Sci. USA* **86**:4166–4170.

244. Leitersdorf, E., Tobin, E. J., Davignon, J., and Hobbs, H. H., 1990, Common low-density lipoprotein receptor mutations in the French Canadian population, *J. Clin. Invest.* **85**:1014–1023.

245. Hobbs, H. H., Brown, M. S., Goldstein, J. L., and Russell, D. W., 1986, Deletion of exon encoding cysteine-rich repeat of low density lipoprotein receptor alters its binding specificity in a subject with familial hypercholesterolemia, *J. Biol. Chem.* **261**:13114–13120.

246. Horsthemke, B., Dunning, A., and Humphries, S., 1987, Identification of deletions in the human low density lipoprotein receptor gene, *J. Med. Gen.* **24**:144–147.

247. Russell, D. W., Lehrman, M. A., Sudhof, T. C., Yamamoto, T., and Davis, C. G., 1987, The LDL receptor in familial hypercholesterolemia: Use of human mutations to dissect a membrane protein, *Cold Spring Harbor Symp. Quant. Biol.* **51**:811–819.

248. Lehrman, M. A., Goldstein, J. L., Russell, D. W., and Brown, M. S., 1987, Duplication of seven exons in LDL receptor gene caused by Alu–Alu recombination in a subject with familial hypercholesterolemia, *Cell* **48**:827–835.

249. Henderson, H. E., Berger, G. M., and Marais, A. D., 1988, A new LDL receptor gene deletion mutation in the South African population, *Hum. Genet.* **80**:371–374.

250. Ma, Y., Betard, C., Roy, M., Davignon, J., and Kessling, A. M., 1989, Identification of a second French Canadian LDL receptor gene deletion and development of a rapid method to detect both deletions, *Clin. Genet.* **36**:219–228.

251. Lehrman, M. A., Goldstein, J. L., Brown, M. S., Russell, D. W., and Schneider, W. J., 1985, Internalization-defective LDL receptor produced by genes with nonsense and frameshift mutations that truncate the cytoplasmic domain, *Cell* **41**:735–743.

252. Lehrman, M. A., Schneider, W. J., Sudhof, T. C., Brown, M. S., Goldstein, J. L., and Russell, D. W., 1985, Mutation in LDL receptor: Alu–Alu recombination deletes exons encoding transmembrane and cytoplasmic domains, *Science* **227**:140–146.

253. Davis, C. G., Lehrman, M. A., Russell, D. W., Anderson, R. G. W., Brown, M. S., and Goldstein, J. L., 1986, The J.D. mutation in familial hypercholesterolemia: Amino acid substitution in cytoplasmic domain impedes internalization of LDL receptor, *Cell* **45**:15–24.

254. Lehrman, M. A., Russell, D. W., Goldstein, J. L., and Brown, M. S., 1987, Alu–Alu recombination deletes splice acceptor sites and produces secreted low density lipoprotein receptor in a subject with familial hypercholesterolemia, *J. Biol. Chem.* **262**:3354–3361.

255. Miyake, Y., Tajima, S., Funahashi, T., and Yamamoto, A., 1989, Analysis of a recycling-impaired mutant of low density lipoprotein receptor in familial hypercholesterolemia, *J. Biol. Chem.* **264**:16584–16590.

256. Aalto-Setala, K., Helve, E., Kovanen, P. T., and Kontula, K. 1989, Finnish type of low density lipoprotein receptor gene mutation (FH-Helsinki) deletes exons encoding and carboxy-terminal part of the receptor and creates an internalization-defective phenotype, *J. Clin. Invest.* **84**:499–505.

257. Davis, C. G., van Driel, I. R., Russell, D. W., Brown, M. S., and Goldstein, J. L., 1987, The low density lipoprotein receptor. Identification of amino acids in cytoplasmic domain required for rapid endocytosis, *J. Biol. Chem.* **262**:4075–4082.

258. Yen, F. T., Mann, C. J., and Bihain, B. E., 1991, Alternate pathway for lipoprotein uptake in fibroblasts from subjects homozygous for familial hypercholesterolemia, *Circulation* **84**(Suppl. II):342 (Abstract).

259. Carrela, M., and Cooper, A. D., 1979, High affinity binding of chylomicron remnants to rat liver plasma membranes, *Proc. Natl. Acad. Sci. USA* **76**:338–342.

260. Hui, D. Y., Innerarity, T. L., and Mahley, R. W., 1981, Lipoprotein binding to canine hepatic membranes. Metabolically distinct apo-E and apoB,E receptors, *J. Biol. Chem.* **256**:5646–5654.

261. Havel, R. J., Goldstein, J. L., and Brown, M. S., 1980, Lipoproteins and lipid transport, in: *Metabolic Control and Disease* (P. K. Bondy and L. L. Rosenberg, eds.), Saunders, Philadelphia, Pennsylvania, pp. 393–494.

262. Rubinsztein, D. C., Cohen, J. C., Berger, G. M., van der Westhuyzen, D. R., Coetzee, G. A., and Gevers, W., 1990, Chylomicron remnant clearance from the plasma is normal in familial hypercholesterolemic homozygotes with defined receptor defects, *J. Clin. Invest.* **86**:1306–1312.

263. Chen, W. J., Goldstein, J. L., and Brown, M. S., 1990, NpxY, sequence often found in cytoplasmic tails, is required for coated pit-mediated internalization of the low density lipoprotein receptor, *J. Biol. Chem.* **265**:3116–3123.

264. Herz, J., Kowal, R. C., Goldstein, J. L., and Brown, M. S., 1990, Proteolytic processing of the 600 kD low density lipoprotein receptor related protein (LRP) occurs in a trans-Golgi compartment, *EMBO J.* **9**:1769–1776.

265. Kowal, R. C., Herz, J., Weisgraber, K. H., Mahley, R. W., Brown, M. S., and Goldstein, J. L., 1990, Opposing effects of apolipoprotein E and C on lipoprotein binding to low density lipoprotein receptor-related protein, *J. Biol. Chem.* **265**:10771–10779.

266. Weisgraber, K. H., Mahley, R. W., Kowal, R. C., Herz, J., Goldstein, J. L., and Brown, M. S., 1990, Apolipoprotein C-I modulates the interaction of apolipoprotein E with beta-migrating very low density lipoprotein (beta-VLDL) and inhibits binding of beta-VLDL to low density lipoprotein receptor-related protein, *J. Biol. Chem.* **265**:22453–22459.

267. Herz, J., Kowal, R. C., Ho, Y. K., Brown, M. S., and Goldstein, J. L., 1990, Low density lipoprotein receptor-related protein mediates endocytosis of monoclonal antibodies in cultured cells and rabbit liver, *J. Biol. Chem.* **265**:21355–21362.

268. Strickland, D. K., Ashcom, J. D., Williams, S., Burgess, W. H., Migliorini, M., Argraves, W. S., 1990, Sequence identity between the alpha2-macroglobulin receptor and low density lipoprotein receptor-related protein suggests that this molecule is a multifunctional receptor, *J. Biol. Chem.* **265**:17401–17404.

269. Kristensen, T., Moestrup, S. K., Gliemann, J., Bendtsen, L., Sand, O., and Sottrup-Jensen, L., 1990, Evidence that the newly cloned low-density-lipoprotein receptor related protein (LRP) is the alpha 2-macroglobulin receptor, *FEBS Lett.* **276**:151–155.

270. Moestrup, S. K., and Gliemann, J., 1989, Purification of the rat hepatic alpha2-macroglobulin receptor as an approximately 440 kD single chain protein, *J. Biol. Chem.* **264**:15574–15577.

271. Ashcom, J. D., Tiller, S. E., Dickerson, K., Cravens, J. L., Argraves, W. S., and Strickland, D. K., 1990, The human alpha2-macroglobulin receptor: Identification of a 420kD cell surface glycoprotein specific for the activated conformation of alpha2-macroglobulin, *J. Cell Biol.* **110**: 1041–1048.

272. Sottrup-Jensen, L., 1989, Alpha2-macroglobulins: Structure, shape, and mechanism of proteinase complex formation, *J. Biol. Chem.* **264**:11539–11542.

272a. Willnow, T. E., Goldstein, J. L., Orth, K., Brown, M. S., and Herz, J., 1992, Low density lipoprotein receptor-related protein and gp330 bind similar ligands, including plasminogen activator-inhibitor complexes and lactoferrin, an inhibitor of chylomicron remnant clearance, *J. Biol. Chem.* **267**:26172–26180.

272b. Chappell, D. A., Fry, G. L., Waknitz, M. A., Iverius, P.-H., Williams, S. E., and Strickland, D. K., 1992, The low density lipoprotein receptor-related protein/α_2-macroglobulin receptor binds and mediates catabolism of bovine milk lipoprotein lipase, *J. Biol. Chem.* **267**:25764–25767.

273. Fogelman, A. M., Schechter, I., Seager, J., Hokom, M., Child, J. S., and Edwards, P. A., 1980, Malondialdehyde alteration of low density lipoprotein leads to cholesteryl ester accumulation in human monocyte-macrophages, *Proc. Natl. Acad. Sci. USA* **77**:2214–2218.

274. Goldstein, J. L., Ho, Y. K., Basu, S. K., and Brown, M. S., 1979, Binding site on macrophages that mediates uptake and degradation of acetylated low density lipoprotein producing massive cholesterol deposition, *Proc. Natl. Acad. Sci. USA* **76**:333–337.

275. Brown, M. S., Goldstein, J. L., Krieger, M., Ho, Y. K., and Anderson, R. G. W., 1979, Reversible accumulation of cholesteryl esters in macrophages incubated with acetylated lipoproteins, *J. Cell Biol.* **82**:597–613.

276. Brown, M. S., and Goldstein, J. L., 1983, Lipoprotein metabolism in the macrophage: Implications for cholesterol deposition in atherosclerosis, *Annu. Rev. Biochem.* **52**:223–261.

277. Brown, M. S., Ho, Y. K., and Goldstein, J. L., 1980, The cholesteryl ester cycle in macrophage foam cells, *J. Biol. Chem.* **255**:9344–9352.

278. Mahley, R. W., 1982, Atherogenic hyperlipoproteinemia: The cellular and molecular biology of plasma lipoproteins altered by dietary fat and cholesterol, in: *Medical Clinics of North America: Lipid Disorders*, Vol. 66 (R. J. Havel, ed.), Saunders, Philadelphia, Pennsylvania, pp. 375–402.

279. Haberland, M. E., and Fogelman, A. M., 1987, The role of altered lipoproteins in the pathogenesis of atherosclerosis, *Am. Heart J.* **113**:573–577.

280. Steinberg, D., Parthasarathy, S., Carew, T. E., Khoo, J. C., and Witztum, J. L., 1989, Beyond cholesterol. Modifications of low density lipoprotein that increase its atherogenicity, *N. Engl. J. Med.* **320**:915–924.

281. Jaakkola, O., Kallioniemi, O. P., and Nikkari, T., 1988, Lipoprotein uptake in primary cell cultures of rabbit atherosclerotic lesions. A fluorescence microscopic and flow cytometric study, *Atherosclerosis* **69**:257–268.

282. Via, D. P., Dresel, H. A., and Gotto, Jr., A. M., 1982, Isolation and characterization of the murine macrophage acetyl LDL receptor, *Circulation* **66**(II):37.

283. Esterbauer, H., Jurgens, G., Quehenberger, O., and Koller, E., 1987, Autoxidation of human low density lipoprotein: Loss of polyunsaturated fatty acids and vitamin E and generation of aldehydes, *J. Lipid Res.* **28**:495–509.

284. Sparrow, C. P., Parthasarathy, S., and Steinberg, D., 1989, A macrophage receptor that recognizes oxidized low density lipoprotein but not acetylated low density lipoprotein, *J. Biol. Chem.* **264**:2599–2604.

285. Arai, H., Kita, T., Yokode, M., Narumiya, S., and Kawai, C., 1989, Multiple receptors for modified low density lipoproteins in mouse peritoneal macrophages: Different uptake mechanisms for acetylated and oxidized low density lipoproteins, *Biochem. Biophys. Res. Commun.* **159**:1375–1382.

286. Via, D. P., Dresel, H. A., Cheng, S. L., and Gotto, Jr., A. M., 1985, Murine macrophage tumors are a source of a 260,000 dalton acetyl low density lipoprotein receptor, *J. Biol. Chem.* **260**:7379–7386.

287. Kodama, T., Reddy, P., Kishimoto, C., and Krieger, M., 1988, Purification and characterization of a bovine acetyl low density lipoprotein receptor, *Proc. Natl. Acad. Sci. USA* **85**:9238–9242.

288. Kodama, T., Freeman, M., Rohrer, L., Zabrecky, J., Matsudaira, P., and Krieger, M., 1990, Type I macrophage scavenger receptor contains alpha-helical and collagen-like coiled coils, *Nature* **343**:531–535.

289. Rohrer, L., Freeman, M., Kodama, T., Penman, M., and Krieger, M., 1990, Coiled-coil fibrous domains mediate ligand binding by macrophage scavenger receptor type II, *Nature* **343**:570–572.

290. Matsumoto, A., Naito, M., Itakura, H., Ikemoto, S., Asaoka, H., Hayakawa, I., Kanamori, H., Aburatani, H., Takaku, F., Suzuki, H., Kobari, Y., Miyai, T., Takahashi, K., Cohen, E. H., Wydro, R., Housman, D. E., and Kodama, T., 1990, Human macrophage scavenger receptors: Primary structure, expression and localization in atherosclerotic lesions, *Proc. Natl. Acad. Sci. USA* **87**:9133–9137.

291. Kodama, T., Doi, T., Matsumoto, A., Higashino, K., Kurihara, Y., Kawabe, Y., Uesugi, S., Imanishi, T., Itakura, H., and Yazaki, Y., 1991, The C-terminal region of a collagen-like domain containing a cluster of basic amino acids mediates binding of modified low density lipoproteins by macrophage scavenger receptors, *Circulation* **84**(Supp. II):229 (Abstract).

292. Innerarity, T. L., Hui, D. Y., Bersot, T. P., and Mahley, R. W., 1986, Type III hyperlipoproteinemia: A focus on lipoprotein receptor–apolipoprotein E2 interactions, *Adv. Exp. Med. Biol.* **201**:273–288.

293. Dejager, S., Friera, A. M., and Pitas, R. E., 1991, Interaction of oxidized LDL with the acetyl LDL receptor expressed by smooth muscle cells and transfected CHO cells, *Circulation* **84** (Suppl. II):568 (Abstract).

294. Stein, O., and Stein, Y., 1973, The removal of cholesterol from Landschutz ascites cells by high-density apolipoprotein, *Biochim. Biophys. Acta* **326**:232–244.

295. Nakai, T., Otto, P. S., Kennedy, D. L., and Whayne, Jr., T. F., 1976, Rat high density lipoprotein subfraction (HDL3) uptake and catabolism by isolated rat liver parenchymal cells, *J. Biol. Chem.* **251**:4914–4921.

296. Drevon, C. A., Berg, T., and Norum, K. R., 1977, Uptake and degradation of cholesteryl ester-labeled rat plasma lipoproteins in purified rat hepatocytes and non-parenchymal liver cells, *Biochim. Biophys. Acta* **487**:122–136.

297. Ose, L., Roken, I., Norum, K. R., Drecon, C. A., and Berg, T., 1981, The binding of high density lipoproteins to isolated rat hepatocytes, *Scand. J. Clin. Invest.* **41**:63–73.

298. Glass, C., Pittman, R. C., Civen, M., and Steinberg, D., 1985, Uptake of high-density lipoprotein-associated apoprotein A-I and cholesterol esters by 16 tissues of the rat *in vivo* and by adrenal cells and hepatocytes *in vitro*, *J. Biol. Chem.* **260**:744–750.

299. Bachorik, P. S., Franklin, Jr., F. A., Virgil, D. G., and Kwiterovich, Jr., P. O., 1985, Reversible

high affinity uptake of apoE-free high density lipoproteins in cultured pig hepatocytes, *Arteriosclerosis* **5**:142–152.

300. Dashti, N., Wolfbauer, G., and Alaupovic, P., 1985, Binding and degradation of human high-density lipoproteins by human hepatoma cell line HepG2, *Biochim. Biophys. Acta* **833**:100–110.

301. Wandel, M., Norum, K. R., Berg, T., and Ose, L., 1981, Binding, uptake and degradation of 125I-labeled high density lipoproteins in isolated non-parenchymal rat liver cells, *Scand. J. Gastroenterol.* **16**:71–80.

302. Oram, J. F., Albers, J. J., and Bierman, E. L., 1980, Rapid regulation of the activity of the low density lipoprotein receptor of cultured human fibroblasts, *J. Biol. Chem.* **255**:475.

303. Oram, J. F., Albers, J. J., Cheung, M. C., and Bierman, E. L., 1981, The effects of subfractions of high density lipoprotein on cholesterol efflux from culture fibroblasts. Regulation of low density lipoprotein receptor activity, *J. Biol. Chem.* **256**:8348–8356.

304. Biesbroeck, R., Oram, M. F., Albers, J. J., and Bierman, E. L., 1983, Specific high-affinity binding of high-density lipoproteins to cultured human skin fibroblasts and arterial smooth muscle cells, *J. Clin. Invest.* **71**:525–539.

305. Miller, N. E., Weinstein, D. B., and Steinberg, D., 1977, Binding, internalization, and degradation of high density lipoprotein by cultured normal human fibroblasts, *J. Lipid Res.* **18**:438–450.

306. Wu, J. D., Butler, J., and Bailey, J. M., 1979, Lipid metabolism in cultured cells. Comparative uptake of low density and high density lipoprotein by normal, hypercholesterolemic, and tumor-transformed human fibroblasts, *J. Lipid Res.* **20**:472–480.

307. Kovanen, P. T., Brown, M. S., and Goldstein, J. L., 1979, Increased binding of low density lipoprotein to liver membranes from rats treated with i-a-ethinyl estradiol, *J. Biol. Chem.* **254**:11367–11373.

308. Hoeg, J. M., Demosky, Jr., S. J., Edge, S. B., Gregg, R. E., Osborne, Jr., J. C., and Brewer, Jr., H. B., 1985, Characterization of a human hepatic receptor for high density lipoproteins, *Arteriosclerosis* **5**:228–237.

309. Fidge, N. H., 1986, Partial purification of a high density lipoprotein-binding protein from rat liver and kidney membranes, *FEBS Lett.* **199**:265–268.

310. Chen, Y. D., Kroemer, F. B., and Reaven, G. M., 1980, Identification of specific high density lipoprotein-binding sites in rat testis and regulation of binding by human chorionic gonadotropin, *J. Biol. Chem.* **255**:9162–9167.

311. Fong, B. S., Rodrigues, P. O., Salter, A. M., Yip, B. P., Despres, J. P., and Angel, A., 1985, Characterization of high density lipoprotein binding to human adipocyte plasma membranes, *J. Clin. Invest.* **75**:1804–1812.

312. Rifici, V. A., and Eder, H. A., 1984, A hepatocyte receptor for high-density lipoproteins specific for apolipoprotein A-I, *J. Biol. Chem.* **259**:13814–13818.

313. Hwang, J., and Memon, K. M. J., 1985, Binding of apolipoprotein A-I and A-II after recombination with phospholipid vesicles to the high density lipoprotein receptor of luteinized rat ovary, *J. Biol. Chem.* **260**:5660–5668.

314. Fidge, N. H., and Nestel, P. J., 1985, Identification of apolipoprotein involved in the interaction of human high density lipoprotein 3 with receptors on cultured cells, *J. Biol. Chem.* **260**:3570–3575.

315. Steinmetz, A., Barbaras, R., Ghalim, N., Clavey, V., Fruchart, J. C., and Ailhaud, G., 1990, Human apolipoprotein A-IV binds to apolipoprotein A-I/A-II receptor sites and promotes cholesterol efflux from adipose cells, *J. Biol. Chem.* **265**:7859–7863.

316. Tabas, I., and Tall, A. R., 1984, Mechanism of the association of HDL3 with endothelial cells, smooth muscle cells, and fibroblasts, *J. Biol. Chem.* **259**:13897–13905.

317. Slotte, J. P., Oram, J .F., and Bierman, E. L., 1987, Binding of high density lipoproteins to cell receptors promotes translocation of cholesterol from intracellular membranes to the cell surface, *J. Biol. Chem.* **262**:12904–12907.

318. Graham, D. L., and Oram, J. F., 1987, Identification and characterization of a high density lipoprotein-binding protein in cell membranes by ligand blotting, *J. Biol. Chem.* **262**:7439–7442.

319. Schmitz, G., Assmann, G., Robenek, H., and Brennhausen, B., 1985, Tangier disease: A disorder of intracellular membrane traffic, *Proc. Natl. Acad. Sci. USA* **82**:6305–6309.

320. Schmitz, G., Robenek, H., Lohmann, U., and Assmann, G., 1985, Interaction of high density lipoproteins with cholesteryl ester-laden macrophages: Biochemical and morphological characterization of cell surface receptor binding, endocytosis and resecretion of high density lipoproteins by macrophages, *EMBO J.* **4**:613–622.

321. Schmitz, G., Niemann, R., Brennhausen, B., Krause, R., and Assmann, G., 1985, Regulation of high density lipoprotein receptors in cultured macrophages: Role of acyl-CoA:cholesterol acyltransferase, *EMBO J.* **4**:2773–2779.

322. Glomset, J. A., and Norum, K. R., 1973, The metabolic role of lecithin:cholesterol acyltransferase: Perspectives from pathology, *Adv. Lipid Res.* **11**:4–65.

323. Glomset, J. A., 1980, High-density lipoproteins in human health and disease, *Adv. Intern. Med.* **25**:91–116.

324. Karlin, J. B., Johnson, W. J., Benedict, C. R., Chacko, G. K., Phillips, M. C., and Rothblat, G. H., 1987, Cholesterol flux between cells and high density lipoprotein. Lack of relationship to specific binding of the lipoprotein to the cell surface, *J. Biol. Chem.* **262**:12557–12564.

325. Johnson, W. J., Chacko, G. K., Phillips, M. C., and Rothblat, G. H., 1990, The efflux of lysosomal cholesterol from cells, *J. Biol. Chem.* **265**:5546–5553.

326. Eisenberg, S., 1984, High density lipoprotein metabolism, *J. Lipid Res.* **25**:1017–1058.

327. Johnson, W. J., Bamberger, M. J., Latta, R. A., Rapp, P. E., Phillips, M. C., and Rothblat, G. H., 1986, The bidirectional flux of cholesterol between cells and lipoproteins: Effects of phospholipid depletion of high density lipoprotein, *J. Biol. Chem.* **261**:5766–5776.

328. Phillips, M. C., Johnson, W. J., and Rothblat, G. H., 1987, Mechanisms and consequences of cellular cholesterol exchange and transfer, *Biochim. Biophys. Acta* **906**:223–276.

329. Monaco, L., Bond, H. M., Howell, K. E., and Cortese, R., 1987, A recombinant apoA-I–protein A hybrid reproduces the binding parameters of HDL to its receptor, *EMBO J.* **6**:3253–3260.

330. Mendel, C. M., Kunitake, S. T., Kane, J. P., and Kempner, E., 1988, Radiation inactivation of binding sites for high density lipoproteins in human fibroblast membranes, *J. Biol. Chem.* **263**:1314–1319.

331. Mendel, C. M., Kunitake, S. T., Hong, K., Erickson, S. K., Kane, J. P., and Kempner, E. S., 1988, Radiation inactivation of binding sites for high-density lipoproteins in human liver membranes, *Biochim. Biophys. Acta* **961**:183–193.

332. Havel, R. J., Shore, V. G., Shore, B., and Bier, D. M., 1970, Role of specific glycopeptides of human serum lipoproteins in the activation of lipoprotein lipase, *Circ. Res.* **27**:595–600.

333. Miller, A. L., and Smith, L. C., 1973, Activation of lipoprotein lipase by apolipoprotein glutamic acid, *J. Biol. Chem.* **248**:3359–3362.

334. Chung, J., and Scanu, A. M., 1977, Isolation, molecular properties, and kinetic characterization of lipoprotein lipase from rat heart, *J. Biol. Chem.* **252**:4202–4209.

335. Krauss, R. M., Windmueller, H. G., Levy, R. I., and Fredrickson, D. S., 1973, Selective measurement of two different triglyceride lipase activities in rat postheparin plasma, *J. Lipid Res.* **14**:286–295.

336. Chait, A., Iverius, P.-H., and Brunzell, J. D., 1982, Lipoprotein lipase secretion by human monocyte-derived macrophages, *J. Clin. Invest.* **69**:490–493.

337. Mahoney, E. M., Khoo, J. C., and Steinberg, D., 1982, Lipoprotein lipase secretion by human monocytes and rabbit alveolar macrophages in culture, *Proc. Natl. Acad. Sci. USA* **79**:1639–1642.

338. Olivecrona, L., and Bengtsson-Olivecrona, G., 1987, in: *Lipoprotein Lipase* (Borensztajn, J., ed.), Evener Press, Chicago, pp. 15–58.

339. Khoo, J. C., Mahoney, E. M., and Witztum, J. L., 1981, Secretion of lipoprotein lipase by macrophages in culture, *J. Biol. Chem.* **256**:7105–7108.

340. Wion, K. L., Kirchgessner, T. G., Lusis, A. J., Schotz, M. C., and Lawn, R. M., 1987, Human lipoprotein lipase complementary DNA sequence, *Science* **235**:1638–1641.

341. Kirchgessner, T. G., Leboef, R. C., Langner, C. A., Zollman, S., Chang, C. H., Taylor, B. A., Schotz, M. C., Gordon, J. I., and Lusis, A. J., 1989, Genetic and developmental regulation of the lipoprotein lipase gene: Loci both distal and proximal to the lipoprotein lipase structural gene control enzyme expression, *J. Biol. Chem.* **264**:1473–1482.

342. Semenkovich, C. F., Chen, S. H., Wims, M., Luio, C. C., Li, W. H., and Chan, I., 1989, Lipoprotein lipase and hepatic lipase mRNA tissue specific expression, developmental regulation, and evolution, *J. Lipid Res.* **30**:423–431.

343. Goldberg, I. J., Soprano, D. R., Wyatt, M. L., Vanni, T. M., Kirchgessner, T. G., and Schotz, M. C., 1989, Localization of lipoprotein lipase mRNA in selected rat tissues, *J. Lipid Res.* **30**:1569–1577.

344. Nilsson-Ehle, P., Carlström, S., and Belfrage, P., 1975, Rapid effect on lipoprotein lipase activity in adipose tissue of humans after carbohydrate and lipid intake, *Scand. J. Clin. Lab. Invest.* **35**:373–378.

345. Pykalisto, O. J., Smith, P. H., and Brunzell, J. D., 1975, Determinations of human adipose tissue lipoprotein lipase. Effect of diabetes and obesity on basal and diet-induced activity, *J. Clin. Invest.* **56**:1108–1117.

346. Lithell, H., Boberg, J., Hellsing, K., Lundqvist, G., and Vessby, B., 1978, Lipoprotein-lipase activity in human skeletal muscle and adipose tissue in the fasting and the fed states, *Atherosclerosis* **30**:89–94.

347. Doolittle, M. H., Ben-Zeev, O., Elovson, J., Martin, D., and Kirchgessner, T. G., 1990, The response of lipoprotein lipase to feeding and fasting. Evidence for posttranslational regulation, *J. Biol. Chem.* **265**:4570–4577.

348. Zinder, O., Hamosh, M., Fleck, T. R. C., and Scow, R. O., 1974, Effect of prolactin on lipoprotein lipase in mammary gland and adipose tissue of rats, *Am. J. Physiol.* **226**:744–748.

349. Eckel, R. H., Fujimoto, W. Y., and Brunzell, J. D., 1979, Gastric inhibitory polypeptide enhanced lipoprotein lipase activity in cultured preadipocytes, *Diabetes* **28**:1141–1142.

350. Wasada, T., McCorkle, K., Harris, V., Kawai, K., Howard, B., and Unger, R. H., 1981, Effect of gastric inhibitory polypeptide on plasma levels of chylomicron triglycerides in dogs, *J. Clin. Invest.* **68**:1106–1112.

351. Raynolds, M. V., Awald, P. D., Gordon, D. F., Gutierrez-Hartmann, A., Rule, D. C., Wood, W. M., and Eckel, R. H., 1990, Lipoprotein lipase gene expression in rat adipocytes is regulated by isoproterenol and insulin through different mechanisms, *Mol. Endocrinol.* **4**:1416–1422.

352. Querfeld, U., Ong, J. M., Prehn, J., Carty, J., Saffari, B., Jordan, S. C., and Kern, P. A., 1990, Effects of cytokines on the production of lipoprotein lipase in cultured human macrophages, *J. Lipid Res.* **31**:1379–1386.

353. Jonasson, I., Hansson, G. K., Bondjers, G., Noe, L., and Etienne, J., 1990, Interferon-gamma inhibits lipoprotein lipase in human monocyte-derived macrophages, *Biochim. Biophys. Acta* **1053**:43–48.

354. Mackay, A. G., Oliver, J. D., and Rogers, M. P., 1990, Regulation of lipoprotein lipase activity and mRNA content in rat epididymal adipose tissue *in vitro* by recombinant tumour necrosis factor, *Biochem. J.* **269**:123–126.

355. Kern, P. A., Ong, J. M., Saffari, B., and Carty, J., 1990, The effects of weight loss on the activity and expression of adipose-tissue lipoprotein lipase in very obese humans, *N. Engl. J. Med.* **322**:1053–1059.

356. Savard, R., and Bouchard, C., 1990, Genetic effects in the response of adipose tissue lipoprotein lipase activity to prolonged exercise. A twin study, *Int. J. Obesity* **14**:771–777.

357. Peterson, J., Bihain, B. E., Bengtsson-Olivecrona, G., Deckelbaum, R. J., Carpentier, Y. A.,

and Olivecrona, T., 1990, Fatty acid control of lipoprotein lipase: A link between energy metabolism and lipid transport, *Proc. Natl. Acad. Sci. USA* **87**:909–913.

358. Pradinesfigueres, A., Barcellinicouget, S., Dani, C., Vannier, C., and Ailhaud, G., 1990, Transcriptional control of the expression of lipoprotein lipase gene by growth hormone in preadipocyte Ob1771 cells, *J. Lipid Res.* **31**:1283–1291.

359. Shimada, K., Gill, P. J., Silbert, J. E., Douglas, W. H., and Fanburg, B. L., 1981, Involvement of cell surface heparin sulfate in the binding of lipoprotein lipase to cultured bovine endothelial cells, *J. Clin. Invest.* **68**:995–1002.

360. Saxena, U., Klein, M. G., and Goldberg, I. J., 1990, Metabolism of endothelial cell-bound lipoprotein lipase. Evidence for heparan sulfate proteoglycan-mediated internalization and recycling, *J. Biol. Chem.* **265**:12880–12886.

360a. Iverius, P-H., and Ostlund-Lindqvist, A-M., 1976, Lipoprotein lipase from bovine milk. Isolation procedure, chemical characterization, and molecular weight analysis, *J. Biol. Chem.* **251**:7791–7795.

361. Hayden, M. R., Ma, Y., Brunzell, J., and Henderson, H. E., 1991, Genetic variants affecting human lipoprotein and hepatic lipases, *Curr. Opin. Lipodol.* **2**:104–109.

362. Kern, P. A., 1991, Lipoprotein lipase and hepatic lipase, *Curr. Opin. Lipidol.* **2**:162–169.

363. Nikkila, E. A., 1982, Familial lipoprotein lipase deficiency and related disorders of chylomicron metabolism, in: *The Metabolic Basis of Inherited Disease*, 5th ed. (J. B. Stanbury, J. B. Wyngaarden, D. S. Fredrickson, J. L. Goldstein, and M. S. Brown, eds.), McGraw-Hill, New York, pp. 622–642.

364. Breckenridge, W. C., Little, J. A., Steiner, G., Chow, A., and Poapst, M., 1978, Hypertriglyceridemia associated with deficiency of apolipoprotein C-II, *N. Engl. J. Med.* **298**:1265–1273.

365. Stalenhoef, A. F., Casparie, A. F., Demacker, P. N., Stouten, J. T., Lutterman, J. A., and van 't Laar, A., 1981, Combined deficiency of apolipoprotein C-II and lipoprotein lipase in familial hyperchylomicronemia, *Metabolism: Clinical and Experimental* **30**:919–926.

366. Deeb, S. S., and Peng, R. L., 1989, Structure of the human lipoprotein lipase gene, *Biochemistry* **28**:4131–4135; Erratum, *Biochemistry* **28**(16):6786 (1989).

367. Sparkes, R. S., Zollman, S., Klisak, I., Kirchgessner, T. G., Komaromy, M. C., Mohandas, T., Schotz, M. C., and Lusis, A. J., 1987, Human genes involved in lipolysis of plasma lipoproteins: Mapping of loci for lipoprotein lipase to 8p22 and hepatic lipase to 15q21, *Genomics* **1**:138–144.

368. Kirchgessner, T. G., Chuat, J. C., Heinzmann, C., Etienne, J., Guilhot, S., Svenson, K., Ameis, D., Pilon, C., d'Auriol, L., and Andalibi, A., 1989, Organization of the human lipoprotein lipase gene and evolution of the lipase gene family, *Proc. Natl. Acad. Sci. USA* **86**:9647–9651.

369. Davis, R. C., Stahnke, G., Wong, H., Doolittle, M. H., Ameis, D., Will, H., and Schotz, M. C., 1990, Hepatic lipase: Site-directed mutagenesis of a serine residue important for catalytic activity, *J. Biol. Chem.* **265**:6291–6295.

370. Semenkovich, C. F., Luo, C. C., Nakanishi, M. K., Chen, S. H., Smith, L. C., and Chan, L., 1990, *In vitro* expression and site-specific mutagenesis of the cloned human lipoprotein lipase gene. Potential N-linked glycosylation site asparagine 43 is important for both enzyme activity and secretion, *J. Biol. Chem.* **265**:5429–5433.

371. Mashes, H., Blanchette-Mackie, E. J., Chernick, S. S., and Scow, R. D., 1990, Synthesis of inactive nonsecretable high mannose-type lipoprotein lipase by cultured brown adipocytes of combined lipase-deficient cld/cld mice, *J. Biol. Chem.* **265**:1628–1638.

372. Davis, R. C., Ben-Zeev, O., Martin, D., and Doolittle, M. H., 1990, Combined lipase deficiency in the mouse. Evidence of impaired lipase processing and secretion, *J. Biol. Chem.* **265**:17960–17966.

373. Ferrans, V. J., Buja, L. M., Roberts, W. C., and Fredrickson, D. S., 1971, The spleen in type I hyperlipoproteinemia. Histochemical, biochemical, microfluorometric and electron microscopic observations, *Am. J. Pathol.* **64**:67–96.

374. Brunzell, J. D., Chait, A., Nikkila, E. A., Ehnholm, C., Huttunen, J. K., and Steiner, G., 1980, Heterogeneity of primary lipoprotein lipase deficiency, *Metabolism: Clinical and Experimental* **29:**624–629.

375. Brunzell, J. D., 1989, Familial lipoprotein lipase deficiency and other causes of the chylomicronemia syndrome, in: *The Metabolic Basis of Inherited Disease* (C. Scriver, A. L. Beaudet, W. S. Sly, and D. Valle, eds.), McGraw-Hill, New York, pp. 1165–1180.

376. Emi, M., Wilson, D. E., Iverius, P. H., Wu, L., Hata, A., Hegele, R., Williams, R. R., and Lalouel, J. M., 1990, Missense mutation (Gly–Glu188) of human lipoprotein lipase imparting functional deficiency, *J. Biol. Chem.* **265:**5910–5916.

377. Babirak, S. P., Iverius, P. H., Fujimoto, W. Y., and Brunzell, J. D., 1989, The detection of the heterozygote state for lipoprotein lipase deficiency, *Arteriosclerosis* **9:**326–334.

378. Wilson, D. E., Emi, M., Iverius, P. H., Hata, A., Wu, L. L., Hillas, E., Williams, R. R., and Lalouel, J. M., 1990, Phenotypic expression of heterozygous lipoprotein lipase deficiency in the extended pedigree of a proband homozygous for a missense mutation, *J. Clin. Invest.* **86:**735–750.

379. Winkler, F. K., D'Arcy, A., and Hunzikas, W., 1990, Structure of human pancreatic lipase, *Nature* **343:**771–774.

380. Langlois, S., Deeb, S., Brunzell, J. D., Kastelein, J. J., and Hayden, M. R., 1989, A major insertion accounts for a significant proportion of mutations underlying human lipoprotein lipase deficiency, *Proc. Natl. Acad. Sci. USA* **86:**948–952.

381. Hata, A., Emi, M., Luc, G., Basdevant, A., Gambert, P., Iverius, P. H., and Lalouel, J. M., 1990, Compound heterozygote for lipoprotein lipase deficiency: Ser–Thr244 and transition in 3′ splice site of intron 2 (AG–AA) in the lipoprotein lipase gene, *Am. J. Hum. Genet.* **47:**721–726.

382. Devlin, R. H., Deeb, S., Brunzell, J., and Hayden, M. R., 1990, Partial gene duplication involving exon–Alu interchange results in lipoprotein lipase deficiency, *Am. J. Hum. Genet.* **46:**112–119.

383. Beg, O. U., Meng, M. S., Skarlatos, S. I., Previato, L., Brunzell, J. D., Brewer, Jr., H. B., and Fojo, S. S., 1990, Lipoprotein lipaseBethesda: A single amino acid substitution (Ala–176–Thr) leads to abnormal heparin binding and loss of enzymic activity, *Proc. Natl. Acad. Sci. USA* **87:**3474–3478.

384. Gotoda, T., Murase, T., Ishibashi, S., Shimario, H., Harada, K., and Yamada, N., 1990, Splicing, nonsense and missense mutation in familial lipoprotein lipase deficiency, *Arteriosclerosis* **10:**833.

385. Emi, M., Hata, A., Robertson, M., Iverius, P. H., Hegele, R., and Lalouel, J. M., 1990, Lipoprotein lipase deficiency resulting from a nonsense mutation in exon 3 of the lipoprotein lipase gene, *Am. J. Hum. Genet.* **47:**107–111.

386. Kantor, M. A., Cullinane, E. M., Sady, S. P., Herbert, P. N., and Thompson, P. D., 1987, Exercise acutely increases high density lipoprotein-cholesterol and lipoprotein lipase activity in trained and untrained men, *Metabolism: Clinical and Experimental* **36:**188–192.

387. Monsalve, M. V., Henderson, H., Roederer, G., Julien, P., Deeb, S., Kastelein, J. J., Peritz, L., Devlin, R., Bruin, T., and Murthy, M. R., 1990, A missense mutation at codon 188 of the human lipoprotein lipase gene is a frequent cause of lipoprotein lipase deficiency in persons of different ancestries, *J. Clin. Invest.* **86:**728–734.

388. Dichek, H. L., Fojo, S. S., Beg, O. U., Skarlatos, S. I., Brunzell, J. D., Cutler, Jr., G. B., and Brewer, Jr., H. B., 1991, Identification of two separate allelic mutations in the lipoprotein lipase gene of a patient with the familial hyperchylomicronemia syndrome, *J. Biol. Chem.* **266:**473–477.

389. Henderson, H. E., Ma, Y., Hassan, M. F., Monsalve, M. V., Marais, A. D., Winkler, K., Gubernator, K., Peterson, J., Brunzell, J. D., and Hayden, M. R., 1991, Amino acid substitution (Ile194–Thr) in exon 5 of the lipoprotein lipase gene causes lipoprotein lipase deficiency in three unrelated probands. Support for a multicentric origin, *J. Clin. Invest.* **87:**2005–2011.

390. Ameis, D., Kobayashi, J., Davis, R. C., Ben-Zeev, O., Malloy, M. J., Kane, J. P., Lee, G., Wong, H., Havel, R. J., and Schotz, M. C., 1991, Familial chylomicronemia (type I hyperlipo-

proteinemia) due to a single missense mutation in the lipoprotein lipase gene, *J. Clin. Invest.* **87**:1165–1170.

391. Deeb, S. S., Reina, M., Peterson, J., Takata, K., Kajiyama, G., and Brunzell, J. D., 1991, Gene mutations in patients with lipoprotein lipase deficiency, *Circulation* **84**(Suppl. II):456 (Abstract).

392. Ishimura-Oka, K., Semenkovich, C. F., Faustinella, F., Goldberg, I. J., Schacter, N., Herbert, P., Thompson, P. D., Smith, L. C., Oka, K., and Chan, L., 1991, Identification of compound heterozygotes for lipoprotein lipase deficiency in three unrelated families, *Circulation* **84** (Suppl. II):376 (Abstract).

392a. Ishimura-Oka, K. Semenkovich, C. F., Faustinella, F., Goldberg, I. J., Shachter, N., Smith, L. C., Coleman, T., Hide, W. A., Brown, W. V., Oka, K., and Chan, L., 1992, A missense (Asp$^{250}\rightarrow$Asn) mutation in the lipoprotein lipase gene in two unrelated families with familial lipoprotein lipase deficiency, *J. Lipid Res.* **33**:745–754.

392b. Ishimura-Oka, K., Faustinella, F., Kihara, S., Smith, L. C., Oka, K., and Chan, L., 1992, A missense mutation (Trp$^{86}\rightarrow$Arg) in exon 3 of the lipoprotein lipase gene: A cause of familial chylomicronemia, *Am. J. Hum. Genet.* **50**:1275–1280.

392c. Faustinella, F., Chang, A., Van Biervliet, J. P., Rosseneu, M., Vinaimont, N., Smith, L. C., Chen, S-H., and Chan, L., 1991, Catalytic triad residue mutation (Asp$^{156}\rightarrow$Gly) causing familial lipoprotein lipase deficiency. Co-inheritance with a nonsense mutation (Ser$^{147}\rightarrow$Ter) in a Turkish family, *J. Biol. Chem.* **266**:14418–14424.

392d. May, Y. H., Bruin, T., Tuzgol, S., Wilson, B. I., Roederer, G., Liu, M. S., Davignon, J., Kastelein, J. J., Brunzell, J. D., and Hayden, M. R., 1992, Two naturally occurring mutations at the first and second bases of codon aspartic acid 156 in the proposed catalytic triad of human lipoprotein lipase. In vivo evidence that aspartic acid 156 is essential for catalysis, *J. Biol. Chem.* **267**:1918–1923.

392e. Sprecher, D. L., Kobayashi, J., Rymaszewski, M., Goldberg, I. J., Harris, B. V., Bellet, P. S., Ameis, D., Yunker, R. L., Black, D. M., and Stein, E. A., 1992, Trp64—Nonsense mutation in the lipoprotein lipase gene, *J. Lipid Res.* **33**:859–866.

392f. Chimienti, G., Capurso, A., Resta, F., and Pepe, G., 1992, A G–C change at the donor splice site of intron 1 causes lipoprotein lipase deficiency in a southern-Italian family, *Biochem. Biophys. Res. Commun.* **187**:620–627.

392g. Hata, A., Ridinger, D. N., Sutherland, S. D., Emi, M., Kwong, L. K., Shuhua, J., Lubbers, A., Guy-Grand, B., Basdevant, A., and Iverius, P. H., 1992, Missense mutations in exon 5 of the human lipoprotein lipase gene. Inactivation correlates with loss of dimerization, *J. Biol. Chem.* **267**:20132–20139.

393. Krauss, R. M., Levy, R. I., and Fredrickson, D. S., 1974, Selective measurement of two lipase activities in postheparin plasma from normal subjects and patients with hyperlipoproteinemia, *J. Clin. Invest.* **54**:1107–1124.

394. Greten, H., DeGrella, R., Klose, G., Rascher, W., de Gennes, J. L., and Gjone, E., 1976, Measurement of two plasma triglyceride lipases by an immunochemical method: Studies in patients with hypertriglyceridemia, *J. Lipid Res.* **17**:203–210.

395. Ostlund-Lindqvist, A. M., 1979, Properties of salt-resistant lipase and lipoprotein lipase purified from human post-heparin plasma, *Biochem. J.* **179**:555–559.

396. Baginsky, M. L., and Brown, W. V., 1979, A new method for the measurement of lipoprotein lipase in postheparin plasma using sodium dodecyl sulfate for the inactivation of hepatic triglyceride lipase, *J. Lipid Res.* **20**:548–556.

397. Breckenridge, W. C., Little, J. A., Alaupovic, P., Wang, C. S., Kuksis, A., Kakis, G., Lindgren, F., and Gardiner, G., 1982, Lipoprotein abnormalities associated with a familial deficiency of hepatic lipase, *Atherosclerosis* **45**:161–179.

398. Kuusi, T., Kinnunen, P. K. J., and Nikkila, E. A., 1979, Hepatic endothelial lipase anti-serum influences rat plasma low and high density lipoproteins in vivo, *FEBS Lett.* **104**:384–388.

399. Twu, J. S., Garfinkel, A. S., and Schotz, M. C., 1984, Hepatic lipase purification and characterization, *Biochim. Biophys. Acta* **792**:330–337.
400. Ben-Zeev, O., Ben-Avram, C. M., Wong, H., Nikazy, J., Shively, J. E., and Schotz, M. C., 1987, Hepatic lipase: A member of a family of structurally related lipases, *Biochim. Biophys. Acta* **919**(1):13–20.
401. Dootlittle, M. H., Wong, H., Davis, R. C., and Schotz, M. C., 1987, Synthesis of hepatic lipase in liver and extrahepatic tissues, *J. Lipid Res.* **28**:1326–1334.
402. Verhoeven, A. J. M., and Jansen, H., 1990, Secretion of hepatic lipase is blocked by inhibition of oligosaccharide processing at the stage of glucosidase, *Int. J. Lipid Res.* **31**:1883–1893.
403. Kuusi, T., Nikkila, E. A., Virtanen, I., and Kinnunen, P. K. J., 1979, Localization of the heparin-releasable lipase *in situ* in the rat liver, *Biochem. J.* **181**:245–246.
404. Jansen, H., VanBerkel, T. J. C., and Hulsmann, W. C., 1978, Binding of liver lipase to parenchymal and non-parenchymal rat liver cells, *Biochem. Biophys. Res. Commun.* **85**:148–152.
405. Datta, S., Luo, C. C., Li, W. H., Van Tuinen, P., Ledbetter, D. H., Brown, M. A., Chen, S. H., Liu, S. W., and Chan, L., 1988, Human hepatic lipase. Cloned cDNA sequence, restriction fragment length polymorphisms, chromosomal localization, and evolutionary relationships with lipoprotein lipase and pancreatic lipase, *J. Biol. Chem.* **263**:1107–1110.
406. Ameis, D., Stahnke, G., Kobayashi, J., McLean, J., Lee, G., Buscher, M., Schotz, M. C., and Will, H., 1990, Isolation and characterization of the human hepatic lipase gene, *J. Biol. Chem.* **265**:6552–6555.
407. Carlson, L. A., Holmquist, L., and Nilsson-Ehle, P., 1986, Deficiency of hepatic lipase activity in post heparin plasma in familial hyperbetatriglyceridemia, *Acta Med. Scand.* **216**:435–447.
408. Connelly, F. W., Maguire, G. F., Lee, M., and Little, J. A., 1990, Plasma lipoprotein in familial hepatic lipase deficiency, *Arteriosclerosis* **10**:40–48.
409. Auwerx, J. H., Babirak, S. P., Hokanson, J. E., Stahnke, G., Will, H., Deeb, S. S., and Brunzell, J. D., 1990, Coexistence of abnormalities of hepatic lipase and lipoprotein lipase in a large family, *Am. J. Hum. Genet.* **46**:470–477.
410. Hegele, R. A., Vezina, C., Moorjani, S., Lupien, P. J., Gagne, C., Brun, L. D., Little, J. A., and Connelly, P. W., 1992, A hepatic lipase gene mutation associated with heritable lipolytic deficiency, *J. Clin. Endocrinol. Metab.* **72**:730–732.
411. Glomset, J. A., 1968, The plasma lecithin:cholesterol acyltransferase reaction, *J. Lipid Res.* **9**:155–167.
412. Fielding, P. E., and Fielding, C. J., 1980, A cholesteryl ester transfer complex in human plasma, *Proc. Natl. Acad. Sci. USA* **77**:3327–3330.
413. Sgoutas, D. S., 1972, Fatty acid specificity of plasma phosphatidylcholine:cholesterol acyltransferase, *Biochemistry* **11**:293–296.
414. Assmann, G., Schmitz, G., Donath, N., and Lekim, D., 1978, Phosphatidylcholine substrate specificity of lecithin:cholesterol acyltransferase, *Scand. J. Clin. Lab. Invest.* **38**(Suppl. 150):16–20.
415. Subbaiah, P. V., Albers, J. J., Chen, C. H., and Bagdade, J. D., 1980, Low density lipoprotein-activated lysolecithin acylation by human plasma lecithin-cholesterol acyltransferase. Identity of lysolecithin acyltransferase and lecithin-cholesterol acyltransferase, *J. Biol. Chem.* **255**:9275–9280.
416. Aron, L., Jones, S., and Fielding, C. J., 1978, Human plasma lecithin-cholesterol acyltransferase. Characterization of cofactor-dependent phospholipase activity, *J. Biol. Chem.* **253**:7220–7226.
417. Sorci-Thomas, M., Babiak, J., and Rudel, L. L., 1990, Lecithin-cholesterol acyltransferase (LCAT) catalyzes transacylation of intact cholesteryl esters. Evidence for the partial reversal of the forward LCAT reaction, *J. Biol. Chem.* **265**:2665–2670.
418. Fielding, C. J., Shore, V. G., and Fielding, P. D., 1972, A protein cofactor of lecithin:cholesterol acyltransferase, *Biochem. Biophys. Res. Commun.* **46**:1943–1949.

419. Soutar, A. K., Garner, C. W., Baker, H. N., Sparrow, J. T., Jackson, R. L., Gotto, A. M., and Smith, L. C., 1975, Effect of the human plasma apolipoproteins and phosphatidylcholine acyl donor on the activity of lecithin:cholesterol acyltransferase, *Biochemistry* **14**:3057–3064.

420. Fukushima, D., Yokoyama, S., Kroon, D. J., Kézdy, F. J., and Kaiser, E. T., 1980, Chain length–function correlation of amphiphilic peptides. Synthesis and surface properties of a tetratetracontapeptide segment of apolipoprotein A-I, *J. Biol. Chem.* **255**:10651–10657.

421. Steinmetz, A., and Utermann, G., 1985, Activation of lecithin:cholesterol acyltransferase by human apolipoprotein A-IV, *J. Biol. Chem.* **260**:2258–2264.

422. Weinberg, R. B., Jordan, M. K., and Steinmetz, A., 1990, Distinctive structure and function of human apolipoprotein variant apoA-IV-2, *J. Biol. Chem.* **265**:18372–18378.

423. Chen, C. H., and Albers, J. J., 1985, Activation of lecithin:cholesterol acyltransferase by apolipoproteins E-2, E-3, and A-IV isolated from human plasma, *Biochim. Biophys. Acta* **836**: 279–285.

424. Jauhiainen, M., Stevenson, K. J., and Dolphin, P. J., 1988, Human plasma lecithin-cholesterol acyltransferase. The vicinal nature of cysteine 31 and cysteine 184 in the catalytic site, *J. Biol. Chem.* **263**:6525–6533.

425. Francone, O. L., and Fielding, C. J., 1991, Effects of site-directed mutagenesis at residues cysteine-31 and cysteine-184 on lecithin-cholesterol acyltransferase activity, *Proc. Natl. Acad. Sci. USA* **88**:1716–1720.

426. Hamilton, R. L., Williams, M. C., Fielding, C. J., and Havel, R. J., 1976, Discoidal bilayer structure of nascent high density lipoproteins from perfused rat liver, *J. Clin. Invest.* **58**: 667–680.

427. Green, P. H. R., Tall, A. R., and Glickman, R. M., 1978, Rat intestine secretes discoid high density lipoproteins, *J. Clin. Invest.* **61**:528–534.

428. Norum, K. R., Gjone, E., and Glomset, J. A., 1989, Familial lecithin:cholesterol acyltransferase deficiency including fish eye disease, in: *The Metabolic Basis of Inherited Disease* (C. R. Scriver, A. L. Beaudet, W. S. Sly, and D. Valle, eds.), McGraw-Hill, New York, pp. 1181–1194.

429. Dory, L., Sloop, C. H., Boquet, L. M., Hamilton, R. L., and Roheim, P. S., 1983, Lecithin:cholesterol acyltransferase-mediated modification of discoidal peripheral lymph high density lipoproteins: Possible mechanism of formation of cholesterol-induced high density lipoproteins (HDLc) in cholesterol-fed dogs, *Proc. Natl. Acad. Sci. USA* **80**:3489–3493.

430. Cheung, M. C., and Wolf, A. C., 1989, *In vitro* transformation of apoA-I-containing lipoprotein subpopulations: Role of lecithin:cholesterol acyltransferase and apoB-containing lipoproteins, *J. Lipid Res.* **30**:499–509.

431. Fielding, C. J., and Fielding, P. E., 1971, Purification and substrate specificity of lecithin-cholesterol acyltransferase from human plasma, *FEBS. Lett.* **15**:355–358.

432. Park, M. S. C., Kudchodkar, B. J., Frohlich, J., Pritchard, H., and Lacko, A. G., 1987, Study of components of reverse cholesterol transport in lecithin:cholesterol acyltransferase deficiency, *Arch. Biochem. Biophys.* **258**:545–554.

433. Fielding, C. J., and Fielding, P. E., 1981, Regulation of human plasma lecithin:cholesterol acyltransferase activity by lipoprotein acceptor cholesteryl ester content, *J. Biol. Chem.* **256**: 2102–2104.

434. Francone, O. L., Gurakar, A., and Fielding, C., 1989, Distribution and functions of lecithin:cholesterol acyltransferase and cholesteryl ester transfer protein in plasma lipoproteins. Evidence for a functional unit containing these activities together with apolipoproteins A-I and D that catalyzes the esterification and transfer of cell-derived cholesterol, *J. Biol. Chem.* **264**: 7066–7072.

435. Miida, T., Fielding, C. J., and Fielding, P. E., 1990, Mechanism of LDL-derived free cholesterol to HDL subfractions in human plasma, *Biochemistry* **29**:10469–10474.

436. Lefevre, M., Goudey-Lefevre, J. C., and Roheim, P. S., 1989, Preferential redistribution of lipoprotein-unassociated apoA-IV to an HDL subpopulation with a high degree of LCAT modification, *Lipids* **24**:1035–1038.

437. Schmitz, G., Bruning, T., Williamson, E., and Nowicka, G., 1990, The role of HDL in reverse cholesterol transport and its disturbances in Tangier disease and HDL deficiency with xanthomas, *Eur. Heart J.* **11**:197–211.

438. Fielding, C. J., 1990, Lecithin:Cholesterol acyltransferase. In: *Advances in Cholesterol Research* (M. Esfahani and J. B. Swaney, eds.), Telford Press, New Jersey, pp. 271–314.

439. Nishida, H. I., Kato, H., and Nishida, T., 1990, Affinity of lipid transfer protein for lipid and lipoprotein particles as influenced by lecithin-cholesterol acyltransferase, *J. Biol. Chem.* **265**:4876–4883.

440. Nichols, A., 1990, Conversions in the origins and metabolism of human plasma HOL. In: *Advances in Cholesterol Research* (M. Esfahani and J. M. Swaney, eds.), Telford Press, New Jersey, pp. 315–365.

441. Simon, J. B., and Boyer, J. L., 1971, Production of lecithin:cholesterol acyltransferase by the isolated perfused rat liver, *Biochim. Biophys. Acta* **218**:549–551.

442. Nordby, G., Berg, T., Nilsson, M., and Norum, K. R., 1976, Secretion of lecithin:cholesterol acyltransferase from isolated rat hepatocytes, *Biochim. Biophys. Acta* **450**:69–77.

443. Warden, C. H., Langner, C. A., Gordon, J. I., Taylor, B. A., McLean, J. W., and Lusis, A. J., 1989, Tissue-specific expression, developmental regulation, and chromosomal mapping of the lecithin:cholesterol acyltransferase gene. Evidence for expression in brain and testes as well as liver, *J. Biol. Chem.* **264**:21573–21581.

444. McLean, J., Fielding, C., Drayna, D., Dieplinger, H., Bear, B., Kohr, W., Henzel, W., and Lawn, R., 1986, Cloning and expression of human lecithin:cholesterol acyltransferase cDNA, *Proc. Natl. Acad. Sci. USA* **83**:2335–2339.

445. McLean, J., Wion, K., Drayna, D., Fielding, C., and Lawn, R., 1986, Human lecithin-cholesterol acyltransferase gene: Complete gene sequence and sites of expression, *Nucleic Acids Res.* **14**:9387–9406.

446. Vergani, C., Catapano, A. L., Roma, P., and Giudici, G., 1983, A new case of familial LCAT deficiency, *Acta Med. Scand.* **214**:173–176.

447. Glomset, J. A., Nichols, A. V., Norum, K. R., King, W., and Forte, T., 1973, Plasma lipoproteins in familial lecithin:cholesterol acyltransferase deficiency: Further studies of very low and low density lipoprotein abnormalities, *J. Clin. Invest.* **52**:1078–1092.

448. Hamilton, R. L., 1978, Hepatic secretion and metabolism of high density lipoproteins, in: *Disturbances in Lipid and Lipoprotein Metabolism* (J. Dietschy, A. M. Gotto, Jr., and J. A. Ontko, eds.), American Physiology Society, Baltimore, Maryland, pp. 155–171.

449. Glomset, J. A., and Norum, K. R., 1973, The metabolic role of lecithin:cholesterol acyltransferase: Perspectives from pathology, *Adv. Lipid Res.* **2**:1–16.

450. Norum, K. R., and Gjone, E., 1968, The influence of plasma from patients with familial plasma lecithin:cholesterol acyltransferase deficiency on the lipid pattern of erythrocytes, *Scand. J. Clin. Lab. Invest.* **22**:94–98.

451. Hovig, T., and Gjone, E., 1973, Familial lecithin:cholesterol acyltransferase deficiency: Ultrastructural aspects of a new syndrome with particular reference to lesions in the kidneys and the spleen, *Acta Pathol. Microbiol. Scand.* **81**:681–697.

452. Alaupovic, P., McConathy, W. J., Curry, M. D., Magnani, H. N., Torsvik, H., Berg, K., and Gjone, E., 1974, Apolipoproteins and lipoprotein families in familial lecithin:cholesterol acyltransferase deficiency, *Scand. J. Clin. Lab. Invest.* **33**(Suppl. 137):83–87.

453. Stokke, K. T., Bjerve, K. S., Blomhoff, J. P., Oystese, B., Flatmark, A., Norum, K. R., and Gjone, E., 1974, Familial lecithin:cholesterol acyltransferase deficiency: Studies on lipid composition and morphology of tissues, *Scand. J. Clin. Lab. Invest.* **33**(Suppl. 137):93–100.

454. Albers, J. J., Adolphson, J. L., and Chen, C. H., 1981, Radioimmunoassay of human plasma lecithin-cholesterol acyltransferase, *J. Clin. Invest.* **67**:141–148.

455. Albers, J. J., and Utermann, G., 1981, Genetic control of lecithin-cholesterol acyltransferase (LCAT): Measurement of LCAT mass in a large kindred with LCAT deficiency, *Am. J. Hum. Genet.* **33**:702–708.

456. Frohlich, J., Hon, K., and McLeod, R., 1982, Detection of heterozygotes for familial lecithin:cholesterol acyltransferase (LCAT) deficiency, *Am. J. Hum. Genet.* **34**:65–72.

457. Frohlich, J., McLeod, R., Pritchard, P. H., Fesmrer, J., and McConathy, W., 1988, Plasma lipoprotein abnormalities in heterozygotes for familiar lecithin:cholesterol acyltransferase deficiency, *Metabolism: Clinical and Experimental* **37**:3–8.

458. Albers, J. J., Chen, C.-H., and Adolphson, J. C., 1981, Familial lecithin:cholesterol acyltransferase: Identification of heterozygotes with half-normal enzyme activity and mass, *Clin. Genet.* **58**:306–309.

459. Taramelli, R., Pontoglio, M., Candini, G., Ottolenghi, S., Dieplinger, H., Catapano, A., Albers, J., Vergani, C., and McLean, J., 1990, Lecithin:cholesterol acyltransferase deficiency: Molecular analysis of a mutated allele, *Hum. Genet.* **85**:195–199.

460. Assmann, G., von Eckardstein, A., and Funke, H., 1991, lecithin:cholesterol acyltransferase deficiency and fish-eye disease, *Curr. Opin. Lipidol.* **2**:110–117.

461. Gotoda, T., Yamada, N., Murase, T., Sakuma, M., Murayama, N., Shimano, H., Kozaki, K., Albers, J. J., Yazaki, Y., and Akanuma, Y., 1991, Differential phenotypic expression by three mutant alleles in familial lecithin:cholesterol acyltransferase deficiency, *Lancet* **338**:778–781.

461a. Klein, H-G., Lohse, P., Duverger, N., Albers, J. J., Rader, D. J., Zech, L. A., Santamarina-Fojo, S., and Brewer, H. B., Jr., 1993, Two different allelic mutations in the lecithin:cholesterol acyltransferase (LCAT) gene resulting in classic LCAT deficiency: LCAT (Tyr[83]→Stop) and LCAT (Tyr[156]→Asn), *J. Lipid Res.* **34**:49–58.

461b. Maeda, E., Naka, Y., Matozaki, T., Sakuma, M., Akanuma, Y., Yoshino, G., and Kasuga, M., 1991, Lecithin:cholesterol acyltransferase (LCAT) deficiency with a missense mutation in exon 6 of the LCAT gene, *Biochem. Biophys. Res. Commun.* **178**:460–466.

461c. Bujo, H., Kusunoki, J., Ogasawara, M., Yamamoto, T., Ohta, Y., Shimada, T., Saito, Y., and Yoshida, S., 1991, Molecular defect in familial lecithin:cholesterol acyltransferase (LCAT) deficiency: A single nucleotide insertion in LCAT gene causes a complete deficient type of the disease, *Biochem. Biophys. Res. Commun.* **181**:933–940.

462. Carlson, L. A., 1989, Fish eye disease. A lesson on plasma cholesterol esterification, in: *High Density Lipoproteins and Atherosclerosis* (N. E. Miller, ed.), Elsevier Science, New York, pp. 95–102.

463. Funke, H., von Eckardstein, A., Pritchard, P. H., Albers, J. J., Kastelein, J. J., Droste, C., and Assmann, G., 1991, A molecular defect causing fish eye disease: An amino acid exchange in lecithin-cholesterol acyltransferase (LCAT) leads to the selective loss of alpha-LCAT activity, *Proc. Natl. Acad. Sci. USA* **88**:4855–4859.

463a. Klein, H.-G., Lohse, P., Pritchard, P. H., Bojanovski, D., Schmidt, H., and Brewer, H. B., Jr., 1992, Two different allelic mutations in the lecithin-cholesterol acyltransferase gene associated with the fish eye syndrome, *J. Clin. Invest.* **89**:499–506.

464. Jarnagin, A. S., Kohr, W., and Fielding, C. J., 1987, Isolation and specificity of a Mr 74,000 cholesteryl ester transfer protein from human plasma, *Proc. Natl. Acad. Sci. USA* **84**:1854–1857.

465. Hesler, C. B., Swenson, T. L., and Tall, A. R., 1987, Purification and characterization of a human plasma cholesteryl ester transfer protein, *J. Biol. Chem.* **262**:2275–2282.

466. Brewster, M. E., Ihm, J., Brainard, J. R., and Harmonay, J. A. K., 1978, Transfer of phosphatidylcholine facilitated by a component of human plasma, *Biochim. Biophys. Acta* **529**:147–159.

467. Albers, J. J., Tollefson, J. H., Chen, C. H., and Steinmetz, A., 1984, Isolation and characterization of human plasma transfer proteins, *Arteriosclerosis* 4:49–58.

468. Nagashima, M., McLean, J. W., and Lawn, R. M., 1988, Cloning and mRNA tissue distribution of rabbit CETP, *J. Lipid Res.* 12:1643–1649.

469. Morton, R. E., 1985, Binding of plasma-derived lipid transfer protein to lipoprotein substrates, *J. Biol. Chem.* 260:12593–12599.

470. Pattnaik, N. M., and Zilversmit, D. B., 1979, Interaction of chol-ester exchange protein with human plasma lipoproteins and phospholipid vesicles, *J. Biol. Chem.* 254:2782–2786.

471. Swenson, T. L., Brocia, R. W., and Tall, A. R., 1988, Plasma cholesteryl ester transfer protein has binding sites for neutral lipids and phospholipids, *J. Biol. Chem.* 263:5150–5157.

472. Hesler, C. B., Tall, A. R., Swenson, T. L., Weech, P. K., Marcel, Y. L., and Milne, R. W., 1988, Monoclonal antibodies to the Mr 74,000 cholesteryl ester transfer protein neutralize all of the cholesteryl ester and triglyceride transfer activities in human plasma, *J. Biol. Chem.* 263:5020–5023.

473. Swenson, T. L., Hesler, C. B., Brown, M. L., Quinet, E., Trotta, P. P., Haslanger, M. F., Gaeta, F. C. A., Marcel, Y. L., Milne, R. W., and Tall, A. R., 1989, Mechanism of cholesteryl ester transfer protein inhibition by a neutralizing monoclonal antibody and mapping of the monoclonal antibody epitope, *J. Biol. Chem.* 264:14318–14320.

474. Brown, M. L., Hesler, C., and Tall, A. R., 1990, Plasma enzymes and transfer proteins in cholesterol metabolism, *Curr. Opin. Lipidol.* 1:122–127.

475. Whitlock, M. E., Swenson, T. L., RamaKrishman, R., Leonard, M. T., Marcel, Y. L., Milne, R. W., and Tall, A. R., 1989, Monoclonal antibody inhibition of cholesteryl ester transfer protein activity in the rabbit. Effects on lipoprotein composition and high density lipoprotein cholesteryl ester metabolism, *J. Clin. Invest.* 84(1):129–137.

476. Bisgaier, C. L., Siebenkas, M. V., Hesler, C. B., Swenson, T. L., Blum, C. B., Marcel, Y. L., Milne, R. W., Glickman, R. M., and Tall, A. R., 1989, Effect of a neutralizing monoclonal antibody to cholesteryl ester transfer protein on the redistribution of apolipoproteins A-IV and E among human lipoproteins, *J. Lipid Res.* 30:1025–1031; Erratum, *J. Lipid Res.* 30(9):1476, 1989.

477. Drayna, D., Jarnagin, A. S., McLean, J., Henzel, W., Kohr, W., Fielding, C., and Lawn, R., 1987, Cloning and sequencing of human cholesteryl ester transfer protein cDNA, *Nature* 327:632–634.

478. Agellon, L. B., Quinet, E. M., Gillette, T. G., Drayna, D. T., Brown, M. L., and Tall, A. R., 1990, Organization of the human cholesteryl ester transfer protein gene, *Biochemistry* 29:1373–1376.

479. Jiang, X. C., Moulin, P., Quinet, E., Goldbert, I. J., Yacoub, L. K., Agellon, L. B., Compton, D., Schnitzer-Polokoff, R., and Tall, A. R., 1991, Mammalian adipose tissue and muscle are major sources of lipid transfer protein mRNA, *J. Biol. Chem.* 266:4631–4639.

480. Quinet, E. M., Agellon, L. B., Kroon, D. A., Marcel, Y. L., Lee, Y., Whitlock, M. F., and Tall, A. R., 1990, Atherogenic diet increases cholesteryl ester transfer protein messenger RNA levels in rabbit liver, *J. Clin. Invest.* 85:357–363.

481. Faust, R. A., and Albers, J. J., 1987, Synthesis and secretion of plasma cholesterol ester transfer protein by human hepatocarcinoma cell line, HepG2, *Arteriosclerosis* 7:267–275.

482. Swenson, T. L., Simmons, J. S., Hesler, C. B., Bisgaier, C., and Tall, A. R., 1987, Cholesteryl ester transfer protein is secreted by HepG2 cells and contains asparagine-linked carbohydrate and sialic acid, *J. Biol. Chem.* 262:16271–16274.

483. Faust, R. A., and Albers, J. J., 1988, Regulated vectorial secretion of cholesteryl ester transfer protein (LTP-I) by the CaCo-2 model of human enterocyte epithelium, *J. Biol. Chem.* 263:8786–8789.

484. Koizumi, J., Mabuchi, H., and Yoshimura, A., 1985, Deficiency of serum cholesteryl-ester transfer activity in patients with familial hyperalphalipoproteinaemia, *Atherosclerosis* 58:175–186.

485. Brown, M. L., Inazu, A., Hesler, C. B., Agellon, L. B., Mann, C., Whitlock, M. E., Marcel, Y. L., Milne, R. W., Koizumi, J., Mabuchi, H., Takeda, R. and Tall, A. R., 1989, Molecular basis of lipid transfer protein deficiency in a family with increased high-density lipoproteins, *Nature* **342**:448–451.

486. Inazu, A., Brown, M. L., Hesler, C. B., Agellon, L. B., Koizumi, J., Takata, K., Maruhama, Y., Mabuchi, H., and Tall, A. R., 1990, Increased high-density lipoprotein levels caused by a common cholesteryl-ester transfer protein gene mutation, *N. Engl. J. Med.* **323**:1234–1238.

487. Yamashita, S., Sprecher, D. L., Sakai, N., Matsuzawa, Y., Tarui, S., and Hui, D. Y., 1990, Accumulation of apolipoprotein E-rich high density lipoproteins in hyperalphalipoproteinemic human subjects with plasma cholesteryl ester transfer protein deficiency, *J. Clin. Invest.* **86:** 688–695.

488. Ikewaki, K., Sakamoto, T., Rader, D., Schaefer, J., Fairwell, T., Kindt, M., Ishikawa, T., Nagano, M., and Brewer, Jr., H. B., 1991, Cholesterol ester transfer protein deficiency: Delayed catabolism of apoA-I leading to elevated plasma HDL levels, *Circulation* **84**(Suppl. II):139 (Abstract).

489. Castelli, W. P., Doyle, J. T., Gordon, T., Hames, C. G., Hjortland, M. C., Hulley, S. B., Kagan, A., and Zukel, W. J., 1977, HDL cholesterol and other lipids in coronary heart disease: The cooperative lipoprotein phenotyping study, *Circulation* **55**:767–772.

490. Heiss, G., and Tyroler, H., 1982, Are apolipoproteins useful for evaluating ischemic heart disease? A brief overview of the literature, in: *Proceedings of the Workshop on Apolipoprotein Quantification* (NIH Publ. No. 83-1266), U.S. Department of Health and Human Services, National Institutes of Health, Bethesda, Maryland, pp. 7–24.

491. Glueck, C. J., Fallat, R. W., Millet, F., and Steiner, M. F., 1975, Familial hyperalphalipoproteinemia, *Arch. Intern. Med.* **135**:1025–1028.

492. Glueck, C. J., Gartside, P., Fallat, R. W., Sielski, J., and Steiner, P. M., 1976, Longevity syndromes: Familial hypobeta- and hyperalphalipoproteinemia, *J. Lab. Clin. Med.* **88**:941–957.

493. Patsch, W., Kuisk, J., Glueck, C., and Schonfeld, G., 1981, Lipoproteins in familial hyperalphalipoproteinemia, *Arteriosclerosis* **1**:156–161.

494. Marotti, K. R., Castle, C. K., Rehberg, E. F., Polites, G., and Melchior, G. W., 1991, The size and plasma concentration of HDL is reduced in CETP transgenic mice, *Circulation* **84**(Suppl. II):253 (Abstract).

495. Hayek, T., Chajek-Shaul, T., Walsh, A., Agellon, L. B., Moulin, P., Tall, A. R., and Breslow, J. L., 1991, Interaction of human apoA-I and CETP genes in transgenic mice results in a profound decrease in HDL cholesterol, *Circulation* **84**(Suppl. II):17 (Abstract).

496. Agellon, L. V., Walsh, A., Hayek, T., Moulin, P., Jiang, X. C., Shelanski, S. A., Breslow, J. L., and Tall, A. R., 1991, Reduced high density lipoprotein cholesterol in human cholesteryl ester transfer protein transgenic mice, *J. Biol. Chem.* **266**:10796–10801.

497. Tollefson, J. H., Ravnik, S., and Albers, J. J., 1988, Isolation and characterization of a phospholipid transfer protein (LTP-II) from human plasma, *J. Lipid Res.* **29**:1593–1602.

498. Morton, R. E., and Zilversmit, D. B., 1981, A plasma inhibitor of triglyceride and cholesteryl ester transfer activities, *J. Biol. Chem.* **256**:11992–11995.

499. Son, Y. S. C., and Zilversmit, D. B., 1984, Purification and characterization of human plasma proteins that inhibit transfer activities, *Biochim. Biophys. Acta* **795**:473–480.

500. Nichide, T., Tollefson, J. H., and Albers, J. J., 1989, Inhibition of lipid transfer by a unique high density lipoprotein subclass containing an inhibitor protein, *J. Lipid Res.* **30**:149–158.

501. Baker, H. N., Gotto, Jr., A. M., and Jackson, R. L., 1975, The primary structure of human plasma high density apolipoprotein glutamine I (apoA-I). II. The amino acid sequence and alignment of cyanogen bromide fragments IV, III, and I, *J. Biol. Chem.* **250**:2725–2738.

502. Brewer, Jr., H. B., Fairwell, T., Larue, A., Ronan, R., Houser, A., and Bronzert, T., 1978, The amino acid sequence of human apoA-I, an apoprotein isolated from high density lipoproteins, *Biochem. Biophys. Res. Commun.* **80**:623–630.

503. Barker, W. C., and Dayhoff, M. O., 1977, Evolution of lipoproteins deduced from protein sequence data, *Comp. Biochem. Physiol. B Comp. Biochem.* **57:**309–315.

504. Fitch, W. M., 1977, Phylogenies constrained by the crossover process as illustrated by human hemoglobins and a thirteen-cycle, eleven-amino-acid repeat in human apolipoprotein A-I, *Genetics* **86:**623–644.

505. McLachlin, A. D., 1977, Repeated helical pattern in apolipoprotein A-I, *Nature* **267:**465–466.

506. Segrest, J. P., Jackson, R. L., Morrisett, J. D., and Gotto, Jr., A. M., 1974, A molecular theory of lipid–protein interactions in the plasma lipoproteins, *FEBS Lett.* **38:**247–253.

507. Fukushima, D., Kupferberg, J. P., Kokoyama, S., Kroon, D. J., Kaiser, E. T., and Kezdy, F. J., 1979, A synthetic amphiphilic helical docosapeptide with the surface properties of plasma apolipoprotein A-I, *J. Am. Chem. Soc.* **101:**3703–3704.

508. Kaiser, E. T., and Kezdy, F. J., 1983, Secondary structures of proteins and peptides in amphiphilic environments, *Proc. Natl. Acad. Sci. USA* **80:**1137–1143.

509. Nakagawa, S. H., Lau, H. S. H., Kezdy, F. J., and Kaiser, E. T., 1985, The use of polymer-bound oximes for the synthesis of large peptides usable in segment condensation: Synthesis of a 44 amino acid amphiphilic peptide model of apolipoprotein A-I, *J. Am. Chem. Soc.* **107:**7087–7092.

510. Anantharamaiah, G. M., Venkatachalapathi, Y. V., Brouillette, C. G., and Segrest, J. P., 1990, Use of synthetic peptide analogues to localize lecithin:cholesterol acyltransferase activating domain in apolipoprotein A, *Arteriosclerosis* **10:**95–105.

511. Breiter, D. R., Kanost, M. R., Benning, M. M., Wesenberg, G., Law, J. H., Wells, M. A., Rayment, I., and Holden, H. M., 1990, Molecular structure of an apolipoprotein determined at 2.5Å resolution, *Biochemistry* **30:**603–608.

512. Wilson, C., Wardell, M. R., Weisgraber, K. H., Mahley, R. W., and Agard, D. A., 1991, Three-dimensional structure of the LDL receptor-binding domain of human apolipoprotein E, *Science* **252:**1817–1822.

513. Marcel, Y. L., Provost, P. R., Kos, H., Rattai, E., Dac, N. V., Fruchart, J. C., and Rassart, E., 1991, The epitopes of apolipoprotein A-I define distinct structural domains including a mobile middle region, *J. Biol. Chem.* **266:**3644–3653.

514. Landschulz, W. H., Johnson, P. F., and McKnight, S. L., 1988, The leucine zipper: A hypothetical structure common to a new class of DNA binding proteins, *Science* **240:**1759–1764.

515. Haddad, I. A., Ordovas, J. M., Fitzpatrick, T., and Karathanasis, S. K., 1986, Linkage, evolution, and expression of the rat apolipoprotein A-I, C-III, and A-IV genes, *J. Biol. Chem.* **261:**13268–13277.

516. Tam, S. P., Archer, T. K., and Deeley, R. G., 1986, Biphasic effects of estrogen on apolipoprotein synthesis in human hepatoma cells: Mechanism of antagonism by testosterone, *Proc. Natl. Acad. Sci. USA* **83:**3111–3115.

517. Masumoto, A., Koga, S., Uchida, E., and Ibayashi, H., 1988, Effects of insulin, dexamethazone and glucagon on the production of apolipoprotein A-I in cultured rat hepatocytes, *Atherosclerosis* **70:**217–223.

518. Lin, R. C., 1988, Effects of hormones on apolipoprotein secreted in cultured rat hepatocytes, *Metabolism: Clinical and Experimental* **37:**745–751.

519. Apostolopoulos, J. J., La Scala, M. J., and Howlett, G. J., 1988, The effect of triiodothyronine on rat apolipoprotein A-I and A-IV gene transcription, *Biochem. Biophys. Res. Commun.* **154:**997–1102.

520. Sorci-Thomas, M., Prack, M. M., Dashti, N., Johnson, F., Rudel, L. L., and Williams, D. L., 1989, Differential effects of dietary fat on the tissue-specific expression of the apolipoprotein A-I gene: Relationship to plasma concentration of high density lipoproteins, *J. Lipid Res.* **30:**1397–1403.

521. Go, M. F., Schonfeld, G., Pfleger, B., Cole, T. G., Sussman, N. L., and Alpers, D. H., 1988,

Regulation of intestinal and hepatic apoprotein synthesis after chronic fat and cholesterol feeding, *J. Clin. Invest.* **81**:1615–1620.

522. Staels, B., Auwerx, J., Chan, L., van Tol, A., Rosseneu, M., and Verhoeven, G., 1989, Influence of development, estrogens, and food intake on apolipoprotein A-I, A-II, and E mRNA in rat liver and intestine, *J. Lipid Res.* **30**:1137–1145.

523. Panduro, A., Lin-Lee, Y. C., Chan, L., and Shafritz, D. A., 1990, Transcriptional and posttranscriptional regulation of apolipoprotein E, A-I, and A-II gene expression in normal rat liver and during several pathophysiologic states, *Biochemistry* **29**:8430–8435.

524. Atmeh, R. F., Shepherd, J., and Packard, C. J., 1983, Subpopulations of apolipoprotein A-I in human high-density lipoproteins. Their metabolic properties and response to drug therapy, *Biochim. Biophys. Acta* **751**:175–188.

525. Skipski, V., 1972, Lipid composition of lipoproteins in normal and diseased states. In: *Blood Lipids and Lipoproteins: Quantitation, Composition, and Metabolism* (G. J. Nelson, ed.), Wiley-Interscience, New York, pp. 471–583.

526. Scanu, A. M., Edelstein, C., and Keim, P., 1975, Serum lipoproteins. In: *The Plasma Proteins*, Vol. 1, 2nd ed. (F. W. Putnam, Ed.), Academic Press, New York, pp. 317–334.

527. Atkinson, D., Smith, H. M., Dickson, J., and Austin, J. P., 1976, Interaction of apoprotein from porcine high-density lipoprotein with dimyristoyl lecithin, *Eur. J. Biochem.* **64**:541–547.

528. Pownall, H. J., Massey, J. B., Kusscrow, S. K., and Gotto, Jr., A. M., 1979, Kinetics of lipid protein interactions: Effect of cholesterol on the association of human plasma high density lipoprotein A-I with L-a-dimyristoylphosphatidylcholine, *Biochemistry* **18**:574–579.

529. Tajima, S., Yokoyama, S.,and Yamamoto, A., 1983, Effect of lipid particle size on association of apolipoproteins with lipid, *J. Biol. Chem.* **258**:10073–10082.

530. Yokoyama, S., Fukushima, D., Kupferberg, J. P., Kёzdy, F. J., and Kaiser, E. T., 1980, The mechanism of activation of lecithin:cholesterol acyltransferase by apolipoprotein A-I and an amphiphilic peptide, *J. Biol. Chem.* **255**:7333–7339.

531. Fielding, C. J., and Fielding, P. E., 1981, Evidence for a lipoprotein carrier in human plasma catalyzing sterol efflux from cultured fibroblasts and its relationship to lecithin:cholesterol acyltransferase, *Proc. Natl. Acad. Sci. USA* **78**:3911–3914.

532. Jonas, A., Wald, J. H., Toohill, K. L., Krul, E. S., and Kёzdy, K. E., 1990, Apolipoprotein A-I structure and lipid properties in homogeneous, reconstituted spherical and discoidal high density lipoproteins, *J. Biol. Chem.* **265**:22123–22129.

533. Segrest, J. P., DeLoof, H., Dohlman, J. G., Brouillette, C. G., and Anantharamaiah, G. M., 1990, Amphipathic helix motif: Classes and properties, *Proteins* **8**:103–117.

534. Sparrow, J. T., and Gotto, Jr., A. M., 1980, Phospholipid binding studies with synthetic apolipoprotein fragments, *Ann. N. Y. Acad. Sci.* **348**:187–211.

535. Utermann, G., Haas, J., Steinmetz, A., Paetzold, R. Rall, Jr., S. C., Weisgraber, K. H., and Mahley, R. W., 1984, Apolipoprotein A-IGiessen (Pro143–Arg). A mutant that is defective in activating lecithin:cholesterol acyltransferase, *Eur. J. Biochem.* **144**:325–331.

536. Jonas, A., von Eckardstein, A., Kёzdy, K. E., Steinmetz, A., and Assmann, G., 1991, Structural and functional properties of reconstituted high density lipoprotein discs prepared with six apolipoprotein A-I variants, *J. Lipid Res.* **32**:97–106.

537. Rall, Jr., S. C., Weisgraber, K. H., Mahley, R. W., Ogawa, Y., Fielding, C. J., Utermann, G., Haas, J., Steinmetz, A., Menzel, H. J., and Assmann, G., 1984, Abnormal lecithin:cholesterol acyltransferase activation by a human apolipoprotein A-I variant in which a single lysine residue is deleted, *J. Biol. Chem.* **259**:10063–10070.

538. Banka, C. L., Smith, R. S., Bonnet, D. J., and Curtiss, L. K., 1990, Localization of an apolipoprotein A-I epitope critical for LCAT activation, *Circulation* **82**:330 (Abstract).

539. Von Eckardstein, A., Funke, H., Walter, M., Altland, K., Benninghoven, A., and Assmann, G., 1990, Structural analysis of human apolipoprotein A-I variants. Amino acid substitutions

are nonrandomly distributed throughout the apolipoprotein A-I primary structure, *J. Biol. Chem.* **265**:8610–8617.

540. Weisgraber, K. H., Bersot, T. P., Mahley, R. W., Franceschini, G., and Sirtori, C. R., 1980, A-I Milano apoprotein. Isolation and characterization of a cysteine-containing variant of the A-I apoprotein from human high density lipoproteins, *J. Clin. Invest.* **66**:901–907.

541. Rall, Jr., S. C., Weisgraber, K. H., Mahley, R. W., Ehnholm, C., Schamaun, O., Olaisen, B., Blomhoff, J. P., and Teisberg, P., 1986, Identification of homozygosity for a human apolipoprotein A-I variant, *J. Lipid Res.* **27**:436–441.

542. Takada, Y., Sasaki, S. Ogata, S., Nakanishi, T., Ikehara, Y., and Arakawa, K., 1990, Isolation and characterization of human apolipoprotein A-I Fukuoka (110 Glu–Lys). A novel apolipoprotein variant, *Biochim. Biophys. Acta* **1043**:169–176.

543. Schonfeld, G., 1990, The genetic dyslipoproteinemias—Nosology update 1990, *Atherosclerosis* **81**:81–93.

544. Assmann, G., Schmitz, G., Funke, H., and von Eckardstein, A., 1990, Apolipoprotein A-I and HDL deficiency, *Curr. Opin. Lipidol.* **1**:110–115.

545. Funke, H., von Eckardstein, A., Pritchard, P. H., Karas, M., Albers, J. J., and Assmann, G., 1991, A frameshift mutation in the human apolipoprotein A-I gene causes high density lipoprotein deficiency, partial lecithin:cholesterol-acyltransferase deficiency, and corneal opacities, *J. Clin. Invest.* **87**:371–376.

546. Turley, S. D., and Dietschy, J. M., 1982, Cholesterol metabolism excretion, in: *The Liver Biology and Pathobiology* (I. Arias, H. Popper, D. Schachter, and D. A. Shafritz, eds.), Raven Press, New York, pp. 468–487.

547. Barbaras, R., Puchois, P., Fruchart, J. C., Pradines-Figueres, A., and Ailhaud, G., 1990, Purification of an apolipoprotein A binding protein from mouse adipose cells, *Biochem. J.* **269**:767–773.

548. Walsh, A., Ito, Y., and Breslow, J. L., 1989, High levels of human apolipoprotein A-I in transgenic mice result in increased plasma levels of small high density lipoprotein (HDL) particles comparable to human HDL3, *J. Biol. Chem.* **264**:6488–6494.

549. Rubin, E. M., Ishida, B. Y., Clift, S. M., and Krauss, R. M., 1991, Expression of human apolipoprotein A-I in transgenic mice results in reduced plasma levels of murine apolipoprotein A-I and the appearance of two new high density lipoprotein size subclasses, *Proc. Natl. Acad. Sci. USA* **88**:434–438.

550. Rubin, E. M., Krauss, R. M., Spangler, E. A., Verstuyft, J. G., and Clift, S. M., 1991, Inhibition of early atherogenesis in transgenic mice by human apolipoprotein AI, *Nature* **353**:265–267.

551. Breslow, J. L., Ross, D., McPherson, J., Williams, H., Kurnit, D., Nussbaum, A. L., Karathanasis, S. K., and Zannis, V. I., 1982, Isolation and characterization of cDNA clones for human apolipoprotein A-I, *Proc. Natl. Acad. Sci. USA* **79**:6861–6865.

552. Karathanasis, S. K., McPherson, J., Zannis, V. I., and Breslow, J. L., 1983, Linkage of human apolipoproteins A-I and C-III genes, *Nature* **304**:371–373.

553. Karathanasis, S. K., 1985, Apolipoprotein multigene family: Tandem organization of human apolipoprotein AI, CIII, and AIV genes, *Proc. Natl. Acad. Sci. USA* **82**:6374–6378.

554. Cheung, P., Kao, F. T., Law, M. L., Jones, C., Puck, T. T., and Chan, L., 1984, Localization of the structural gene for human apolipoprotein A-I on the long arm of human chromosome 11, *Proc. Natl. Acad. Sci. USA* **81**:508–511.

555. Franceschini, G., Sirtori, C. R., Capurso, A., Weisgraber, K. H., and Mahley, R. W., 1980, Decreased high density lipoprotein cholesterol levels with significant lipoprotein modification and without clinical atherosclerosis in an Italian family, *J. Clin. Invest.* **66**:892–900.

556. Franceschini, G., Baio, M., Calabresi, L., Sirtori, C. R., and Cheung, M. C., 1990, Apolipoprotein A-I Milano. Partial lecithin:cholesterol acyltransferase deficiency due to low levels of a functional enzyme, *Biochim. Biophys. Acta* **1043**:1–6.

557. Weisgraber, K. H., Rall, Jr., S. C., Bersot, T. P., Mahley, R. W., Franceschini, G., and Sirtori, C. R., 1983, Apolipoprotein A-I Milano. Detection of normal A-I in affected subjects and evidence for a cysteine for arginine substitution in the variant A-I, *J. Biol. Chem.* **258:**2508–2513.

558. Utermann, G., Feussner, G., Franceschini, G., Haas, J., and Steinmetz, A., 1982, Genetic variants of group A apolipoproteins. Rapid methods for screening and characterization without ultracentrifugation, *J. Biol. Chem.* **257:**501–507.

559. Menzel, H. J., Assmann, G., Rall, Jr., S. C., Weisgraber, K. H., and Mahley, R. W., 1984, Human apolipoprotein A-I polymorphism. Identification of amino acid substitutions in three electrophoretic variants of the Munster-3 type, *J. Biol. Chem.* **259:**3070–3076.

560. Rall, S. C., Weisgraber, K. H., Mahley, R. W., Ehnholm, O., Schamaun, D., Olaisen, B., Blomhoff, J. P., and Teisberg, P., 1986, Identification of homozygosity for a human apolipoprotein A-I variant, *J. Lipid Res.* **27:**436–441.

561. Strobl, W., Jabs, H. U., Hayde, M., Holzinger, T., Assmann, G., and Widhalm, K., 1988, Apolipoprotein A-I (Glu 198–Lys): A mutant of the major apolipoprotein of high-density lipoproteins occurring in a family with dyslipoproteinemia, *Pediatr. Res.* **24:**222–228.

562. Nichols, W. C., Dwulet, F. E., Liepnieks, J., and Benson, M. D., 1988, Variant apolipoprotein AI as a major constituent of a human hereditary amyloid, *Biochem. Biophys. Res. Commun.* **156:**762–768.

563. Nichols, W. C., Gregg, R. E., Brewer, Jr., H. B., and Benson, M. D., 1990, A mutation in apolipoprotein A-I in the Iowa type of familial amyloidotic polyneuropathy, *Genomics* **8:** 318–323.

564. Jeenah, M., Kessling, A., Miller, N., and Humphries, S., 1990, G to A substitution in the promoter region of the apolipoprotein AI gene is associated with elevated serum apolipoprotein AI and high density lipoprotein cholesterol concentrations, *Mol. Biol. Med.* **7:**233–241.

565. Matsunaga, T., Hiasa, Y., Yanagi, H., Maeda, T., Hattori, N., Yamakawa, K., Yamanouchi, Y., Tanaka, I., Obara, T., and Hamaguchi, H., 1991, Apolipoprotein A-I deficiency due to a codon 84 nonsense mutation of the apolipoprotein A-I gene, *Proc. Natl. Acad. Sci. USA* **88:**2793–2797.

566. Breslow, J. L., 1989, Familial disorders of high density lipoprotein metabolism, in: *The Metabolic Basis of Inherited Disease* (C. R. Scriver, A. I. Beaudet, W. S. Sly, and D. Valle, eds.), McGraw-Hill, New York, pp. 1251–1266.

567. Von Eckardstein, A., Funke, H., Henke, A., Altland, K., Benninghoven, A., and Assmann, G., 1989, Apolipoprotein A-I variants. Naturally occurring substitutions of proline residues affect plasma concentration of apolipoprotein A-I, *J. Clin. Invest.* **84:**1722–1730.

568. Schaefer, E. J., Ordovas, J. M., Law, S. W., Ghiselli, G., Kashyap, M. L., Srivastava, L. S., Heaton, W. H., Albers, J. J., Connor, W. E., Lindgren, F. T., Lemeshev, Y., Segrest, J. P., and Brewer, Jr., H. B., 1985, Familial apolipoprotein A-I and C-III deficiency, variant II, *J. Lipid Res.* **26:**1089–1101.

569. Ordovas, J. M., Cassidy, D. K., Civeira, F., Bisgaier, C. L., and Schaefer, E. J., 1989, Familial apolipoprotein A-I, C-III, and A-IV deficiency and premature atherosclerosis due to deletion of a gene complex on chromosome 11, *J. Biol. Chem.* **264:**16339–16342.

570. Karathanasis, S. K., Ferris, E., and Haddad, I. A., 1987, DNA inversion within the apolipoproteins AI/CIII/AIV-encoding gene cluster of certain patients with premature atherosclerosis, *Proc. Natl. Acad. Sci. USA* **84:**7198–7202.

571. Antonarakis, S. E., Oettgen, P., Chakravarti, A., Halloran, S. L., Hudson, R. R., Feisee, L., and Karathanasis, S. K., 1988, DNA polymorphism haplotypes of the human apolipoprotein APOA1-APOC3-APOA4 gene cluster, *Hum. Genet.* **80:**265–273.

572. Fredrickson, D. S., Altrocchi, P. H., Avioli, L. V., Goodman, D. W. S., and Goodman, H. C., 1961, Tangier disease, *Ann. Intern. Med.* **55:**1016–1031.

573. Heinen, R. J., Herbert, P. N., Fredrickson, D. S., Forte, T., and Lindgren, F. T., 1978, Properties of the plasma very low and low density lipoproteins in Tangier disease, *J. Clin. Invest.* **61:** 120–132.

574. Hofman, H. N., and Fredrickson, D. S., 1965, Tangier disease (familial high density lipoprotein deficiency). Clinical and genetic features in two adults, *Am. J. Med.* **39:**582–593.

575. Assmann, G., Schmitz, G., and Brewer, Jr., H. B., 1989, Familial high density lipoprotein deficiency, in: *The Metabolic Basis of Inherited Disease* (C. R. Scriver, A. I. Beaudet, W. S. Sly, and D. Valle, eds.), McGraw-Hill, New York, pp. 1267–1282.

576. Kostner, G., Holasek, A., Schoenborn, W., and Fuhrmann, W., 1972, Immunochemische untersuchung und analytische isoelektrische forussiereines patienten mit Tangier Krankheit, *Clin. Chim. Acta* **38:**155–162.

577. Henderson, L. O., Herbert, P. N., Fredrickson, D. S., Heinen, R. J., and Easterling, J. C., 1978, Abnormal concentration and anomalous distribution of apolipoprotein A-I in Tangier disease, *Metabolism: Clinical and Experimental* **27:**165–174.

578. Glickman, R. M., Green, P. H. R., Lees, R. S., and Tall, A., 1978, Apoprotein A-I synthesis in normal intestinal mucosa and in Tangier disease, *N. Engl. J. Med.* **299:**1424–1427.

579. Schaefer, E. J., Blum, C. B., Levy, R. I., Jenkins, L. L., Alaupovic, P., Foster, D. M., and Brewer, Jr., H. B., 1978, Metabolism of high-density lipoprotein apolipoproteins in Tangier disease, *N. Engl. J. Med.* **299:**905–910.

580. Assmann, G., Capurso, A., Smootz, E., and Wellner, U., 1978, Apoprotein A metabolism in Tangier disease, *Atherosclerosis* **30:**321–332.

581. Zannis, V. I., Lees, A. M., Lees, R. S., and Breslow, J. L., 1982, Abnormal apoA-I isoprotein composition in patients with Tangier disease, *J. Biol. Chem.* **257:**4978–4986.

582. Zannis, V. I., Karathanasis, S. K., Forbes, G. M., and Breslow, J. L., 1986, Intra- and extracellular modifications of apolipoproteins, in: *Methods in Enzymology*, Vol. 128 (J. P. Segrest and J. J. Albers, eds.), Academic Press, Orlando, Florida, pp. 690–712.

583. Law, S. W., and Brewer, Jr., H. B., 1985, Tangier disease: The complete mRNA sequence encoding for preproapoA-I, *J. Biol. Chem.* **260:**12810–12814.

584. Brewer, Jr., H. B., Lux, S. E., Ronan, R., and John, K. M., 1972, Amino acid sequence of human apoLp-Gln-II (apoA-II), an apolipoprotein isolated from the high-density lipoprotein complex, *Proc. Natl. Acad. Sci. USA* **69:**1304–1308.

585. Weisgraber, K. H., and Mahley, R. W., 1978, Apoprotein (E-A-II) complex of human plasma lipoproteins. I. Characterization of this mixed disulfide and its identification in a high density lipoprotein subfraction, *J. Biol. Chem.* **253:**6281–6288.

586. Gordon, J. I., Budelier, K. A., Sims, H. F., Edelstein, C., Scanu, A. M., and Strauss, A. W., 1983, Biosynthesis of human preproapolipoprotein A-II, *J. Biol. Chem.* **258:**14054–14059.

587. Gordon, J. I., Bisgaier, C. L., Sims, H. F., Sachdev, O. P., Glickman, R. M., and Strauss, A. W., 1984, Biosynthesis of human preproapolipoprotein A-IV, *J. Biol. Chem.* **259:**468–474.

588. Albers, J. J., and Aladjem, F., 1971, Precipitation of 125I-labeled lipoproteins with specific polypeptide antisera. Evidence for two populations with differing polypeptide compositions in human high density lipoproteins, *Biochemistry* **10:**3436–3442.

589. Assmann, G., Herbert, P. N., Fredrickson, D. S., and Forte, T., 1977, Isolation and characterization of an abnormal high density lipoprotein in Tangier disease, *J. Clin. Invest.* **60:**242–252.

590. Gustafson, A., McConathy, W. J., Alaupovic, P., Curry, M. D., and Persson, B., 1979, Identification of lipoprotein families in a variant of human plasma apolipoprotein A deficiency, *Scand. J. Clin. Lab. Invest.* **39:**377–387.

591. Bekaert, E. D., Kallel, R., Bouma, M. E., Lontie, J. F., Mebazaa, A., Malmendier, C. L., and Ayrault-Jarrier, M., 1989, Plasma lipoproteins in infantile visceral leishmaniasis: Deficiency of apolipoproteins A-I and A-II, *Clin. Chim. Acta* **184:**181–191.

592. Jackson, R. L., Morrisett, J. D., Pownall, H. J., and Gotto, Jr., A. M., 1973, Human high

density lipoprotein, apolipoprotein glutamine II. The immunochemical and lipid-binding properties of apolipoprotein glutamine II derivatives, *J. Biol. Chem.* **248:**5218–5224.

593. Jackson, R. L., Gotto, Jr., A. M., Lux, S. E., John, K. M., and Fleischer, S., 1973, Human high density lipoprotein: Interaction of the cyanogen bromide fragments from apolipoprotein glutamine II (A-II) with phosphatidylcholine, *J. Biol. Chem.* **248:**8449.

594. Mao, S. J., Jackson, R. L., Gotto, Jr., A. M., and Sparrow, J. T., 1981, Mechanism of lipid–protein interaction in the plasma lipoproteins: Identification of a lipid-binding site in apolipoprotein A-II, *Biochemistry* **20:**1676–1680.

595. Massey, J. B., Gotto, Jr., A. M., and Pownall, H. J., 1980, Dynamics of lipid–protein interactions. Interaction of apolipoprotein A-II from human plasma high density lipoproteins with dimyristoylphosphatidylcholine, *J. Biol. Chem.* **255:**10167–10173.

596. Massey, J. B., Rohde, M. F., Van Winkle, W. B., Gotto, Jr., A. M., and Pownall, H. J., 1981, Physical properties of lipid–protein complexes formed by the interaction of dimyristoylphosphatidylcholine and human high-density apolipoprotein A-II, *Biochemistry* **20:**1569–1574.

597. Lagocki, P. A., and Scanu, A. M., 1980, *In vitro* modulation of the apolipoprotein composition of high density lipoprotein, *J. Biol. Chem.* **255:**3701–3706.

598. Vadiveloo, P. K., and Fidge, N. H., 1990, Studies on the interaction between apolipoprotein A-II-enriched HDL3 and cultured bovine aortic endothelial (BAE) cells, *Biochim. Biophys. Acta* **1045:**135–141.

599. Fong, B. S., Salter, A. M., Jimenez, J., and Angel, A., 1987, The role of apolipoprotein A-I and apolipoprotein A-II in high-density lipoprotein binding to human adipocyte plasma membranes, *Biochim. Biophys. Acta* **920:**105–113.

600. Shinomiya, M., Sasaki, N., Barnhart, R. L., Shirai, K., and Jackson, R. L., 1982, Effect of apolipoproteins on the hepatic lipase-catalyzed hydrolysis of human plasma high density lipoprotein2-triacylglycerols, *Biochim. Biophys. Acta* **713:**292–299.

601. Jahn, C. E., Osborne, Jr., J. C., Schaefer, E. J., and Brewer, Jr., H. B., 1983, Activation of enzymatic activity of hepatic lipase by apolipoprotein A-II, *Eur. J. Biochem.* **131:**25–29.

602. Kubo, M., Matsuzawa, J., Yokoyama, S., Tajima, S., Ishikawa, K., and Tarui, S., 1982, Mechanism of inhibition of hepatic triglyceride lipase from human postheparin plasma by apolipoproteins A-I and A-II, *J. Biochem. (Tokyo)* **1982:**865–870.

603. Chung, J., Abano, D. A., Fless, G. M., and Scanu, A. M., 1979, Isolation, properties, and mechanism of *in vitro* action of lecithin:cholesterol acyltransferase from human plasma, *J. Biol. Chem.* **254:**7456–7464.

604. Carson, S. D., 1987, Tissue factor (coagulation factor III) inhibition by apolipoprotein A-II, *J. Biol. Chem.* **262:**718–721.

605. Knott, T. J., Priestley, L. M., Urdea, M., and Scott, J., 1984, Isolation and characterisation of a cDNA encoding the precursor for human apolipoprotein AII, *Biochem. Biophys. Res. Commun.* **120:**734–740.

606. Lackner, K. J., Law, S. W., and Brewer, Jr., H. B., 1984, Human apolipoprotein A-II: Complete nucleic acid sequence of preproapoA-II, *FEBS Lett.* **175:**159–164.

607. Sharpe, C. R., Sidoli, A., Shelley, C. S., Lucero, M. A., Shoulders, C. C., and Baralle, F. E., 1984, Human apolipoproteins AI, AII, CII and CIII. cDNA sequences and mRNA abundance, *Nucleic Acids Res.* **12:**3917–3932.

608. Tsao, Y. K., Wei, C. F., Robberson, D. L., Gotto, Jr., A. M., and Chan, L., 1985, The structure of the human apolipoprotein C-II Gene: Electron microscopic analysis of RNA:DNA hybrids, complete nucleotide sequence, and identification of 5' homologous sequences among apolipoprotein genes, *J. Biol. Chem.* **260:**15222–15231.

609. Shelley, C. S., Sharpe, C. R., Baralle, F. E., and Shoulders, C. C., 1985, Comparison of the human apolipoprotein genes. ApoA-II presents a unique functional intron–exon junction, *J. Mol. Biol.* **186:**43–51.

610. Moore, M. N., Kao, F. T., Tsao, Y. K., and Chan, L., 1984, Human apolipoprotein A-II: Nucleotide sequence of a cloned cDNA, and localization of its structural gene on human chromosome 1, *Biochem. Biophys. Res. Commun.* **123:**1–7.

611. Deeb, S. S., Takata, K., Peng, R. L., Kajiyama, G., and Albers, J. J., 1990, A splice-junction mutation responsible for familial apolipoprotein A-II deficiency, *Am. J. Hum. Genet.* **46:** 822–827.

612. Yonezu, T., Toda, M., Yamagishi, H., Higuchi, K., and Takeda, T., 1989, Structural organization of the gene encoding apolipoprotein A-II in an amyloidotic strain of senescence-accelerated mouse, *Gene* **84:**187–191.

613. Lusis, A. J., 1988, Genetic factors affecting blood lipoproteins: The candidate gene approach, *J. Lipid Res.* **29:**397–429.

614. Doolittle, M. H., LeBoeuf, R. C., Warden, C. H., Bee, L. M., and Lusis, A. J., 1990, A polymorphism affecting apolipoprotein A-II translational efficiency determines high density lipoprotein size and composition, *J. Biol. Chem.* **265:**16380–16388.

615. Schultz, J. R., Gong, E. L., McCall, M. R., Nichols, A. V., Clift, S. M., and Rubin, E. M., 1992, Expression of human apolipoprotein A-II and its effect on high density lipoproteins in transgenic mice, *J. Biol. Chem.* **267:**21630–21636.

616. Swaney, J. B., Reese, H., and Eder, H. A., 1974, Polypeptide composition of rat high density lipoprotein: Characterization by SDS–gel electrophoresis, *Biochem. Biophys. Res. Commun.* **59:**513–519.

617. Weisgraber, K. H., Bersot, T. P., and Mahley, R. W., 1978, Isolation and characterization of an apoprotein from the $d<1.006$ lipoproteins of human and canine lymph homologous with the rat A-IV apoprotein, *Biochem. Biophys. Res. Commun.* **85:**287–292.

618. Beisiegel, U., and Utermann, G., 1979, An apolipoprotein homolog of rat apolipoprotein A-IV in human plasma. Isolation and partial characterization, *Eur. J. Biochem.* **93:**601–608.

619. Utermann, G., and Beisiegel, U., 1979, Apolipoprotein A-IV: A protein occurring in human mesenteric lymph chylomicrons and free in plasma. Isolation and quantification, *Eur. J. Biochem.* **99:**333–343.

620. Green, P. H., Glickman, R. M., Riley, J. W., and Quinet, E., 1980, Human apolipoprotein A-IV. Intestinal origin and distribution in plasma, *J. Clin. Invest.* **65:**911–919.

621. Bisgaier, C. L., Sachdev, O. P., Megna, L., and Glickman, R. M., 1985, Distribution of apolipoprotein A-IV in human plasma, *J. Lipid Res.* **26:**11–25.

622. Elshourbagy, N. A., Walker, D. W., Boguski, M. S., Gordon, J. I., and Taylor, J. M., 1986, The nucleotide and derived amino acid sequence of human apolipoprotein A-IV mRNA and the close linkage of its gene to the genes of apolipoproteins A-I and CIII, *J. Biol. Chem.* **261:**1998–2002.

623. Karathanasis, S. K., Yunis, I., and Zannis, V. I., 1986, Structure, evolution and tissue-specific synthesis of human apolipoprotein A-IV, *Biochemistry* **25:**3962–3970.

624. Weinberg, R. B., and Spector, M. S., 1985, Human apolipoprotein A-IV. Displacement from the surface of triglyceride-rich particles by HDL2-associated C-apoproteins, *J. Lipid Res.* **26:** 26–37.

625. Weinberg, R. B., and Spector, M. S., 1985, Structural properties and lipid binding of human apolipoprotein A-IV, *J. Biol. Chem.* **260:**4914–4921.

626. Weinberg, R. B., 1987, Differences in the hydrophobic properties of discrete alpha-helical domains of rat and human apolipoprotein A-IV, *Biochim. Biophys. Acta* **918:**299–303.

627. Weinberg, R. B., and Jordan, M. K., 1990, Effects of phospholipid on the structure of human apolipoprotein A-IV, *J. Biol. Chem.* **265:**8081–8086.

628. Staels, B., van Tol, A., Verhoeven, G., and Auwerx, J., 1990, Apolipoprotein A-IV messenger ribonucleic acid abundance is regulated in a tissue-specific manner, *Endocrinology* **126:**2153–2163.

629. Weinberg, R. B., Dantzker, C., and Patton, C. S., 1990, Sensitivity of serum apolipoprotein A-IV levels to changes in dietary fat content, *Gastroenterology* **98:**17–24.

630. Lenich, C., Brecher, P., Chobanian, A., and Zannis, V. I., 1987, Distribution of apolipoprotein mRNA in rabbit tissues and gene expression in response to fat feeding in normal and diabetic rabbits, *Fed. Proc.* **46:**2095.

631. Apfelbaum, T. F., Davidson, N. O., and Glickman, R. M., 1987, Apolipoprotein A-IV synthesis in rat intestine: Regulation by dietary triglyceride, *Am. J. Physiol.* **252:**G662–G666.

632. Goldberg, I. J., Scheraldi, C. A., Yacoub, L. K., Saxena, U., and Bisgaier, C. L., 1990, Lipoprotein ApoC-II activation of lipoprotein lipase. Modulation by apolipoprotein A-IV, *J. Biol. Chem.* **265:**4266–4272.

633. Stein, O., Stein, Y., Lefevre, M., and Roheim, P. S., 1986, The role of apolipoprotein A-IV in reverse cholesterol transport studied with cultured cells and liposomes derived from an ether analog of phosphatidylcholine, *Biochim. Biophys. Acta* **878:**7–13.

634. Savion, N., and Gamliel, A., 1988, Binding of apolipoprotein A-I and apolipoprotein A-IV to cultured bovine aortic endothelial cells, *Arteriosclerosis* **8:**178–186.

635. Weinberg, R. B., and Patton, C. S., 1990, Binding of human apolipoprotein A-IV to human hepatocellular plasma membranes, *Biochim. Biophys. Acta* **1044:**255–261.

636. Karathanasis, S. K., Oettgen, P., Haddad, I. A., and Antonarakis, S. E., 1986).Structure, evolution and polymorphisms of the human apolipoprotein A4 gene (apoA4), *Proc. Natl. Acad. Sci. USA* **83:**8456–8461.

637. Menzel, H. J., Kladetzky, R. G., and Assmann, G., 1982, One-step screening method for the polymorphism of apolipoproteins A-I, A-II, and A-IV, *J. Lipid Res.* **23:**915–922.

638. Menzel, H. J., Kövary, P. M., and Assmann, G., 1982, Apolipoprotein A-IV polymorphism in man, *Hum. Genet.* **62:**349–352.

639. Kamboh, M. I., and Ferrell, R. E., 1987, Genetic studies of human apolipoproteins. I. Polymorphism of apolipoprotein A-IV, *Am. J. Hum. Genet.* **41:**119–127.

640. Bisgaier, C. L., Lee, E. S., and Glickman, R. M., 1987, A method to screen apolipoprotein polymorphism in whole plasma: Description of apolipoprotein A-IV variants in dyslipidemias and a reassessment of apolipoprotein A-I in Tangier disease, *Biochim. Biophys. Acta* **918:** 242–249.

641. Menzel, H. J., Boerwinkle, E., Schrangl-Will, S., and Utermann, G., 1988, Human apolipoprotein A-IV polymorphism: Frequency and effect on lipid and lipoprotein levels, *Hum. Genet.* **79:** 368–372.

642. De Knijff, P., Rosseneu, M., Beisiegel, U., De Keersgieter, W., Frants, R. R., and Havekes, L. M., 1988, Apolipoprotein A-IV polymorphism and its effect on plasma lipid and apolipoprotein concentrations, *J. Lipid Res.* **29:**1621–1627.

643. Lukka, M., Metso, J., and Ehnholm, C., 1988, Apolipoprotein A-IV polymorphism in the Finnish population: Gene frequencies and description of a rare allele, *Hum. Hered.* **38:** 359–362.

644. Lohse, P., and Brewer, Jr., H. B., 1991, Genetic polymorphism of apolipoprotein A-IV, *Curr. Opin. Lipidol.* **2:**90–95.

645. Asakawa, J., Takahashi, N., Rosenblum, B. B., and Neel, J. V., 1985, Two dimensional gel studies of genetic variation in the plasma proteins of Amerindians and Japanese, *Hum. Genet.* **70:**222–230.

646. Schamaun, O., Olaisen, B., Teisberg, P., Gedde-Dahl, Jr., T., and Ehnholm, C., 1985, Genetic studies of apolipoprotein A-IV by two-dimensional electrophoresis, in: *Protides of the Biological Fluids* (H. Peeters, ed.), Pergamon Press, New York, pp. 471–474.

647. Menzel, H. J., Sigurdsson, G., Boerwinkle, E., Schrangl-Will, S., Dieplinger, H., and Utermann, G., 1990, Frequency and effect of human apolipoprotein A-IV polymorphism on lipid and lipoprotein levels in an Icelandic population, *Hum. Genet.* **84:**344–346.

648. Menzel, H. J., Dwyer, K., Sandholzer, C., Schrangl-Will, S., Dieplinger, H., Hoye, E., Lackner, C., and Utermann, G., 1990, New electrophoretic methods for the detection of apolipoprotein variants, in: *Biotechnology of Dyslipoproteinemias: Applications in Diagnosis and Control* (C. Lenfant, A. Albertini, R. Paoletti, and A. L. Catapano, eds.), Raven Press, New York, pp. 171–178.

649. Akiyama, K., 1989, Population study of apolipoprotein A-IV polymorphism and report of a new variant in Japanese, *Hum. Hered.* **39**:302–304.

650. Lohse, P., Kindt, M. R., Rader, D. J., and Brewer, Jr., H. B., 1990, Genetic polymorphism of human plasma apolipoprotein A-IV is due to nucleotide substitutions in the apolipoprotein A-IV gene, *J. Biol. Chem.* **265**:10061–10064.

651. Lohse, P., Kindt, M. R., Rader, D. J., and Brewer, Jr., H. B., 1990, Human plasma apolipoproteins A-IV-0 and A-IV-3. Molecular basis for two rare variants of apolipoprotein A-IV-1, *J. Biol. Chem.* **265**:12734–12739.

652. Eichner, J. E., Kuller, L. H., Ferrell, R. E., and Kamboh, M. I., 1989, Phenotypic effects of apolipoprotein structural variation on lipid profiles: II. Apolipoprotein A-IV and quantitative lipid measures in the healthy women study, *Genet. Epidemiol.* **6**:493–499.

653. Weinberg, R. B., Jordan, M. K., and Steinmetz, A., 1990, Distinctive structure and function of human apolipoprotein variant ApoA-IV-2, *J. Biol. Chem.* **265**:18372–18378.

654. Hardman, D. A., and Kane, J. P., 1986, Isolation and characterization of apolipoprotein B-48, in: *Methods in Enzymology*, Vol. 128 (J. P. Segrest and J. J. Albers, eds.), Academic Press, Orlando, Florida, pp. 262–272.

655. Cladaras, C., Hadzopoulou-Cladaras, M., Nolte, R. T., Atkinson, D., and Zannis, V. I., 1986, The complete sequence and structural analysis of human apolipoprotein B-100: Relationship between apoB-100 and apoB-48 forms, *EMBO J.* **5**:3495–3507.

656. Law, S. W., Grant, S. M., Higuchi, K., Hospattankar, A., Lackner, K., Lee, N., and Brewer, Jr., H. B., 1986, Human liver apolipoprotein B-100 cDNA: Complete nucleic acid and derived amino acid sequence, *Proc. Natl. Acad. Sci. USA* **83**:8142–8146.

657. Yang, C. Y., Chen, S. H., Gianturco, S. H., Bradley, W. A., Sparrow, J. T., Tanimura, M., Li, W. H., Sparrow, D. A., DeLoof, H., and Rosseneu, M., 1986, Sequence, structure, receptor-binding domains and internal repeats of human apolipoprotein B-100, *Nature* **323**:738–742.

658. Knott, T. J., Pease, R. J., Powell, L. M., Wallis, S. C., Rall, Jr., S. C., Innerarity, T. L., Blackhart, B., Taylor, W. H., Marcel, Y., Milne, R., Johnson, D., Fuller, M., Lusis, A. J., McCarthy, B. J., Mahley, R. W., Levy-Wilson, B., and Scott, J., 1986, Complete protein sequence and identification of structural domains of human apolipoprotein B, *Nature* **323**:734–839.

659. Blackhart, B. D., Ludwig, E. M., Pierotti, V. R., Caiati, L., Onasch, M. A., Wallis, S. C., Powell, L., Pease, R., Knott, T. J., Chu, M. L., Mahley, R. W., Scott, J., McCarthy, B. J., and Levy-Wilson, B., 1986, Structure of the human apolipoprotein B gene, *J. Biol. Chem.* **261**:15364–15367.

660. Ludwig, E. H., Blackhart, B. D., Pierotti, V. R., Caiati, L., Fortier, C., Knott, T., Scott, J., Mahley, R. W., Levy-Wilson, B., McCarthy, B. J., 1987, DNA sequence of the human apolipoprotein B gene, *DNA* **6**:363–372.

661. Cardin, A. D., Witt, K. R., Barhart, C. L., and Jackson, R. L., 1982, Sulfhydryl chemistry and solubility properties of human plasma apolipoprotein B, *Biochemistry* **21**:4503–4511.

662. Yang, C. Y., Kim, T. W., Weng, S. A., Lee, B., Yang, M., and Gotto, Jr., A. M., 1990, Isolation and characterization of sulfhydryl and disulfide peptides of human apolipoprotein B-100, *Proc. Natl. Acad. Sci. USA* **87**:5523–5527.

663. Fisher, W. R., and Gurin, S., 1964, Structure of lipoproteins: Covalently bound fatty acids, *Science* **143**:362–363.

664. Krishnaiah, K. V., Walker, L. F., Borensztajn, J., Schonfeld, G., and Getz, G. S., 1980, Apolipoprotein B variant derived from rat intestine, *Proc. Natl. Acad. Sci. USA* **77**:3806–3810.

665. Teng, B., Verp, M., Salomon, J., and Davidson, N. O., 1990, Apolipoprotein B messenger RNA editing is developmentally regulated and widely expressed in human tissues, *J. Biol. Chem.* **265**:20616–20620.

666. Demmer, L. A., Levin, M. S., Elovson, J., Reuben, M. A., Lusis, A. J., and Gordon, J. I., 1986, Tissue-specific expression and developmental regulation of the rat apolipoprotein B gene, *Proc. Natl. Acad. Sci. USA* **83**:8102–8106.

667. Dashti, N., Williams, D. L., and Alaupovic, P., 1989, Effects of oleate and insulin on the production rates and cellular mRNA concentrations of apolipoproteins in HepG2 cells, *J. Lipid Res.* **30**:1365–1373.

668. Pullinger, C. R., North, J. D., Teng, B. B., Rifici, V. A., Ronhild de Brito, A. E., and Scott, J., 1989, The apolipoprotein B gene is constitutively expressed in HepG2 cells: Regulation of secretion by oleic acid, albumin, and insulin, and measurement of the mRNA half-life, *J. Lipid Res.* **30**:1065–1077.

669. Gibbons, G. F., and Pullinger, C. R., 1987, Regulation of hepatic very low density lipoprotein secretion in rats fed on a diet high in unsaturated fats, *Biochem. J.* **243**:487–492.

670. Wong, S. H., Fisher, E. A., and Marsh, J. B., 1989, Effects of eicosapentaenoic and docosahexaenoic acids on apoprotein B mRNA and secretion of very low density lipoprotein in HepG2 cells, *Arteriosclerosis* **9**:836–841.

671. Kwong, T. C., Sparks, J. D., Pryce, D. J., Cianci, J. F., and Sparks, C. E., 1989, Inhibition of apolipoprotein B net synthesis and secretion from cultured rat hepatocytes by the calcium-channel blocker diltiazem, *Biochem. J.* **263**:411–415.

672. Sparks, J. D., Sparks, C. E., and Miller, L. L., 1989, Insulin effects on apolipoprotein B production by normal, diabetic and treated-diabetic rat liver and cultured rat hepatocytes, *Biochem. J.* **261**:83–88.

673. Kosykh, V. A., Surguchov, A. P., Podres, E. A., Novikov, D. K., Sudarickov, A. B., Berestetskaya, Yu. V., Repin, V. S., and Smirnov, V. N., 1988, VLDL apoprotein secretion and apo-B mRNA level in primary culture of cholesterol-loaded rabbit hepatocytes, *FEBS Lett.* **232**:103–106.

674. Hughes, T. E., Ordovas, J. M., and Schaefer, E. J., 1988, Regulation of intestinal apolipoprotein B synthesis and secretion by Caco-2 cells. Lack of fatty acid effects and control by intracellular calcium ion, *J. Biol. Chem.* **263**:3425–3431.

675. Powell, L. M., Wallis, S. C., Pease, R. J., Edwards, Y. H., Knott, T. J., and Scott, J., 1987, A novel form of tissue-specific RNA processing produces apolipoprotein-B48 in intestine, *Cell* **50**:831–840.

676. Chen, S. H., Habib, G., Yang, C. Y., Gu, Z. W., Lee, B. R., Weng, S. A., Silberman, S. R., Cai, S. J., Deslypere, J. P., Rosseneu, M., Gotto, Jr., A. M., Li, W. H., and Chan, L., 1987, Apolipoprotein B-48 is the product of a messenger RNA with an organ-specific in-frame stop codon, *Science* **238**:363–366.

677. Driscoll, D. M., Wynne, J. K., Wallis, S. C., and Scott, J., 1989, An *in vitro* system for the editing of apolipoprotein B mRNA, *Cell* **58**:519–525.

678. Garcia, Z. C., Poksay, K. S., Bostrom, K., Johnson, D. F., Balestra, M. E., and Schechter, I., 1992, Characterization of apolipoprotein B mRNA editing from rabbit intestine, *Arteriosclerosis Thrombosis* **12**:172–179.

679. Greeve, J., Navaratnam, N., and Scott, J., 1991, Characterization of the apolipoprotein B mRNA editing enzyme: No similarity to the proposed mechanism of RNA editing in kinetoplastid protozoa, *Nucleic Acids Res.* **19**:3569–3576.

680. Shah, R., Knott, T., LeGros, J., Navaratnam, N., Greeve, J., and Scott, J., 1991, A 9-nucleotide motif is essential for the editing of apolipoprotein (apo) B mRNA, *Circulation* **84**(Supp. II):225 (Abstract).

681. Wu, J. H., Semenkovich, C. F., Chen, S. H., Li, W. H., and Chan, L., 1990, Apolipoprotein B

mRNA editing. Validation of a sensitive assay and developmental biology of RNA editing in the rat, *J. Biol. Chem.* **265**:12312–12316.

682. Glickman, R. M., Rogers, M., and Glickman, J. N., 1986, Apolipoprotein B synthesis by human liver and intestine *in vitro*, *Proc. Natl. Acad. Sci. USA* **83**:5296–5302.

683. Davidson, N. O., Carlos, R. C., and Lukaszewicz, A. M., 1990, Apolipoprotein B mRNA editing is modulated by thyroid hormone analogs but not growth hormone administration in the rat, *Mol. Endocrinol.* **4**:779–785.

684. Davidson, N. O., Powell, L. M., Wallis, S. C., and Scott, J., 1988, Thyroid hormone modulates the introduction of a stop codon in rat liver apolipoprotein B messenger RNA, *J. Biol. Chem.* **263**:13482–13485.

685. Sjoberg, A., Oscarsson, J., Bostrom, K., Innerarity, T. L., Eden, S., and Olofsson, S. O., 1991, Growth hormone participates in the regulation of the editing of apolipoprotein B mRNA in the rat liver, *Circulation* **84**(Suppl. II):226 (Abstract).

686. Baum, C. L., Teng, B. B., and Davidson, N. O., 1990, Apolipoprotein B messenger RNA editing in the rat liver. Modulation by fasting and refeeding a high carbohydrate diet, *J. Biol. Chem.* **265**:19263–19270.

687. Davis, R. A., and Boogaerts, J. R., 1982, Intrahepatic assembly of very low density lipoproteins. Effect of fatty acids on triacylglycerol and apolipoprotein synthesis, *J. Biol. Chem.* **257**: 10908–10913.

688. Suita-Mangano, P., Janero, D. R., and Lane, M. D., 1982, Association and assembly of triglyceride and phospholipid with glycosylated and unglycosylated apoproteins of very low density lipoprotein in the intact liver cell, *J. Biol. Chem.* **257**:11463–11467.

689. Kane, J. P., and Havel, R. J., 1989, Disorders of the biogenesis and secretion of lipoproteins containing the B apolipoproteins, in: *The Metabolic Basis of Inherited Disease* (C. R. Scriver, A. I. Beaudet, W. S. Sly, and Valle, D., eds.), McGraw-Hill, New York, pp. 1139–1164.

690. Parthasarathy, S., and Barnett, J., 1990, Phospholipase A2 activity of low density lipoprotein: Evidence for an intrinsic phospholipase A2 activity of apoprotein B-100, *Proc. Natl. Acad. Sci. USA* **87**:9741–9745.

691. Marcel, Y. L., Innerarity, T. L., Spilman, C., Mahley, R. W., Protter, A. A., and Milne, R. W., 1987, Mapping of human apolipoprotein B antigenic determinants, *Arteriosclerosis* **7**:166–175.

692. Milne, R. W., Thëolis, Jr., R., Maurice, R., Pease, R. J., Weech, P. K., Rassart, E., Fruchart, J. C., Scott, J., and Marcel, Y. L., 1989, The use of monoclonal antibodies to localize the low density lipoprotein receptor-binding domain of apolipoprotein B, *J. Biol. Chem.* **264**:19754–19760.

693. Pease, R. J., Milne, R. W., Jessup, W. K., Law, A., Provost, P., Fruchart, J. C., Dean, R. T., Marcel, Y. L., and Scott, J., 1990, Use of bacterial expression cloning to localize the epitopes for a series of monoclonal antibodies against apolipoprotein B100, *J. Biol. Chem.* **265**:553–568.

694. Rall, S. C., Weisgraber, K. H., and Mahley, R. W., 1981, Human apolipoprotein E: The complete amino acid sequence, *J. Biol. Chem.* **257**:4171–4178.

695. Soria, L. F., Ludwig, E. H., Clarke, H. R., Vega, G. L., Grundy, S. M., and McCarthy, B. J., 1989, Association between a specific apolipoprotein B mutation and familial defective apolipoprotein B-100, *Proc. Natl. Acad. Sci. USA* **86**:587–591.

696. Hirose, N., Blankenship, D. T., Krivanek, M. A., Jackson, R. L., and Cardin, A. D., 1987, Isolation and characterization of four heparin-binding cyanogen bromide peptides of human plasma apolipoprotein B, *Biochemistry* **26**:5505–5512.

697. Weisgraber, K. H., and Rall, Jr., S. C., 1987, Human apolipoprotein B-100 heparin-binding sites, *J. Biol. Chem.* **262**:11097–11103.

698. Innerarity, T. L., Friedlander, E. J., Rall, Jr., S. C., Weisgraber, K. H., and Mahley, R. W., 1983, The receptor-binding domain of human apolipoprotein E. Binding of apolipoprotein E fragments, *J. Biol. Chem.* **258**:12341–12347.

699. Weisgraber, K. H., Innerarity, T. L., Harder, K. J., Mahley, R. W., Milne, R. W., Marcel, Y. L., and Sparrow, J. T., 1983, The receptor-binding domain of human apolipoprotein E, *J. Biol. Chem.* **258**:12348–12354.

700. Krul, E. S., Wagner, R. D., Parhofer, K. G., Barrett, H., and Schonfeld, G., 1991, Apolipoprotein B-75: A truncated form of apolipoprotein B with increased affinity for the LDL receptor, *Circulation* **84**:II-340.

701. Young, S. G., Peralta, F. P., Dubois, B. W., Curtiss, L. K., Boyles, J. K., and Witztum, J. L., 1987, Lipoprotein B37, a naturally occurring lipoprotein containing the amino-terminal portion of apolipoprotein B100, does not bind to the apolipoprotein B,E (low density lipoprotein) receptor, *J. Biol. Chem.* **262**:16604–16611.

702. Dunning, A. M., Houlston, R., Frostegard, J., Revill, J., Nilsson, J., Hamsten, A., Talmud, P., and Humphries, S., 1991, Genetic evidence that the putative receptor binding domain of apolipoprotein B (residues 3130 to 3630) is not the only region of the protein involved in interaction with the low density lipoprotein receptor, *Biochim. Biophys. Acta* **1096**:231–237.

703. Walsh, M. T., and Atkinson, D., 1990, Calorimetric and spectroscopic investigation of the unfolding of human apolipoprotein B, *J. Lipid Res.* **31**:1051–1062.

704. Schiffer, M., and Edmundson, H. R., 1967, Use of helical wheels to represent the structures of proteins and to identify segments with helical potential, *Biophys. J.* **7**:121–135.

705. Cohn, J. S., Wagner, D. A., Cohn, S. D., Millar, J. S., and Schaefer, E. J., 1990, Measurement of very low density and low density lipoprotein apolipoprotein (Apo) B-100 and high density lipoprotein Apo A-I production in human subjects using deuterated leucine. Effect of fasting and feeding, *J. Clin. Invest.* **85**:804–811.

706. Chatterton, J. E., Phillips, M. L., Curtiss, L. K., Milne, R. W., Marcel, Y. L., and Schumaker V. N., 1991, Mapping of apolipoprotein B on the low density lipoprotein surface by immunoelectron microscopy, *J. Biol. Chem.* **266**:5955–5962.

707. Law, S. W., Lackner, K. J., Hospattankar, A. V., Anchors, J. M., Sakaguchi, A. Y., Naylor, S. L., and Brewer, Jr., H. B., 1985, Human apolipoprotein B-100: Cloning, analysis of liver mRNA, and assignment of the gene to chromosome 2, *Proc. Natl. Acad. Sci. USA* **82**:8340–8344.

708. Deeb, S. S., Disteche, C., Motulsky, A. G., Lebo, R. B., and Kan, Y. W., 1986, Chromosome localization of the human apolipoprotein B gene and detection of homologous RNA in monkey intestine, *Proc. Natl. Acad. Sci. USA* **83**:419–422.

709. Hegele, R. A., Huang, L. S., Herbert, P. N., Blum, C. B., Buring, J. E., Hennekens, C. H., and Breslow, J. L., 1986, Apolipoprotein B gene DNA polymorphisms associated with myocardial infarction, *N. Engl. J. Med.* **315**:1509–1515.

710. Xu, C., Nanjee, N., Tikkanen, M. J., Huttunen, J. K., Pietinen, P., Butler, R., Angelico, F., Del Ben, M., Mazzarella, B., Antonio, R., Miller, N. G., Humphries, S., and Talmud, P. J., 1989, Apolipoprotein B amino acid 3611 substitution from arginine to glutamine creates the Ag(h/i) epitope: The polymorphism is not associated with differences in serum cholesterol and apolipoprotein B levels, *Hum. Genet.* **82**:322–326.

711. Huang, L. S., de Graaf, J., and Breslow, J. L., 1988, ApoB gene *Msp*I RFLP in exon 26 changes amino acid 3611 from Arg to Gln, *J. Lipid Res.* **29**:63–67.

712. McCarthy, B. J., 1991, Polymorphisms and markers associated with apolipoprotein B, *Curr. Opin. Lipidol.* **2**:81–85.

713. Ladias, J. A., Kwiterovich, Jr., P. O., Smith, H. H., Miller, M., Bachorik, P. S., Forte, T., Lusis, A. J., and Antonarakis, S. E., 1989, Apolipoprotein B-100 Hopkins (arginine4019–tryptophan, A new apolipoprotein B-100 variant in a family with premature atherosclerosis and hyperapobetalipoproteinemia, *J. Am. Med. Assoc.* **262**:1980–1988.

713a. Boerwinkle, E., Chen, S-H., Visvikis, S., Hanis, C. L., Siest, G., and Chan, L., 1991, Signal peptide-length variation in human apolipoprotein B gene. Molecular characteristics and association with plasma glucose levels, *Diabetes* **40**:1539–1544.

713b. Huang, L. S., and Breslow, J. L., 1987, A unique AT-rich hypervariable minisatellite 3' to the apoB gene defines a high information restriction fragment length polymorphism, *J. Biol. Chem.* **262**:8952–8955.

713c. Boerwinkle, E., Xiong, W., Fourest, E., and Chan, L., 1989, Rapid typing of tandemly repeated hypervariable loci by the polymerase chain reaction: Application to the apolipoprotein B 3' hypervariable region, *Proc. Natl. Acad. Sci. USA* **86**:212–216.

714. Ma, Y., Schumaker, V. N., Butler, R., and Sparkes, R. S., 1987, Two DNA restriction fragment length polymorphisms associated with Ag(t/z) and Ag(g/c) antigenic sites of human apolipoprotein B, *Arteriosclerosis* **7**:301–305.

715. Young, S. G., and Hubl, S. T., 1989, An ApaLI restriction site polymorphism is associated with the MB19 polymorphism in apolipoprotein B, *J. Lipid Res.* **30**:443–449.

716. Wang, X., Schlapfer, P., Ma, Y., Butler, R., Elovson, J., and Schumaker, V. N., 1988, Apolipoprotein B: The Ag(al/d) immunogenic polymorphism coincides with a T-to-C substitution at nucleotide 1981, Creating an AluI restriction site, *Arteriosclerosis* **8**:429–435.

717. Wu, M. J., Butler, E., Butler, R., and Schumaker, V. N., 1991, identification of the base substitution responsible for the Ag(x/y) polymorphism of apolipoprotein B-100, *Arteriosclerosis Thrombosis* **11**:379–384.

718. Rapacz, J., Chen, L., Butler-Brunner, E., Wu, M. J., Hasler-Rapacz, J. O., Butler, R., and Schumaker, V. N., 1991, Identification of the ancestral haplotype for apolipoprotein B suggests an African origin of *Homo sapiens sapiens* and traces their subsequent migration to Europe and the Pacific, *Proc. Natl. Acad. Sci. USA* **88**:1403–1406.

719. Cooper, R. A., and Gulbrandsen, C. L., 1971, The relationship between serum lipoproteins and red cell membranes in abetalipoproteins: Deficiency of lecithin:cholesterol acyltransferase, *J. Lab. Clin. Med.* **78**:323–335.

720. Zannis, V. I., and Breslow, J. L., 1982, Apolipoprotein E, *Mol. Cell. Biol.* **42**:3–20.

721. Lackner, K. J., Monge, J. C., Gregg, R. E., Hoeg, J. M., Triche, T. J., Law, S. W., and Brewer, Jr., H. B., 1986, Analysis of the apolipoprotein B gene and messenger ribonucleic acid in abetalipoproteinemia, *J. Clin. Invest.* **78**:1707–1712.

722. Bouma, M. E., Beucler, I., Pessah, M., Heinzmann, C., Lusis, A. J., Naim, H. Y., Ducastelle, T., Leluyer, B., Schmitz, J., and Infante, R., 1990, Description of two different patients with abetalipoproteinemia: Synthesis of a normal-sized apolipoprotein B-48 in intestinal organ culture, *J. Lipid Res.* **31**:1–15.

723. Talmud, P. J., Lloyd, J. K., Muller, D. P., Collins, D. R., Scott, J., and Humphries, S., 1988, Genetic evidence from two families that the apolipoprotein B gene is not involved in abetalipoproteinemia, *J. Clin. Invest.* **82**:1803–1806.

724. Huang, L. S., Janne, P. A., de Graaf, J., Cooper, M, Deckelbaum, R. J., Kayden, H., Breslow, J. L., and Decklebaum, R. J., 1990, Exclusion of linkage between the human apolipoprotein B gene and abetalipoproteinemia, *Am. J. Hum. Genet.* **46**:1141–1148; Erratum, *Am. J. Hum. Genet.* **47**(1):172, 1990.

725. Ross, R. S., Gregg, R. E., Law, S. W., Monge, J. C., Grant, S. M., Higuchi, K., Triche, T. J., Jefferson, J., and Brewer, Jr., H. B., 1988, Homozygous hypobetalipoproteinemia: A disease distinct from abetalipoproproteinemia at the molecular level, *J. Clin. Invest.* **81**:590–595.

726. Menzel, H. J., Dieplinger, H., Lackner, C., Hoppichler, F., Lloyd, J. K., Muller, D. R., Labeur, C., Talmud, P. J., and Utermann, G., 1990, Abetalipoproteinemia with an ApoB-100-lipoprotein(a) glycoprotein complex in plasma. Indication for an assembly defect, *J. Biol. Chem.* **265**:981–986.

726a. Wetterau, J. R., Aggerbeck, L. P., Bouma, M.-E., Eisenberg, C., Munck, A., Hermier, M.,

Schmitz, J., Gay, G., Rader, D. J., and Gregg, R. E., 1992, Absence of microsomal triglyceride transfer protein in individuals with abetalipoproteinemia, *Science* **258**:999–1001.

726b. Wetterau, J. R., and Zilversmit, D. B., 1985, Purification and characterization of microsomal triglyceride and cholesteryl ester transfer protein from bovine liver microsomes, *Chem. Phys. Lipids* **38**:205–222.

727. Mars, H., Lewis, L. A., Robertson, Jr., A. L., Butkus, A., and Williams, Jr., C. H., 1969, Familial hypo-B-lipoproteinemia: A genetic disorder of lipid metabolism with nervous system involvement, *Am. J. Med.* **46**:886–900.

728. Ricket, G., Durepaire, H., Hartmann, L., Ollier, M. P., Polonovski, J., and Maitrot, B., 1969, Hypolipoprotiemie familiale asymptomatique predominant sur les beta-lipoproteines, *Presse Med. (Paris)* **77**:2045–2048.

729. Levy, R. I., Langer, T., Gotto, A. M., and Fredrickson, D. S., 1970, Familial hypobetalipoproteinemia, a defect in lipoprotein synthesis, *Clin. Res.* **18**:539.

730. Tamir, I., Levtow, O., Lotan, D., Lequin, C., Heldenberg, D., and Werbin, B., 1976, Further observations on familial hypobetalipoproteinemia, *Clin. Genet.* **9**:149–155.

731. Fosbrooke, A., Choksey, S., and Wharton, B., 1978, Familial hypo-B-lipoproteinemia, *Arch. Dis. Child.* **48**:729–732.

732. Steinberg, D., Grundy, S. M., Mok, H. Y. I., Turner, J. D., Weinstein, J. J., Brown, W. V., and Albers, J. J., 1979, Metabolic studies in an unusual case of asymptomatic familial hypobetalipoproteinemia with hypoalphalipoproteinemia and fasting chylomicronemia, *J. Clin. Invest.* **64**:292–301.

733. Forsyth, C. C., Lloyd, J. K., and Fosbrooke, A. S., 1965, A-beta-lipoproteinemia, *Arch. Dis. Child.* **40**:47–52.

734. Khachadurian, A. K., Freyha, R., Shammas, M. M., and Baghdassarian, S. A., 1971, A-beta-lipoproteinemia and colour blindness, *Arch. Dis. Child.* **46**:871–873.

735. Kostner, G., Holasek, A., Bohlmann, H. G., and Thiede, H., 1974, Investigation of serum lipoproteins and apoproteins in abetalipoproteinaemia, *Clin. Sci. Mol. Med. Suppl.* **46**:457–468.

736. Young, S. G., Bertics, S. J., Curtiss, L. K., and Witztum, J. L., 1987, Characterization of an abnormal species of apolipoprotein B, apolipoprotein B-37, associated with familial hypobetalipoproteinemia, *J. Clin. Invest.* **79**:1831–1841.

737. Young, S. G., Bertics, S. J., Curtiss, L. K., Dubois, B. W., and Witztum, J. L., 1987, Genetic analysis of a kindred with familial hypobetalipoproteinemia. Evidence for two separate gene defects: One associated with an abnormal apolipoprotein B species, apolipoprotein B-37; and a second associated with low plasma concentrations of apolipoprotein B-100, *J. Clin. Invest.* **79**:1842–1851.

738. Young, S. G., Northey, S. T., and McCarthy, B. J., 1988, Low plasma cholesterol levels caused by a short deletion in the apolipoprotein B gene, *Science* **241**:591–593.

739. Huang, L. S., Ripps, M. E., Korman, S. H., Deckelbaum, R. J., and Breslow, J. L., 1989, Hypobetalipoproteinemia due to an apolipoprotein B gene exon 21 deletion derived by Alu–Alu recombination, *J. Biol. Chem.* **264**:11394–11400.

740. Collins, D. R., Knott, T. J., Pease, R. J., Powell, L. M., Wallis, S. C., Robertson, S., Pullinger, C. R., Milne, R. W., Marcel, Y. L., and Humphries, S. E., 1988, Truncated variants of apolipoprotein B cause hypobetalipoproteinaemia, *Nucleic Acids Res.* **16**:8361–8375.

741. Young, S. G., Hubl, S. T., Smith, R. S., Snyder, S. M., and Terdiman, J. F., 1990, Familial hypobetalipoproteinemia caused by a mutation in the apolipoprotein B gene that results in a truncated species of apolipoprotein B (B-31). A unique mutation that helps to define the portion of the apolipoprotein B molecule required for the formation of buoyant, triglyceride-rich lipoproteins, *J. Clin. Invest.* **85**:933–942.

742. Talmud, P., King-Underwood, L., Krul, E., Schonfeld, G., and Humphries, S., 1989, The molecular basis of truncated forms of apolipoprotein B in a kindred with compound hetero-zygous hypobetalipoproteinemia, *J. Lipid Res.* **30**:1773–1779.

743. Krul, E. S., Kinoshita, M., Talmud, P., Humphries, S. E., Turner, S., Goldberg, A. C., Cook, K., Boerwinkle, E., and Schonfeld, G., 1989, Two distinct truncated apolipoprotein B species in a kindred with hypobetalipoproteinemia, *Arteriosclerosis* **9**:856–868.

744. Young, S. G., Hubl, S. T., Chappell, D. A., Smith, R. S., Claiborne, F., Snyder, S. M., and Terdiman, J. F., 1989, Familial hypobetalipoproteinemia associated with a mutant species of apolipoprotein B (B-46), *N. Engl. J. Med.* **320**:1604–1610.

745. Malloy, M. J., Kane, J. P., Hardman, D. A., Hamilton, R. L., and Dalal, K. B., 1981, Normotriglyceridemic abetalipoproteinemia. Absence of the B-100 apolipoprotein, *J. Clin. Invest.* **67**:1441–1450.

746. Hardman, D. A., Pullinger, C. R., Kane, J. P., and Malloy, M. J., 1989, Molecular defect in normotriglyceridemic abetalipoproteinemia, *Circulation* **80**:II-466.

747. Wagner, R., Krul, E. S., Tang, J. J., Parhofer, K. G., Garlock, K., Talmud, P., and Schonfeld, G., 1991, ApoB-45.8, a truncated apolipoprotein found primarily in VLDL, is associated with a nonsense mutation in the apoB gene and hypobetalipoproteinemia, *J. Lipid Res.* **32**:1001–1011.

748. Pullinger, C. R., Hillas, E., and Hardman, D. A., 1990, A mutation in the apolipoprotein B gene results in a truncated variant, apoB-61, in VLDL and LDL in a kindred with hypobetalipopro-teinemia, *Circulation* **82**:III-424.

749. Welty, F. K., Hubl, S. T., Pierotti, V. R., and Young, S. G., 1991, A truncated species of apolipoprotein B (B67) in a kindred with familial hypobetalipoproteinemia, *J. Clin. Invest.* **87**:1748–1754.

750. Tennyson, G. E., Gabelli, C., Baggio, G., Bilato, C., and Brewer, Jr., H. B., 1990, Molecular defect in the apolipoprotein B gene in a patient with hypobetalipoproteinemia and three distinct apoB species, *Clin. Res.* **38**:2 (Abstract).

751. Parhofer, K. G., Daugherty, A., Kinoshita, M., and Schonfeld, G., 1990, Enhanced clearance from plasma of low density lipoproteins containing a truncated apolipoprotein, apoB-89, *J. Lipid Res.* **31**:2001–2007.

752. Vega, G. L., and Grundy, S. M., 1986, *In vivo* evidence for reduced binding of low density lipoproteins to receptors as a cause of primary moderate hypercholesterolemia, *J. Clin. Invest.* **78**:1410–1414.

753. Innerarity, T. L., Weisgraber, K. H., Arnold, K. S., Mahley, R. W., Krauss, R. M., Vega, G. L., and Grundy, S. M., 1987, Familial defective apolipoprotein B-100: Low density lipoproteins with abnormal receptor binding, *Proc. Natl. Acad. Sci. USA* **84**:6919–6923.

754. Innerarity, T. L., Mahley, R. W., Weisgraber, K. H., Bersot, T. P., Krauss, R. M., Vega, G. L., Grundy, S. M., Friedl, W., Davignon, J., and McCarthy, B. J., 1990, Familial defective apolipoprotein B-100: A mutation of apolipoprotein B that causes hypercholesterolemia, *J. Lipid Res.* **31**:1337–1349.

755. Breslow, J. L., Zannis, V. I., SanGiacomo, T. R., Third, J. L., Tracy, T., and Glueck, C. J., 1982, Studies of familial type III hyperlipoproteinemia using as a genetic marker the apoE phenotype E2/2, *J. Lipid Res.* **23**:1224–1235.

756. Kannel, W. B., Castelli, W. P., and Gordon, T., 1979, Cholesterol in the prediction of atherosclerotic disease: New perspectives based on the Framingham Study, *Ann. Intern. Med.* **90**:85–91.

757. Sniderman, A., Shapiro, S., Marpole, D., Skinner, B., Teng, B., and Kwiterovich, Jr., P. O., 1980, Association of coronary atherosclerosis with hyperapobetalipoproteinemia [increased protein but normal cholesterol levels in human plasma low density (beta) lipoproteins], *Proc. Natl. Acad. Sci. USA* **77**:604–608.

758. Sniderman, A., Shapiro, S., Marpole, D., Skinner, B., Teng, B., and Kwiterovich, Jr., P. O., 1980, Enhanced drug-metabolizing capacity within liver adjacent to human and rat liver tumors, *Proc. Natl. Acad. Sci. USA* **77**:601–603.

759. Jackson, R. L., Sparrow, J. T., Baker, H. N., Morrisett, J. D., Taunton, O. D., and Gotto, Jr., A. M., 1974, The primary structure of apolipoprotein-serine, *J. Biol. Chem.* **249**:5308–5313.

760. Shulman, R. S., Herbert, P. N., Wehrly, K., and Fredrickson, D. S., 1975, The complete amino acid sequence of CI (apo Lp=Ser)), an apolipoprotein from human very low density lipoprotein, *J. Biol. Chem.* **250**:182–190.

761. Knott, T. J., Robertson, M. E., Priestley, L. M., Urdea, M., Wallis, S., and Scott, J., 1984, Characterization of mRNAs encoding the precursor for human apolipoprotein CI, *Nucleic Acids Res.* **12**:3909–3915.

762. Lauer, S. J., Walker, D., Elshourbagy, N. A., Reardon, C. A., Levy-Wilson, B., and Taylor, J. M., 1988, Two copies of the human apolipoprotein C-I gene are linked closely to the apolipoprotein E gene, *J. Biol. Chem.* **263**:7277–7286.

763. Sparrow, J. T., and Gotto, Jr., A. M., 1982, Apolipoprotein/lipid interactions: Studies with synthetic polypeptides, *CRC Crit. Rev. Biochem.* **13**:87–107.

764. Jackson, R. L., Morrisett, J. D., Sparrow, J. T., Segrest, J. P., Pownall, H. J., Smith, L. C., Hoff, H. F., and Gotto, Jr., A. M., 1974, The interaction of apolipoprotein-serine with phosphatidylcholine, *J. Biol. Chem.* **249**:5314–5320.

765. Forte, T., Gong, E., and Nichols, A. V., 1974, Interaction by sonication of C-apolipoproteins with lipid: An electron microscopic study, *Biochim. Biophys. Acta* **337**:169–183.

766. Sparrow, J. T., Pownall, H. J., Sigler, G. F., Smith, L. C., Soutar, A. K., and Gotto, Jr., A. M., 1977, The mechanism of phospholipid binding by the plasma apolipoproteins, in: *Peptides—Proceedings of the Fifth American Peptide Symposium* (M. Goodman and J. Meienhofer, eds.), Wiley, New York, pp. 149–152.

767. Windler, E., Chao, Y., and Havel, R. J., 1980, Determinants of hepatic uptake of triglyceride-rich lipoproteins and their remnants in the rat, *J. Biol. Chem.* **255**:5475–5480.

768. Shelburne, F., Hanks, J., Meyers, W., and Quarfordt, S., 1980, Effect of apoproteins on hepatic uptake of drug emulsions in the rat, *J. Clin. Invest.* **65**:652–658.

769. Windler, E., Chao, Y., and Havel, R. J., 1980, Regulation of the hepatic uptake of triglyceride-rich lipoproteins in the rat, *J. Biol. Chem.* **255**:8303–8307.

770. Quarfordt, S. H., Michalopoulos, G., and Schirmer, B., 1982, The effect of human C apolipoproteins on the *in vitro* hepatic metabolism of triglyceride emulsions in the rat, *J. Biol. Chem.* **257**:14642–14647.

771. Simonet, W. S., Bucay, N., Pitas, R. E., Lauer, S. J., and Taylor, J. M., 1991, Multiple tissue-specific elements control the apolipoprotein E/C-I gene locus in transgenic mice, *J. Biol. Chem.* **266**:8651–8654.

772. Poensgen, J., 1990, Apolipoprotein C-I inhibits the hydrolysis by phospholipase A2 of phospholipids in liposomes and cell membranes, *Biochim. Biophys. Acta* **1042**:188–192.

773. Jackson, R. I., Baker, H. N., Gilliam, E. B., and Gotto, A. M., 1977, Primary structure of very low density apolipoprotein C-II of human plasma, *Proc. Natl. Acad. Sci. USA* **74**:1942.

774. Mantulin, W. W., Rohde, M. F., Gotto, Jr., A. M., and Pownall, H. J., 1980, The conformational properties of human plasma apolipoprotein C-II: A spectroscopic study, *J. Biol. Chem.* **255**:8185–8191.

775. Morrisett, J. D., Jackson, R. L., and Gotto, Jr., A. M., 1977, Lipid–protein interactions in the plasma lipoproteins, *Biochim. Biophys. Acta* **427**:93.

776. Scow, R. O., and Egelrud, T., 1976, Hydrolysis of chylomicron phosphatidylcholine *in vitro* by lipoprotein lipase, phospholipase A2 and phospholipase C, *Biochim. Biophys. Acta* **431**:538–549.

777. Fielding, C. J., and Fielding, P. E., 1977, The activation of lipoprotein lipase by lipase co-protein (apo C-2), *Exposes Ann. Biochim. Med.* **33**:165–172.

778. Fielding, C. J., and Fielding, P. E., 1976, Mechanism of salt-mediated inhibition of lipoprotein lipase, *J. Lipid Res.* **17**:248.

779. Musliner, T. A., Church, E. C., Herbert, P. N., Kingston, M. J., and Shulman, R. S., 1977, Lipoprotein lipase cofactor activity of a carboxyl-terminal peptide of apolipoprotein C-II, *Proc. Natl. Acad. Sci. USA* **74**:5358–5362.

780. Kinnunen, P. K. J., Jackson, R. L., Smith, L. C., Gotto, Jr., A. M., and Sparrow, J. T., 1977, Activation of lipoprotein lipase by native and synthetic fragments of human plasma apolipoprotein C-II, *Proc. Natl. Acad. Sci. USA* **74**:4848–4851.

781. Smith, L. C., Kinnunen, P. K. J., Jackson, R. L., Gotto, A. M., and Sparrow, J. T., 1978, Activation of lipoprotein lipase by native and synthetic peptide fragments of apolipoprotein C-II, in: *Proceedings of International Conference on Atherosclerosis* (L. A. Carlson, R. Paoletti, and G. Weber, eds.), Raven Press, New York, pp. 269–273.

782. Connelly, P. W., Maguire, G. F., and Little, J. A., 1987, Apolipoprotein CII St. Michael, Familial apolipoprotein CII deficiency associated with premature vascular disease, *J. Clin. Invest.* **80**: 1597–1606.

783. Cox, D. W., Wills, D. E., Quan, F., and Ray, P. N., 1988, A deletion of one nucleotide results in functional deficiency of apolipoprotein CII (apo CII Toronto), *J. Med. Gen.* **25**:649–652.

784. Connelly, P. W., Maguire, G. F., Hofmann, T., and Little, J. A., 1987, Structure of apolipoprotein C-II Toronto, a nonfunctional human apolipoprotein, *Proc. Natl. Acad. Sci. USA* **84**: 270–273.

785. Holtfreter, C., and Stoffel, W., 1988, Expression of normal and mutagenized apolipoprotein CII in procaryotic cells. Structure–function relationship, *Biol. Chem. Hoppe-Seyler* **369**:1045–1054.

786. Fojo, S. S., Law, S. W., and Brewer, Jr., H. B., 1984, Human apolipoprotein C-II: Complete nucleic acid sequence of preapolipoprotein C-II, *Proc. Natl. Acad. Sci. USA* **81**:6354–6357.

787. Myklebost, O., Williamson, B., Markham, A. F., Myklebost, S. R., Rogers, J., Woods, D. E., and Humphries, S. E., 1984, The isolation and characterization of cDNA clones for human apolipoprotein CII, *J. Biol. Chem.* **259**:4401–4404.

788. Jackson, C. L., Bruns, G. A., and Breslow, J. L., 1984, Isolation and sequence of a human apolipoprotein CII cDNA clone and its use to isolate and map to human chromosome 19 the gene for apolipoprotein CII, *Proc. Natl. Acad. Sci. USA* **81**:2945–2949.

789. Wei, C. F., Tsao, Y. K., Robberson, D. L., Gotto, Jr., A. M., Brown, K., and Chan, L., 1985, The structure of the human apolipoprotein C-II gene, *J. Biol. Chem.* **260**:15211–15221.

790. Fojo, S. F., Law, S., Baggio, G., Gregg, R. E., and Brewer, Jr., H. B., 1984, Complete nucleic acid sequence of preapolipoprotein C-II and an analysis of the apolipoprotein (apo) C-II gene in apoC-II deficient patients, *Circulation* **70**:II-8.

791. Baggio, G., Manzato, E., Gabelli, C., Fellin, R., Martini, S., Baldo Enzi, G., Verlato, F., Baiocchi, M. R., Sprecher, D. L., Kashyap, M. L., Brewer, Jr., H. B., and Crepaldi, G., 1986, Apolipoprotein C-II deficiency syndrome, *J. Clin. Invest.* **77**:520–527.

792. Fojo, S. S., Beisiegel, U., Beil, U., Higuchi, K., Bojanovski, M., Gregg, R. E., Greten, H., and Brewer, Jr., H. B.,·1988, Donor splice site mutation in the apolipoprotein (Apo) C-II gene (Apo C-II Hamburg) of a patient with Apo C-II deficiency, *J. Clin. Invest.* **82**:1489–1494.

793. Fojo, S. S., Stalenhoef, A. F., Marr, K., Gregg, R. E., Ross, R. S., and Brewer, Jr., H. B., 1988, A deletion mutation in the ApoC-II gene (ApoC-II Nijmegen) of a patient with a deficiency of apolipoprotein C-II, *J. Biol. Chem.* **263**:17913–17916.

794. Menzel, H. J., Kane, J. P., Malloy, M. J., and Havel, R. J., 1986, A variant primary structure of apolipoprotein C-II in individuals of African descent, *J. Clin. Invest.* **77**:595–601.

795. Fojo, S. S., de Gennes, J. L., Chapman, J., Parrott, C., Lohse, P., Kwan, S. S., Truffert, J., and Brewer, Jr., H. B., 1989, An initiation codon mutation in the apoC-II gene (apoC-II Paris) of a patient with a deficiency of apolipoprotein C-II, *J. Biol. Chem.* **264**:20839–20842.

795a. Xiong, W., Li, W-H., Posner, I., Yamamura, T., Yamamoto, A., Gotto, A. M. Jr., and Chan, L., 1991, No severe bottleneck during human evolution: Evidence from two apolipoprotein CII deficiency alleles, *Am. J. Hum. Genet.* **48:**383–389.

796. Shulman, R. S., Herbert, P. N., Fredrickson, D. S., Wehrly, K., and Brewer, Jr., H. B., 1974, Isolation and alignment of the tryptic peptides of alanine apolipoprotein, an apolipoprotein from human plasma very low density lipoproteins, *J. Biol. Chem.* **249:**4969–4974.

797. Morrisett, J. D., David, J. S. K., Pownall, H. J., and Gotto, Jr., A. M., 1973, Interaction of an apolipoprotein (apolp-alanine) with phosphatidylcholine, *Biochemistry* **12:**1290–1299.

798. Sparrow, J. T., Pownall, H. J., Hsu, F. J., Blumental, L. E., Culwell, A. R., and Gotto, Jr., A. M., 1977, Lipid binding by fragments of apolipoprotein C-III-1 obtained by thrombin cleavage, *Biochemistry* **16:**5427–5431.

799. Anne, K. C., Gallagher, J. G., Gotto, Jr., A. M., and Morrisett, J. D., 1977, Physical properties of the dimyristoylphosphatidylcholine vesicle and of complexes formed by its interaction with apolipoprotein C-III, *Biochemistry* **16:**2151–2156.

800. Laggner, P., Gotto, Jr., A. M., and Morrisett, J. D., 1979, Structure of the dimyristoyl-phosphatidylcholine vesicle and the complex formed by its interaction with apolipoprotein CIII: X-ray single angle scattering studies, *Biochemistry* **18:**164–171.

801. Brown, W. V., and Baginsky, M. L., 1972, Inhibition of lipoprotein lipase by an apoprotein of human very low density lipoprotein, *Biochem. Biophys. Res. Commun.* **46:**375–382.

802. Krauss, R. M., Herbert, P. N., Levy, R. I., and Fredrickson, D. S., 1973, Further observations on the activation and inhibition of lipoprotein lipase by apolipoproteins, *Circ. Res.* **33:**403–411.

803. Ito, Y., Azrolan, N., O'Connell, A., Walsh, A., and Breslow, J. L., 1990, Hypertriglyceridemia as a result of human apo CIII gene expression in transgenic mice, *Science* **249:**790–793.

804. Karathanasis, S. K., Zannis, V. I., and Breslow, J. L., 1985, Isolation and sequence of human apoC-III cDNA clones specifying two different alleles, *J. Lipid Res.* **26:**451–456.

805. Protter, A. A., Levy-Wilson, B., Miller, J., Bencen, G., White, T., and Seilhamer, J. J., 1984, Isolation and sequence analysis of the human apolipoprotein CIII gene and the intergenic region between the apoAI and apoCIII genes, *DNA* **3:**449–456.

806. Schonfeld, G., George, P. K., Miller, J., Reilly, P., and Witztum, J., 1979, Apolipoprotein C-II and C-III levels in hyperlipoproteinemia, *Metabolism Clin. Exp.* **28:**1001–1010.

807. Kashyap, M. L., Hynd, B. A., Robinson, K., and Gartside, P. S., 1981, Abnormal preponderance of sialylated apolipoprotein CIII in triglyceride rich lipoproteins in type V hyperlipoproteinemia, *Metabolism: Clinical and Experimental* **30:**111–118.

808. Maeda, H., Uzawa, H., and Kamei, R., 1981, Unusual familial lipoprotein C-III associated with apolipoprotein C-III-O preponderance, *Biochim. Biophys. Acta* **665:**578–585.

809. Maeda, H., Hashimoto, R. K., Ogura, T., Hiraga, S., and Uzawa, H., 1987, Molecular cloning of a human apoC-III variant: Thr 74–Ala 74 mutation prevents O-glycosylation, *J. Lipid Res.* **28:**1405–1409.

810. Von Eckardstein, A., Holz, H., Sandkamp, M., Weng, W., Funke, H., and Assmann, G., 1991, Apolipoprotein C-III(Lys58–Glu). Identification of an apolipoprotein C-III variant in a family with hyperalphalipoproteinemia, *J. Clin. Invest.* **87:**1724–1731.

811. Das, H. K., McPherson, J., Bruns, G. A. P., Karathanasis, S. K., and Breslow, J. L., 1985, Isolation, characterization, and mapping to chromosome 19 of the human apolipoprotein E gene, *J. Biol. Chem.* **260:**6240–6246.

812. Paik, Y. K., Chang, D. J., Reardon, C. A., Davies, G. E., Mahley, R. W., and Taylor, J. M., 1985, Nucleotide sequence and structure of the human apolipoprotein E gene, *Proc. Natl. Acad. Sci. USA* **82:**3445–3451.

813. Wallis, S. C., Rogne, S., Gill, L., Markham, A., Edge, M., Woods, D. Williamson, R., and Humphries, S., 1983, The isolation of cDNA clones for human apolipoprotein E and the detection of apoE RNA in hepatic and extra-hepatic tissues, *EMBO J.* **2:**2369–2373.

814. Zannis, V. I., McPherson, J., Goldberger, G., Karathanasis, S. K., and Breslow, J. L., 1984, Synthesis, intracellular processing and signal peptide of human apoE, *J. Biol. Chem.* **259:** 5495–5499.

815. Wetterau, J. R., Aggerbeck, L. P., Rall, Jr., S. C., and Weisgraber, K. H., 1988, Human apolipoprotein E3 in aqueous solution. I. Evidence for two structural domains, *J. Biol. Chem.* **263:**6240–6248.

816. Aggerbeck, L. P., Wetterau, J. R., Weisgraber, K. H., Wu, C. S. C., and Lindgren, F. T., 1988, Human apolipoprotein E3 in aqueous solution. II. Properties of the amino- and carboxyl-terminal domains, *J. Biol. Chem.* **263:**6249–6258.

817. Weisgraber, K. H., 1990, Apolipoprotein E distribution among human plasma lipoproteins: Role of the cysteine–arginine interchange at residue 112, *J. Lipid Res.* **31:**1503–1511.

818. Roth, R. I., Jackson, R. L., Pownall, H. J., and Gotto, Jr., A. M., 1977, Interaction of plasma "arginine-rich" apolipoprotein with dimyristoylphosphatidylcholine, *Biochemistry* **16:**5030–5036.

819. Derksen, A., and Small, D. M., 1989, Interaction of ApoA-1 and ApoE-3 with triglyceride-phospholipid emulsions containing increasing cholesterol concentrations. Model of triglyceride-rich nascent and remnant lipoproteins, *Biochemistry* **28:**900–906.

820. Elshourbagy, N. A., Liao, W. S., Mahley, R. W., and Taylor, J. M., 1985, Apolipoprotein E mRNA is abundant in the brain and adrenals, as well as in the liver, and is present in other peripheral tissues of rats and marmosets, *Proc. Natl. Acad. Sci. USA* **82:**203–208.

821. Lin-Lee, Y. C., Tanaka, Y., Lin, C. T., and Chan, L., 1981, Effects of an atherogenic diet on apolipoprotein E biosynthesis in the rat, *Biochemistry* **20:**6474–6480.

822. Mangeney, M., Cardot, P., Lyonnet, S., Coupe, C., Benarous, R., Munnich, A., Girard, J., Chambaz, J., and Bereziat, G., 1989, Apolipoprotein-E-gene expression in rat liver during development in relation to insulin and glucagon, *Eur. J. Biochem.* **181:**225–230.

823. Kim, M. H., Nakayama, R., Manos, P., Tomlinson, J. E., Choi, E., Ng, J. D., and Holten, D., 1989, Regulation of apolipoprotein E synthesis and mRNA by diet and hormones, *J. Lipid Res.* **30:**663–671.

824. Driscoll, D. M., Mazzone, T., Matsushima, T., and Getz, G. S., 1990, Apoprotein E biosynthesis in the cholesterol-fed guinea pig, *Arteriosclerosis* **10:**31–39.

825. Lenich, C. M., Chobanian, A. V., Brecher, P., and Zannis, V. I., 1991, Effect of dietary cholesterol and alloxan-diabetes on tissue cholesterol and apolipoprotein E mRNA levels in the rabbit, *J. Lipid Res.* **32:**431–438.

826. Basheeruddin, K., Stein, P., Strickland, S., and Williams, D. L., 1987, Expression of the murine apolipoprotein E gene is coupled to the differentiated state of F9 embryonal carcinoma cells, *Proc. Natl. Acad. Sci. USA* **84:**709–713.

827. Zechner, R., Moser, R., Newman, T. C., Fried, S. K., and Breslow, J. L., 1991, Apolipoprotein E gene expression in mouse 3T3-L1 adipocytes and human adipose tissue and its regulation by differentiation and lipid content, *J. Biol. Chem.* **266:**10583–10588.

828. Wyne, K. L., Schreiber, J. R., Larsen, A. L., and Getz, G. S., 1989, Rat granulosa cell apolipoprotein E secretion. Regulation by cell cholesterol, *J. Biol. Chem.* **264:**16530–16536.

829. Basu, S. K., Brown, M. S., Ho, Y. K., Havel, R. J., and Goldstein, J. L., 1981, Mouse macrophages synthesize and secrete a protein resembling apolipoprotein E, *Proc. Natl. Acad. Sci. USA* **78:**7545–7549.

830. Basu, S. K., Ho, Y. K., Brown, M. S., Bilheimer, D. W., Anderson, R. G., and Goldstein, J. L., 1982, Biochemical and genetic studies of the apoprotein E secreted by mouse macrophages and human monocytes, *J. Biol. Chem.* **257:**9788–9795.

831. Mazzone, T., Gump, H., Diller, P., and Getz, G. S., 1987, Macrophage free cholesterol content regulates apolipoprotein E synthesis, *J. Biol. Chem.* **262:**11657–11662.

832. Werb, Z., and Chin, J. R., 1983, Endotoxin suppresses expression of apoprotein E by mouse macrophages *in vivo* and in culture, *J. Biol. Chem.* **258**:10642–10648.

833. Driscoll, D. M., and Getz, G. S., 1984, Extrahepatic synthesis of apolipoprotein E, *J. Lipid Res.* **25**:1368–1374.

834. Hussain, M. M., Bucher, N. L. R., Faris, B., Franzblau, C., and Zannis, V. I., 1988, Tissue-specific post-translational modification of rat apoE synthesis of sialated apoE forms by neonatal rat aortic smooth muscle cells, *J. Lipid Res.* **29**:915–923.

835. Friedman, G., Liu, L. M., Friedman, S. L., and Boyles, J. K., 1991, Apolipoprotein E is secreted by cultured lipocytes of the rat liver, *J. Lipid Res.* **32**:107–114.

836. Murakami, M., Ushio, Y., Morino, Y., Ohta, T., and Matsukado, Y., 1988, Immunohistochemical localization of apolipoprotein E in human glial neoplasms, *J. Clin. Invest.* **82**:177–188; Erratum, *J. Clin. Invest.* **83**(1):347, 1989.

837. Pitas, R. E., Boyles, J. K., Lee, S. H., Foss, D., and Mahley, R. W., 1987, Astrocytes synthesize apolipoprotein E and metabolize apolipoprotein E-containing lipoproteins, *Biochim. Biophys. Acta* **917**:148–161.

838. Boyles, J., Pitas, R. E., Wilson, E., Mahley, R. W., and Taylor, J. M., 1985, Apolipoprotein E associated with astrocytic glia of the central nervous system and with nonmyelinating glia of the central nervous system and with nonmyelinating glia at the peripheral nervous system, *J. Clin. Invest.* **76**:1501–1513.

839. Mahley, R. W., 1988, Apolipoprotein E: Cholesterol transport protein with expanding role in cell biology, *Science* **240**:622–630.

840. Muller, H. W., Ignatius, M. J., Hanagen, D. H., and Shooter, E. M., 1986, Expression of specific sheath cell proteins during peripheral nerve growth and regeneration in mammals, *J. Cell Biol.* **102**:393–402.

841. Ignatius, M. J., Gebicke-Harter, P. J., Skene, J. H. P., Schilling, J. W., Weisgraber, K. H., Mahley, R. W., and Shooter, E. M., 1986, Expression of apolipoprotein E during nerve degeneration and regeneration, *Proc. Natl. Acad. Sci. USA* **83**:1125–1129.

842. Boyles, J. K., Notterpek, L. M., and Anderson, L. J., 1990, Accumulation of apolipoproteins in the regenerating and remyelinating mammalian peripheral nerve. Identification of apolipoprotein D, apolipoprotein A-IV, apolipoprotein E, and apolipoprotein A-I, *J. Biol. Chem.* **265**:17805–17815.

843. Gordon, V., Innerarity, T. L., and Mahley, R. W., 1983, Formation of cholesterol- and apoprotein E-enriched high density lipoproteins *in vitro*, *J. Biol. Chem.* **258**:6202–6212.

844. Krul, E. S., Dolphin, P. J., and Rubinstein, D., 1981, Secretion of nascent lipoproteins by isolated rat hepatocytes, *Can. J. Biochem.* **59**:676–686.

845. Krul, E. S., and Dolphin, P. J., 1982, Secretion of nascent lipoproteins by isolated hepatocytes from hypothyroid and hypothyroid, hypercholesterolemic rats, *Biochim. Biophys. Acta* **713**:609–621.

846. Zannis, V. I., Ordovas, J. M., Cladaras, C., Cole, F. S., Forbes, G., and Schaefer, E. J., 1985, mRNA and apolipoprotein E synthesis abnormalities in peripheral blood monocyte macrophages in familial apolipoprotein E deficiency, *J. Biol. Chem.* **24**:12891–12894.

847. Lalazar, A., Weisgraber, K. H., Rall, Jr., S. C., Giladi, H., Innerarity, T. L., Levanon, A. Z., Boyles, J. K., Amit, B., Gorecki, M., and Mahley, R. W., 1988, Site-specific mutagenesis of human apolipoprotein E. Receptor binding activity of variants with single amino acid substitutions, *J. Biol. Chem.* **263**:3542–3545.

848. Lalazar, A., and Mahley, R. W., 1989, Human apolipoprotein E. Receptor binding activity of truncated variants with carboxyl-terminal deletions, *J. Biol. Chem.* **264**:8447–8450.

849. DeLoof, H., Rosseneu, M., Brasseur, R., and Ruysschaert, J. M., 1986, Use of hydrophobicity profiles to predict receptor binding domains on apolipoprotein E and the low density lipoprotein apolipoprotein B-E receptor, *Proc. Natl. Acad. Sci. USA* **83**:2295–2299.

850. Weisgraber, K. H., Rall, Jr., S. C., Mahley, R. W., Milne, R. W., Marcel, Y. L., and Sparrow, J. T., 1986, Human apolipoprotein E, *J. Biol. Chem.* **261**:2068–2076.

851. Cardin, A. D., Hirose, N., Blankenship, D. T., Jackson, R. L., and Harmony, J. A. K., 1986, Binding of a high reactive heparin to human apolipoprotein E: Identification of two heparin-binding domains, *Biochem. Biophys. Res. Commun.* **134**:783–789.

852. Pitas, R. E., Boyles, J. K., Lee, S. H., Hui, D., and Weisgraber, K. H., 1987, Lipoproteins and their receptors in the central nervous system, *J. Biol. Chem.* **262**:14352–14360.

853. Reyland, M. E., Gwynne, J. T., Forgez, P., Prack, M. M., and Williams, D. L., 1991, Expression of the human apolipoprotein E gene suppresses steroidogenesis in mouse Y1 adrenal cells, *Proc. Natl. Acad. Sci. USA* **88**:2375–2379.

854. Ehnholm, C., Mahley, R. W., Chappell, D. A., Weisgraber, K. H., Ludwig, E., and Witztum, J. L., 1984, Role of apolipoprotein E in the lipolytic conversion of beta-very low density lipoproteins to low density lipoproteins in type III hyperlipoproteinemia, *Proc. Natl. Acad. Sci. USA* **81**:5566–5570.

855. Lin-Lee, Y. C., Kao, F. T., Cheung, P., and Chan, L., 1985, Apolipoprotein E gene mapping and expression: Localization of the structural gene to human chromosome 19 and expression of apoE mRNA in lipoprotein and non-lipoprotein-producing tissues, *Biochemistry* **24**:3751–3756.

856. Utermann, G., Jaeschke, M., and Menzel, J., 1975, Familial hyperlipoproteinemia type III: Deficiency of a specific apolipoprotein (apoE-III) in the very low density lipoproteins, *FEBS Lett.* **56**:352–355.

857. Zannis, V. I., Just, P. W., and Breslow, J. L., 1981, Human apolipoprotein E isoprotein subclasses are genetically determined, *Am. J. Hum. Genet.* **33**:11–24.

858. Zannis, V. I., Breslow, J. L., Utermann, G., Mahley, R. W., Weisgraber, K. H., Havel, R. J., Goldstein, J. L., Brown, M. S., Schonfeld, G., Hazzard, W. R., and Blum, C. B., 1982, Proposed nomenclature of apoE isoprotein genotypes and phenotypes, *J. Lipid Res.* **23**:911–914.

859. Zannis, V. I., and Breslow, J. L., 1981, Human very low density lipoprotein apolipoprotein E isoprotein polymorphism is explained by genetic variation and posttranslational modification, *Biochemistry* **20**:1033–1041.

860. Zannis, V. I., 1986, Genetic polymorphism in human apolipoprotein E, in: *Methods in Enzymology*, Vol. 128 (J. P. Segrest and J. J. Albers, eds.), Academic Press, Orlando, Florida, pp. 823–851.

861. Davignon, J., Gregg, R. E., and Sing, C. F., 1988, Apolipoprotein E polymorphism and atherosclerosis, *Arteriosclerosis* **8**:1–21.

862. Utermann, G., Pruin, N., and Steinmetz, A., 1979, Polymorphism of apolipoprotein E. III. Effect of a single polymorphic gene locus on plasma lipid levels in man, *Clin. Genet.* **15**:63–72.

863. Uterman, G., Steinmetz, A., and Weber, W., 1982, Genetic control of human apolipoprotein E polymorphism: Comparison of one- and two-dimensional techniques of isoprotein analysis, *Hum. Genet.* **60**:344–351.

864. Menzel, H. J., Kladetzky, R. G., and Assmann, G., 1983, Apolipoprotein E polymorphism and coronary artery disease, *Arteriosclerosis* **3**:310–315.

865. Sing, C. F., and Davignon, J., 1985, Role of the apolipoprotein E polymorphism in determining normal plasma lipid and lipoprotein variation, *Am. J. Hum. Genet.* **37**:268–285.

866. Ehnholm, C., Lukka, M., Kuush, T., Nikklia, E., and Utermann, G., 1986, Apolipoprotein E polymorphism in the Finnish population: Gene frequencies and relation to lipoprotein concentrations, *J. Lipid Res.* **27**:227–235.

867. Eto, M., Watanabe, K., Iwashima, Y., Morikawa, A., Oshima, E., Sekiguchi, M., and Ishii, K., 1986, The effects of apolipoprotein E alleles (E2, E3 and E4) on plasma lipids in middle-aged subjects, *J. Jpn. Atheroscler. Soc.* **14**:491–495.

868. Boerwinkle, E. Visvikis, S., Welsh, D., Steinmetz, J., Hanash, S. M., and Sing, C. F., 1987, The use of measured genotype information in the analysis of quantitative phenotypes in man. II. The

role of the apolipoprotein E polymorphism in determining levels, variability and covariability of cholesterol, betalipoprotein and triglycerides in a sample of unrelated individuals, *Am. J. Med. Genet.* **27**:567–582.

869. Breslow, J. L., and Zannis, V. I., 1986, Genetic variation in apolipoprotein E and type III hyperlipoproteinemia, *Arterioscler. Rev.* **14**:119–141.

870. Ordovas, J. M., Litwack-Klein, L., Wilson, P. W. F., Schaefer, M. M., and Schaefer, E. J., 1987, Apolipoprotein E isoform phenotyping methodology and population frequency with identification of apoE1 and apoE5 isoforms, *J. Lipid Res.* **28**:371–380.

871. Utermann, G., 1985, Genetic polymorphism of apolipoprotein E-impact on plasma lipoprotein metabolism. In: *Diabetes, Obesity and Hyperlipidemias* (G. Crepaldi, A. Tiengo, and G. Baggio, eds.), Excerpta Medica, Amsterdam, pp. 1–28.

872. Utermann, G., 1985, Apolipoprotein E mutants, hyperlipidemia and arteriosclerosis, *Adv. Exp. Med. Biol.* **183**:173–188.

873. Assmann, G., Schmitz, G., Menzel, H. J., and Schultz, H., 1984, Apolipoprotein E polymorphism and hyperlipidemia, *Clin. Chem.* **30**:641–643.

874. Olaisen, B., Teisberg, P., and Gedde-Dahl, Jr., T., 1982, The locus for apolipoprotein E (apoE) is linked to the complement component C3 (C3) locus on chromosome 19 in man, *Hum. Genet.* **62**:233–236.

875. Lussier-Cacan, S., Bouthillier, D., and Davignon, J., 1985, ApoE allele frequency in primary endogenous hypertriglyceridemia (type IV) with and without hyperapobetalipoproteinemia, *Arteriosclerosis* **5**:639–643.

876. Eto, M., Watanabe, K., Iwashima, Y., Morikawa, A., Oshima, E., Sekiguchi, M., and Ishii, K., 1986, Apolipoprotein E polymorphism and hyperlipemia in type II diabetics, *Diabetes* **35**: 1374–1382.

877. Leren, T. P., Borresen, A. L., Berg, K., Hjermann, P., and Leren, P., 1985, Increased frequency of the apolipoprotein E-4 isoform in male subjects with multifactorial hypercholesterolemia, *Clin. Genet.* **27**:458–462.

878. Utermann, G., Kindermann, I., Kaffarnik, H., and Steinmetz, A., 1984, Apolipoprotein E phenotypes and hyperlipidemia, *Human Genetics* **65**:232–236.

879. Utermann, G., Hardewig, A., and Zimmer, E., 1984, ApoE phenotypes in patients with myocardial infarction, *Hum. Genet.* **65**:237–241.

880. Kameda, K., Matsuzawa, Y., Kubo, M., Ishikawa, K., Maejima, I., Yamamura, T., Yamamoto, A., and Tarui, S., 1984, Increased frequency of lipoprotein disorders similar to type III hyperlipoproteinemia in survivors of myocardial infarction in Japan, *Basic Res. Cardiol.* **51**: 241–249.

881. Lenzen, H. J., Assmann, G., Buchwalsky, R., and Schulte, H., 1986, Association of apolipoprotein E polymorphism, low density lipoprotein cholesterol, and coronary artery disease, *Clin. Chem.* **32**:778–781.

882. Hazzard, W. R., Prote, Jr., D., and Bierman, E. L., 1972, Abnormal lipid composition of very low density lipoproteins in diagnosis of broad beta disease (type III hyperlipoproteinemia), *Metabolism: Clinical and Experimental* **21**:1009–1019.

883. Fredrickson, D. S., Goldstein, J. L., and Brown, M. S., 1978, The familial hyperlipoproteinemias. In: *The Metabolic Basis of Inherited Disease* (J. B. Stanbury, J. Wyngaarden, and D. S. Fredrickson, eds.), McGraw-Hill, New York, pp. 604–655.

884. Mahley, R. W., and Rall, Jr., S. C., 1989, Type III hyperlipoproteinemia (dysbetalipoproteinemia): The role of apolipoprotein E in normal and abnormal lipoprotein metabolism, in: *The Metabolic Basis of Inherited Disease* (C. R. Scriver, A. I. Beaudet, W. S. Sly, and D. Valle, eds.), McGraw-Hill, New York, pp. 1195–1211.

885. Fredrickson, D. S., Morganroth, J., and Levy, R. I., 1975, Type III hyperlipoproteinemia: An analysis of two contemporary definitions, *Ann. Intern. Med.* **82**:150–157.

886. Morrison, J. A., Kelly, K., Horvitz, R., Khoury, P., Laskarzewski, P. M., Mellies, M. J., and Glueck, C. J., 1982, Parent–offspring and sib–sib lipid and lipoprotein associations during and after sharing of household environments. The Princeton school district family study, *Metabolism* **31**:158–167.

887. Breslow, J. L., and Zannis, V. I., 1988, Genetic variation in apolipoprotein E and type III hyperlipoproteinemia, *Arterioscler. Rev.* **14**:119–142.

888. Breslow, J. L., Zannis, V. I., SanGiacomo, R. R., Third, J. L. H. C., Tracy, T., and Glueck, C. J., 1982, Studies of familial type III hyperlipoproteinemia using as a genetic marker the apoE phenotype E2/2, *J. Lipid Res.* **23**:1224–1235.

889. Gregg, R. E., Zech, L. A., Schaefer, E. J., and Brewer, Jr., H. B., 1981, Type III hyperlipoproteinemia: Defective metabolism of an abnormal apolipoprotein E, *Science* **211**:584–586.

890. Havel, R. J., Chao, Y., Windler, E. E., Kotite, L., and Guo, L. S., 1980, Isoprotein specificity in the hepatic uptake of apolipoprotein E and the pathogenesis of familial dysbetalipoproteinemia, *Proc. Natl. Acad. Sci. USA* **77**:4349–4353.

891. Gregg, R. E., Zech, L. A., Schaefer, E. J., Stark, D., Wilson, D., and Brewer, Jr., H. B., 1986, Abnormal *in vivo* metabolism of apolipoprotein E4 in humans, *J. Clin. Invest.* **78**:815–821.

892. Schneider, W. J., Kovanen, P. T., Brown, M. S., Goldstein, J. L., Utermann, G., Weber, W., Havel, R. J., Kotite, L., Kane, J. P., Innerarity, T. L., and Mahley, R. W., 1981, Familial dysbetalipoproteinemia. Abnormal binding of mutant apoprotein E to low density lipoprotein receptors of human fibroblasts and membranes from liver and adrenal of rats, rabbits, and cows, *J. Clin. Invest.* **68**:1075–1085.

893. Rall, Jr., S. C., Weisgraber, K. H., Innerarity, T. L., and Mahley, R. W., 1982, Structural basis for receptor binding heterogeneity of apolipoprotein E from type III hyperlipoproteinemic subjects, *Proc. Natl. Acad. Sci. USA* **79**:4696–4703.

894. Weisgraber, K. H., Innerarity, T. L., and Mahley, R. W., 1982, Abnormal lipoprotein receptor-binding activity of the human E apoprotein due to cysteine–arginine interchange at a single site, *J. Biol. Chem.* **257**:2518–2521.

895. Innerarity, T. L., Weisgraber, K. H., Arnold, K. S., Rall, Jr., S. C., and Mahley, R. W., 1984, Normalization of receptor binding of apolipoprotein E2. Evidence for modulation of the binding site conformation, *J. Biol. Chem.* **259**:7261–7267.

896. Havel, R. J., and Kane, J. L., 1973, Primary dysbetalipoproteinemia: Predominance of a specific apoprotein species in triglyceride-rich lipoproteins, *Proc. Natl. Acad. Sci. USA* **70**:2015–2019.

897. Hazzard, W. R., and Bierman, E. L., 1976, Delayed clearance of chylomicron remnants following vitamin-A-containing oral fat loads in broad-B disease (type III hyperlipoproteinemia), *Metabolism: Clinical and Experimental* **25**:777–801.

898. Curry, M. D., McConathy, W. J., Alaupovic, P., Ledford, J. H., and Popovic, M., 1976, Determination of human apolipoprotein E by electroimmunoassay, *Biochim. Biophys. Acta* **439**:413–425.

899. Kushwaha, R. S., Hazzard, W. R., Wahl, P. W., and Hoover, J. J., 1977, Type III hyperlipoproteinemia: Diagnosis in whole plasma by apolipoprotein-E immunoassay, *Ann. Intern. Med.* **87**:509–516.

900. Chait, A., Hazzard, W. R., Albers, J. J., Kushwaha, R. P., and Brunaell, J. D., 1978, Impaired very low density lipoprotein and triglyceride removal in broad beta disease: Comparison with endogenous hypertriglyceridemia, *Metabolism: Clinical and Experimental* **27**:1055–1066.

901. Mahley, R. W., Innerarity, T. L., Rall, Jr., S. C., Weisgraber, K. H., and Taylor, J. M., 1990, Apolipoprotein E: Genetic variants provide insights into its structure and function, *Curr. Opin. Lipidol.* **1**:87–95.

902. Bersot, T. P., Innerarity, T. L., Mahley, R. W., and Havel, R. J., 1983, Cholesteryl ester accumulation in mouse peritoneal macrophages induced by beta-migrating very low density

lipoproteins from patients with atypical dysbetalipoproteinemia, *J. Clin Invest* 72(3):1024–1033.

903. Havel, R. J., Kotite, L., Kane, J. P., Tun, P., and Bersot, T., 1983, Atypical familial dysbetalipoproteinemia associated with apolipoprotein phenotype E3/3, *J. Clin. Invest.* 72:379–387.

904. Wardell, M. R., Brennan, S. O., Janus, E. D., Fraser, R., and Carrell, R. W., 1987, Apolipoprotein E2-Christchurch (136 Arg–Ser, New variant of human apolipoprotein E in a patient with type III hyperlipoproteinemia, *J. Clin. Invest.* 80:483–490.

905. Rall, Jr., S. C., Newhouse, Y. M., Clarke, H. R. G., Weisgraber, K. H., McCarthy, B. J., Mahley, R. W., and Bersot, T. P., 1989, Type III hyperlipoproteinemia associated with apolipoprotein E phenotype E3/3: Structure and genetics of an apolipoprotein E3 variant, *J. Clin. Invest.* 83:1095–1101.

906. Mann, W. A., Gregg, R. E., Sprecher, D. L., and Brewer, Jr., H. B., 1989, Apolipoprotein E-1Harrisburg: A new variant of apolipoprotein E dominantly associated with type III hyperlipoproteinemia, *Biochim. Biophys. Acta* 1005:239–244.

907. Wardell, M. R., Weisgraber, K. H., Havekes, L. M., and Rall, Jr., S. C., 1989, Apolipoprotein E3-Leiden contains a seven-amino acid insertion that is a tandem repeat of residues 121–127, *J. Biol. Chem.* 264:21205–21210.

908. Van den Maagdenberg, A. M., de Knijff, P., Stalenhoef, A. F., Gevers Leuven, J. A., Havekes, L. M., and Frants, R. R., 1989, Apolipoprotein E*3-Leiden allele results from a partial gene duplication in exon 4, *Biochem. Biophys. Res. Commun.* 165:851–857.

909. Smit, M., de Knijff, P., van der Kooij-Meijs, E., Groenendijk, C., van den Maagdenberg, A. M., Gevers Leuven, J. A., Stalenhoef, A. F., Stuyt, P. M., Frants, R. R., and Havekes, L. M., 1990, Genetic heterogeneity in familial dysbetalipoproteinemia. The E2(Lys146–Gln) variant results in a dominant mode of inheritance, *J. Lipid Res.* 31:45–53.

910. Wardell, M. R., Rall, Jr., S. C., Brennan, S. O., Nye, E. R., George, P. M., Janus, E. D., and Weisgraber, K. H., 1990, Apolipoprotein E2-Dunedin (228 Arg replaced by Cys): An apolipoprotein E2 variant with normal receptor-binding activity, *J. Lipid Res.* 31:535–543.

911. McLean, J. W., Elshourbagy, N. A., Chang, D. J., Mahley, R. W., and Taylor, J. M., 1984, Human apolipoprotein E mRNA. cDNA cloning and nucleotide sequencing of a new variant, *J. Biol. Chem.* 259:6498–6504.

912. Weisgraber, K. H., Rall, Jr., S. C., Innerarity, T. L., and Mahley, R. W., 1984, A novel electrophoretic variant of human apolipoprotein E: Identification and characterization of apolipoprotein E1, *J. Clin. Invest.* 73:1024–1033.

913. Steinmetz, A., Assefbarkhi, N., Eltze, C., Ehlenz, K., Funke, H., Pies, A., Assmann, G., and Kaffarnik, H., 1990, Normolipemic dysbetalipoproteinemia and hyperlipoproteinemia type III in subjects homozygous for a rare genetic apolipoprotein E variant (apoE1), *J. Lipid Res.* 31:1005–1013.

914. Wardell, M. R., Rall, Jr., S. C., Schaefer, E. J., Kane, J. P., and Weisgraber, K. H., 1991, Two apolipoprotein E5 variants illustrate the importance of the position of additional positive charge on receptor-binding activity, *J. Lipid Res.* 32:521–528.

915. Maeda, H., Nakamura, H., Kobori, S., Okada, M., Mori, H., Niki, H., Ogura, T., and Hiraga, S., 1989, Identification of human apolipoprotein E variant gene: Apolipoprotein E7 (Glu244,245–Lys244,245), *J. Biochem.* 105:51–54.

916. Tajima, S., Yamamura, T., Menju, M., and Yamamoto, A., 1989, Analysis of apolipoprotein E7 (apolipoprotein E-Suita) gene from a patient with hyperlipoproteinemia, *J. Biochem.* 105:249–253.

917. Lohse, P. Mann, W. A., Stein, E. A., and Brewer, Jr., H. B., 1991, Apolipoprotein E-4Philadelphia (Glu13–Lys,Arg145–Cys). Homozygosity for two rare point mutations in the apolipoprotein E gene combined with severe type III hyperlipoproteinemia, *J. Biol. Chem.* 266:10479–10484.

918. Mailly, F., Xu, C. F., Xhignesse, M., Lussier-Cacan, S., Talmud, P. J., Davignon, J., Humphries,

S. E., and Nestruck, A. C., 1991, Characterization of a new apolipoprotein E5 variant detected in two French-Canadian subjects, *J. Lipid Res.* **32**:613–620.

919. Tajima, S., Yamamura, T., and Yamamoto, A., 1988, Analysis of apolipoprotein E5 gene from a patient with hyperlipoproteinemia, *J. Biochem.* **104**:48–52.

920. Maeda, H., Nakamura, H., Kobori, S., Okada, M., Niki, H., Ogura, T., and Hiraga, S., 1989, Molecular cloning of a human apolipoprotein E variant: E5 (Glu3–Lys3), *J. Biochem.* **105**: 491–493.

921. Dong, L. M., Yamamura, T., and Yamamoto, A., 1990, Enhanced binding activity of an apolipoprotein E mutant, APO E5, to LDL receptors on human fibroblasts, *Biochem. Biophys. Res. Commun.* **168**:409–414.

922. Ghiselli, G., Schaefer, E. J., Gascon, P., and Breser, Jr., H. B., 1981, Type III hyperlipoproteinemia associated with apolipoprotein E deficiency, *Science* **214**:1239–1241.

923. Schaefer, E. J., Gregg, R. E., Ghiselli, G., Forte, T. M., Ordovas, J. M., Zech, L. A., and Brewer, B. H., 1986, Familial apolipoprotein E deficiency, *J. Clin. Invest.* **78**:1206–1219.

924. Cladaras, C., Hadzopoulou-Cladaras, M., Felber, B. K., Pavlakis, G., and Zannis, V. I., 1987, The molecular basis of a familial apoE deficiency. An acceptor splice site mutation in the third intron of the deficient apoE gene, *J. Biol. Chem.* **262**:2310–2315.

925. Lohse, P., Brewer, B., Meng, M. S., LaRosa, J., and Brewer, Jr., H. B., 1989, Familial apolipoprotein (apoE) E deficiency: Identification of a new kindred with a premature stop codon in the apoE gene, *Circulation* **80**:277.

925a. Plump, A. S., *et al.*, 1992, Severe hypercholesterolemia and atherosclerosis in apolipoprotein E-deficient mice created by homologous recombination in ES cells, *Cell* **71**:343–353.

925b. Zhang, S. H., *et al.*, 1992, Spontaneous hypercholesterolemia and arterial lesions in mice lacking apolipoprotein E, *Science* **258**:468–471.

926. Fredrickson, D. S., Levy, R. I., and Lees, R. S., 1967, Fat transport in lipoproteins—An integrated approach to mechanisms and disorders, *N. Engl. J. Med.* **276**:34–44.

927. Fredrickson, D. S., and Levy, R. I., 1972, Familial hyperlipoproteinemia, in: *The Metabolic Basis of Inherited Disease* (J. B. Stanbury, J. B. Wyngaarden, and D. S. Fredrickson, eds.), McGraw-Hill, New York, pp. 545–614.

928. Marien, K. J., Hulsmans, H. A., and vanGent, C. M., 1974, On a family with coexistence of phenotypes II and III hyperlipoproteinemia, *Acta Med. Scand.* **196**:149–153.

929. Morganroth, J., Levy, R. I., and Fredrickson, D. S., 1975, The biochemical, clinical, and genetic features of type III hyperlipoproteinemia, *Ann. Intern. Med.* **82**:158–174.

930. Utermann, G., Vogelberg, K. H., Steinmetz, A., Schoenborn, W., Pruin, N., Jaeschke, M., Hees, M., and Canzler, H., 1979, Polymorphism of apolipoprotein E. II. Genetics of hyperlipoproteinemia type III, *Clin. Genet.* **15**:37–62.

931. Hazzard, W. R., Warnick, G. R., Utermann, G., and Albers, J. J., 1981, Genetic transmission of isoapolipoprotein E phenotypes in a large kindred: Relationship to dysbetalipoproteinemias and hyperlipidemia, *Metabolism: Clinical and Experimental* **30**:79–89.

932. Rall, Jr., S. C., Weisgraber, K. H., Innerarity, T. L., and Mahley, R. W., 1983, Identical structure and receptor binding defects in apolipoprotein E2 in hypo-, normo-, and hypercholesterolemic dysbetalipoproteinemia, *J. Clin. Invest.* **71**:1023–1031.

933. Kostner, G. M., 1974, Studies of the composition and structure of human serum lipoproteins: Isolation and partial characterization of apolipoprotein A-III, *Biochim. Biophys. Acta* **336**: 383–395.

934. Curry, M. D., McConathy, W. J., and Alaupovic, P., 1977, Quantitative determination of human apolipoprotein D by electroimmunoassay and radial immunodiffusion, *Biochim. Biophys. Acta* **491**:232–241.

935. McConathy, W. J., and Alaupovic, P., 1976, Studies on the isolation and partial characterization of apolipoprotein D and lipoprotein D of human plasma, *Biochemistry* **15**:515–520.

936. McConathy, W. J., and Alaupovic, P., 1973, Isolation and partial characterization of apolipo-

protein D: A new protein moiety of the human plasma lipoprotein system, *FEBS Lett.* **37**: 178–182.

937. Drayna, D., McLean, J. W., Wion, K. L., Trent, J. M., Drabkin, H. A., and Lawn, R. M., 1987, Human apolipoprotein D gene: Gene sequence, chromosome localization, and homology to the a2-globulin superfamily, *DNA* **6**:199–204.

938. Boyles, J. K., Notterpek, L. M., Wardell, M. R., and Rall, Jr., S. C., 1990, Identification, characterization and tissue distribution of apolipoprotein D in the rat, *J. Lipid Res.* **31**:2243–2256.

939. Albers, J. J., Lin, J. T., and Roberts, G. P., 1979, Effect of human plasma apolipoproteins on the activity of purified lecithin:cholesterol acyltransferase, *Artery* **5**:61–75.

940. Kostner, G., 1974, Studies on the cofactor requirements for lecithin:cholesterol acyltransferase, *Scand. J. Clin. Invest.* **33**:19–21.

941. Peitsch, M., and Boguski, M. S., 1990, Is apolipoprotein D a mammalian bilin-binding protein? *New Biol.* **2**:197–206.

942. Utermann, G., 1989, The mysteries of lipoprotein(a), *Science* **246**:904–910.

943. Utermann, G., 1990, Lipoprotein (a): A genetic risk factor for premature coronary heart disease, *Curr. Opin. Lipidol.* **1**:404–410.

944. Scanu, A. M., 1991, Update on lipoprotein (a), *Curr. Opin. Lipidol.* **2**:253–258.

945. Berg, K., 1963, A new serum type system in man: The Lp system, *Acta Pathol. Microbiol. Scand.* **59**:369–382.

946. Scanu, A., 1990, Lipoprotein(a): Heterogeneity and biological relevance, *J. Clin. Invest.* **85**: 1709–1715.

947. Gaubatz, J. W., Heideman, C., Gotto, Jr., A. M., Morrisett, J. D., and Dahlen, G. H., 1982, Human plasma lipoprotein (a): Structural properties, *J. Biol. Chem.* **258**:4582–4589.

948. Utermann, G., and Weber, W., 1983, Protein composition of Lp(a) lipoprotein from human plasma, *FEBS Lett.* **154**:357–361.

949. McLean, J. W., Tomlinson, J. E., Kuang, W. J., Eaton, D. L., Chen, E. Y., Fless, G. M., Scanu, A. M., and Lawn, R. M., 1987, cDNA sequence of human apolipoprotein(a) is homologous to plasminogen, *Nature* **330**:132–137.

950. Kraft, H. G., Menzel, H. J., Hoppichler, F., Vogel, W., and Utermann, G., 1989, Changes of genetic apolipoprotein phenotypes causes by liver transplantations: Implications for apolipoprotein synthesis, *J. Clin. Invest.* **83**:137–142.

951. Henriksson, P., Angelin, B., and Berglund, L., 1991, Hormonal effects on serum Lp (a) levels: Marked reduction during estrogen treatment in males with prostatic cancer, *Circulation* **84** (Suppl. II):484 (Abstract).

952. Salonen, E. M., Jauhiainen, M., Zardi, L., Vaheri, A., and Ehnholm, C., 1989, Lipoprotein(a) binds to fibronectin and has a serine proteinase activity capable of cleaving it, *EMBO J.* **8**:4035–4040.

953. Krempler, F., Kostner, G. M., Roscher, A., Bolzano, K., and Sandhofer, F., 1984, The interaction of human apoB-containing lipoproteins with mouse peritoneal macrophages: A comparison of Lp(a) with LDL, *J. Lipid Res.* **25**:283–287.

954. Bersot, T. P., Innerarity, T. L., Pitas, R. E., Rall, Jr., S. C., Weisgraber, K. H., and Mahley, R. W., 1986, Fat feeding in humans induces lipoproteins of density less than 1.006 that are enriched in apolipoprotein[a] and that cause lipid accumulation in macrophages, *J. Clin. Invest.* **77**: 622–630.

955. Miles, L. A., Hoover-Plow, J. L., Levin, E. G., Curtiss, L. K., Baynham, P. J., Fless, G. M., Scanu, A. M., and Plow, E. F., 1991, Interaction of Lp (a) with plasminogen receptors, *Circulation* **84**(Suppl. II):566 (Abstract).

955a. Lawn, R. M., Wade, D. P., Hammer, R. E., Chiesa, G., Verstuyft, J. G., and Rubin, E. M., 1992, Atherogenesis in transgenic mice expressing human apolipoprotein(a), *Nature* **360**:670–672.

956. Scanu, A. M., 1991, Lp(a) as a marker of coronary heart disease, *Clin. Cardiol.* **14**:35–39.
957. Marz, W., Aygoren, E., Trommlitz, M., Scharrer, I., and Gross, W., 1990, Lipoprotein(a): Risikoindicator by thromboembolischem erkrankungen, *Kolin-Wochenschr.* **68**:111–112.
958. Tranchesi, B., Marankao, R., Cobbaert, C., Vanhove, P., and Verstraete, M., 1990, Lack of association between raised serum lipoprotein(a) and thrombolysis, *Lancet* **336**:1587–1588.
959. Alessi, M. C., Parra, H. J., Joly, P., Vu-Dac, N., Bard, J. M., Fruchart, J. C., and Juhan-Vague, I., 1990, The increased plasma Lp(a): B lipoprotein particle concentration in angina pectoris is not associated with hypofibrinolysis, *Clin. Chim. Acta* **188**:119–127.
960. Loscalzo, J., 1990, Lipoprotein(a). A unique risk factor for atherothrombotic disease, *Arteriosclerosis* **10**:671–679.
961. Mbewu, A. D., and Durrington, P. N., 1990, Lipoprotein(a): Structure, properties and possible involvement in thrombogenesis and atherogenesis, *Atherosclerosis* **85**:1–14.
962. Edelberg, J. M., Gonzalez-Gronow, M., and Pizzo, S. W., 1989, Lipoprotein(a) inhibits streptokinase mediated activation of human plasminogen, *Biochemistry* **28**:2370–2374.
963. Harpel, P. C., Gordon, B. R., and Parker, T. S., 1989, Plasmin catalyzes binding of lipoprotein(a) to immobilized fibrinogen and fibrin, *Proc. Natl. Acad. Sci. USA* **86**:3847–3851.
964. Etingin, O. R., Hajjar, D. P., Hajjar, K., Harpel, P. C., and Nachman, R. L., 1991, Lipoprotein(a) regulates plasminogen activator inhibitor-1 expression in endothelial cells. A potential mechanism of thrombogenesis, *J. Biol. Chem.* **266**:2459–2465.
965. Armstrong, V. W., Neubauer, C., Schutz, E., and Tebbe, U., 1990, Lack of association between raised serum Lp(a) concentrations and unsuccessful thrombolysis after acute myocardial infarction, *Lancet* **336**:1077.
966. Utermann, G., Hoppichler, F., Dieplinger, H., Seed, M., Thompson, G. R., and Boerwinkle, E., 1989, Defects in the low density lipoprotein receptor gene affect lipoprotein (a) levels: Multiplicative interaction of two gene loci associated with premature atherosclerosis, *Proc. Natl. Acad. Sci. USA* **86**:4171–4174.
967. Hofmann, S. L., Eaton, D. L., Brown, M. S., McConathy, W. J., Goldstein, J. L., and Hammer, R. E., 1990, Overexpression of human low density lipoprotein receptors leads to accelerated catabolism of Lp(a) lipoprotein in transgenic mice, *J. Clin. Invest.* **85**:1542–1547.
968. Williams, K. J., Fless, G. M., Petrie, K., Snyder, M. L., Brocia, R. W., and Swenson, T. L., 1991, Lipoprotein lipase enhances cellular catabolism of lipoprotein (a), *Circulation* **84**(Suppl II):566 (Abstract).
969. Krempler, F., Kostner, G. M., Foscher, A., Haslauer, F., Bolzano, K., and Sandhofer, F., 1983, Studies on the role of specific cell surface receptors in the removal of lipoprotein (a) in man, *J. Clin. Invest.* **71**:1431–1441.
970. Drayna, D. T., Hegele, R. A., Hass, P. E., Emi, M., and Wu, L. L., 1988, Genetic linkage between lipoprotein(a) phenotype and a DNA polymorphism in the plasminogen gene, *Genomics* **3**:230–236.
971. Frank, S. L., Klisak, I., Sparkes, R. S., Mohandas, T., Tomlinson, J. E., McLean, J. W., Lawn, R. M., and Lusis, A. J., 1988, The apolipoprotein(a) gene resides on human chromosome 6q26–27, in close proximity to the homologous gene for plasminogen, *Hum. Genet.* **79**:352–356.
972. Lindahl, G., Gersdorf, E., Menzel, H. J., Duba, C., Cleve, H., Humphries, S., and Utermann, G., 1989, The gene for the Lp(a)-specific glycoprotein is closely linked to the gene for plasminogen on chromosome 6, *Hum. Genet.* **81**:149–152.
973. Gavish, D., Azrolan, N., and Breslow, J. L., 1989, Plasma Ip(a) concentration is inversely correlated with the ratio of Kringle IV/Kringle V encoding domains in the apo(a) gene, *J. Clin. Invest.* **84**:2021–2027.
974. Hixson, J. E., Britten, M. L., Scott Manis, G., and Rainwater, D. L., 1989, Apolipoprotein(a) [apo(a)] glycoprotein isoforms result from size differences in apo(a) mRNA in baboons, *J. Biol. Chem.* **264**:6013–6016.

975. Lindahl, G., Gersdorf, E. Menzel, H. J., Seed, M., Humphries, S., and Utermann, G., 1990, Variation in the size of human apolipoprotein(a) is due to a hypervariable region in the gene, *Hum. Genet.* **84**:563–567.

976. Utermann, G., Menzel, H. J., Kraft, H. G., Duba, H. C., Kemmler, H. G., and Seitz, C., 1987, Lp(a) glycoprotein phenotypes. Inheritance and relation to Lp(a)-lipoprotein concentrations in plasma, *J. Clin. Invest.* **80**:458–465.

977. Utermann, G., Duba, C., and Menzel, H. J., 1988, Genetics of the quantitative Lp(a) lipoprotein trait: II Inheritance of Lp(a) glycoprotein phenotypes, *Hum. Genet.* **78**:47–50.

978. Utermann, G., Kraft, H. G., Menzel, H. J., Hopferwieser, T., and Seitz, C., 1988, Genetics of quantitative Lp(a) lipoprotein trait: I. Relation of Lp(a) glycoprotein phenotypes to Lp(a) lipoprotein concentrations in plasma, *Hum. Genet.* **78**:41–46.

979. Rader, D., Cain, W., Zech, L., Kindt, M., Usher, D., and Brewer, Jr., H. B., 1991, Lp(a): Plasma levels in individuals with the same apo(a) isoprotein are determined by differences in production rates, *Circulation* **84**(Suppl. II):565 (Abstract).

980. Seed, M., Hoppichler, F., Reaveley, D., McCarthy, S., Thompson, G. R., Boerwinkle, E., and Utermann, G., 1990, Relation of serum lipoprotein(a) concentration and apolipoprotein(a) phenotype to coronary heart disease in FH, *N. Engl. J. Med.* **322**:1494–1499.

981. Wiklund, O., Angelin, B., Olofsson, S. O., Eriksson, M., Fager, G., Berglund, L., and Bondjers, G., 1990, Apolipoprotein(a) and ischaemic heart disease in familial hypercholesterolaemia [see comments], *Lancet* **335**:1360–1363.

982. Armstrong, V. W., Cremer, P., Eberle, E., Manke, A., Schuzef, F., Wieland, H., Kreuzer, H., and Seidel, D., 1986, The association between serum Lp(a) concentrations and angiographically assessed coronary atherosclerosis dependence on serum LDL levels, *Atherosclerosis* **61**: 249–257.

983. Dahlen, G. H., Guyton, J. R., Attar, M., Farmer, J. A., and Kautz, J. A., 1986, Association of levels of lipoprotein Lp(a), plasma lipids, and other proteins with coronary artery disease documented by angiography, *Circulation* **74**:758–765.

984. Rhoads, G. G., Dahlen, G., Berg, K., Morton, N. E., and Dannenberg, A. L., 1986, Lp(a) lipoprotein as a risk factor for myocardial infarction, *J. Am. Med. Assoc.* **256**:2540–2544.

985. Olofsson, S. O., McConathy, W. J., and Alaupovic, P., 1978, Isolation and partial characterization of a new acidic apolipoprotein (apolipoprotein F) from high density lipoproteins of human plasma, *Biochemistry* **17**:1032–1036.

986. Ayrault-Jarrier, M., Alix, J. F., and Polonovsky, J., 1978, Une nouvelle proteine des lipoproteines du serum humain: Isolement et caracterisation partielle d'une apolipoproteine G, *Biochimie* **60**:65–70.

987. Polz, E., and Kostner, G. M., 1979, The binding of beta 2-glycoprotein-I to human serum lipoproteins: Distribution among density fractions, *FEBS Lett.* **102**:183–186.

988. Nakaya, Y., Schaefer, E. J., and Brewer, Jr., H. B., 1980, Activation of human post heparin lipoprotein lipase by apolipoprotein H (beta 2-glycoprotein I), *Biochem. Biophys. Res. Commun.* **95**:1168–1172.

989. Lee, N. S., Brewer, Jr., H. B., and Osborne, Jr., J. C., 1983, B2-glycoprotein I. Molecular properties of an unusual apolipoprotein, apolipoprotein H, *J. Biol. Chem.* **258**:4765–4770.

990. Polz, E., and Kostner, G. M., 1979, Binding of B2-glycoprotein 1 to intralipid: Determination of the dissociation constant, *Biochem. Biophys. Res. Commun.* **90**:1305–1312.

991. Sata, T., Havel, R. J., Kotite, L., and Kane, J. L., 1976, New protein in human blood plasma, rich in proline, with lipid-binding properties, *Proc. Natl. Acad. Sci. USA* **73**:1063–1067.

992. De Silva, H. V., Stuart, W. D., Duvic, C. R., Wetterau, J. R., Ray, M. J., Ferguson, D. B., Albers, H. W., Smith, W. R., and Harmony, J. A. K., 1990, A 70kD apolipoprotein designated apoJ is a marker for subclasses of human plasma high density lipoproteins, *J. Biol. Chem.* **265**: 13240–13247.

993. Rosenthal, C. J., and Franklin, E. C., 1975, Variation with age and disease of an amyloid A protein-related serum component, *J. Clin. Invest.* **55**:746–753.

994. Benditt, E. P., and Eriksen, N., 1977, Amyloid protein SAA is associated with high density lipoprotein from human serum, *Proc. Natl. Acad. Sci. USA* **74**:4025–4028.

995. Bausserman, L. L., Herbert, P. N., and McAdam, K. P. W. L., 1980, Heterogeneity of human serum amyloid A proteins, *J. Exp. Med.* **152**:641–656.

996. Hoffman, J. S., and Benditt, E. P., 1982, Secretion of serum amyloid protein and assembly of serum amyloid protein-rich high density lipoprotein in primary mouse hepatocyte culture, *J. Biol. Chem.* **257**:10518–10522.

997. Eriksen, N., and Benditt, E. P., 1980, Isolation and characterization of the amyloid-related apoprotein (SAA) from human high density lipoprotein, *Proc. Natl. Acad. Sci. USA* **77**:6860–6864.

998. Parmelee, D. C., Titani, K., Ericsson, L. H., Eriksen, N., Benditt, E. P., and Walsh, K. A., 1982, Amino acid sequence of amyloid-related apoprotein (apoSAA1) from human high-density lipoprotein, *Biochemistry* **21**:3298–3303.

999. Dwulet, F. E., Wallace, D. K., and Benson, M. D., 1988, Amino acid structures of multiple forms of amyloid-related serum protein SAA from a single individual, *Biochemistry* **27**:1677–1682.

1000. Woo, P., Sipe, J., Dinarello, C. A., and Colten, H. R., 1987, Structure of a human serum amyloid A gene and modulation of its expression in transfected L cells, *J. Biol. Chem.* **262**:15790–15795.

1001. Kluve-Beckerman, B., Dwulet, F. E., and Benson, M. D., 1988, Human serum amyloid A. Three hepatic mRNAs and the corresponding proteins in one person, *J. Clin. Invest.* **82**:1670–1675.

1002. Beach, C. M., DeBeer, M. C., Sipe, J. D., Loose, L. D., and DeBeer, F. C., 1991, Mouse serum amyloid A protein: Complete amino acid sequence and mRNA and analysis of a new isoform, *Biochem. J.* **283**:673–678.

1003. Steinkasserer, A., Weiss, E. H., Schwaeble, W., and Linke, R. P., 1990, Heterogeneity of human serum amyloid A protein. Five different variants from one individual demonstrated by cDNA sequence analysis, *Biochem. J.* **268**:187–193.

1004. Sellar, G. C., DeBeer, M. C., Lelias, J. M., Snyder, P. W., Glickman, L. T., Felsburg, P. J., and Whitehead, A. S., 1991, Dog serum amyloid A protein, *J. Biol. Chem.* **266**:3500–3510.

1005. Lowell, G. A., Potter, D. A., Stearman, R. S., and Morrow, J. F., 1986, Structure of the murine serum amyloid A gene family, *J. Biol. Chem.* **261**:8442–8452.

1006. Zannis, V. I., Kardassis, D., Ogami, K., Hadzopoulou-Cladaras, M., and Cladaras, C., 1990, Transcriptional regulation of the human apolipoprotein genes, in: *Advances in Experimental Medicine and Biology* (C. L. Malmendier, P. Alaupovic, and H. B. Brewer, Jr., eds.), Plenum Press, New York, pp. 1–23.

1007. Brinster, R. L., Chen, H. Y., Trumbauer, M. E., Yagle, M. K., and Palmiter R. D., 1985, Factors affecting the efficiency of introducing foreign DNA into mice by microinjecting eggs, *Proc. Natl. Acad. Sci. USA* **82**:4438–4442.

1008. Taylor, J. M., Simonet, W. S., Bucay, N., Lauer, S. J., and deSilva, H. V., 1991, Expression of the human apolipoprotein E/apolipoprotein C-I gene locus in transgenic mice, *Curr. Opin. Lipidol.* **2**:73–80.

1009. Kardassis, D., Hadzopoulou-Cladaras, M., Ramji, D. P., Cortese, R., Zannis, V. I., and Cladaras, C., 1990, Characterization of the promoter elements required for hepatic and intestinal transcription of the human apoB gene: Definition of the DNA-binding site of a tissue-specific transcriptional factor, *Mol. Cell. Biol.* **10**:2653–2659.

1010. Kardassis, D., Zannis, V. I., and Cladaras, C., 1992, Organization of the regulatory elements and nuclear activities participating in the transcription of the human apolipoprotein B gene, *J. Biol. Chem.* **267**:2622–2632.

1011. Kardassis, D., Zannis, V. I., and Cladaras, C., 1990, Purification and characterization of the nuclear factor BA1. A transcriptional activator of the human apoB gene, *J. Biol. Chem.* **265:** 21733–21740.

1012. Carlsson, P., and Bjursell, G., 1989, Negative and positive promoter elements contribute to tissue specificity of apolipoprotein B expression, *Gene* **77:**113–121.

1013. Das, H. K., Leff, T., and Breslow, J. L., 1988, Cell type-specific expression of the human apoB gene is controlled by two cis-acting regulatory regions, *J. Biol. Chem.* **263:**11452–11458.

1014. Brooks, A. R., Blackhart, B. D., Haubold, K., and Levy-Wilson, B., 1991, Characterization of tissue-specific enhancer elements in the second intron of the human apolipoprotein B gene, *J. Biol. Chem.* **266:**7848–7859.

1015. Paulweber, B., Brooks, A. R., Nagy, B. P., and Levy-Wilson, B., 1991, Identification of a negative regulatory region 5′ of the human apolipoprotein B promoter, *J. Biol. Chem.* **266:** 21956–21961.

1016. Miyajima, N., Kadowaki, Y., Fukushige, S., Shimizu, S., Semba, K., Yamanashi, Y., Matsubara, K., Toyoshima, K., and Yamamoto, T., 1988, Identification of two novel members of *erbA* superfamily by molecular cloning: The gene products of the two are highly related to each other, *Nucleic Acids Res.* **16:**11057–11074.

1017. Sladek, F. M., Zhong, W., Lai, E., and Darnell, Jr., J. E., 1990, Liver enriched transcription factor HNF-4 is a novel member of the steroid hormone receptor superfamily, *Genes Dev.* **47:** 2353–2365.

1018. Ogami, K., Hadzopoulou-Cladaras, M., Cladaras, C., and Zannis, V. I., 1990, Promoter elements and factors required for hepatic and intestinal transcription of the human ApoCIII gene, *J. Biol. Chem.* **265:**9808–9815.

1019. Reue, K., Leff, T., and Breslow, J. L., 1988, Human apolipoprotein CIII gene expression is regulated by positive and negative cis-acting elements and tissue-specific protein factors, *J. Biol. Chem.* **263:**6857–6864.

1020. Leff, T., Reue, K., Melian, A., Culver, H., and Breslow, J. L., 1989, A regulatory element in the ApoCIII promoter that directs hepatic specific transcription binds to proteins in expressing and nonexpressing cell types, *J. Biol. Chem.* **264:**16132–16137.

1021. Ladias, J. A., and Karathanasis, S. K., 1991, Regulation of the apolipoprotein AI gene by ARP-1, a novel member of the steroid receptor superfamily, *Science* **251:**561–565.

1022. Ogami, K., Kardassis, D., Cladaras, C., and Zannis, V. I., 1991, Purification and characterization of a heat stable nuclear factor CIIIB1 involved in the regulation of the human ApoC-III gene, *J. Biol. Chem.* **266:**9640–9646.

1023. Widom, R. L., Ladias, J. A., Kouidou, S., and Karathanasis, S. K., 1991, Synergistic interactions between transcription factors control expression of the apolipoprotein AI gene in liver cells, *Mol. Cell. Biol.* **11:**677–687.

1024. Papazafiri, P., Ogami, K., Ramji, D. P., Nicosia, A., Monaci, P., Cladaras, C., and Zannis, V. I., 1991, Promoter elements and factors involved in hepatic transcription of the human ApoA-I gene positive and negative regulators bind to overlapping sites, *J. Biol. Chem.* **266:**5790–5797.

1025. Rottman, J. N., Widom, R. L., Nadal-Ginard, B., Mahdavi, V., and Karathanasis, S. K., 1991, A retinoic acid-responsive element in the apolipoprotein AI gene distinguishes between two different retinoic acid response pathways, *Mol. Cell. Biol.* **11:**3814–3820.

1026. Lucero, M. A., Sanchez, D., Ochoa, A. R., Brunel, F., Cohen, G. N., Baralle, F. E., and Zakin, M. M., 1989, Interaction of DNA-binding proteins with the tissue-specific human apolipoprotein-AII enhancer, *Nucleic Acids Res.* **17:**2283–2300.

1027. Chambaz, J., Cardot, P., Pastier, D., Zannis, V. I., and Cladaras, C., 1991, Promoter elements and factors required for hepatic transcription of the human ApoA-II gene, *J. Biol. Chem.* **266:** 11676–11685.

1028. Cardot, P., Chambaz, J., Cladaras, C., and Zannis, V. I., 1992, Regulation of the human apoA-II gene by the synergistic action of factors binding to the proximal and distal regulatory elements, *J. Biol. Chem.* **266**:24460–24470.

1029. Smith, J. D., Melian, A., Leff, T., and Breslow, J. L., 1988, Expression of the human apolipoprotein E gene is regulated by multiple positive and negative elements, *J. Biol. Chem.* **263**:8300–8308.

1030. Chang, D. J., Paik, Y. K., Leren, T. P., Walker, D. W., Howlett, G. J., and Taylor, J. M., 1990, Characterization of a human apolipoprotein E gene enhancer element and its associated protein factors, *J. Biol. Chem.* **265**:9496–9504.

1031. Ladias, J. A. A., Hadzopoulou-Cladaras, M., Kardassis, D., Cardot, P., Cheng, J., Zannis, V. I., and Cladaras, C., 1992, Transcriptional regulation of human apolipoprotein genes apoB, apoCIII, and apoAII by members of the steroid hormone receptor superfamily HNF-4, ARP-1, EAR-2, and EAR-3, *J. Biol. Chem.* **267**:15849–15860.

1032. Frain, M., Swart, G., Monaci, P., Nicosia, A., Stampfli, S., Frank, R., and Cortese, R., 1989, The liver-specific transcription factor LF-B1 contains a highly diverged homeobox DNA binding domain, *Cell* **59**:145–157.

1033. Lauer, S. J., Simonet, W. S., Bucay, N., deSilva, H. V., and Taylor, J. M., 1991, Tissue-specific expression of the human apolipoprotein A-IV gene in transgenic mice, *Circulation* **84**(Suppl. II):17 (Abstract).

1033a. Walsh, A., Azrolan, N., Wang, K., Marcigliano, A., O'Connell, A., and Breslow, J. L., 1993, Intestinal expression of the human apoA-I gene in transgenic mice is controlled by a DNA region 3′ to the gene in the promoter of the adjacent convergently transcribed apoCIII gene, *J. Lipid Res.* (in press).

1034. Cladaras, C., Ogami, K., Kardassis, D., Hadzopoulou-Cladaras, M., and Zannis, V. I., 1991, Recognition of proximal human apoCIII promoter by NFY* type, C/EBP, and a family of heat stable activities, *Circulation* **84**:109 (Abstract).

1035. Mansour, S. L., Thomas, K. R., and Capecchi, M. R., 1988, Disruption of the proto-oncogene int-2 in mouse embryo-derived stem cells: A general strategy for targeting mutations to non-selectable genes, *Nature* **336**:348–352.

1036. Dorin, J. R., Inglis, J. D., and Porteous, D. J., 1989, Selection for precise chromosomal targeting of a dominant marker by homologous recombination, *Science* **243**:1357–1360.

1037. Schwartzberg, P.L., Goff, S. P., and Robertson, E. J., 1989, Germ-line transmission of a c-*abl* mutation produced by targeted gene disruption in ES cells, *Science* **246**:799–803.

1038. Jasin, M., and Berg, P., 1988, Homologous integration in mammalian cells without target gene selection, *Genes Dev.* **2**:1353–1363.

1039. Johnson, R. S., Sheng, M., Greenberg, M. E., Kolodner, R. D., Papaioannou, V. E., and Spiegelman, B. M., 1989, Targeting of nonexpressed genes in embryonic stem cells via homologous recombination, *Science* **245**:1234–1236.

1040. Williamson, R., Lee, D., Hagaman, J., and Maeda, N., 1992, Marked reduction of high density lipoprotein cholesterol in mice genetically modified to lack apolipoprotein A-I, *Proc. Natl. Acad. Sci. USA* **89**:7134–7138.

1041. Piedrahita, J. A., *et al.*, 1992, Generation of mice carrying a mutant apolipoprotein E gene inactivated by gene targeting in embryonic stem cells, *Proc. Natl. Acad. Sci. USA* **89**:4471–4475.

1042. Homanics, G. E., Smith, T. J., Zhang, S. H., Lee, D., Young, S. G., and Maeda, N., 1993, Targeted modification of the apolipoprotein B gene results in hypobetalipoproteinemia and developmental abnormalities in mice, *Proc. Natl. Acad. Sci. USA* (in press).

Chapter 4

Genetic Aspects of Cancer

Audrey D. Goddard* and Ellen Solomon

Somatic Cell Genetics
Imperial Cancer Research Fund
London, England WC2A 3PX

INTRODUCTION

Gain-of-Function Mutations: Dominantly Acting Oncogenes

The two basic premises upon which investigation of the malignant phenotype is based are that malignancy results via alterations in the genetic material of the cell, and that the accumulation of these alterations occurs in a multistep fashion. Early indication that normal cellular genes were the targets for mutation came from the observation that the acutely transforming retroviruses contained abnormal copies of cellular genes (the protooncogenes), the presence of which was required for the transforming activity of the viruses [reviewed in J. M. Bishop (1983)]. Over three dozen protooncogenes have been identified through transfection by retroviruses, gene transfection, and homology to retroviral oncogenes (J. M. Bishop, 1985; Rhim, 1988). These activated oncogenes tend to be dominant, their activity overriding that of the normal allele. The multistage hypothesis was supported by *in vitro* studies demonstrating that the conversion of primary cells from the normal to the tumorigenic phenotype requires a minimum of two cooperating oncogenes (Land *et al.*, 1983) and from epidemiological studies of human cancer

Present address: Department of Molecular Biology, Genentech, Inc., South San Francisco, California 94080.

Advances in Human Genetics, Volume 21, edited by Henry Harris and Kurt Hirschhorn. Plenum Press, New York, 1993.

which suggested that two to seven steps were required before cancer was manifested *in vivo* (Armitage and Doll, 1954; Ashley, 1969b; Nordling, 1953).

Loss-of-Function Mutations

In addition to the dominant oncogenes, which acquire novel or aberrant functions through specific mutations, the recessive oncogenes or tumor suppressor genes, whose loss of function is important, also appear to be the targets of alterations in many malignant processes. The presence of these loci, which must be inactivated for malignancy to proceed, was first suggested by two independent lines of evidence.

Familial Predisposition to Malignancy

The existence of tumor suppressor genes was suggested by reports of families affected with an inherited genetic predisposition to one or multiple forms of cancer. Such families are rare and typically exhibit an autosomal dominant mode of inheritance (Knudson, 1986; Li, 1990). The two most common conditions that result in a hereditary predisposition to tumor formation are von Recklinghausen neurofibromatosis (NF-1) and familial adenomatous polyposis (FAP) (Knudson, 1986). It was studies of retinoblastoma (RB), however, a rare intraocular tumor of early childhood, which suggested that an inherited mutation could contribute to the neoplastic process (Knudson, 1971).

RB occurs in both dominantly inherited and nonfamilial (sporadic) forms. Mathematical modeling based upon the distribution of age at first diagnosis for heritable and nonheritable RD led Knudson (1971) to propose that two rate-limiting mutational events are required for RB development. The first mutation could be either germinal (in heritable RB) or somatic (in nonfamilial cases) and the second event somatic in both cases. This "two-hit" hypothesis also fits the epidemiological data for other cancers observed in both familial and sporadic forms, including neuroblastoma, pheochromocytoma, and perhaps Wilms' tumor (WT) (Knudson and Strong, 1972a,b; Breslow *et al.*, 1988).

The germline interstitial deletions of chromosome 13 in the 3% of patients who develop RB in association with mild congenital anomalies (Yunis *et al.*, 1981) suggested that loss of one allele of a gene could be the first mutation. Considering that only a minute proportion of the predisposed cells progress to the malignant state, the predisposing lesions were proposed to be recessive "loss-of-function" mutations in genes such as the tumor suppressors suggested by the experiments demonstrating hybrid suppression of tumorigenicity (see p. 323). In the predisposed cells the recessive nature of the inherited mutations allows them to remain latent until the remaining normal allele is lost. Thus, some familial cancers were suggested to result from recessive mutations which increase the risk of malignancy above that of the normal population by producing a subset of cells requiring one less mutation to initiate malignancy.

This model has been extended to cancers which require multiple cumulative mutations. Like RB, colorectal carcinoma occurs in both familial and sporadic forms; the epidemiological data for colorectal cancer, however, suggest the accumulation of approximately six independent mutational events (Armitage and Doll, 1954) as opposed to two events in RB. Ashley (1969a) compared the age of diagnosis for sporadic colorectal cancer with that in patients with familial disease and concluded that one or two fewer mutations were necessary in those patients carrying an inherited mutation. The similarity between the observations with colorectal cancer, RB, and other tumors suggested that inactivation of a single allele of a regulatory locus with subsequent somatic inactivation of the remaining allele, with or without the requirement for subsequent rate-limiting alterations in the genome, is a common mechanism in the development of human malignancy.

Tumor Suppression in Somatic Cell Hybrids

The suppression of the transformed phenotype in somatic cell hybrids provides the second line of evidence for tumor suppressor loci. Somatic cell hybrids formed by the fusion of tumor cells with normal cells exhibit a normal phenotype (Harris et al., 1969; Stanbridge, 1976). This repression of tumorigenicity is often followed by the reemergence of the tumorigenic phenotype as chromosomes derived from the normal parental cell are lost (Klinger, 1982). Hybrid suppression of tumorigenicity suggests that the normal parental cell contributes "suppressor" genes, absent or inactive in the malignant parent, which are able to regulate some aspects of the transformed phenotype. The repression of the tumorigenic phenotype in hybrids formed between malignant cells derived from tissues of different histological types (for example, carcinoma × sarcoma or carcinoma × melanoma) suggests that more than one "suppressor" locus is involved in the control of malignancy (Weissman and Stanbridge, 1983). In support of this, Stoler and Bouck (1985) determined that human chromosome 1 contained genetic information enabling normal human fibroblasts to suppress the transformed phenotype of carcinogen-transformed baby hamster kidney cells. Suppression of the tumorigenic potential of HeLa cells required the presence of a normal chromosome 11 (Kaelbling and Klinger, 1986; Saxon et al., 1986; Srivatsan et al., 1986).

The loss of normal gene function is a key step in the expression of malignancy. Hybrids formed between normal fibroblasts and malignant cells containing either an activated oncogene, such as *HRAS*, or an expressed viral genome also result in the suppression of tumorigenicity even under conditions in which the expression of oncogenic sequences is maintained (Geiser *et al.*, 1986; Meyer *et al.*, 1974). A normal genetic background is not permissive for these "dominant" changes to induce neoplastic growth. Thus, some normal regulatory function must first be impaired before the effects of these "dominant" changes can be manifested.

CHROMOSOMAL LOCALIZATION OF RECESSIVE ONCOGENES

The study of hybrid suppression of malignancy and inherited predisposition to cancer demonstrated the presence of normal loci in the human genome which function to maintain the balance among cell death, differentiation, and division. Inactivation of both alleles at these regulatory loci perturbs this equilibrium and eventually leads to the development of cancer. Recent technical developments have provided the means to localize precisely these genes to particular chromosomal regions, and the application of current technologies has made possible gene cloning based solely on the knowledge of chromosomal position.

Heritable Cancer Syndromes and Genetic Linkage

The collection of informative pedigrees segregating the disease is an initial step in determining the chromosomal interval harboring the predisposing gene. Any trait, such as the expression of a predisposition to cancer, can be followed through a family and linkage to other markers assessed. Until the beginning of the last decade, mapping genes within the human genome by meiotic linkage was limited due to the lack of polymorphic markers to which genetic linkage could be established and by the small number of affected families available for analysis. Early studies relied upon the use of expression of protein polymorphisms. Such an approach led to the linkage of the gene predisposing to RB with that for the enzyme esterase D (Sparkes *et al.*, 1983).

Linkage to protein polymorphisms or to the expression of genetically controlled phenotypes is severely limited by the sparsity of such markers. The advent of restriction fragment length polymorphisms (RFLPs), which rely upon cloned DNA segments detecting DNA sequence differences that result in the loss or gain of single restriction enzyme recognition and cleavage sites (Kan and Dozy,

1978), and variable number tandem repeats (VNTRs), which are generated by variation in the number of copies of a short repeat element located between two adjacent restriction sites (Nakamura *et al.*, 1987), revolutionized linkage analysis. More recently short di-, tri-, and tetranucleotide repeats (microsatellites) have been discovered to be highly polymorphic (Weber and May, 1989) and easily detected by the polymerase chain reaction (PCR) technique (White *et al.*, 1989). All three classes of DNA polymorphism can be found near and even within genes, the latter providing the potential for complete, unambiguous linkage to the gene causing the disease. The high degree of DNA sequence polymorphism and the isolation of probes which allows detection of these differences have resulted in the construction of a linkage map completely covering the human genome as predicted by Solomon and Bodmer (1979) and Botstein *et al.* (1980).

Like any other inherited disease, when a cancer occurs in a familial form, DNA polymorphisms can be used to determine the chromosome to which the disease is linked. Table I summarizes the chromosomal locations of inherited cancer syndromes determined by linkage analysis. In the absence of any clues to the chromosomal location of a gene, the task of searching the whole genome becomes very labor intensive. Coupling linkage analysis with other gene-mapping techniques such as the search for disease-associated cytogenetic abnormalities and loss of constitutional DNA heterozygosity can narrow the search to a small region of the genome.

TABLE I.   Inherited Cancer-Predisposing Loci Mapped by Linkage Analysis

Syndrome	Location[a]
Beckwith–Wiedemann syndrome	11p15
Bilateral acoustic neurofibromatosis (neurofibromatosis type 2)	22q11–q13.1
Breast cancer, early-onset	17q12–q21
Breast–ovarian carcinoma	17q12–q21
Dysplastic naevus syndrome	1p36
Familial adenomatous polyposis	5q21–q22
Gorlins syndrome	9q22.3–q31
Malignant melanoma/dysplastic naevus syndrome	1p36
Malignant melanoma	9p13–p22
Multiple endocrine neoplasia type 1	11q13
Multiple endocrine neoplasia type 2A	10q11.2
Neurofibromatosis type 1	17q11.2
Retinoblastoma	13q14
VonHippel–Lindau disease	3p25–p26
WAGR syndrome	11p13

[a]For references not cited in the text see Frézal and Schinzel (1991), Gailani *et al.* (1992), Farndon *et al.* (1992), Reis *et al.* (1992), and Cannon-Albright *et al.* (1992).

Chromosomal Abnormalities in Cancer

Karyotypic abnormalities provide essential information in mapping the predisposing lesions in cancers, especially for those in which no inherited form has been identified or for which appropriate families are not available. A specific chromosomal abnormality or a limited number of aberrations associated with the condition may indicate potential candidate regions in which gene(s) affected by the predisposing lesion reside.

Constitutional Abnormalities

Mutations of the same genes are believed to be involved in the development of both inherited and sporadic tumors. The consistent constitutional chromosomal abnormalities (deletions and translocations) observed in the rare patients present-ing with congenital abnormalities in association with malignancy often implicate specific regions of the genome that contain genes important in the etiology of cancer. Examples of this include the interstitial deletions of chromosome 13q in RB patients with congenital abnormalities (Yunis *et al.*, 1981), translocations involving chromosome 17q11.2 in NF-1 patients (Ledbetter *et al.*, 1989; Rey *et al.*, 1987), and deletions of 5q in FAP (Herrera *et al.*, 1986; Hockey *et al.*, 1989) (Table IIA).

Tumor-Specific Abnormalities

The majority of cancer patients, however, do not present with visible constitutional karyotypic changes. Cytogenetic analysis of tumors has revealed the association of specific chromosomal aberrations with particular types of cancer, resulting in the identification of the specific chromosomal regions involved in a number of cancers (summarized in Table IIB).

Often the data collated from the various approaches for mapping the initiating lesion agree, providing a great degree of confidence in the mapping. For example, deletions of 13q were observed in both the constitutional cells of RB patients presenting with other developmental abnormalities and in the tumor cells of spontaneously occurring RB tumors (Balaban *et al.*, 1982; Kusnetsova *et al.*, 1982; Yunis *et al.*, 1981). Subsequent linkage analysis demonstrated that the RB susceptibility gene *(RB1)* was located in 13q14 (Sparkes *et al.*, 1983). Similarly, the chromosomal location of a gene predisposing to a familial form of colorectal cancer (Gardner's syndrome; GS) was suggested by a constitutional deletion of

TABLE II.   Chromosomal Location of Primary
Mutations Implied by Cytogenetic Abnormalities

Disease	Chromosomal region[a]
A. Constitutional abnormalities	
Colorectal carcinoma	5q15–q22
Neurofibromatosis type 1	17q11.2
Renal cell carcinoma	3p11–p14
Retinoblastoma	13q14
Beckwith–Wiedemann syndrome/WT	11p15
WAGR syndrome/WT	11p13
B. Sporadic abnormalities	
Leukemia	5q22–q31
Melanoma	1p11–p22, 9p21
Meningioma	Monosomy 22
Neuroblastoma	1p
Non-small cell lung carcinoma	9q22–pter
Small cell lung carcinoma	3p14–p23

[a]References not cited in the text are as follows: meningioma
(Zang, 1982); neuroblastoma (Gilbert et al., 1984); non-
SCLC (Lukeis et al., 1990).

5q15–q22 in one patient (Herrera et al., 1986) and confirmed by genetic linkage
(Bodmer et al., 1987; Leppert et al., 1987; Meera Khan et al., 1988).

ALLELE LOSS

The Knudson two-hit model of carcinogenesis (Knudson, 1971) proposed that
a mutation transmitted as a dominant trait may be recessive at the cellular level and
lead to cancer when the normal allele is lost or inactivated. This mechanism is also
applicable to malignancies developing as the result of the accumulation of
alterations at multiple loci, some of which must be activated and some inactivated.

The chromosomal mechanisms which can lead to somatic inactivation of loci
in tumors (Fig. 1) suggest another method of determining the location of genes
whose function must be lost for malignancy to develop: assay for tumor-specific
loss of genetic material. Allele loss studies require the identification of locus-
specific polymorphic markers (RFLPs, VNTRs, microsatellites) in the DNA of
constitutional cells. The detection of only one variant in the tumor (homozygous or
hemizygous) when two forms are present in constitutional cells (heterozygous) is

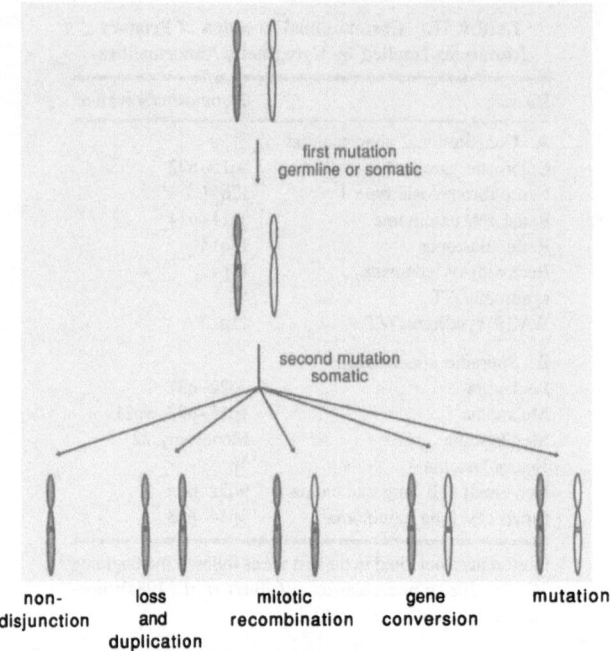

Fig. 1. Mechanisms which can inactivate a tumor suppressor locus during the development of a tumor. The first mutation inactivates one allele at the locus, and can be either a germline event (as in inherited cases) or a somatic event (in sporadic tumors). The second mutation is always a somatic occurrence. The most common mechanisms leading to LOH are chromosomal nondisjunction resulting in simple loss, loss of the chromosome carrying the normal allele followed by duplication of the mutant chromosome, and mitotic recombination. The tumors which do not exhibit LOH probably have undergone a more localized lesion, such as gene conversion, point mutations, or small deletions or insertions. [Adapted from Cavenee *et al.* (1983).]

indicative of loss of genetic material. Studies of chromosome-specific loss of heterozygosity (LOH) in tumor cells affecting a limited genomic region have supported gene localizations suggested by constitutional and tumor-specific deletions or linkage analysis. LOH has been used to suggest the presence of tumor suppressor genes in a wide variety of tumors (summarized in Table III). These studies detect not only loci which predispose to cancer when mutant alleles are inherited, but also those that are inactivated during the progression of the tumor.

Loci predisposing to a number of tumors, such as RB, renal cell carcinoma (RCC), and bilateral acoustic neurofibromatosis (BANF), consistently map to a single chromosomal region in each case by the three methodologies (linkage; cytogenetic abnormality; allele loss) as discussed. The data for WT and multiple endocrine neoplasia type 2A (MEN2A) demonstrate that assessment of LOH does not always point to the site of a predisposing germinal mutation. The consistent loss of heterozygosity in WT at 11p13–p15 and that in MEN2A-associated tumors at 1p33–p35 point to regions carrying tumor suppressor loci; the LOH does not necessarily correlate, however, with the linkage of the predisposing genes in all familial cases. Familial MEN2A maps to 10p11 (Mathew *et al.*, 1987; Simpson *et al.*, 1987) and the familial form of WT has been excluded from the 11p13–p15 region in four pedigrees (Grundy *et al.*, 1988; Huff *et al.*, 1988; Schwartz *et al.*, 1991). In colorectal carcinoma three distinct regions, 5q, 17p, and 18q, show allele loss, although only the *APC* gene on 5q is thought to be associated with hereditary predisposition to disease in FAP and GS families. Predisposition to WT is now known to be associated with at least three different loci (see p. 342), and the same is probably true for colorectal carcinoma.

THE POSITION-DEPENDENT APPROACH TO GENE CLONING

Positional cloning is an approach to isolating genes based solely on their chromosomal location. Once the chromosomal region containing the gene has been defined, sequences from the region are isolated and used to search for gross molecular changes associated with the disease. The ultimate aim is to identify the gene affected by the mutations and determine its function. As the Human Genome Mapping initiative gains momentum, the identification of candidate genes in the region and assessment of their involvement in the disease will be simplified.

Generation of Random Probes

The exclusion of known genes and probes mapped to the region from involvement in the disease necessitates the saturation of the area with new probes. The general approach is to isolate clones from libraries of genomic DNA constructed from a source enriched for the region of interest. A number of different methodologies are available for generating DNA enhanced in a portion of the human genome: flow-sorted chromosomes (Lalande *et al.*, 1984); somatic cell hybrids containing a whole chromosome or a derivative translocation chromosome (Cavenee *et al.*, 1984); somatic cell hybrids containing chromosomal fragments such as chromosome-mediated gene transfectants (Porteous, 1987; Pritchard and Goodfellow, 1987) or radiation hybrids (Cox *et al.*, 1989; Goss and Harris, 1975;

TABLE III. Allele Loss in Tumors

Disease	Chromosome region[a]
Adrenocortical carcinoma	*11p15 (BWS)*; 13q12–q21 (*RB1*); **17** (*TP53*)
Bilateral acoustic neurofibromatosis/ neurofibromatosis 2	
Acoustic neuroma	*22q11.2–q12*
Meningioma	*22q*
Bladder carcinoma	**9q; 11p; 17p13** (*TP53*)
Brain tumors	9p; **10**; 17p11–pter (*TP53*); **22q**
Astrocytomas	
Glioblastoma multiforme	
Gliomas	
Rhabdoid brain tumor	
Breast carcinoma	1p; 1q; 3p21–p25 (*VHL*?q); 6q13–q21; 11p15 (*BWS*?); 13q12–q22 (*RB1*); 16q; **17p13.1** (*TP53*); **17p13.3**; *17q (BRCA1*?); **18q21–qter** (*DCC*?); 22q
Colorectal carcinoma	1p(35–pter); *5q21–q22 (APC)*; **17p13** (*TP53*); **18q21.3–qter** (*DCC*); 22q
Esophageal cancer	**5q** (*APC?, MCC*?); 13q (*RB1*?); 17p (*TP53*)
Gastric carcinoma	1; 5q (*APC*?); 13q (*RB1*?); **17p13** (*TP53*?)
Gorlin syndrome	9q31
Basal cell carcinoma	
Ovarian fibroma	9q31
Hemangioblastoma	*3p21–pter (VHL)*
Hepatoblastoma	*11p15 (BWS?)*
Hepatocellular carcinoma	1; **4q11–q32**; 5q (*APC*?); 10q; 13q12–q22 (*RB1*?) 14q32; **16q22.3–23.2**; **17p13** (*TP53*)
Lung adenocarcinoma (non-small cell)	**3p21–p25**; 11p15 (*BWS*?); 13q12–q32 (*RB1*) 17p13 (*TP53*); 18
Lung carcinoma (small-cell)	**3p21–p25**; 11p15 (*BWS*?); **13q12–q222** (*RB1*); *17p13 (TP53)*
Melanoma	1p36; 1p22–p31;6q; 9p21
Multiple endocrine neoplasia type 1	
Insulinoma	*11q13*
Parathyroid adenoma	*11q13*
Multiple endocrine neoplasia type 2	
Thyroid medullary carcinoma	1p; 1p; 11; 22q
Pheochromocytoma	1p; 1p; 11; 22q
Nasopharyngeal carcinoma	**3p**
Neuroblastoma	1p36; 14q
Neurofibromatosis 1	**17p12–p13** (*TP53*); *17q (NF1)*
Neurofibrosarcoma	
Osteosarcoma	*13q12–q22 (RB1)*; **17p13** (*TP53*); 18q2.13qter (*DCC*?)
Ovarian carcinoma	3p; 6q; 11p; 17p13 (*TP53*?); 17q (*BRACA1*?)
Prostatic carcinoma	**8p**; 10q(22); 16q22–q24
Peripheral neuroepithelioma	17p13

TABLE III. *(Continued)*

Disease	Chromosome region[a]
Renal cell carcinoma	**3p12–p14**; *3p21.3*; *3p21–pter* (*VHL*); 5q (*APC*); 10q; 13q12–q22 (*RB1*?); 18 (*DCC*?)
Retinoblastoma	*13q14* (*RB1*)
Rhabdomyosarcoma	*11p15.5* (*BWS*); 17p13 (*TP53*)
Testicular carcinoma	3p21–pter; 11p15 (*BWS*?)
Thyroid follicular carcinoma	**3p**
Uterine cervical carcinoma	**3p**
Wilms' tumor	*11p13* (*WT1*); *11p15* (*BWS*)

[a]Boldfaced type indicates greater than 50% allele loss in the tumor. Italics indicates a primary genetic event. The genetic loci implicated by allele loss are indicated (where names have been designated) in italics and parentheses. *RB1*, *APC*, *DCC*, *WT1*, *NF1*, and *TP53* have been cloned, while the *VHL*, *BWS*, and *BRCA1* loci have yet to be isolated. A question mark implies a suggestion that the locus may be the target of the allele loss. See Seizinger *et al.* (1991), Boyton *et al.* (1992), Gailani *et al.* (1992), and Fountain *et al.* (1992) for references not cited in the text.

Graw *et al.*, 1988); and microdissected chromosomes (Ludecke *et al.*, 1989). Another approach is to generate probes by Alu-PCR from DNA sources enriched for the region of interest (Guzzetta *et al.*, 1991).

A variety of different library types have proven effective for the generation of random probes (Smith *et al.*, 1987). Linking libraries exploit the observation that CpG islands, in which methylation-sensitive rare-cutting restriction enzyme sites are clustered, are not randomly distributed in the genome, but are associated with the 5' ends of some genes (Gardiner-Garden and Frommer, 1987). Linking libraries are generated with rare-cutting enzymes and therefore are associated with CpG islands and genes. End-clone or jumping libraries consist of the ends of large restriction fragments generated by rare-cutting enzymes cloned without the enormous amount of intervening DNA. Each clone consists of sequences from two adjacent CpG islands. Partial-digest genomic libraries are generated from partially digested DNA and the clones contain 20–50 kilobases (kb) of random DNA fragments. DNA fragments of greater than 1 million base pairs (Mbp) can be stably maintained in yeast in the form of yeast artificial chromosomes (YACs). YAC vectors contain a centromere, two telomeres, and an autonomous replication sequence which enable them to survive and replicate in the yeast cell (Burke *et al.*, 1987; Coulson *et al.*, 1988).

Verification that the random probes are derived from the appropriate region can be accomplished either through *in situ* hybridization of cloned DNA to chromosome spreads using fluorescein-labeled probes (Landegrent *et al.*, 1987) or

mapping panels consisting of somatic cell hybrids retaining different chromosomal subfragments. Both approaches allow mapping of probes to particular
regions of the chromosome.

Physical Mapping

The mapping of both the disease locus and random probes to a narrow
interval is followed by the physical analysis of patient and tumor DNA for clues
to the precise location of the gene. Pulsed field gel electrophoresis (PFGE)
(Anand, 1986) can be used to look for rearrangements over large (up to 2–3 Mbp)
stretches of DNA, while conventional Southern blots cover distances in the range
of 1–50 kbp. The observation of rearrangements localizes the gene to a more
precisely defined region.

The search through a candidate region for suitable genes requires efficient
methods for cloning continuous DNA fragments while looking for transcribed
sequences. This can be accomplished for short distances (50–300 kbp) by walking
and jumping along the chromosome by isolating overlapping cosmid and phage
clones or end clones and linking clone pairs. If larger distances have to be covered,
fragments can be isolated in YAC vectors (Schlessinger, 1990). Potential exon
sequences can be identified as phylogenetically conserved nonrepetitive DNA
fragments. It is possible to look for evidence of transcription by using an
evolutionarily conserved probe on northern blots and cDNA libraries or through a
PCR exon trapping strategy (Fearon et al., 1990).

Evaluation of Candidate Genes

To be a plausible candidate a probe must fulfill certain criteria. It must detect
rearrangements in tumor DNA and/or RNA and be associated with transcribed
sequences. Confirmation that a candidate is the predisposing gene requires the
demonstration of somatic mutations of the gene unique to tumor material and
germline mutation transmitted to affected individuals in families segregating the
disease. In general, Southern and northern blots detect only a small percentage of
mutations in tumors. Techniques which detect more subtle mutations, such as
RNase protection (Myers et al., 1985b), single-strand conformational polymorphisms (Orita et al., 1989), denaturing gradient gel electrophoresis (Myers
et al., 1985a), chemical cleavage (Cotton et al., 1988), and direct sequencing of
PCR amplified complementary DNA (cDNA) or genomic DNA, are more effective
in detecting the elusive mutations that often inactivate genes.

The ultimate proof that any candidate is truly the predisposing gene is a functional test. Reintroduction of the cloned gene into the appropriate tumor should result in some phenotypic changes suggestive of tumor suppression, as has been done with the *RB1* and *TP53* loci (see pp. 334, 342, 351). Alternatively, mutation of the normal gene should result in the predisposition to malignant transformation (see pp. 341, 354).

MOLECULARLY CHARACTERIZED RECESSIVE ONCOGENES

With the advent of genetic localization of genes predisposing to cancer and positional cloning it has become possible to characterize recessive oncogenes at the molecular and functional levels. Tumor suppressor loci have been defined as "genetic element(s) whose loss or inactivation allows a cell to display one or another phenotype of neoplastic deregulation" (Weinberg, 1991). Functionally, these loci are expected to be components of cellular pathways transducing negative-growth regulatory signals controlling the delicate balances among cell proliferation, differentiation, and death. Biochemically, they could act at any stage in the transduction of the antiproliferative signal from the membrane to the nucleus, including extracellular interactions, membrane-associated signal transduction, and, in the nucleus, regulating transcription or cell division. These predictions are fulfilled by the recessive oncogenes isolated to date, as will be discussed in the remainder of this review. A number of other loci which are involved in these processes have been identified through the presence of consistent chromosomal abnormalities in leukemias and lymphomas. For a recent discussion of this, the reader is referred to Solomon *et al.* (1991).

The Retinoblastoma Susceptibility Locus (RB1)

Genetics

Retinoblastoma (RB) provided the first suggestion that inactivation of a specific gene could be important in the development of a human cancer and it was the first tumor in which these molecular events were described. Involvement of the 13q14 region in the etiology of RB was first inferred by the observation of constitutional interstitial deletions of 13q14 in patients presenting with both RB and congenital abnormalities. The importance of this region was further bolstered by the observation of interstitial deletions of, or monosomy for, chromosome 13 in a small proportion of sporadic RB tumors (Balaban *et al.*, 1982; Gardner *et al.*, 1982; Kusnetsova *et al.*, 1982). The regional assignment of the polymorphic enzyme esterase D (EsD) to chromosome band 13q14 (Sparkes *et al.*, 1980) and demonstration of linkage of hereditary RB and EsD (Sparkes *et al.*, 1983) provided confirmation that the locus for hereditary RB was located in this chromosomal region.

Once polymorphic markers for chromosome 13 became available, analysis of normal and tumor cells from RB patients who were heterozygous for EsD (Godbout *et al.*, 1983) and RFLPs specific to chromosome 13 (Cavenee *et al.*, 1983; Dryja *et al.*, 1984) showed that the second event in 70% of RB tumors was a somatic event which made the original mutation homozygous. Reduction to homozygosity occurred via two main mechanisms: loss of the chromosome carrying the normal allele followed by duplication of the mutant chromosome and mitotic recombination (Cavenee *et al.*, 1983) (see Fig. 1).

Although RB follows an autosomal dominant mode of inheritance, it is recessive at the cellular level. Both alleles of *RB1* must be inactivated or deleted before neoplastic development begins. The recessive nature of the mutation is not in disagreement with a dominant pattern of inheritance, since the mutation rate for the second event and the number of potential target cells in the developing retina are such that an individual with a germline mutation will usually acquire the second hit in multiple cells (Hethcote and Knudson, 1978).

Isolation of *RB1*

A cDNA copy of the messenger RNA (mRNA) produced from the *RB1* gene was isolated after a random probe was shown to lie in close proximity to *RB1*. The probe was deleted in a low percentage of RB tumors (Dryja *et al.*, 1986; Friend *et al.*, 1986; Fung *et al.*, 1987; W.-H. Lee *et al.*, 1987a) and a neighboring evolutionarily conserved genomic fragment was found to contain an exon of a gene encoding a novel protein.

Only a small proportion of the alterations in the *RB1* gene can be detected by Southern and northern blotting (Goddard *et al.*, 1988). However, close to 60% of tumors tested for abnormalities in *RB1* mRNA had alterations in the size, amount, or sequence of the transcript (Dunn *et al.*, 1988; Friend *et al.*, 1986, 1987; Fung *et al.*, 1987; Goddard *et al.*, 1988; W.-H. Lee *et al.*, 1987a,b), and at the protein level almost all RB tumors exhibit abnormalities (W.-H. Lee *et al.*, 1987b).

Functional proof of the authenticity of this candidate gene came from the reintroduction of the *RB1* cDNA, via retroviral transfer, into cell lines lacking detectable *RB1* product. The presence of a functional *RB1* allele resulted in the partial restoration of cell division regulation, morphological differentiation, and reduction of tumorigenic potential (Bookstein et al., 1990; H-J. S. Huang et al., 1988; Takahashi et al., 1991; Xu et al., 1991), thus demonstrating that it acts to suppress tumorigenicity in the appropriate environment.

RB1 gene expression is not limited to the developing retina. Fetal and adult human and rat tissues and a variety of human tumors express a normal size transcript, albeit at varying levels (Goddard et al., 1988; W.-H. Lee et al., 1987a). The broad tissue distribution implies that the RB1 product has an important role in the development and maintenance of more than simply the embryonic retina. Although the expression of RB1 is ubiquitous, inheritance of a mutant allele predisposes to a limited repertoire of malignancies other than RB, including osteogenic sarcoma (OS), soft tissue sarcomas, and, less commonly, melanomas, brain tumors, epithelial tumors, and tumors of the peripheral nervous system (Friend et al., 1987). The observation of high frequencies of RB1 inactivation in OS (Friend et al., 1986, 1987; Fung et al., 1987; H-J.S. Huang et al., 1988; W. H. Lee et al., 1987b; Toguchida et al., 1988), ductal breast carcinoma (Friend et al., 1987; E.Y.-H.P. Lee et al., 1988; T'Ang et al., 1988), small cell lung carcinoma (SCLC) (Harbour et al., 1988; Yokota et al., 1987), bladder carcinoma (Friend et al., 1987; Horowitz et al., 1989), soft tissue sarcoma (Friend et al., 1987; Mendoza et al., 1988), and leukemias and lymphomas (Furukawa et al., 1991) implies that RB1 may function as a suppressor in a broad range of tumor types. In some cancers, such as RB and OS, inactivation of RB1 initiates the malignant process, while in other tumors aberrations in RB1 may contribute to the progression of the tumor.

The *RB1* Gene Product and Regulation of the Cell Cycle

The RB1 gene product (pRB) appears to be involved in the regulation of a number of cellular processes, including cell cycle and transcriptional regulation, the disruption of which leads to cellular transformation. pRB is a nuclear protein with DNA-binding activity (W.-H. Lee et al., 1987b) which undergoes cyclical changes in the degree of its phosphorylation during the cell cycle (Buchkovich et al., 1989; Chen et al., 1989; DeCaprio et al., 1989; Ludlow et al., 1990; Mihara et al., 1989). It is predominantly unphosphorylated in quiescent (G_0) cells, and becomes phosphorylated at the G_1/S boundary. pRB remains phosphorylated throughout G_2 and M and is dephosphorylated as the cell reenters G_1 or G_0. The phosphorylation of pRB is accomplished, at least in part, by the master cell-cycle kinase p34[cdc2] (Lin et al., 1991) and results in an altered affinity of pRB for nuclear structures (Mittnacht and Weinber, 1991; Templeton et al., 1991) (Fig. 2).

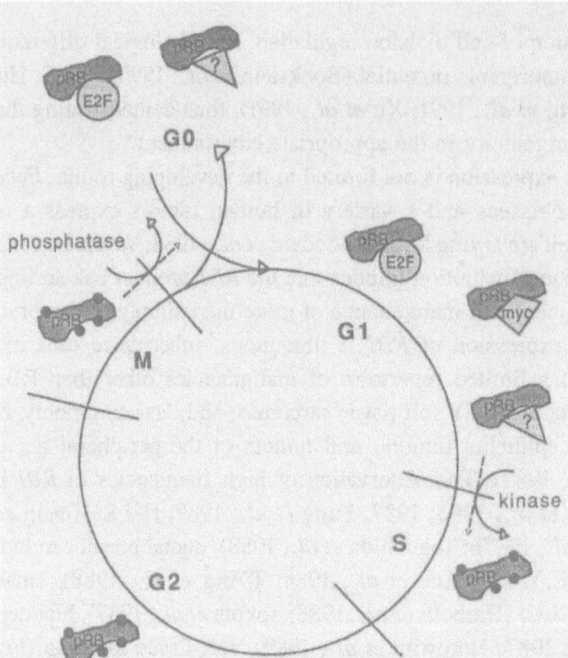

Fig. 2. pRB function is regulated during the cell cycle through changes in phosphorylation. In both G_0 and G_1, pRB is unphosphorylated and binds to normal cellular proteins, presumably modifying their function. pRB has been shown to bind directly to E2F and myc; this binding may occur as part of a larger, yet undefined protein complex. pRB is phosphorylated (indicated in the figure by black dots on the pRB protein) at the G_1/S boundary, at least in part by the p34^{cdc2} kinase. Protein–protein interactions are disrupted by phosphorylation and the bound proteins are released from the pRB complex. It is not known if phospho-pRB binds to other proteins during the S, G_2, and M phases. Just prior to or at the M/G_1 boundary, the RB protein is dephosphorylated.

The discovery that the nuclear oncoproteins of the DNA tumor viruses adenovirus (E1a), simian virus SV40 (large T antigen), and human papilloma virus (E7) interact with unphosphorylated pRB (DeCaprio *et al.*, 1988; Egan *et al.*, 1989; Imai *et al.*, 1991; Ludlow *et al.*, 1989, 1990; Whyte *et al.*, 1988) through regions of the oncoproteins known to be essential for transformation suggested that viruses act by sequestering pRB in nonfunctional complexes. Thus, loss of pRB function through mutational inactivation or binding to viral oncoproteins liberates the cell from pRB constraints by inactivating a growth-suppressing pathway.

The region of pRB necessary for oncoprotein binding is almost always disrupted in pRB mutants isolated from tumors (Hu *et al.*, 1990), indicating that this region is required for interaction with normal cellular proteins through which pRB exerts its effects. At least seven nuclear proteins (Defeo-Jones *et al.*, 1991; S. Huang *et al.*, 1991; Kaelin *et al.*, 1991) interact with pRB through this region. One target has been identified as the transcription factor E2F/DRTF. Unphosphorylated pRB binds to E2F/DRTF during G_1 (Bagchi *et al.*, 1991; Bandara and La Thangue, 1991; Bandara *et al.*, 1991; Chellappan *et al.*, 1991; Chittenden *et al.*, 1991). The same factor is bound to cyclin A during the S phase (Chellappan *et al.*, 1991; Mudryj *et al.*, 1991). Cyclin A is required for progression through the cell cycle and undergoes cyclic accumulation and destruction during each cycle of the cell. Both pRB and cyclin A participate in cell-cycle regulation, pRB as a negative regulator and cyclin A as a positive regulator. One mechanism through which both exhibit their functions appears to be the formation of complexes with a transcription factor, thereby potentially modulating its activity.

The targets of negative transcriptional regulation by pRB include TGF-β1 (Kim *et al.*, 1991; Kimchi *et al.*, 1988), co-*fos* (Robbins *et al.*, 1990), and c-*myc* (Moses *et al.*, 1990). Both c-*fos* and c-*myc* have E2F-binding sites in their promoters (Mudryj *et al.*, 1990, 1991; Wagner and Green, 1991), implying that pRB represses transcription of these genes by sequestering E2F in an inactive complex. pRB regulation of c-*myc* and N-*myc* is more complex, since pRB and the *myc* proteins can interact directly (Rustgi *et al.*, 1991).

In summary, pRB regulates passage through the cell cycle by sequestering a number of nuclear proteins, including E2F/DRTF and c-*myc*, in potentially inactive complexes. The normal cell-cycle-associated phosphorylation of pRB releases these proliferation-associated proteins from the complexes and thus from inhibition. It is not known if phosphorylated pRB interacts with normal cellular proteins to effect other functions. Yet uncharacterized nuclear proteins may bind unphosphorylated pRB, and thus represent further potential regulatory functions for pRB.

The ubiquitous role of pRB in the regulation of the cell cycle is enigmatic, considering that inheritance of a mutant allele predisposes to RB, and, to a lesser extent, osteogenic and soft tissue sarcomas. It is possible that there are redundant growth regulatory pathways in most cell types, especially those undergoing rapid or continuous proliferation. In retinoblasts and other cells susceptible to transformation in RB patients, the pRB-dependent pathway may be the sole regulatory mechanism functioning, and inactivation of pRB is therefore sufficient to permit the unregulated growth of the cell. In other tissues redundant pathways would have to be disrupted before deregulated proliferation could occur. Alternatively, loss of pRB could be lethal in most normal cells. Only cells in which loss is not lethal (retinoblasts and other cells susceptible to transformation in RB patients, and partially transformed cells) can benefit from reduced proliferative control due to loss of pRB. An increased understanding of pRB function and the variations in the mechanisms regulating cell proliferation in different cell types is necessary to discriminate between these possibilities.

Colorectal Carcinoma

Genetics

Colorectal carcinoma differs from RB in that more than one locus is known to be mutated during the progression of the malignant clone. FAP and GS patients inherit dominant mutations at the *APC* locus and develop precancerous hyperproliferative lesions. Extracolonic manifestation of FAP can include gastric and intestinal polyps, desmoid tumors, and osteomas. The predisposing gene in these families has been mapped to the long arm of chromosome 5. A case report of a patient affected by GS and congenital abnormalities with constitutional interstitial deletion of 5q, originally described as 5q13–q15 or 5q15–q22, (Herrera *et al.*, 1986) prompted a detailed analysis of this chromosome in FAP families. Linkage was demonstrated between the 5q21–q22 region and the *APC* locus (Bodmer *et al.*, 1987; Leppert *et al.*, 1987; Meera Khan *et al.*, 1988).

The 5q21–q22 region is, however, rarely the target of visible somatic deletions in colorectal carcinomas. Its involvement in the etiology of sporadic colorectal cancer was demonstrated by allele loss studies. Forty percent of colorectal carcinomas show LOH at markers on 5q (Solomon *et al.*, 1987). 5q is not the only region consistently affected by LOH in colorectal cancer. Some degree of loss was observed with every probe tested (Vogelstein *et al.*, 1989); three regions of the genome were most commonly affected, however, 5q, 17p, and 18q (Vogelstein *et al.*, 1988). Constitutional heterozygosity in 17p and 18q is lost in over 75% of tumors (Vogelstein *et al.*, 1989). Less frequent loss on 1q, 4p, 6p, 6q, 8p, 9p, 9q, 21, and 22 may represent the contribution of rare mutations in yet unidentified tumor suppressor loci in these chromosomes.

Neoplastic development in colorectal cancer is characterized by defined stages, including hyperplastic epithelium, early-stage adenomatous polyps, late-stage polyps, and noninvasive and invasive carcinomas. Associated with this progression is the accumulation of genetic changes. Since 5q allele loss is not observed in benign polyps from FAP patients, it appears that the loss of the remaining normal allele at the *APC* locus is secondary to the initial hyperplastic development (Solomon *et al.*, 1987). One mutant allele at the *APC* locus is sufficient for adenoma formation to begin. LOH on 5q has been observed in the dysplastic but still benign adenomas (Rees *et al.*, 1989; Vogelstein *et al.*, 1988).

Deletion of sequences from 18q occurred rarely in early adenomas (11–13%), more commonly in late-stage adenomas (47%), and frequently at the carcinoma stage (73%). The allelic loss of chromosome 17p sequences was restricted to carcinomas (75%) (Vogelstein *et al.*, 1988). In addition, the acquisition of mutations activating the dominantly acting oncogene *RAS* appears to be associated with the transition from early- to late-stage adenomas (Bos *et al.*, 1987; Forrester *et al.*, 1987; Vogelstein *et al.*, 1988). Thus, conversion of normal colonic epithelium into a malignant state requires inactivation of at least three loci and the activation of at least one dominantly acting oncogene.

The 5q, 17p, and 18q regions have been identified through routes which imply that they contain tumor suppressor genes. The presence of suppressor loci on two of these chromosomes has been confirmed by the reintroduction of a normal chromosome 5 or chromosome 18 into a colon carcinoma cell line and the observation of altered morphology, altered growth characteristics, and reduced tumorigenicity in the transformed cells (Tanaka *et al.*, 1991).

The *APC* Gene

Two candidate genes for the *APC* locus (*APC* and *MCC*) were isolated by delimiting the most common region of allele loss in sporadic colorectal carcinomas followed by searching for gross rearrangements in tumor DNA and constitutional DNA from affected FAP individuals with probes from the region (Joslyn *et al.*, 1991; Kinzler *et al.*, 1991a; 1991b). Candidate genes expressed in normal colonic mucosa were identified by phylogenetic conservation and cDNA-PCR-based exon-connection strategies (Fearon *et al.*, 1990), and were subjected to extensive structural and sequence analysis to search for mutations in tumors and in FAP and GS patients (Groden *et al.*, 1991; Kinzler *et al.*, 1991b; Nishisho *et al.*, 1991).

Mutations in the *APC* gene have been observed in both sporadic and familial colorectal carcinomas and in the germ line of FAP and GS patients (Groden *et al.*, 1991; Nishisho *et al.*; 1991). Inheritance of a new *APC* mutation in an FAP kindred (Groden *et al.*, 1991) supports the designation of this locus as *APC*, as does the observation that a nonsense mutation in the murine homologue of *APC* (*mAPC*) is responsible for multiple intestinal neoplasia in mice (Su *et al.*, 1992). The majority of FAP patients carry mutations in *APC*. Miyoshi *et al.* (1992) identified disruptions in the coding region of the *APC* gene in 67% of FAP kindreds using a combined PCR and RNase protection protocol. More than 90% of the alterations observed resulted in truncation of the protein (Cottrell *et al.* 1992; Miyoshi *et al.*, 1992). Identical *APC* mutations can predispose to either GS or FAP, implying that the final phenotype exhibited by the patient is influenced by other factors, either genetic or environmental. *APC* mutation may also be associated with the sporadic form of the extracolonic neoplasms observed in FAP patients.

APC encodes a large (2843-amino acid) putative protein of unknown function. The homology of the *APC* protein product with other proteins and the presence of known structural motifs have provided limited insight into its potential roles. It shares three short stretches of amino acid sequence homology with myosins and keratins which are predicted to form coiled-coil structures. Coiled-coil domains often mediate protein–protein interactions. Other motifs identified in the protein include heptad repeats, homology to *ral*-2 and the muscarinic acetylcholine receptor (a G-protein), and 20-amino acid repeats (Groden *et al.*, 1991; Kinzler *et al.*, 1991a). The carboxy terminus of the *APC* protein presumably provides a functional domain required for tumor suppression, since a high percentage of mutant alleles give rise to truncated protein products lacking this region (Miyoshi *et al.*, 1992). The much rarer missense mutations (Miyoshi *et al.*, 1992) may indicate other functional domains.

Mutations in a second gene in the region, *MCC* (mutated in colorectal cancer), have only been identified as sporadic events in colorectal carcinoma. The open reading frame in *MCC* codes for a conceptual protein of 829 amino acids which exhibits general homology to structural proteins and which has the potential to form multiple discontinuous coiled-coil domains separated by hinge regions (Bourne, 1991; Kinzler *et al.*, 1991a). The pattern of deletions observed in sporadic tumors suggests that both *APC* and *MCC* may be involved in progression of colorectal carcinoma, but only germline mutations in *APC* predispose individuals to the disease. Nishisho *et al.* (1991) proposed that both loci encode proteins which function as components of the same biochemical pathway and disruption of either function contributes to the development of colorectal cancer. Alternatively, random mutations in *MCC* could become fixed in a tumor cell population by coselection along with the biologically relevant *APC* mutation.

The *DCC* Gene Is Inactivated in the Later Stages of Colon Cancer

The LOH on the long arm of chromosome 18 in the 18q21–qter region correlated with the more malignant phenotype of the late-stage adenomas and carcinomas. This suggested the inactivation of a gene in this region in the progression from the benign to the malignant phenotype. The same strategy that yielded *APC* and *MCC* also facilitated cloning of a large gene from this region of chromosome 18q (Fearon *et al.*, 1990). This locus, *DCC* (deleted in colorectal cancer), undergoes allelic loss in 71% of colorectal carcinomas. It also is expressed at much lower level in 88% of colon tumors when compared to normal colonic mucosa and is disrupted by somatic mutations, including deletions, insertions, and point mutations, in 13% of colorectal carcinomas.

DCC encodes a conceptual 190-kilodalton (kD) transmembrane phosphoprotein with homology to the immunoglobulin superfamily in general and to the neural cell adhesion molecules (NCAMs) in particular (Fearon *et al.*, 1990). CAMs play a role in cell–cell adhesion and communication, two functions which are disrupted in malignant cells. *DCC* may encode a signal transducing receptor whose loss in the later stage of colon cancer releases the cell from normal negative regulatory signals by disrupting recognition of cell–cell communication or of a diffusible macromolecule. Alternatively it may function in cell-to-cell or cell-to-matrix adhesion in another manner. Disruption of *DCC* expression in Rat-1 cells through expression of a *DCC*-specific antisense vector reduced the ability of the cells to adhere to the substratum *in vitro* and the Rat-1 clones expression antisense *DCC* were tumorigenic *in vivo* (Narayanan *et al.*, 1992). Thus, *DCC* appears to function as a tumor suppressor by regulating normal cell adhesion.

Lynch syndrome II or hereditary nonpolyposis colorectal carcinoma (HNPCC), an autosomal dominant predisposition to multiple forms of cancer including colorectal carcinoma, was initially linked to the Kidd blood group in 18q21–qter (Lynch *et al.*, 1985; Mecklin, 1987); the *DCC* locus (and 60 centimorgans of chromosome 18q DNA in the region of *DCC*), however, was excluded as the HNPCC susceptibility locus (Peltomäki *et al.*, 1991). Although it is possible that *DCC* may be involved in the development of colon carcinoma as either a late event or an initiating mutation, evidence for an early role has not been obtained. *DCC* is expressed in a broad range of cell types and chromosome 18q LOH is seen in many tumors (Table III). Drawing on similarity to *RB1*, it is possible that *DCC* inactivation could play a role in the progression of many human cancers.

The *TP53* Gene

Involvement of the tumor-suppressor gene *TP53*, encoding the tumor-associated antigen p53 (for details see p. 351), was suggested by LOH affecting chromosome 17p in over 75% of colon carcinomas (Vogelstein *et al.*, 1988). More precisely, the region containing *TP53*, 17p12–p13.3, had the highest levels of allelic deletion (Baker *et al.*, 1989). The majority of colon carcinomas which have undergone loss of heterozygosity around *TP53* retain only a mutated form of this gene. Missense mutations affecting highly conserved regions of *TP53* are present in more than 50% of colon carcinomas (Baker *et al.*, 1989; Nigro *et al.*, 1989). The lack of allele loss on 17p and low levels of reactivity to mutant-specific anti-p53 antibodies in benign polyps from FAP patients and premalignant lesions from sporadic adenomas (Rodrigues *et al.*, 1990) suggest that the early stages of development in colon cancer are not dependent on repression of p53 activity.

If p53 normally functions to regulate negatively proliferation in normal colonic epithelium, inactivation would lead to a growth advantage. Reintroduction of wild-type *TP53* into colon carcinoma cell lines carrying *TP53* mutations blocked their progress through the cell cycle (Baker *et al.*, 1990b). Thus, low levels of normal p53 inhibited one component of the malignant phenotype in colon carcinoma, namely deregulated proliferation, supporting its role as a tumor suppressor.

Wilms' Tumor

The Genetics of Wilms' Tumor Implicates Three Predisposing Loci

Wilms' tumor (WT) is a rare pediatric tumor of the embryonic kidney. It is similar to RB in that it occurs in both inherited and sporadic forms, and the kidneys can be bilaterally or unilaterally affected. Approximately 1% of WT occurs as an autosomal dominant condition (Breslow *et al.*, 1988); most patients, however, with a predisposing germline mutation arise as the result of new mutations in the population. The earlier age of presentation of inherited WT, its bilateral nature, and constitutional chromosomal deletion of 11p13 in a minority of patients fit with the predictions of Knudson's two-hit hypothesis (Knudson and Strong, 1972b).

Despite the initial similarity to RB, the genetics of WT appears to be much more complex. In addition to the familial form of the disease, a predisposition to WT is observed in two syndromes characterized by genitourinary malformations. Wilms' tumor in association with aniridia (malformation of the irises), genitourinary abnormalities, mental retardation, and constitutional deletion of the 11p13 region characterizes the WAGR syndrome (Miller *et al.*, 1964; Riccardi *et al.*, 1978). In Denys–Drash syndrome (Denys *et al.*, 1967; Drash *et al.*, 1970; Miller *et al.*, 1964) WT occurs in association with pseudohermaphroditism and progressive renal failure. WT is observed in yet a third syndrome. In Beckwith–Wiedemann syndrome (BWS) rearrangements of the 11p15 region, including trisomy and paternal uniparental disomy of 11p15, lead to multiorgan developmental abnormalities and to an increased risk of WT, rhabdomyosarcoma, hepatoblastoma, and adrenal carcinoma (Henry *et al.*, 1991; Sotelo-Avila and Gooch, 1976). Familial BWS is a rare autosomal dominant trait showing incomplete penetrance and is linked to 11p15 (Koufos *et al.*, 1989; Ping *et al.*, 1989). WAGR syndrome and BWS imply that either loss of a locus at 11p13 or duplication of a second locus at 11p15 may predispose to WT.

Early allele loss studies demonstrated that up to 50% of sporadic WTs had LOH for markers on 11p (Fearon *et al.*, 1984; Henry *et al.*, 1989; Koufos *et al.*, 1984, 1989; Orkin *et al.*, 1984; Reeve *et al.*, 1984, 1989), supporting the hypothesis that a gene or genes on 11p contribute to the development of WT. Reintroduction of a fragment of chromosome 11, including all of the short arm, into a cell line derived from a WT resulted in a complete loss of tumorigenic potential (Weissman *et al.*, 1987), verifying that chromosome 11 does contain genetic information which is capable of suppressing expression of the fully malignant phenotype of some transformed cells.

The genetics of WT is complicated further by linkage studies indicating that the predisposing mutation in four families segregating WT as an autosomal dominant trait is not linked to either the 11p13 or 11p15 regions (Grundy *et al.*, 1988; Huff *et al.*, 1988; Schwartz *et al.*, 1991). The genetic complexity of WT suggests that at least three loci can contribute to the development of the tumor and predispose to its development when inherited in an aberrant form. Introduction of a normal chromosome 11 did not completely reverse the transformed phenotype (Weissman *et al.*, 1987), a finding expected if other genetic alterations also contribute to the development of the tumor.

The 11p13 WT Gene (*WT1*) Is Involved in Urogenital Differentiation

The 11p13 region commonly deleted in WAGR syndrome is thought to contain a cluster of genes involved in normal organ development since its deletion leads to a constellation of abnormalities. Although numerous genes and anonymous probes were initially mapped to 11p13, few were close enough to the WAGR region to be of use in isolating a candidate gene for WT (Glaser *et al.*, 1986; van Heyningen *et al.*, 1985). Random probes from chromosome 11 libraries were isolated and markers mapping to 11p13 were identified with somatic cell hybrid panels containing chromosome 11p deletions and translocations (Bickmore *et al.*, 1989; Compton *et al.*, 1988; Davis *et al.*, 1988; Gessler *et al.*, 1989b; Wadey *et al.*, 1990).

Long-range restriction maps constructed by PFGE defined the WAGR region in much greater detail, narrowing the location of several of the WAGR genes, including the WT-predisposing locus (*WT1*) (Bickmore *et al.*, 1989; Compton *et al.*, 1988, 1990; Gessler *et al.*, 1989a). In order to isolate a candidate for *WT1*, a cosmid from the 11p13 region (Glaser *et al.*, 1989) deleted in all WAGR deletions tested (Call *et al.*, 1990) was mapped to the small region defined by PFGE to contain the gene (Rose *et al.*, 1990). A second strategy exploited the observation of several CpG islands in the PFG map of the WT region (Bonetta *et al.*, 1990; Gessler *et al.*, 1990).

The *WT1* locus produces four transcripts reflecting the presence or absence of two alternatively spliced exons (Haber *et al.*, 1991) and encodes a nuclear protein (WT1) with a structure reminiscent of a transcription factor. WT1 contains four zinc fingers and a proline- and glutamine-rich domain; both these motifs are found in proteins exhibiting sequence-specific DNA binding and transactivation activity. The WT1 zinc finger domain shows significant sequence homology to the EGR and Krox family of transcription factors (Call *et al.*, 1990; Gessler *et al.*, 1990), which are activated during the G_0 to G_1 transition in cultured cells in response to growth stimulation (Chavrier *et al.*, 1988). The WT1 zinc finger domain can also recognize and bind to the same DNA sequence as EGR-1 (Rauscher *et al.*, 1990). Binding of WT1 to the EGR sequences represses transcriptional activation by EGR-1 (Madden *et al.*, 1991); the physiological significance of these cotransfection experiments remains to be determined.

The expression profile of *WT1* fits that of a gene involved in the regulation of development of the fetal kidney and gonads (Pritchard-Jones *et al.*, 1990; van Heyningen *et al.*, 1990). The gene is most highly expressed in fetal kidney, spleen, and gonads, with much lower levels of mRNA observed in fetal brain and infant and adult kidney (Call *et al.*, 1990; A. Huang *et al.*, 1990; Pritchard-Jones *et al.*, 1990). *In situ* mRNA hybridization of human fetal kidney revealed that *WT1* was expressed in a very specific manner (Pritchard-Jones *et al.*, 1990) suggesting a role in organogenesis of the kidney and other tissues. In the cells expressing *WT1*, the presence of the protein could counteract mitogenic signaling, perhaps mediated by EGR-1, causing the cells to commit to terminal differentiation.

Mutations in *WT1*

The varying presentation of WT suggested heterogeneity at the genetic level and predicted that some WTs would have mutations in *WT1* while others would arise independently of alteration of this gene. Similarly, only a subset of patients with heritable WT should be expected to carry germline mutations of *WT1*. These predictions have largely been supported by experimental observations. Although the exact proportion of WTs carrying *WT1* mutations is unknown, it appears that less than 10% have deletions of *WT1* (Cowell *et al.*, 1991; Gessler *et al.*, 1990; Haber *et al.*, 1990; A. Huang *et al.*, 1990; Pritchard-Jones *et al.*, 1990; Ton *et al.*, 1991) and in those patients with heritable WT due to mutations in *WT1*, the tumors often, but not always, undergo reduction to homozygosity for the mutant allele (Huff *et al.*, 1991; Pelletier *et al.*, 1991b; Haber *et al.*, 1990).

Although no linkage or cytogenetic evidence linked Denys–Drash syndrome to 11p13, the similarities between the expression profile of *WT1* and the spectrum of developmental abnormalities in the syndrome suggested that *WT1* was a good candidate for mutation in this disease. Mutations in *WT1* in Denys–Drash patients confirmed this hypothesis (Pelletier *et al.*, 1991a); the molecular nature of the mutations, however, was restricted to a set of point mutations affecting exons 8 and 9 (encoding zinc fingers II and III, respectively), which reduced the efficiency with which the protein bound the EGR-1 consensus recognition sequence. Comparison of the WAGR and Denys–Drash syndromes suggests that a dysfunctional WT1 protein can produce more severe abnormalities than a reduction in the level of the protein due to deletion of a single allele. The amino acid substitutions characterized in Denys–Drash patients could behave as: (1) dominant-negative mutations by sequestering the normal WT1 protein, or another polypeptide required for WT1 function, in inactive complexes (Herskowitz, 1987); (2) a gain-of-function mutation altering the specificity of DNA binding such that novel genes are regulated by the mutant protein (Pelletier *et al.*, 1991a); or (3) a dominant, incompletely penetrant, haploid insufficiency resulting from lower levels of WT1 function. WTs from Denys–Drash patients lose the remaining normal allele at the *WT1* locus, indicating that the loss of normal WT1 function can be sufficient for tumorigenesis (Pelletier *et al.*, 1991a). Considering, however, that WTs exist in which one mutant *WT1* allele has been observed in the presence of a normal allele (Haber *et al.*, 1990), it is not clear that *WT1* always functions as a tumor suppressor in the classic recessive manner.

The developmental effects of reducing WT1 dosage in WAGR syndrome or acquiring specific mutations as in Denys–Drash syndrome demonstrate the link between development and oncogenesis, and indicate the role played by WT1 in directing normal organogenesis and regulating cell differentiation (and thereby cell division). Pelletier *et al.* (1991b) have documented reduction to homozygosity for a *WT1* mutation in a granulosa cell tumor (a nonkidney malignancy) in a Denys–Drash patient. It should be noted that ovarian granulosa cells normally express *WT1*. Thus, both the developmental and tumor suppressor aspects of WT1 may not be limited to the developing kidney and inactivation of WT1 may be involved in other tumor types associated with Denys–Drash syndrome.

11p15 and Insulin-Like Growth Factor II

As described above, the association between BWS, WT, and 11p15 abnormalities implicates another chromosome 11p locus in the etiology of a subset of WT cases. The allele loss affecting 11p15 in WTs (Henry *et al.*, 1989; Koufos *et al.*, 1989; Reeve *et al.*, 1989) is unusual in that when tumor-specific loss is observed it is virtually always the maternally derived allele which is lost (Henry *et al.*, 1991). This observation is suggestive of an imprinting effect in which the maternally and paternally derived alleles of a gene or genes in the region are not functionally equivalent. Interestingly, one gene which maps to this region and is known to be differentially imprinted depending on whether it is inherited through the maternal or paternal genome encodes insulin-like growth factor II (IGF-II). In mouse embryos, the maternal gene is transcribed at very low levels, while the paternal allele is expressed to a much greater degree (DeChiara *et al.*, 1991). The loss of maternally derived sequences from the 11p15 region coupled with a duplication of the paternal counterpart could result in higher levels of IGF-II and developmental abnormalities. IGF-II is highly expressed in WT and enhanced levels of expression could provide the tumor with a growth advantage. Loss of maternal 11p15 material coupled with duplication of the paternal sequences would be manifested as LOH at 11p15 in WT, and this LOH would therefore be indicative of an imprinting effect rather than the location of a tumor suppressor gene. Further analysis is needed of IGF-II and other loci in the region of common allele loss in 11p15 before this mechanism can be confirmed or refuted.

Models for WT Development

WT is genetically heterogeneous, with mutation of at least three loci implicated in the formation of the tumor. It is unlikely that accumulation of mutations at all three loci is required for WT to develop, given the observation of WTs lacking *WT1* mutations. Also, the early age at which WT presents argues against the requirement for numerous genetic alterations. Interruption of one of parallel but distinct regulatory pathways represented by the three loci may be sufficient for tumorigenesis in the fetal kidney, or alternatively the three loci may function in the same regulatory hierarchy and inactivation of each locus individually would be functionally equivalent (Francke, 1990). It seems likely that *WT1* mutations will be limited to WTs, genitourinary tumors, and other malignancies arising in the limited number of tissues expressing *WT1*.

Neurofibromatosis Type I

Linkage to 17q11.2 and Generation of a PFGE Map

Mutations predisposing to von Recklinghausen neurofibromatosis (neurofibromatosis type I; NF-1) are common in all human populations. One in 3500 individuals has this dominantly inherited condition and is affected by multiple benign abnormalities such as café-au-lait spots, nodular abnormalities of the iris, and skin neurofibromas. Tumors with malignant potential observed in these patients include neurofibromas, schwannomas, CNS tumors, pheochromo-cytomas, and neurofibrosarcomas (Ponder, 1990). Like RB, this disease results from a germline mutation either inherited from an affected parent or generated through a new germinal mutation. The high penetrance of expression of the inherited mutations coupled with an extreme variability in the severity of the phenotype within families suggests a complex biology behind the manifestations of the disease.

Genetic linkage localized the NF-1-predisposing locus (*NF1*) to the peri-centric region of chromosome 17 (17q11.2) (Barker *et al.*, 1987; Collins *et al.*, 1989b; Seizinger *et al.*, 1987). The characterization of two apparently balanced constitutional translocations in NF-1 patients involving 17q11.2 (Ledbetter *et al.*, 1989; Rey *et al.*, 1987) strengthened the genetic assignment of *NF1* and provided a valuable resource in identifying the sequences responsible for NF-1. The trans-locations also suggested that gross rearrangement of the *NF1* locus is one mechanism by which mutation could predispose to NF-1.

If *NF1* is a tumor suppressor gene, LOH on proximal 17q should be observed in NF-1-associated tumors. A significant frequency of allele loss has been observed in malignant pheochromocytomas and neurofibrosarcomas from NF-1 patients (Glover *et al.*, 1991; Menon *et al.*, 1990; Xu *et al.*, 1992). Allele loss is not restricted to 17q, however, and occasionally only involves 17p, suggesting a role for *TP53* in the progression of malignant tumors in NF-1 patients (Collins *et al.*, 1989a; Glover *et al.*, 1991).

The order of loci in proximal 17q determined by detailed genetic linkage analysis and the identification of the closest linked probes were confirmed by physical mapping techniques. The closest linked loci failed to detect the transloca-tion breakpoints on PFGE (Fountain *et al.*, 1989a); therefore new DNA markers in this region were identified from linking (Fountain *et al.*, 1989b), cosmid (O'Con-nell *et al.*, 1989; Yagle *et al.*, 1990), and jumping libraries (Wallace *et al.*, 1990a). Long-range PFGE maps encompassing the translocation region were constructed (Fountain *et al.*, 1989b; Yagle *et al.*, 1990). This region was analyzed for expressed sequences, but the early candidate genes identified via this strategy, those for EVI-1, EVI-2 (Cawthon *et al.*, 1990), and oligodendrocyte-myelin glycoprotein (Viskochil *et al.*, 1990), proved not to be interrupted by the translocations (O'Connell *et al.*, 1990; Wallace *et al.*, 1990b).

Mutations in *NF1* Inactivate a rasGAP Protein

Two groups simultaneously isolated the *NF1* gene and showed that it encoded a large transcript (12–13 kb) that is commonly expressed (Cawthon *et al.*, 1990; Viskochil *et al.*, 1990; Wallace *et al.*, 1990b). The locus defined by the cDNAs was interrupted by both translocations, and 1 of 35 NF-1 patients had a new mutation detectable with a partial cDNA probe on Southern blots (Wallace *et al.*, 1990b). Application of the single strand conformation polymorphism (SSCP) technique to the analysis of five exons of the gene in 72 NF-1 patients detected six variant alleles, two of which were sequenced and shown to result in missense and nonsense mutations, respectively (Cawthon *et al.*, 1990).

The variation in the penetrance of different *NF1* mutations and differences in their phenotypic manifestation may reflect particular types of mutations in *NF1*, i.e., deletions removing the genes nested within *NF1* or mutations also affecting neighboring genes. The variable severity of the phenotype associated with a specific *NF1* mutation within a family points to involvement of secondary genetic factors in the development of the disease (Easton *et al.*, in press).

Analysis of the putative protein encoded by the *NF1* cDNA (Buchberg *et al.*, 1990; Xu *et al.*, 1990b) revealed significant homology between a part of the protein and the catalytic domains of the yeast IRA proteins (inhibitory regulators of the *ras*-cAMP pathway) and mammalian p120^{GAP} (p21^{ras} GTPase-activating protein) (Trahey and McCormick, 1987). The *NF1* protein (neurofibromin or p280^{NF1}) has more extensive homology to yeast IRA1 and IRA2 than it does with p120^{GAP}, the latter homology being restricted solely to the catalytic domain (the GAP-related domain, GRD). A portion of neurofibromin containing the putative catalytic domain was able to compensate for *ira1* and *ira2* mutants of *Saccharomyces cerivisiae* and stimulate the GTPase activity of both yeast RAS2 and mammalian p21^{ras} *in vitro*, but was not able to stimulate that of p21^{ras} proteins carrying oncogenic point mutations blocking hydrolysis of GTP to GDP (Ballester *et al.*, 1990; Bollag and McCormick, 1991; Martin *et al.*, 1990; Xu *et al.*, 1990a). Regulation of neurofibromin's GAP activity may occur through alternative splicing of an exon which introduces 21 amino acids to the center of the GRD. Neurofibromin lacking these additional amino acids is predominantly expressed in undifferentiated cells, whereas the isoform with the insertion in the GRD is more abundant in differentiated cells (Nishi *et al.*, 1991).

p21^{ras} is a membrane-associated protein involved in signal transduction processes affecting regulation of growth. When activated, p21^{ras} exchanges GDP for GTP, thereby switching from an inactive to an active state. A weak intrinsic GTPase activity hydrolyses GTP into GDP, returning the protein to the inactive form. The GAP proteins may serve two functions in this *ras* cycle [reviewed in A. Hall (1990)] (Fig. 3). First, they may downregulate p21^{ras}·GTP levels by enhancing the GTPase activity of p21^{ras} (Trahey and McCormick, 1987). Second, the GAP proteins may be targets of activated p21^{ras}·GTP since p120^{GAP} binds to the effector-binding domain of p21^{ras} (Adari *et al.*, 1988; Cales *et al.*, 1988) and the

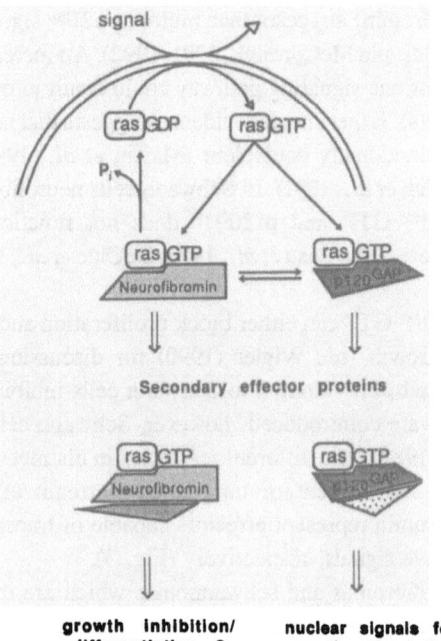

Fig. 3. The functions of the two *ras* GAP proteins neurofibromin and p120^{GAP} may differ in cells derived from the neural crest. p21^{ras} exchanges GDP for GTP in response to an extracellular signal, then the activated p21^{ras}·GTP complex interacts with neurofibromin or p120^{GAP}. In Schwann cells, it appears that neurofibromin both enhances the intrinsic GTPase activity of p21^{ras} and mediates the transduction of a growth-inhibitory or differentiative signal to the cell. Both these activities are reduced in cells carrying one mutant *NFI* allele and absent from tumors in which both alleles have been inactivated. p120^{GAP} may be functioning as an effector molecule transducing a growth-promoting signal. Both p21^{ras}·GTP·neurofibromin and p21^{ras}·GTP·p210^{GAP} are likely to act as effector complexes through the interaction with other cellular proteins [reviewed in A. Hall (1992)]; thus the message they send to the nucleus could depend upon the profile of secondary effectors expressed in the cell. [Adapted from Bollag and McCormick (1992).]

p21^{ras}·GTP·p120^{GAP} complex can interact with downstream targets (Yatani *et al.*, 1991). GAPs release their own downstream signal and then enhance p21^{ras}·GTP GTPase activity, thus regulating the levels of the effector complex. Thus, inactivation of neurofibromin would result in a phenotype similar to that of the activating *ras* mutations with p21^{ras} being locked into an active state. In addition to increasing the levels of p21^{ras}·GTP, loss of neurofibromin would remove one pathway through which p21^{ras}·GTP communicates its signal to the cell. The identification of more than one ubiquitously expressed p21^{ras}-specific GAP (p120^{GAP} and neurofibromin) suggests that multiple p21^{ras} signalling routes are present in the cell (Bollag and McCormick, 1991, 1992). An increase in p21^{ras}·GTP coupled with the loss of one signaling pathway could result in overstimulation of the remaining pathway(s). Experimental evidence suggests that neurofibromin and p120^{GAP} are neither functionally equivalent (Martin *et al.*, 1990) nor activated by the same signals (Han *et al.*, 1991). In Schwann cells neurofibromin appears to regulate levels of p21^{ras}·GTP, and p120^{GAP} does not function as a GTPase-activating protein in these cells (Basu *et al.*, 1992; DeClue *et al.*, 1992), supporting this model.

High levels of p21^{ras}·GTP can either block proliferation and induce differentiation or stimulate growth [see Wigler (1990) for discussion]. For example, introduction of activated p21^{ras} alone into Schwann cells inhibits their proliferation; if p53 and p21^{ras} are cointroduced, however, Schwann cells become transformed (Ridley *et al.*, 1988). The different responses in distinct cell types may be accomplished by the recruitment of unique downstream effectors. Perhaps p120^{GAP} and neurofibromin represent effectors capable of transmitting the mitogenic and differentiative signals, respectively (Fig. 3).

The benign neurofibromas and schwannomas which are the most common manifestation of germline mutations in *NF1* are focal lesions which may be monoclonal in origin (Skuse *et al.*, 1991) and remain heterozygous for the *NF1* region. Thus, half-normal levels of neurofibromin may result in a growth advantage (Ponder, 1990), perhaps through the overstimulation of a mitogenic pathway mediated by an effector other than neurofibromin in response to a local signal. Pheochromocytomas, on the other hand, show allele loss on proximal 17q (Xu *et al.*, 1992). A number of routes may be available for the manifestation of *NF1* mutations, including dosage effects or subsequent genetic changes, such as loss of the second *NF1* allele or mutation of other genes, such as *TP53* (Menon *et al.*, 1990; Nigro *et al.*, 1989).

NF1 Mutations in Other Tumor Types

Like *RB1*, *NF1* is broadly expressed (Buchberg *et al.*, 1990). *NF1* mutations could potentially affect the phenotype of multiple cell types, especially given the broad distribution of *ras* mutations in human cancer (Bos, 1989) and the functional association between p21^{ras} and neurofibromin. The slightly increased risk of cancer in NF-1 (B. H. Cohen and Rothner, 1989; Sorensen *et al.*, 1986) suggests that the analysis of *NF1* in sporadic cases of the same tumors may reveal somatic changes. Mutations of the *RB1* gene predispose to a highly specific subset of cancers when inherited, but somatic inactivation of the gene is now known to be involved in the progression of many malignancies as a necessary but non-rate-limiting step. A similar mechanism may be functioning with *NF1*, as is implied by the recent observation of a specific amino acid substitution in the neurofibromin GRD in NF-1 patients and three types of sporadic tumors, colorectal carcinoma, myelodysplastic syndrome, and astrocytoma (Li *et al.*, 1992).

Tumor-Associated Antigen p53

Loss of Function of p53 Is Important in Malignancy

The nuclear phosphoprotein p53 (encoded by the chromosome 17p locus *TP53*) was first identified in complexes with the oncoprotein large T in SV40-transformed cells (Lane and Crawford, 1979; Linzer and Levine, 1979). Early reports suggested p53 could immortalize adult rodent cells (Jenkins *et al.*, 1984) and cooperate with activated *ras* to transform embryonic rodent fibroblasts (Eliyahu *et al.*, 1984; Parada *et al.*, 1984). It now appears, however, that *TP53* has more of the characteristics of a gene whose loss of function is important in transformation as opposed to one which is mutationally activated.

Normal p53 appears to regulate cell division as a component of a process which suppresses deregulated growth. This assumption is supported by a number of experimental observations. First, recent reevaluation of p53 transforming activity in transfection assays (Finlay *et al.*, 1988; Hinds *et al.*, 1989) suggested that wild-type p53 cannot cooperate with *ras* to transform primary rodent cells; only mutated forms have this ability. The early studies had inadvertently been carried out with mutant *TP53* clones (Green, 1989). Second, like pRB, p53 also forms stable complexes with SV40, adenovirus, and human papillomavirus type 16 and 18 oncoproteins (Crawford *et al.*, 1981; Lane and Harlow, 1982; Lane and Crawford, 1979; Linzer and Levine, 1979; McCormick *et al.*, 1981; Sarnow *et al.*, 1982). The finding that the nuclear oncoproteins of at least three DNA tumor viruses participate in transformation by binding, and presumably inactivating, the pRB suggests that the other cellular proteins bound by these transforming agents may be similarly inactivated. Third, complete deletion and internal mutations inactivate *TP53* in a Moloney murine leukemia virus-transformed cell line (Wolf and Rotter, 1984), a proportion of Friend virus-induced erythroleukemias (Hicks and Mowat, 1988; Mowat *et al.*, 1985; Munroe *et al.*, 1988), human OS cells (Masuda *et al.*, 1987), and human colorectal carcinomas (Baker *et al.*, 1989, 1990b). Fourth, reintroduction of wild-type p53 into human colorectal carcinoma cell lines (Baker *et al.*, 1990a) and a human OS cell line (Chen *et al.*, 1990) suppressed the expression of the transformed phenotype.

The apparent dominant characteristic of p53 mutants in the transfection/ transformation assay, so much at odds with expectations for a recessive oncogene, may be due to the formation of inactive p53 oligomers between the mutant and wild-type proteins (Green, 1989). p53 could thus behave in a dominant-negative fashion (Herskowitz, 1987) in the presence of normal protein. The cellular heat shock protein HSP70 is also irreversibly bound up in these inactive p53 oligomers (Clarke et al., 1988; Finlay et al., 1988; Hinds et al., 1987; Pinhasi-Kimhi et al., 1986; Stürzbecher et al., 1987). Both wild-type p53 and HSP70 become locked into an inactive complex, depriving the cell of normal levels of these proteins. In a similar manner, viral oncoprotein binding to p53 sequesters the protein in inactive complexes.

p53 Is Commonly Mutated in Human Cancer

The TP53 gene on chromosome 71p is perhaps the most common target of mutations in human cancers. LOH affecting 17p (see Table III) has been observed in colorectal carcinoma (Baker et al., 1989; Fearon et al., 1987; Monpezat et al., 1988), SCLC (Yokota et al., 1987), breast cancer (MacKay et al., 1988; Sato et al., 1990), glioblastoma (James et al., 1988), astrocytomas (James et al., 1989), bladder carcinomas (Olumi et al., 1990; Tsai et al., 1990), and hepatocellular carcinoma (Fujimori et al., 1991; Slagle et al., 1991). TP53 is rarely subjected to gross mutations of the sort detectable by Southern and northern blotting techniques. Analysis of TP53 coding sequence in the tumors which exhibit allele loss on 17p revealed mutations clustered in four of the five highly conserved regions (II–V) of the gene (Soussi et al., 1987) (Fig. 4). TP53 mutations have been observed in more than 30 tumor types, including carcinomas (bladder, breast, brain, cervix, colon, esophagus, stomach, liver, ovary, and pancreas), sarcomas (osteosarcoma, neurofibrosarcoma, leiomyosarcoma, rhabdomyosarcoma, and cholangiosarcoma), leukemias, lymphomas, and myelodysplastic syndromes (Caron de Fromentel and Soussi, 1992; Hollstein et al., 1991).

Hepatocellular carcinoma is unique in that the mutations in TP53 are not distributed over the four conserved regions as in other tumor types, but are biased toward one codon (codon 249) in the gene (Bressac et al., 1991; Hsu et al., 1991). It may be that the involvement of hepatitis B virus (HBV) and aflotoxin in the etiology of the hepatocellular carcinomas (HCCs) studied in some way selects for a particular mutant form of p53, as has been suggested by the preliminary observation that non-HBV, non-aflotoxin-related HCCs do not carry this mutation (Bressac et al., 1991). Selection for a specific mutant may be an indication that it is an activating lesion, conferring a gain of function as with a dominantly acting

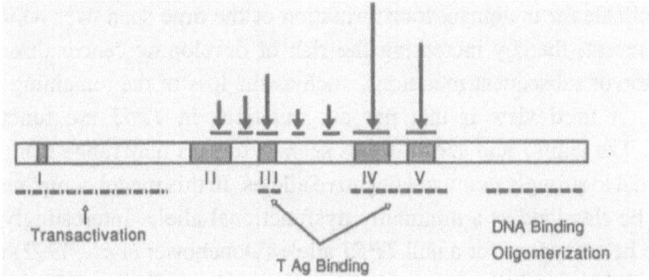

Fig. 4. Schematic map of p53, showing the evolutionarily conserved domains and functional motifs. Eighty-one percent of the mutations fall within the indicated regions (the arrow height represents the relative frequency of mutation in each region). The approximate boundaries of the transactivation, T Ag-binding, DNA-binding, oligomerization, and nuclear localizing motifs are indicated.

oncogene (Harris, 1991), or it may represent requirement for a highly specific change, induced by aflotoxin, to allow the biological interaction between HBV and mutant p53.

Germline *TP53* Mutations and the Cancer-Family Syndrome

While somatic mutations in *TP53* are implicated in the progression of a host of human malignancies, germline mutations predispose to a limited subset of cancers in families affected by Li–Fraumeni syndrome (Malkin *et al.*, 1990; Srivastava *et al.*, 1990). Li–Fraumeni families inherit an autosomal dominant predisposition to cancer of the breast, brain, bone, soft tissues, hematopoietic system, and adrenal cortex (Marx, 1990). Traditional linkage analysis could not be applied to these families due to their rareness, small sizes, death of affected members, and incomplete penetrance. Since *TP53* is the target of somatic mutations in the sporadic forms of many of these tumors, Malkin *et al.* (1990) and Srivastava *et al.* (1990) looked for p53 mutations in Li–Fraumeni families. Inherited mutations were found clustered in the small region encoding amino acids 245–258 (in conserved region V). It is interesting that the expression of Li–Fraumeni mutations is tolerated by normal cells and they do not appear to act in a dominant negative manner (Srivastava *et al.*, 1992). Vogelstein (1990) proposed that *TP53* mutations are only one step in the process of malignant transformation, and whether the first mutation occurs early (in the germ line) or late (as a somatic mutation) is not so important as the fact that it does occur. Another interpretation is that in certain tissues reduction in normal p53 activity may increase the number of cells available for malignant transformation or the time span over which such cells are present, thereby increasing the risk of developing cancer through the accumulation of subsequent mutations, such as the loss of the remaining normal p53 allele. A third view is that not all mutations in *TP53* are functionally equivalent. The amino acid substitutions seen in tumors may range from mildly dysfunctional to strongly dominant negative alleles. In this model complete loss of p53 could be classified as a minimally dysfunctional allele. Interestingly, transgenic mice heterozygous for a null *TP53* allele (Donehower *et al.*, 1992) showed only a moderate increase in tumor formation compared to wild-type mice. Transgenic mice homozygous for the null allele developed normally but showed a dramatic increase in early-onset tumors. It appears that loss of p53 accentuates the strain-specific predispositions to cancer in mice.

p53 Is Involved in Cell Cycle Regulation

Microinjection of anti-p53 antibodies into mouse 3T3 cells blocks cell proliferation (Mercer *et al.*, 1982, 1984), indicating that p53 is involved in cell cycle control. p53 has subsequently been shown to be regulated during the cell cycle by both changes in protein levels and phosphorylation and is similar to pRB in that it becomes phosphorylated in S phase (Bischoff *et al.*, 1990). Phosphorylation of pRB is thought to allow progression from G_1 to S phase; p53 may function in a similar fashion. Absolute levels of p53 are low after mitosis and increase during G_1 (Bischoff *et al.*, 1990), also supporting a role in cell cycle regulation. A possible role for p53 in controlling DNA polymerase alpha interaction with other components of the replication complex has been proposed (Marshall, 1991), but there is no direct evidence for this.

p53 may also function in transcriptional regulation. A transcription activation domain has been found in the amino terminus of p53 (Fields and Jang, 1990; Raycroft *et al.*, 1990) and transforming mutations in p53 can inactivate this transactivation activity (Raycroft *et al.*, 1990). p53 binds preferentially to specific DNA sequences (El-Deiry *et al.*, 1992) and can stimulate expression of a reporter gene adjacent to a p53–DNA binding element (Kern *et al.*, 1992). Mutant p53 proteins lack this activity.

The characterized mutations in p53 are found in four of the conserved domains of the protein and are commonly base-pair substitutions, small deletions or insertions, and splice mutants (Caron de Fromentel and Soussi, 1992) (Fig. 4). Most of the mutations affect exons 5–8, which encode the central domains of the protein. The acidic transactivation domain at the amino terminus and the carboxy-terminal DNA-binding, oligomerization, and nuclear-localizing domains are rarely mutated. The consequences of the mutations in the central region are not known, since the function of this region remains largely undefined. However, this region binds SV40 large T antigen and it is possible that mutations in this region interfere with interaction between p53 and a cellular protein as is observed with mutations in the "RB pocket" of pRB. Given the selection for specific mutations in Li–Fraumeni syndrome and hepatocellular carcinoma, it is unlikely that all mutations affect p53 function in the same way. p53 appears to have multiple functions, and different alterations could have differential effects on protein activity. If p53 acts as part of a larger protein complex for at least some of its functions, mutant p53 could bind to, and sequester in inactive complexes, more than just normal p53, providing one explanation for the observation that the presence of some mutant p53 proteins can be more transforming than complete deletion of the protein itself.

TUMOR SUPPRESSOR GENES CLOSE TO ISOLATION

As more cancers are subjected to detailed cytogenetic analysis, linkage analysis, and loss of constitutional heterozygosity studies, and as the number of anonymous probes and genes mapped increases, the position-dependent approach to isolating or identifying tumor suppressor genes will begin to yield results at an increasing rate. The identification of chromosomal regions harboring newly suspected tumor suppressor genes is rapidly expanding the search for and isolation of this class of genes.

A number of the chromosomal regions indicated by these approaches have been previously implicated in diverse tumor types and are known to contain a tumor suppressor gene. It is important to determine whether a previously characterized gene or a novel locus is involved. The 13q14 region, for example, is commonly the target for allele loss in retinoblastoma, osteosarcoma, breast carcinoma, small cell lung carcinoma, gastric carcinoma, hepatocellular carcinoma, and renal cell carcinoma (see Table III). The *RB1* gene has been demonstrated to be the target for mutational inactivation in most of these malignancies as either an initiating or progression-associated event. Similarly, allele loss on the short arm of chromosome 17 (17p13) is indicative of homozygous mutation of the *TP53* gene in a broad range of tumors. Although the short arm of chromosome 11 undergoes allele loss in many tumors, *WT1* may be involved in a small subset of these, including WT and ovarian carcinoma, as is indicated by its limited tissue distribution. Other loci on 11p are presumably responsible for the broad spectrum of tumors exhibiting 11p allele loss.

The 5q– Syndrome

The cytogenetic deletions of 5q22–q31 observed in a number of hematologi-cal conditions (refractory anemias, acute nonlymphocytic leukemia, treatment-induced leukemias, lymphoproliferative disorders, and chronic myeloprofilerative disorders) (Van den Berghe *et al.*, 1985) pinpoint a distinct locus from that involved in colorectal carcinoma. This region of 5q contains many growth factor and growth factor receptor genes, including interleukins 3, 4, and 5, macrophage-colony stimulating factor (CSF1), granulocyte/macrophage-colony stimulating factor (CSF2), and the CSF1 receptor (CSF1R) (D. T. Bishop and Westbrook, 1990; Wasmuth *et al.*, 1989). CSF1 and its receptor (CSF1R) are involved in myeloid lineage development. The CSF1R gene has been found to be deleted from the normal chromosome 5 in 4 of 10 patients with myelodysplasia and a 5q– chromosome (Boultwood *et al.*, 1991), suggesting that loss of either CSF1R or a closely linked locus is important in the etiology of these conditions. Thus, the loss of a single allele of one or more critical growth factors or growth factor receptors may be sufficient to perturb normal development and lead to malignant growth, or, alternatively, interstitial deletion accompanied by a cytogenetically invisible alteration may inactivate a key suppressor locus.

Breast Carcinoma

The predisposition to both early-onset familial breast carcinoma and familial breast–ovarian carcinoma was recently linked to chromosome 17q12–q21 (J. M. Hall *et al.* 1990; Narod *et al.*, 1991). Approximately 40% of early-onset breast cancer families and 100% of breast–ovarian cancer kindreds are linked to this region (Easton *et al.*, in press). Breast cancer is a heterogeneous disease, and, not surprisingly, multiple genetic loci are implicated in predisposition to breast cancer (see Tables I and III). Inherited *TP53* mutations increase the risk of developing breast cancer in Li–Fraumeni families; linkage to a locus (*BRCA1*) on 17q12–q21 is observed in some early-onset breast cancer families (J. M. Hall *et al.*, 1990) but not in others (Skolnick *et al.*, 1990); and some late-onset breast cancer families may possibly be linked to the estrogen receptor gene (Zuppan *et al.*, 1991). Thus, multiple loci appear to contribute to the predisposition of malignant transfor-mation of breast tissue. A number of genes assist progression in breast cancer. Mutations of *RB1* and *TP53*, and presumably a number of unidentified loci, occur as secondary changes.

The presence of the tumor suppressor locus *BRCA1* in the 17q12–q21 region and its involvement in the etiology of both familial and sporadic breast cancer is suggested by the linkage of familial breast and ovarian cancer to this region and allele loss in sporadic breast cancers centering around the 17q11.2–q12 region (Bowcock *et al.*, in press; Futreal *et al.*, 1992; Sato *et al.*, 1990). A number of biologically plausible candidate genes map to the chromosome interval defined by the initial linkage reports, including *HER2* (human epidermal growth factor, amplified or overexpressed in a significant proportion of breast cancers), *THRA1* (thyroid hormone receptor), *RARA* (retinoic acid receptor alpha), *17HSD1* and *17HSD2* (hydroxysteroid dehydrogenases), *WNT3* (human homologue of a murine mammary tumor virus integration site), *HOX2* (homeobox gene involved in differentiation), *NGFR* (nerve growth factor receptor), *PHB* (prohibitin, an anti-proliferative protein), and *NME1* (expression of this gene is associated with metastasis) (Editorial, 1991; J. M. Hall *et al.*, 1990; Sato *et al.*, 1992). More precise linkage (Hall *et al.*, 1992; Bowcock *et al.*, in press) and LOH studies (Futreal *et al.*, n.d.) have excluded a number of these candidates. These studies do not decisively define the *BRCA1* region; the common region appears, however, to encompass the chromosomal region from the D17S250 to D17S588 markers.

The complexity of the genetic events involved in development of breast carcinoma is further highlighted by the *PHB* gene. This gene encodes an anti-proliferative activity and was considered to be a candidate for *BRCA1*. Finer mapping of the *BRCA1* locus, however, excluded *PHB* from the list of possible candidates. Analysis of *PHB* in sporadic breast carcinomas with an early age of onset or having undergone LOH for 17q markers revealed mutations in 4 of 23 tumors in an evolutionarily conserved exon of the gene (Sato *et al.*, 1992). Although *PHB* is not thought to be *BRCA1*, it does appear to be mutated during the development of sporadic breast cancers.

Chromosome 3p Contains Multiple Recessive Oncogenes

Three distinct regions of the short arm of chromosome 3 are suspected of containing recessive oncogenes involved in the genesis of multiple human cancers. In renal cell carcinoma (RCC), LOH of 3p12–p14, 3p21.3, and 3p21–pter occurs frequently (Seizinger *et al.*, 1991); a translocation in a family with hereditary renal cancer disrupts 3p14.2 (A. J. Cohen *et al.*, 1979; Wang and Perkins, 1984); sporadic renal cell carcinomas are commonly deleted for the 3p11–p14 region (Kovacs *et al.*, 1988); and inheritance of mutations at the von Hippel–Landau (VHL) disease locus (mapped to 3p24–p25) predisposes to renal cell carcinomas, hemangioblastomas, and pheochromocytomas (Seizinger *et al.*, 1988; Tory *et al.*, 1989). Malignancies not associated with VHL disease also commonly undergo LOH for distal 3p. In small cell lung carcinoma (SCLC), loci in the 3p21–p25 region are consistently deleted or reduced to homozygosity (Brauch *et al.*, 1990; Ibsen *et al.*, 1987; Kok *et al.*, 1987; Whang-Peng *et al.*, 1982a,b). This region contains the VHL gene, suggesting either a possible role for the VHL gene in lung carcinoma or the presence of a second recessive oncogene in this region. The pattern of LOH in non-SCLC may be different from that observed for SCLC (Brauch *et al.*, 1990; Kok *et al.*, 1987) and more similar to that reported for carcinoma of the uterine cervix (Yokota *et al.*, 1989). LOH affecting distal 3p is observed in both breast and testicular carcinoma, albeit at a lower frequency than in RCC and lung carcinomas (Seizinger *et al.*, 1991).

Although no tumor suppressor gene has been identified in this region by the reverse genetics approach, two genes mapped to the region can be considered candidates. A receptor protein with tyrosine phosphatase activity (*PTPG*) has been mapped to the smallest region of allele loss observed in lung carcinomas in the 3p21 region (La Forgia *et al.*, 1991). Phosphatases could function to regulate growth through specific dephosphorylation of phosphotyrosine residues in response to growth inhibitory signals, thereby counterbalancing growth factor-induced tyrosine phosphorylation. The loss of such a function could leave a cell in a permanently stimulated state and thus provide release from growth control. A second candidate was isolated through the identification of a CpG island in the vicinity of a polymorphic locus in 3p21 (Erlandsson *et al.*, 1990) lost in 76% of RCCs (Kovacs *et al.*, 1988) and all SCLCs (Brauch *et al.*, 1990). The *RIK* gene encodes an acylpeptide hydrolase (Erlandsson *et al.*, 1991) which is highly expressed in kidney and shows severely reduced expression in RCCs. The enzyme hydrolyzes short N-acetylated peptides of less than 15 amino acids in length, increasing their rate of degradation, and it is conceivable that lost or reduced activity could lead to deregulated growth by allowing the accumulation of an acylated growth-stimulatory peptide (Erlandsson *et al.*, 1991). The involvement of either *PTPG* or *RIK* in carcinogenesis has not been supported by identification of somatic mutations in tumors, germline mutations in familial patients, or functional studies involving reintroduction of the gene into cell lines lacking normal product. Thus, it is possible that loss of both of these genes in RCC and SCLC may reflect coincidental loss occurring along with the inactivation of an undiscovered recessive oncogene linked to these genes.

CONCLUSIONS

The protooncogenes encode proteins with functions as diverse as growth factors, receptors, cytoplasmic signal transducers, and nuclear factors. The activating mutations in these genes result in constitutively active proteins which stimulate the cell to proliferate in the absence of normal mitogenic signals. The roles played by tumor suppressors are as diverse as those filled by the protooncogenes. Conversely, the tumor suppressors appear to regulate unrestricted growth by transducing inhibitory growth signals or limiting the efficacy of stimulatory signals. Investigation of the mechanisms through which the tumor suppressor genes act has resulted in functional links between these two classes of oncogenic sequences. Through its interaction with the transcription factor E2F, pRB regulates the expression of two known protooncogenes (c-*fos* and c-*myc*) expressed as an early response to mitogenic signals, and the *WT1* product may also inhibit the expression of early response genes through binding to the same DNA sequence as the EGR-1 transcription factor. Neurofibromin potentiates p21^{ras} through two possible routes, by stimulating p21^{ras} GTPase activity and by acting as a downstream effector molecule. DCC may function as a transmembrane receptor or in cell–cell contact.

The division between dominant and recessive mutations becomes blurred in some cases. For example, the oligomerization of p53 can lead to the formation of nonfunctional complexes between mutant and wild-type proteins. Other tumor suppressor products with the potential for protein–protein interactions may also act in this dominant-negative fashion. The coiled-coil domains in the *APC* and *MCC* gene products are suggestive of protein–protein interactions and some mutations in these genes may act in a dominant-negative fashion.

Genes with tumor suppressor activity have been identified through approaches other than those described in this review. For example, gene transfer of human cDNA or genomic clones into *ras*-transformed rodent cells has induced morphologic revertants and permitted isolation of a gene encoding a putative protein with structure similar to *ras* (Noda *et al.*, 1989). The presence of yet other tumor suppressor loci is suggested by the study of *Drosophila* neoplasms. At least 25 distinct loci have been found to cause tissue-specific neoplasms in the mutant state [reviewed in Gateff (1982)]. The mutations in the *giant (2) lethal larvae (l(2)gl)* gene cause malignant neuroblastomas of the presumptive adult optic neuroblasts of the larvae and tumors of the imaginal disc. The human homologues of *l(2)gl* and other putative *Drosophila* tumor suppressor loci will be good candidates for tumor suppressors in the mammalian system.

ACKNOWLEDGMENTS. We would like to thank Veronica van Heyningen, Jeremy Squire, and Bruce Ponder for advice and comments during the preparation of this manuscript.

REFERENCES

Adari, H., Lowly, D. R., Willumsen, B. M., Der, C. J., and McCormick, F., 1988, Guanosine triphosphatase activating protein (GAP) interacts with the p21 *ras* effector binding domain, *Science* 240:518–521.

Anand, R., 1986, Pulsed field gel electrophoresis: A technique for fractionating large DNA molecules, *Trends Genet.* 2:278–283.

Armitage, P., and Doll, R., 1954, The age distribution of cancer and a multi-stage theory of carcinogenesis, *Br. J. Cancer* 8:1–12.

Ashley, D. J. B., 1969a, Colonic cancer arising in polyposis coli, *J. Med. Genet.* 6:376–378.

Ashley, D. J. B., 1969b, The two "hit" theory of carcinogenesis, *Br. J. Cancer* 23:313–328.

Bagchi, S., Weinmann, R., and Raychaudhuri, P., 1991, The retinoblastoma protein copurifies with E2F-I, an E1A-regulated inhibitor of the transcription factor E2F, *Cell* 65:1063–1072.

Baker, S. J., Fearon, E. R., Nigro, J. M., Hamilton, S. R., Preisinger, A. C., Jessup, J. M., van Tuinen, P., Ledbetter, D. H., Barker, D. F., Nakamura, Y., White, R., and Vogelstein, B., 1989, Chromosome 17 deletions and p53 gene mutations in colorectal carcinomas, *Science* 244: 217–221.

Baker, S. J., Markowitz, S., Fearon, E. R., Willson, J. K., and Vogelstein, B., 1990a, Suppression of human colorectal carcinoma cell growth by wild-type p53, *Science* 249:912–915.

Baker, S. J., Preisinger, A. C., Jessup, J. M., Paraskeva, C., Markowitz, S., Willson, J. K., Hamilton, S., and Vogelstein, B., 1990b, p53 gene mutations occur in combination with 17p allelic deletions as late events in colorectal tumorigenesis, *Cancer Res.* 50:7717–7722.

Balaban, G., Gilbert, F., Nichols, W., Meadows, A. T., and Shields, J., 1982, Abnormalities of chromosome #13 in retinoblastomas from individuals with normal constitutional karyotypes, *Cancer Genet. Cytogenet.* 6:213–221.

Ballester, R., Marchuk, D., Boguski, M., Saulino, A., Letcher, R., Wigler, M., and Collins, F., 1990, The NF1 locus encodes a protein functionally related to mammalian GAP and yeast IRA proteins, *Cell* 63:851–859.

Bandara, L., and La Thangue, N. B., 1991, Adenovirus E1a prevents the retinoblastoma gene product from complexing with a cellular transcription factor, *Nature* 351:494–497.

Bandara, L. R., Adamczewski, J. P., Hunt, T., and La Thangue, N. B., 1991, Cyclin A and the retinoblastoma gene product complex with a common transcription factor, *Nature* 352:249–251.

Barker, D., Wright, E., Nguyen, K., Cannon, L., Fain, P., Goldgar, D., Bishop, D. T., Carey, J., Baty, B., Kivlin, J., Willard, H., Waye, J. S., Greig, G., Leinwand, L., Nakamura, Y. P., O'Connell, P., Leppert, M., Lalouel, J.-M., White, R., and Skolnick, M., 1987, Gene for von Recklinghausen neurofibromatosis is in the pericentromeric region of chromosome 17, *Science* 236:1100–1102.

Basu, T. N., Gutmann, D. H., Fletcher, J. A., Glover, T. W., Collins, F. S., and Downward, J., 1992, Aberrant regulation of *ras* proteins in malignant cells from type 1 neurofibromatosis patients, *Nature* 356:713–715.

Bickmore, W. A., Porteous, D. J., Christie, S., Seawright, A., Fletcher, J. M., Maule, J. C., Couillin, P., Junien, C., Hastie, N. D., and van Heyningen, V., 1989, CpG islands surround a DNA segment located between translocation breakpoints associated with genitourinary dysplasia and aniridia, *Genomics* 5:685–693.

Bischoff, J. R., Friedman, P. N., Marshak, D. R., Prives, C., and Beach, D., 1990, Human p53 is phosphorylated by p60-cdc2 and cyclin B-cdc2, *Proc. Natl. Acad. Sci. USA* 87:4766–4770.

Bishop, D. T., and Westbrook, C., 1990, Report of the committee on the constitution of chromosome 5, *Cytogenet. Cell. Genet.* 55:111–117.

Bishop, J. M., 1983, Cellular oncogenes and retroviruses, *Annu. Rev. Biochem.* 52:301–354.

Bishop, J. M., 1985, Viral oncogenes, *Cell* 42:23–38.

Bodmer, W. F., Bailey, C. J., Bodmer, J., Bussey, H. J. R., Ellis, A., Gorman, P., Lucibello, F. C., Murday, V. A., Rider, S. H., Scambler, P., Sheer, D., Solomon, E., and Spurr, N. K., 1987, Localization of the gene for familial adenomatous polyposis on chromosome 5, *Nature* **328**:614–616.

Bollag, G., and McCormick, F., 1991, Differential regulation of rasGAP and neurofibromatosis gene product activities, *Nature* **351**:576–579.

Bollag, G., and McCormick, F., 1992, NF is enough of GAP, *Nature* **356**:663–664.

Bonetta, L., Kuehn, S. E., Huang, A., Law, D. J., Kalikin, L. M., Koi, M., Reeve, A. E., Brownstein, B. H., Yeger, H., Williams, B. R. G., and Fenberg, A. P., 1990, Wilms tumor locus on 11p13 defined by multiple CpG island-associated transcripts, *Science* **250**:994–997.

Bookstein, R., Shew, J. Y., Chen, P. L., Scully, P., and Lee, W.-H., 1990, Suppression of tumorigenicity of human prostate carcinoma cells by replacing a mutated *RB* gene, *Science* **247**: 712–715.

Bos, J. L., 1989, *ras* oncogenes in human cancer: A review, *Cancer Res.* **49**:4682–4689.

Bos, J. L., Fearon, E. R., Hamilton, S. R., Verlaan-de Vries, M., van Boom, J. H., van der Eb, A. J., and Vogelstein, B., 1987, Prevalence of *ras* gene mutations in human colorectal cancers, *Nature* **327**:293–297.

Botstein, D., White, R. L., Skolnick, M., and Davis, R. W., 1980, Construction of a genetic linkage map in man using restriction fragment length polymorphisms, *Am. J. Hum. Genet.* **32**:314–331.

Boultwood, J., Rack, K., Kelly, S., Madden, J., Sakaguchi, A. Y., Wang, L.-M., Oscier, D. G., Buckle, V. J., and Wainscoat, J. S., 1991, Loss of both CSF1R (FMS) alleles in patients with myelodysplasia and a chromosome 5 deletion, *Proc. Natl. Acad. Sci. USA* **88**:6176–6180.

Bourne, H. R., 1991, Consider the coiled coil . . . , *Nature* **351**:188–190.

Bowcock, A. M., Anderson, L. A., Black, D. M., Lawrence, S. O., Hall, J. M., Solomon, E., and King, M.-C., (in press). Polymorphorism in THRA1 and D17S181 flank a 5 cM interval for the breast-ovarian cancer gene on chromosome 17q, *Am. J. Hum. Genet.*

Boyton, R. F., Blount, P. L., Yin, J., Brown, V. L., Huang, Y., Tong, Y., McDaniel, T., Newkirk, C., Resau, J. H., Raskind, W. H., Haggitt, R. C., Reid, B. J., and Meltzer, S. L., 1992, Loss of heterozygosity involving the *APC* and *MCC* genetic loci occurs in the majority of human esophageal cancers, *Proc. Natl. Acad. Sci. USA* **89**:3385–3388.

Brauch, H., Tory, K., Kotler, F., Gazdar, A. F., Pettengill, O. S., Johnson, B., Graziano, S., Winton, T., Buys, C. H. C. M., Sorenson, G. D., Poiesz, B. J., Minna, J. D., and Zbar, B., 1990, Molecular mapping of deletion sites in the short arm of chromosome 3 in human lung cancer, *Genes Chromosomes Cancer* **1**:240–246.

Breslow, N., Beckwith, J. B., Ciol, M., and Shaples, K., 1988, Age distribution of Wilms' tumor: Report from the National Wilms' Tumor Study, *Cancer Res.* **48**:1653–1657.

Bressac, B. Kew, M., Wands, J., and Ozturk, M., 1991, Selective G to T mutations of p53 gene in hepatocellular carcinoma from southern Africa, *Nature* **350**:429–431.

Buchberg, A. M., Cleveland, L. S., Jenkins, N. A., and Copeland, N. G., 1990, Sequence homology shared by neurofibromatosis type-1 gene and IRA-1 and IRA-2 negative regulators of the RAS cyclic AMP pathway, *Nature* **347**:291–294.

Buchkovich, K., Duffy, L. A., and Harlow, E., 1989, The retinoblastoma protein is phosphorylated during specific phases of the cell cycle, *Cell* **58**:1097–1105.

Burke, D. T., Carle, G. F., and Olson, M. V., 1987, Cloning of large segments of exogenous DNA into yeast by means of artificial chromosome vectors, *Science* **236**:806–812.

Cales, C., Hancock, J. F., Marshall, C. J., and Hall, A., 1988, The cytoplasmic protein GAP is implicated as the target for regulation by the *ras* gene product, *Nature* **332**:548–551.

Call, K. M., Glaser, T., Ito, C. Y., Buckler, A. J., Pelletier, J., Haber, D. A., Rose, E. A., Kral, A., Yeger, H., Lewis, W. H., Jones, C., and Houseman, D. E., 1990, Isolation and characterization of a zinc finger polypeptide gene at the human chromosome 11 Wilms' tumor locus, *Cell* **60**: 509–520.

Cannon-Albright, L. A., Goldgar, D. E., Meyerm, L. J., Lewis, C. M., Anderson, D. E., Fountain, J. W., Hegi, M. E., Wiseman, R. W., Petty, E. M., Bale, A. E., Olopade, O. I., Diaz, M. O., Kwiatkowski, D. J., Piepkorn, M. W., Zone, J. J., and Skolnick, M. H., 1992, Assignment of a locus for familial melanoma, MLM, to chromosome 9p13–p22, *Science* **258:**1148–1152.

Caron de Fromentel, C., and Soussi, T., 1992, TP53 tumor suppressor gene: A model for investigating human mutagenesis, *Genes Chromosomes Cancer* **4:**1–15.

Cavenee, W. K., Dryja, T. P., Phillips, R. A., Benedict, R., Gallie, B. L., Murphree, A. L., Strong, L. C., and White, R. L., 1983, Expression of recessive alleles by chromosomal mechanisms in retinoblastoma, *Nature* **305:**779–784.

Cavenee, W., Leach, R., Mohandas, T., Pearson, P., and White, R., 1984, Isolation and regional localization of DNA segments revealing polymorphic loci from human chromosome 13, *Am. J. Hum. Genet.* **36:**10–24.

Cawthon, R. M., Weiss, R., Xu, G. F., Viskochil, D., Culver, M. Stevens, J., Robertson, M., Dunn, D., Gesteland, R., O'Connell, P., and White, R., 1990, A major segment of the neurofibromatosis type 1 gene: cDNA sequence, genomic structure, and point mutations, *Cell* **62:**193–201.

Chavrier, P., Zerial, M., Lemaire, P., Almendral, J., Bravo, R., and Charnay, P., 1988, A gene encoding a protein with zinc fingers is activated during G0/G1 transition in cultured cells, *EMBO J* **7:**2–35.

Chellappan, S. P., Hiebert, S., Mudryj, M., Horowitx, J. M., and Nevins, J. R., 1991, The E2F transcription factor is a cellular target for the RB protein, *Cell* **65:**1053–1061.

Chen, P.-L., Scully, P., Shew, J.-Y., Wang, J. Y. J., and Lee, W.-H., 1989, Phosphorylation of the retinoblastoma gene product is modulated during the cell cycle and cellular differentiation, *Cell* **58:**1193–1198.

Chen, P. L., Chen, Y. M., Bookstein, R., and Lee, W. H., 1990, Genetic mechanisms of tumor suppression by the human p53 gene, *Science* **256:**1576–1580.

Chittenden, T., Livingston, D. M., and Kaelin, W. G. J., 1991, The T/E1A-binding domain of the retinoblastoma product can interact selectively with sequence-specific DNA-binding protein, *Cell* **65:**1073–1082.

Clarke, C. F., Cheng, K., Frey, A. B., Stein, R., Hinds, P. W., and Levine, A. J., 1988, Purification of complexes of nuclear oncogene p53 with rat and *Escherichia coli* heat shock proteins: *In vitro* dissociation of hsc70 and dnaK from murine p53 by ATP, *Mol. Cell. Biol.* **8:**1206–1215.

Cohen, A. J., Li, F. P., Berg, S., Marchetto, D. J., Tsai, S., Jacobs, S. C., and Brown, R. S., 1979, Hereditary renal-cell carcinoma associated with a chromosomal translocation, *N. Engl. J. Med.* **301:**592–595.

Cohen, B. H., and Rothner, A. D., 1989, Incidence, types, and management of cancer in patients with neurofibromatosis, *Oncology* **3:**23–30.

Collins, F. S., O'Connell, P., Ponder, B. A., and Seizinger, B. R., 1989a, Progress towards identifying the neurofibromatosis (NF1) gene, *Trends Genet.* **5:**217–221.

Collins, F. S., Ponder, B. A. P., Seizinger, B. R., and Epstein, C. J., 1989b, Editorial: The von Recklinghausen neurofibromatosis region on chromosome 17—Genetic and physical maps come into focus, *Am. J. Hum. Genet.* **44:**1–5.

Compton, D. A., Weil, M. M., Jones, C., Riccardi, V. M., Strong, L. C., and Saunders, G. F., 1988, Long range physical map of the Wilms' tumor–aniridia region on human chromosome 11, *Cell* **55:** 827–836.

Compton, D. A., Weil, M. M., Bonetta, L., Huang, A., Jones, C., Yeger, H., Williams, B. R., Strong, L. C., and Saunders, G. F., 1990, Definition of the limits of the Wilms tumor locus on human chromosome 11p13, *Genomics* **6:**309–315.

Cotton, R. G. H., Rodrigues, N. R., and Campbell, R. D., 1988, Reactivity of cytosine and thymidine in single-base-pair mismatches with hydroxylamine and osmium tetroxide and its application to the study of mutations, *Proc. Natl. Acad. Sci. USA* **85:**4397–4401.

Cottrell, S., Bicknell, D., and Bodmer, W., 1992, Detection of mutations in part of the APC gene, *Lancet* **340:**626–630.

Coulson, A., Waterston, R., Kiff, J., Sulston, J., and Kohara, Y., 1988, Genome linking with yeast artificial chromosomes, *Nature* 335:184–186.

Cowell, J. K., Wadey, R. B., Haber, D. A., Call, K. M., Houseman, D. E., and Pritchard, J., 1991, Structural rearrangements of the WT1 gene in Wilms' tumour cells, *Oncogene* 6:595–599.

Cox, D. R., Pritchard, C. A., Uglum, E., Casher, D., Kobori, J., and Myers, R. M., 1989, Segregation of the Huntington disease region of the human chromosome 4 in a somatic cell hybrid, *Genomics* 4:397–407.

Crawford, L. C., Pim, D. C., Gurney, E. G., Goodfellow, P., and Taylor-Papadimitriou, J., 1981, Detection of a common feature in several human tumor cell lines—A 53000-dalton protein, *Proc. Natl. Acad. Sci. USA* 78:41–45.

Davis, L. M., Byers, M. G., Fukushima, Y., Qin, S. Z., Nowak, N. J., Scoggin, C., and Shows, T. B., 1988, Four new DNA markers are assigned to the WAGR region of 11p13: Isolation and regional assignment of 112 chromosome 11 anonymous DNA segments, *Genomics* 3:264–271.

DeCaprio, J. A., Ludlow, J. W., Figge, J., Shew, J.-Y., Huang, C.-M., Lee, W.-H., Marsilio, E., Paucha, E., and Livingston, D. M., 1988, SV40 large tumor antigen forms a specific complex with the product of the retinoblastoma susceptibility gene, *Cell* 54:275–283.

DeCaprio, J. A., Ludlow, J. W., Lynch, D., Furukawa, Y., Griffen, J., Piwnica-Worms, H., Huang, C.-M., and Livingston, D. M., 1989, The product of the retinoblastoma susceptibility gene has properties of a cell cycle regulatory element, *Cell* 58:1085–1095.

DeChiara, T. M., Robertson, E. J., and Efstratiadis, A., 1991, Paternal imprinting of the mouse insulin-like growth factor II gene, *Cell* 64:849–859.

DeClue, J. E., Papageorge, A. G., Fletcher, J. A., Diehl, S. R., Ratner, N., Vass, W. C., and Lowy, D. R., 1992, Abnormal regulation of mammalian p21^{ras} contributes to malignant tumor growth in von Recklinghausen (type 1) neurofibromatosis, *Cell* 69:265–273.

Defeo-Jones, D., Huang, P. S., Jones, R. E., Haskell, K. M., Vuocolo, G. A., Hanobik, M. G., Hube, H. E., and Oliff, A., 1991, Cloning of cDNAs for cellular proteins that bind to the retinoblastomas gene product, *Nature* 352:251–254.

Denys, P., Malvaux, P., van den Berghe, H., Tanghe, W., and Proesmans, W., 1967, Association d'un syndrome anatomo-pathologique de pseudohermaphrodisme masculin, d'une tumeur de Wilms, d'une nephropathie parenchymateuse et d'un mosaicisme XX/XY, *Arch. Fr. Pediatr.* 24:729–739.

Donehower, L. A., Harvey, M., Slagle, B. L., McArthur, M. J., Montgomery, C. A. J., Butel, J. S., and Bradley, A., 1992, Mice deficient for p53 are developmentally normal but susceptible to spontaneous tumours, *Nature* 356:215–221.

Drash, A., Sherman, F., Hartmann, W. H., and Blizzard, R. M., 1970, A syndrome of pseudo-hermaphroditism, Wilms' tumor, hypertension and degenerative renal disease, *J. Pediatr.* 76:585–593.

Dryja, T. P., Cavenee, W., White, R., Rapaport, J. M., Petersen, R., Albert, D. M., and Bruns, G. A. P., 1984, Homozygosity of chromosome 13 in retinoblastoma, *N. Engl. J. Med.* 310:550–553.

Dryja, T. P., Rapaport, J. M., Epstein, J., Goorin, A. M., Weichselbaum, R., Koufos, A., and Cavenee, W. K., 1986, Chromosome 13 homozygosity in osteosarcoma without retinoblastoma, *Am. J. Hum. Genet.* 38:59–66.

Dunn, J. M., Phillips, R. A., Becker, A. J., and Gallie, B. L., 1988, Identification of germline and somatic mutations affecting the retinoblastoma gene, *Science* 241:1979–1800.

Easton, D. F., Bishop, D. T., Ford, D., Crockford, G. P., and the Breast Cancer Linkage Consortium, (in press). Genetic linkage analysis in familial breast and ovarian cancer—Results from 214 families, *Am. J. Hum. Genet.*

Easton, D. F., Ponder, M. A., Huson, S. M., and Ponder, B. A. J., (n.d.). An analysis of variation in expression of neurofibromatosis type 1: Evidence for modifying genes, *Am. J. Hum. Genet.* (submitted).

Editorial., 1991, Breast cancer genetics: Some clues, more questions, *Lancet* 337:329–331.

Egan, C., Bayley, S. T., and Branton, P. E., 1989, Binding of the Rbl protein to E1A products is required for adenovirus transformation, *Oncogene* 4:383–388.

El-Deiry, W. S., Kern, S. E., Pietenpol, J. A., Kinzler, K. W., and Vogelstein, B., 1992, Definition of a consensus binding site for p53, *Nature Genet.* 1:45–49.

Eliyahu, D., Raz, A., Gruss, P., Givol, D., and Oren, M., 1984, Participation of p53 cellular tumor antigen in transformation of normal embryonic cells, *Nature* 312:646–649.

Erlandsson, R., Bergerheim, U. S. R., Boldog, F., Marcsek, Z., Kunimi, K., Lin, B. Y.-T., Ingvarsson, S., Castresana, J. S., Lee, W.-H., Lee, E., Klein, G., and Sümegi, J., 1990, A gene near the D3F15S2 site on 3p is expression in normal human kidney but not or only at a severely reduced level in 11 of 15 primary renal cell carcinomas (RCC), *Oncogene* 5:1207–1211.

Erlandsson, R., Boldog, F., Persson, B. Zabarovsky, E. R., Allikmets, R. L., Sümegi, J., Klein, G., and Jörnvall, H., 1991, The gene from the short arm of chromosome 3, at D3F15S2, frequently deleted in renal cell carcinoma encodes acylpeptidase hydrolase, *Oncogene* 6:1293–1295.

Farndon, P. A., Del Masro, R. G., Evans, D. G. R., and Kilpatrick, M. W., 1992, Location of gene for Gorlin syndrome, *Lancet* 339:581–582.

Fearon, E. R., Vogelstein, B., and Feinberg, A. P., 1984, Somatic deletion and duplication of genes on chromosome 11 in Wilms' tumors, *Nature* 309:176–178.

Fearon, E. R., Hamilton, S. R., and Vogelstein, B., 1987, Clonal analysis of human colorectal tumours, *Science* 238:193–197.

Fearon, E. R., Cho, K. R., Nigro, J. M., Kern, S. E., Simons, J. W., Ruppert, J. M., Hamilton, S. R., Preisinger, A. C., Thomas, G., Kinzler, K. W., and Vogelstein, B., 1990, Identification of a chromosome 18q gene that is altered in colorectal cancers, *Science* 247:49–56.

Fields, S., and Jang, S. K., 1990, Presence of a potent transcription activating sequence in the p53 protein, *Science* 249:1046–1049.

Finlay, C. A., Hinds, P. W., Tan, T.-H., Eliyahu, D., Oren, M., and Levine, A. J., 1988, Activating mutations by p53 produce a gene product that forms an hsc70-p53 complex with an altered half life, *Mol. Cell. Biol.* 8:531–539.

Forrester, K., Almoguera, C., Han, K., Grizzle, W. E., and Perucho, M., 1987, Detection of high incidence of K-ras oncogenes during human colon tumorigenesis, *Nature* 327:298–303.

Fountain, J. W., Wallace, M. R., Brereton, A. M., O'Connell, P., White, R. L., Rich, D. C., Ledbetter, D. H., Leach, R. J., Fournier, R. E., Menon, A. G., Gusella, J. F., Barker, D., Stephens, K., and Collons, F. S., 1989a, Physical mapping of the von Recklinghausen neurofibromatosis region on chromosome 17, *Am. J. Hum. Genet.* 44:58–67.

Fountain, J. W., Wallace, M. R., Bruce, M. A., Seizinger, B. R., Menon, A. G., Gusella, J. F., Michels, V. V., Schmidt, M. A., Dewald, G. W., and Collins, F. S., 1989b, Physical mapping of a translocation breakpoint in neurofibromatosis, *Science* 244:1085–1087.

Fountain, J. W., Karayiorgou, M., Ernstoff, M. S., Kirkwood, J. M., Vlock, D. R., Titus-Ernstoff, L., Bouchard, B., Vijayasaradhi, S., Houghton, A. N., Lahti, J., Kidd, V. J., Housman, D. E., and Dracopoli, N. C., 1992, Homozygous deletions within human chromosome band 9p21 in melanoma, *Proc. Natl. Acad. Sci USA* 89:10557–10561.

Francke, U., 1990, A gene for Wilms tumour? *Nature* 343:692–694.

Frézal, J., and Schinzel, A., 1991, Report of the committee on clinical disorders, chromosome aberrations and uniparental disomy, *Cytogenet. Cell Genet.* 58:986–1052.

Friend, S. H., Bernards, R., Rogelj, S., Weinberg, R. A., Rapaport, J. M., Alberts, D. M., and Dryja, T. P., 1986, A human DNA segment with properties of the gene that predisposes to retinoblastoma and osteosarcoma, *Nature* 323:643–646.

Friend, S. H., Horowitz, J. M., Gerber, M. R., Wang, X-F., Bogenmann, E., Li, F. P., and Weinberg, R. A., 1987, Deletions of a DNA sequence in retinoblastomas and mesenchymal tumors: Organization of the sequence and its encoded protein, *Proc. Natl. Acad. Sci. USA* 84:9059–9063.

Fujimori, M., Tokino, T. Hino, O., Kitagawa, T., Imamura, T., Okamoto, E., Mitsunobu, M.,

Ishikawa, T., Nakagama, H., Harada, H., Yagura, H., Matsubara, K., and Nakamura, Y., 1991, Allelotype study of primary hepatocellular carcinoma, *Cancer Res.* **51**:89–93.

Fung, Y.-K. T., Murphree, A. L., T'Ang, A., Qian, J., Hinrichs, S. H., and Benedict, W. F., 1987, Structural evidence for the authenticity of the human retinoblastoma gene, *Science* **236**:1657–1661.

Furukawa, Y., DeCaprio, J. A., Belvin, M., and Griffen, J. D., 1991, Heterogeneous expression of the product of the retinoblastoma susceptibility gene in primary human leukemia cells, *Oncogene* **6**: 1343–1346.

Futreal, P. A., Söderkvist, P., Marks, J. R., Inglehart, J. D., Cochran, C., Barrett, J. C., and Wiseman, R. W., 1992, Detection of frequent allelic loss on proximal chromosome 17q in sporadic breast carcinoma using microsatellite length polymorphisms, *Cancer Res.* **52**:2624–2627.

Gailani, M. R., Bale, S. J., Leffell, D. J., DiGiovanna, J. J., Peck, G. L., Poliak, S., Drum, M. A., Pastakia, B., McBride, O. W., Kase, R., Greene, M., Mulvihill, J. J., and Bale, A. E., 1992, Developmental defects in Gorlin syndrome related to a putative tumor suppressor gene on chromosome 9, *Cell* **69**:111–117.

Gardiner-Garden, M., and Frommer, M., 1987, CpG islands in vertebrate genomes, *J. Mol. Biol.* **196**: 261–282.

Gardner, H. A., Gallie, B. L., Knight, L. A., and Phillips, R. A., 1982, Multiple karyotypic changes in retinoblastoma tumor cells: Presence of normal chromosome No. 13 in most tumors, *Cancer Genet. Cytogenet.* **6**:201–211.

Gateff, E., 1982, Cancer, genes, and development: The *Drosophila* case, *Adv. Cancer Res.* **37**:33–74.

Geiser, A. G., Der, C. J., Marshall, C. J., and Stanbridge, E. J., 1986, Suppression of tumorigenicity with continued expression of the c-Ha-*ras* oncogene in EJ bladder carcinoma–human fibroblast hybrid cells, *Proc. Natl. Acad. Sci. USA* **83**:5209–5213.

Gessler, M., Simola, K. O., and Bruns, G. A., 1989a, Cloning of breakpoints of a chromosome translocation identifies the AN2 locus, *Science* **244**:1575–1578.

Gessler, M. Thomas, G. H., Couillin, P., Junien, C., McGillivray, B. C., Hayden, M., Jaschek, G., and Bruns, G. A., 1989b, A deletion map of the WAGR region on chromosome 11, *Am. J. Hum. Genet.* **44**:486–495.

Gessler, M., Poustka, A., Cavenee, W., Neve, R. L., Orkin, S. H., and Bruns, G. A., 1990, Homozygous deletion in Wilms tumours of a zinc-finger gene identified by chromosome jumping, *Nature* **343**:774–778.

Gilbert, F., Feder, M., Balaban, G., Brangman, D., Lurie, D. K., Podolsky, R., Rinaldt, V., Vinikoor, N., and Weisband, J., 1984, Human neuroblastomas and abnormalities of chromosomes 1 and 17, *Cancer Res.* **44**:5444–5449.

Glaser, T., Lewis, W. H., Bruns, G. A., Watkins, P. C., Rogler, C. E., Shows, T. B., Powers, V. E., Willard, H. F., Goguen, J. M., Simola, K. O., and Houseman, D. E., 1986, The β-subunit of follicle-stimulating hormone is deleted in patients with aniridia and Wilms' tumour, allowing a further definition of the WAGR locus, *Nature* **321**:882–887.

Glaser, T., Houseman, D. E., Lewis, W. H., Gerhard, D., and Jones, C., 1989, A fine structure map of human chromosome 11p: Analysis of the J1 series hybrids, *Somatic Cell Mol. Genet.* **15**:477–501.

Glover, T. W., Stein, C. K., Legius, E., Andersen, L. B., Brereton, A., and Johnson, S., 1991, Molecular and cytogenetic analysis of tumors in von Recklinghausen neurofibromatosis, *Genes Chromosomes Cancer* **3**:62–70.

Godbout, R., Dryja, T. P., Squire, J., Gallie, B. L., and Phillips, R. A., 1983, Somatic inactivation of genes on chromosome 13 is a common event in retinoblastoma, *Nature* **304**:451–453.

Goddard, A. D., Balakier, H., Canton, M., Dunn, J., Squire, J., Reyes, E., Becker, A., Phillips, R. A., and Gallie, B. L., 1988, Infrequent genomic rearrangement and normal expression of the putative RB1 gene in retinoblastoma tumors, *Mol. Cell. Biol.* **8**:2082–2088.

Goss, S. J., and Harris, H., 1975, New method for mapping genes in human chromosomes, *Nature* **255**: 680–683.

Graw, S., Davidson, J., Gusella, J., Watkins, P., Tanzi, R., Neve, R., and Patterson, D., 1988, Irradiation-induced human chromosome 21 hybrids, *Somatic Cell Genet.* **14**:233–242.

Green, M. R., 1989, When the products of oncogenes and anti-oncogens meet, *Cell* **56**:1–3.

Groden, J., Thliveris, A., Samowitz, W., Carlson, M., Gelbert, L., Albertsen, H,. Joslyn, G., Stevens, J., Spirio, L., Robertson, K., Sargeant, L., Krapcho, K., Wolff, E., Burt, R., Hughes, J. P., Warrington, J., McPherson, J., Wasmuth, J., Le Pasilier, D., Abderrahim, H., Cohen, D., Leppert, M., and White, R., 1991, Identification and characterization of the familial adenomatous polyposis coli gene, *Cell* **66**:589–600.

Grundy, P., Koufos, A., Morgan, K., Li, F. P., Meadows, A. T., and Cavenee, W. C., 1988, Familial predisposition to Wilms' tumour does not map to the short arm of chromosome 11, *Nature* **33**: 374–376.

Guzzetta, V., Montes de Oca-Luna, R., Lupski, J. R., and Patel, P. I., 1991, Isolation of region-specific and polymorphic markers from chromosome 17 by restricted *Alu* polymerase chain reaction, *Genomics* **9**:31–36.

Haber, D. A., Buckler, A. J., Glaser, T., Call, K. M., Pelletier, J., Sohn, R. L., Douglass, E. C., and Houseman, D. E., 1990, An internal deletion within an 11p13 zinc finger gene contributes to the development of Wilms' tumor, *Cell* **61**:1257–1269.

Haber, D. A., Sohn, R. L., Buckler, A. J., Pelletier, J., Call, K. M., and Houseman, D. E., 1991, Alternative splicing an genomic structure of the Wilms tumor gene *WT1*, *Proc. Natl. Acad. Sci. USA* **88**:9618–9622.

Hall, A., 1990, The cellular functions of small GTP-binding proteins, *Science* **249**:635–640.

Hall, A., 1992, Signal transduction through small GTPases—A tale of two GAPS, *Cell* **69**:389–391.

Hall, J. M., Lee, M. K., Newman, B., Morrow, J. E., Anderson, L. A., Huey, B., and King, M.-C., 1990, Linkage of early-onset familial breast cancer to chromosome 17q21, *Science* **250**:1684–1689.

Hall, J. M., Friedman, L., Guenther, C., Lee, M. K., Weber, J. L., Black, D. M., and King, M-C., 1992, Closing in on a breast cancer gene on chromosome 17q, *Am. J. Hum. Genet.* **50**: 1235–1242.

Han, J.-W., McCormick, F., and Macara, I. G., 1991, Regulation of Ras-GAP and the neurofibromatosis-I gene product by eicosanoids, *Science* **252**:576–579.

Harbour, J. W., Lai, S.-L. L., Whang-Peng, J., Gazdar, A. F., Minna, J. D., and Kaye, F. J., 1988, Abnormalities in structure and expression of the human retinoblastoma gene in SCLC, *Science* **241**:353–357.

Harris, A. L., 1991, Telling change of base, *Nature* **350**:377–378.

Harris, H., Miller, O. J., Klein, G., Worst, P. and Tachibana, T., 1969, Suppression of malignancy by cell fusion, *Nature* **223**:363–368.

Henry, I., Grandjouan, S., Coullin, P., Barichard, F., Huerre, J. C., Glaser, T., Philip, T., Lenoir, G., Chaussain, J. L., and Junien, C., 1989, Tumor-specific loss of 11p15.5 alleles in del11p13 Wilms tumor and in familial adrenocortical carcinoma, *Proc. Natl. Acad. Sci. USA* **86**:3247–3251.

Henry, I., Bonaiti-Pellié, C., Chehensse, V., Beldjord, C., Schwartz, C., Utermann, G., and Junien, C., 1991, Uniparental paternal disomy in a genetic cancer-predisposing syndrome, *Nature* **351**: 665–667.

Herrera, L., Katati, S., Gibas, L., Pierrzak, E, and Sandberg, A. A., 1986, Gardner syndrome in a man with an interstitial deletion of 5q, *Am. J. Med. Genet.* **25**:473–476.

Herskowitz, I., 1987, Functional inactivation of genes by dominant negative mutations, *Nature* **329**: 219–222.

Hethcote, H. W., and Knudson, A. G., 1978, Model for the incidence of embryonal cancers: Application to retinoblastoma, *Proc. Natl. Acad. Sci. USA* **75**:2453–2457.

Hicks, G. G., and Mowat, M., 1988, Integration of Friend murine leukemia virus into both alleles of p53 oncogene in an erythroleukemic cell line, *J. Virol.* **62**:4752–4755.

Hinds, P. W., Finlay, C. A., Frey, A. B., and Levine, A. J., 1987, Immunological evidence for the association of p53 with a heat shock protein, hsc70, in p53-plus-*ras*-transformed cell lines, *Mol. Cell. Biol.* **7**:2863–2869.

Hinds, P., Finlay, C., and Levine, A. J., 1989, Mutation is required to activate the p53 gene for cooperation with the ras oncogene and transformation, *J. Virol.* **63**:739–746.

Hockey, K. A., Mulcahy, M. T., Montgomery, P., and Levitt, S., 1989, Chromosome deletion of 5q and familial adenomatous polyposis, *J. Med. Genet.* **26**:61–62.

Hollstein, M., Sidransky, D., Vogelstein, B., and Harris, C. C., 1991, p53 mutations in human cancers, *Science* **253**:49–53.

Horowitz, J. M., Yandell, D. W., Park, S.-H., Canning, S., Whyte, P., Buchkovich, K., Harlow, E., Weinberg, R. A., and Dryja, T. P., 1989, Point mutational inactivation of the retinoblastoma antioncogene, *Science* **243**:937–940.

Hsu, I. C., Metcalf, R. A., Sun, T., Welsh, J. A., Wang, N. J., and Harris, C. C., 1991, Mutational hotspot in the p53 gene in human hepatocellular carcinomas, *Nature* **350**:427–428.

Hu, Q., Dyson, N., and Harlow, E., 1990, The regions of the retinoblastoma protein needed for binding to adenovirus E1A or SV40 large T antigen are common sites for mutations, *EMBO J.* **9**:1147–1155.

Huang, A., Campbell, C. E., Bonetta, L., McAndrews-Hill, M. S., Chilton-MacNeill, S., Coppes, M. J., Law, D. J., Feinberg, A. P., Yeger, H., and Williams, B. R. G., 1990, Tissue, developmental, and tumor-specific expression of divergent transcripts in Wilms tumor, *Science* **250**:991–997.

Huang, H-J. S., Yee, J.-K., Shew, J.-Y., Chen, P.-L., Bookstein, R., Friedmann, T., Lee, E. Y.-H. P., and Lee, W.-H., 1988, Suppression of the neoplastic phenotype by replacement of the RB gene in human cancer cells, *Science* **242**:1563–1566.

Huang, S., Lee, W-H., and Lee, E.Y.-H. P., 1991, A cellular protein that competes with SV40 T antigen for binding to the retinoblastoma gene product, *Nature* **350**:160–162.

Huff, V., Compton, D. A., Chao, L.-Y., Strong, L. C., Geiser, C. F., and Saunders, G. F., 1988, Lack of linkage of familial Wilms' tumour to chromosomal band 11p13, *Nature* **336**:377–378.

Huff, V., Miwa, H., Haber, D. A., Call, K. M., Houseman, D. E., Strong, L. C., and Saunders, G. F., 1991, Evidence for WT1 as a Wilms tumor (WT) gene: Intragenic germinal deletion in bilateral WT, *Am. J. Hum. Genet.* **48**:997–1003.

Ibsen, J. M., Waters, J. J., Twentyman, P. R., Bleehen, N. M., and Rabbitt, P. H., 1987, Oncogene amplification and chromosomal abnormalities in small cell lung cancer, *J. Cell. Biochem.* **33**: 267–288.

Imai, Y., Matsushima, Y., Sugimura, T., and Terada, M., 1991, Purification and characterization of human papillomavirus type 16 E7 protein with preferential binding capacity to the underphosphorylated form of retinoblastoma gene product, *J. Virol.* **65**:4966–4972.

James, C. D., Carlbom, E., Dumanski, J. P., Hansen, M., Nordenskjold, M., Collins, P., and Cavenee, W. K., 1988, Clonal genomic alterations in glioma malignancy stages, *Cancer Res.* **48**:5546–5551.

James, C. D., Carlbom, E., Nordenskjold, M., Collins, V. P., and Cavenee, W. K., 1989, Mitotic recombination of chromosome 17 in astrocytomas, *Proc. Natl. Acad. Sci. USA* **86**:2858–2862.

Jenkins, J. R., Rudge, K., and Currie, G. A., 1984, Cellular immortalization by cDNA clone encoding the transformation-associated phosphoprotein p53, *Nature* **312**:651–654.

Joslyn, G., Carlson, M., Thliveris, A., Albertsen, H., Gelbert, L., Samowitz, W., Groden, J., Stevens, J., Spirio, L., Robertson, M., Sargeant, L., Krapcho, K., Wolff, E., Burt, R., Hughes, J. P., Warrington, J., McPherson, J., Wasmuth, J., Le Pasilier, D., Abderrahim, H., Cohen, D., Leppert, M., and White, R., 1991, Identification of deletion mutations and three new genes at the familial polyposis locus, *Cell* **66**:601–613.

Kaelbling, M., and Klinger, H. P., 1986, Suppression of tumorigenicity in somatic cell hybrids. III. Cosegregation of human chromosome 11 of a normal cell and suppression of tumorigenicity in intraspecies hybrids of normal diploid × malignant cells, *Cytogenet. Cell. Genet.* **41**:65–70.

Kaelin, W. G. J., Pallas, D. C., DeCaprio, J. A., Kaye, F. J., and Livingston, D. M., 1991, Identification of cellular proteins that can interact specifically with the T/E1A-binding region of the retinoblastoma gene product, *Cell* **64**:521–532.

Kan, Y. W., and Dozy, A. M., 1978, Polymorphism of DNA sequence adjacent to human beta-globin structural gene: Relationship to sickle mutation, *Proc. Natl. Acad. Sci. USA* **75**:5631–5635.

Kern, S. E., Pietenpol, J. A., Thiagalingam, S., Seymour, A., Kinzler, K. W., and Vogelstein, B., 1992, Oncogenic forms of p53 inhibit p53-regulated gene expression, *Science* **256**:827–830.

Kim, S.-J., Lee, H.D., Robbins, P. D., Busam, K., Sporn, M. B., and Robberts, A. B., 1991, Regulation of transforming growth factor beta 1 gene expression by the product of the retinoblastoma-susceptibility gene, *Proc. Natl. Acad. Sci. USA* **88**:3052–3056.

Kimchi, A., Wang, X.-F., Weinberg, R. A., Cheifetz, S., and Massague, J., 1988, Absence of TGF-beta receptors and growth inhibitory responses in retinoblastoma cells, *Science* **240**: 196–199.

Kinzler, K. W., Nilbert, M. C., Su, L.-K., Vogelstein, B., Bryan, T. M., Levy, D. B., Smith, K. J., Preisinger, A. C., Hedge, P., McKechnie, D., Finniear, R., Markham, A., Groffen, J., Boguski, M. S., Altschul, S. F., Horii, A., Ando, H., Miyoshi, Y., Miki, Y., Nishisho, I., and Nakamura, Y., 1991a, identification of FAP locus genes from chromosome 5q21, *Science* **253**:661–665.

Kinzler, K. W., Nilbert, M. C., Vogelstein, B., Bryan, T. M., Levy, D. B., Smith, K. J., Preisinger, A. C., Hamilton, S. R., Hedge, P., Markham, A., Carlson, M., Joslyn, G., Groden, J., White, R., Miki, Y., Miyoshi, Y., Nishisho, I., and Nakamura, Y., 1991b, Identification of a gene located at chromosome 5q21 that is mutated in colorectal cancers, *Science* **251**:1366–1370.

Klinger, H. P., 1982, Suppression of tumorigenicity, *Cytogenet. Cell Genet.* **32**:68–84.

Knudson, A. G., 1971, Mutation and cancer: Statistical study of retinoblastoma, *Proc. Natl. Acad. Sci. USA* **68**:820–823.

Knudson, A. G., 1986, Genetics of human cancer, *Annu. Rev. Genet.* **20**:231–251.

Knudson, A. G., and Strong, L. C., 1972a, Mutation and cancer: Neuroblastoma and pheochromocytoma, *Am. J. Hum. Genet.* **24**:514–532.

Knudson, A. G., and Strong, L. C., 1972b, Mutation and cancer: A model for Wilms' tumor of the kidney, *J. Natl. Cancer Inst.* **48**:313–324.

Kok, K., Osinga, J., Carritt, B., Davis, M. B., van der Hout, A. H., van der Veen, A. Y., Landsvater, R. M., de Leij, L. F. M. H., Berendsen, H. H., Postmus, P. E., Poppema, S., and Buys, C. H. C. M., 1987, Deletion of a DNA sequence at the chromosomal region 3p21 in all major types of lung cancer, *Nature* **330**:578–581.

Koufos, A., Hansen, M. F., Lampkin, B. C., Workman, M. L., Copeland, N. G., Jenkins, N. A., and Cavenee, W. K., 1984, Loss of alleles at loci on human chromosome 11 during genesis of Wilms' tumour, *Nature* **309**:170–172.

Koufos, A., Grundy, P., Morgan, K., Aleck, K. A., Hadro, T., Lampkin, B. C., Kalbakji, A., and Cavenee, W. C., 1989, Familial Weidemann–Beckwith syndrome and a second Wilms tumor locus both map to 11p15.5, *Am. J. Hum. Genet.* **44**:711–719.

Kovacs, G., Erlandsson, R., Boldog, F., Ingvarsson, S., Müller-Brechlin, R., Klein, G., and Sümegi, J., 1988, Consistent chromosome 3p deletion and loss of heterozygosity in renal cell carcinoma, *Proc. Natl. Acad. Sci. USA* **85**:1571–1575.

Kusnetsova, L. E., Prigogina, E. L., Progosianz, H. E., and Belkina, B. M., 1982, Similar chromosomal abnormalities in several retinoblastomas, *Hum. Genet.* **61**:201–204.

La Forgia, S., Morse, B., Levy, J., Barnea, G., Cannizzaro, L. A., Li, F., Nowell, P. C., Boghosian-Sell, L., Glick, J., Weston, A., Harris, C. C., Drabkin, H., Patterson, D., Croce, C. M., Schlessinger, J., and Huebner, K., 1991, Receptor protein-tyrosine phosphatase gamma is a candidate tumor suppressor gene at human chromosome region 3p21, *Proc. Natl. Acad. Sci. USA* **88**:8036–5040.

Lalande, M., Dryja, T. P., Schreck, R. R., Shipley, J., Flint, A., and Latt, S. A., 1984, Isolation of human chromosome 13-specific DNA sequences cloned from flow sorted chromosomes and potentially linked to the retinoblastoma locus, *Cancer Genet. Cytogenet.* **13**:283–295.

Land, H., Parada, L. F., and Weinberg, R. A., 1983, Tumorigenic conversion of primary embryo fibroblasts requires at least two cooperating oncogenes, *Nature* **304**:596–602.

Landegrent, J. E., Jansen in de Wal, J., Dirks, J., Baas, F., and van der Ploeg, M., 1987, Use of whole cosmid cloned genomic sequences for chromosomal localization by non-radioactive *in situ* hybridization, *Hum. Genet.* **77**:366–370.

Lane, D., and Harlow, E., 1982, Two different viral transforming proteins bind the same host tumor antigen, *Nature* **298**:517.

Lane, D. P., and Crawford, L. V., 1979, T antigen bound to a host protein in SV40-transformed cells, *Nature* **278**:261–263.

Ledbetter, D. H., Rich, D. C., O'Connell, P., Leppert, M., and Carey, J. C., 1989, Precise localization of NF1 to 17q11.2 by balanced translocation, *Am. J. Hum. Genet.* **44**:20–24.

Lee, E. Y.-H. P., To, H., Shew, J.-Y., Bookstein, R., Scully, P., and Lee, W.-H., 1988, Inactivation of the retinoblastoma susceptibility gene in human breast cancers, *Science* **241**:218–221.

Lee, W.-H., Bookstein, R., Hong, F., Young, L.-J., Shew, J.-Y., and Lee, E. Y.-H.P., 1987a, Human retinoblastoma susceptibility gene: Cloning, identification, and sequence, *Science* **235**:1394–1399.

Lee, W-H., Shew, J.-Y., Hong, F. D., Sery, T. W., Donoso, L. A., Young, L.-J., Bookstein, R., and Lee, E. Y.-H.P., 1987b, The retinoblastoma susceptibility gene encodes a nuclear phosphoprotein associated with DNA binding activity, *Nature* **329**:642–645.

Leppert, M., Dobbs, M., Scambler, P., O'Connel, P., Nakamura, Y., Stauffer, D., Woodward, S., Burt, R., Hughes, J., Gardner, E., Lathrop, M., Wasmuth, J., Lalouel, M., and White, R., 1987, The gene for familial polyposis coli maps to the long arm of chromosome 5, *Science* **238**:1411–1413.

Li, F. P., 1990, Familial cancer syndromes and clusters, *Curr. Probl. Cancer* **14**:73–114.

Li, Y., Bollag, G., Clark, R., Stevens, J., Conroy, L., Fults, D., Ward, K., Friedman, E., Samowitz, W., Robertson, M., Bradley, P., McCormick, F., White, R., and Cawthorn, R., 1992, Somatic mutations in the neurofibromatosis 1 gene in human tumors, *Cell* **69**:275–281.

Lin, B. T., Gruenwald, S., Morla, A. O., Lee, W. H., and Wang, J. Y., 1991, Retinoblastoma cancer suppressor gene product is a substrate of the cell cycle regulator cdc2 kinase, *EMBO J.* **10**:857–864.

Linzer, D. I. H., and Levine, A. J., 1979, Characterization of a 54K dalton cellular SV40 tumor antigen present in SV40-transformed cells and uninfected embryonal carcinoma cells, *Cell* **17**:43–52.

Ludecke, H.-J., Senger, G., Claussen, U., and Horsthemke, B., 1989, Cloning defined regions of the human genome by microdissection of banded chromosomes and enzymatic amplification, *Nature* **338**:348–350.

Ludlow, J. W., DeCaprio, J. A., Huang, C.-M., Lee, W.-H., Paucha, E., and Livingston, D. M., 1989, SV40 large T antigen binds preferentially to an underphosphorylated member of the retinoblastoma susceptibility gene family product, *Cell* **56**:57–65.

Ludlow, J. W., Shon, J., Pipas, J. M., Livingston, D. M., and DeCaprio, J. A., 1990, The retinoblastomas susceptibility gene product undergoes cell cycle-dependent dephosphorylation and binding to and release from SV40 large T, *Cell* **60**:387–396.

Lukeis, R., Irving, L., Garson, M., and Hasthorpe, S., 1990, Cytogenetics of non-small cell lung cancer: Analysis of consistent non-random abnormalities, *Genes Chrom. Cancer* **2**:116–124.

Lynch, H. T., Kimberling, W., Albano, W. A., Lynch, J. F., Biscone, K., Schuelke, G. S., Sandberg, A. V., Lipkin, M., Deschner, E. E., Mikol, Y. B., Elston, R. C., Bailey-Wilson, J. E., and Shannon-Danes, B., 1985, Hereditary nonpolyposis colorectal cancer (Lynch syndromes 1 and 11) 1. Clinical description of resource, *Cancer* **56**:934–938.

MacKay, J., Steele, C. M., Elder, P. A., Forrest, A. P. M., and Evans, H. J., 1988, Allele loss on short arm of chromosome 17 in breast cancers, *Lancet* **ii**:1384–1385.

Madden, S. L., Cook, D. M., Morris, J. F., Gashler, A., Sukhatme, V. P., and Rauscher, F. J., 1991,

Transcriptional repression mediated by the WT1 Wilms-tumor gene-product, *Science* **253**:1550–1553.

Malkin, D., Li, F. P., Strong, L. C., Fraumeni, J. F., Nelson, C. E., Kim, D. H., Kassel, J., Gryka, M. A., Bischoff, F. Z., Tainsky, M. A., and Friend, S. H., 1990, Germ line p53 mutations in a familial syndrome of breast cancer, sarcomas, and other neoplasms, *Science* **250**:1233–1238.

Marshall, C. J., 1991, Tumor suppressor genes, *Cell* **64**:313–326.

Martin, G. A., Viskochil, D., Bollag, G., McCabe, P. C., Crosier, W. J., Haubruck, H., Conroy, L., Clark, R., O'Connell, P., Cawthon, R. M., Innis, M. A., and McCormick, F., 1990, The GAP-related domain of the neurofibromatosis type 1 gene product interacts with ras p21, *Cell* **63**:843–849.

Marx, J., 1990, Genetic defect identified in rare cancer syndrome, *Science* **250**:1209.

Masuda, H., Miller, C., Koeffler, H. P., Battifora, H., and Cline, M. J., 1987, Rearrangement of the p53 gene in human osteogenic sarcoma, *Proc. Natl. Acad. Sci. USA* **84**:7716–7719.

Mathew, C. G. P., Chin, K. S., Easton, D. F., Thorpe, K., Carteer, C., Liou, G. I., Fong, S.-L., Bridges, C. D. B., Haak, H., Kruseman, A. C. N., Schifter, S., Hansen, H. H., Telenius, H., Telenius-Berg, M., and Ponder, B. A. J., 1987, A linked genetic marker for multiple endocrine neoplasia type 2A on chromosome 10, *Nature* **328**:527–528.

McCormick, F., Clark, R. Harlow, E., and Tjian, R., 1981, SV40 antigen binds specifically to a cellular 53K protein *in vitro*, *Nature* **292**:63–65.

Mecklin, J.-P., 1987, Frequency of hereditary colorectal carcinoma, *Gastroenterology* **93**:1021–1025.

Meera Khan, P., Tops, C. M. J., v. d. Broek, M., Breukel, C., Wijnen, J. T., Oldenberg, M., v. d. Bos, J., van Leeuwen-Cornelisse, I. S. J., Vasen, H. F. A., Griffioen, G., Verspaget, H. M., den Hartog Jager, F. C. A., and Lamers, C. B. H. W., 1988, Close linkage of a highly polymorphic marker (D5S37) to familial adenomatous polyposis (FAP) and confirmation of FAP localization on chromosome 5q21–22, *Hum. Genet.* **79**:183–185.

Mendoza, A. E., Shew, J.-Y., Lee, E. Y.-H. P., Bookstein, R., and Lee, W.-H., 1988, A case of synovial sarcoma with abnormal expression of the human retinoblastoma susceptibility gene, *Hum. Pathol.* **19**:487–488.

Menon, A. G., Anderson, K. M., Riccardi, V. M., Chung, R. Y., Whaley, J. M., Yandell, D. W., Farmer, G. E., Freiman, R. N., Lee, J. K., Li, F. P., Barker, D. F., Ledbetter, D. H., Klelcer, A., Martuza, R., Gusella, J. F., and Seizinger, B. R., 1990, Chromosome 17p deletions and p53 mutations associated with the formation of malignant neurofibromas in von Recklinghausen neurofibromatosis, *Proc. Natl. Acad. Sci. USA* **87**:5435–5439.

Mercer, W. E., Nelson, D., Deleo, A. B., Old, L. J., and Baserga, R., 1982, Microinjection of monoclonal antibody to protein p53 inhibits serum-induced DNA synthesis in 3T3 cells, *Proc. Natl. Acad. Sci. USA* **79**:6309–6312.

Mercer, W. E., Avignolo, C., and Baserga, R., 1984, Role of the p53 protein in cell proliferation as studied by microinjection of monoclonal antibodies, *Mol. Cell. Biol.* **4**:276–281.

Meyer, G., Berebbi, M., and Klein, G., 1974, Expression of polyoma-induced antigens in low malignant hybrids derived from fusion of a polyoma-induced tumor with a fibroblast line, *Nature* **249**:47–49.

Mihara, K., Cao, X.-R., Yen, A., Chandler, S., Driscoll, B., Murphree, A. L., T'Ang, A., and Fung, Y.-K. T., 1989, Cell cycle-dependent regulation of phosphorylation of the human retinoblastoma gene product, *Science* **246**:1300–1303.

Miller, R. W., Fraumeni, J. F. J., and Manning, M. D., 1964, Association of Wilms' tumor with aniridia, hemihypertrophy and other congenital malformations, *N. Engl. J. Med.* **270**:922–927.

Mittnacht, S., and Weinber, R. A., 1991, G1/S phosphorylation of the retinoblastoma protein is associated with an altered affinity for the nuclear compartment, *Cell* **65**:381–393.

Miyoshi, Y., Ando, H., Nagese, H., Nishisho, I., Horii, A., Miki, Y., Mori, T., Utsunomiya, J., Baba, S., Petersen, G., Hamilton, S. R., Kinzler, K. W., Vogelstein, B., and Nakamura, Y., 1992, Germline mutations of the APC gene in 53 familial adenomatous polyposis patients, *Proc. Natl. Acad. Sci. USA* **89:**4452–4456.

Monpezat, J.-P., Delattre, O., Bernard, A., Grunwald, D., Remvikos, Y., Muleris, M., Salmon, R. J., Frelat, G., Dutrillaux, B., and Thomas, G., 1988, Loss of alleles on chromosome 18 and on the short arm of chromosome 17 in polypoid colorectal carcinomas, *Int. J. Cancer* **41:**404–408.

Moses, H. L., Yang, E. Y., and Pietenpol, J. A., 1990, TGF-β stimulation and inhibition of cell proliferation: New mechanistic insights, *Cell* **63:**245–247.

Mowat, M., Cheng, A., Kimura, N., Bernstein, A., and Benchimol, S., 1985, Rearrangement of the cellular p53 gene in erythroleukaemic cells transformed by Friend virus, *Nature* **314:**633–636.

Mudryj, M., Hiebert, S. W., and Nevins, J. R., 1990, A role for the adenovirus inducible E2F transcription factor in a proliferation dependent signal transduction pathway, *EMBO J.* **9:**2179–2184.

Mudryj, M., Devoto, S. H., Hiebert, S. W., Hunter, T., Pines, J., and Nevins, J. R., 1991, Cell cycle regulation of the E2F transcription factor involves an interaction with cyclin A, *Cell* **65:**1243–1253.

Munroe, D. G., Rovinski, B., Bernstein, A., and Benchimol, S., 1988, Loss of a highly conserved domain on p53 as a result of a gene deletion during Friend virus-induced erythroleukemia, *Oncogene* **2:**621–624.

Myers, R. M., Fischer, S. G., Maniatis, T., and Lerman, L. S., 1985a, Modification of the melting properties of duplex DNA by attachment of a GC-rich DNA sequence as determined by denaturing gradient gel electrophoresis, *Nucleic Acids Res.* **13:**3111–3145.

Myers, R. M., Larin, Z., and Maniatis, T., 1985b, Detection of single base substitutions by ribonuclease cleavage at mismatches in RNA:DNA duplexes, *Science* **230:**1242–1246.

Nakamura, Y., Leppert, M., O'Connell, P., Wolff, R., Hom, T., Culver, M., Marin, C., Fujimoto, E., Hoff, M., Kumlen, E., and White, R., 1987, Variable number of tandem repeat (VNTR) markers for human gene mapping, *Science* **235:**1616–1622.

Narod, S. A., Feunteun, J., Lynch, H. T., Watson, P., Conway, T., Lynch, J., and Lenoir, G. M., 1991, Familial breast–ovarian cancer locus on chromosome 17q12–23, *Lancet* **338:**82–83.

Narayanan, R,. Lawlor, K. G., Schaapveld, R. Q. J., Cho, K. R., Vogelstein, B., Tran, P. B.-V., Osborne, M. P., and Telang, N. T., 1992, Antisense RNA to the putative tumor-suppressor gene DCC transforms Rat-1 cells, *Oncogene* **7:**553–561.

Nigro, J. M., Baker, S. J., Preisinger, A. C., Jessup, J. M., Hostetter, R., Cleary, K., Bigner, S. H., Davidson, N., Baylin, S., Devilee, P., Glover, T., Collins, F. S., Weston, A., Modali, R., Haris, C. C., and Vogelstein, B., 1989, Mutations in the p53 gene occur in diverse human tumour types, *Nature* **342:**705–708.

Nishi, T., Lee, P., Okra, K., Levin, V. A., Tanase, S., Morino, Y., and Saya, H., 1991, Differential expression of 2 types of the neurofibromatosis type-1 (nf1) gene transcripts related to neuronal differentiation, *Oncogene* **6:**1555–1559.

Nishisho, I., Nakamura, Y., Miyoshi, Y., Miki, Y., Ando, H., Horii, A., Koyama, K., Utsunomiya, J., Baba, S., Hedge, P., Markham, A., Krush, A. J., Petersen, G., Hamilton, S. R., Nilbert, M. C., Levy, D. B., Bryan, T. M., Preisinger, A. C., Smith, K. J., Su, L.-K., Kinzler, K. W., and Vogelstein, B., 1991, Mutations of chromosome 5q21 genes in FAP and colorectal cancer patients, *Science* **253:**665–669.

Noda, M., Kitayam, H., Matsuzaki, T., Sugimoto, Y., Okayama, H., Bassin, R. H., and Ikawa, Y., 1989, Detection of genes with a potential for suppressing the transformed phenotype associated with activated *ras* genes, *Proc. Natl. Acad. Sci. USA* **86:**162–166.

Nordling, C. O., 1953, A new theory on the cancer-inducing mechanism, *Br. J. Cancer* **7:**68–72.

O'Connell, P., Leach, R. J., Ledbetter, D. H., Cawthon, R. M., Culver, M., Eldridge, J. R., Frej, A. K., Holm, T. R., Wolff, E., Thayer, M. J., Schafer, A. J., Fountain, J. W., Wallace, M. R., Collins, F. S., Skolnick, M. H., Rich, D. C., Fournier, R. E. K., Baty, B. J., Carey, J. C., Leppert, M. F.,

Lathrop, G. M., Lalouel, J.-M., and White, R., 1989, Fine structure DNA mapping studies of the chromosomal region harboring the genetic defect in neurofibromatosis type I, *Am. J. Hum. Genet.* **44**:51–57.

O'Connell, P., Viskochil, D., Buchberg, A. M., Fountain, J., Cawthon, R., Culver, M., Stevens, J., Rich, D. C., Ledbetter, D. H., Wallace, M., Carey, J. C., Jenkins, N. A., Copeland, N. G., Collins, F. S., and White, R., 1990, The human homolog of murine Evi-2 lies between two von Recklinghausen neurofibromatosis translocations, *Genomics* **7**:547–554.

Olumi, A. F., Tsai, Y. C., Nichols, P. W., Skinner, D. G., Cain, D. R., Bender, L. I., and Jones, P. A., 1990, Allelic loss of chromosome 17p distinguishes high grade from low grade transitional cell carcinomas of the bladder, *Cancer Res.* **50**:7081–7083.

Orita, M., Iwahana, H., Kanazawa, H., Hayashi, K., and Sekiya, T., 1989, Detection of polymorphisms of human DNA by gel electrophoresis as single-strand conformation polymorphisms, *Proc. Natl. Acad. Sci. USA* **86**:2766–2770.

Orkin, S. H., Goldman, D. S., and Sallan, S. E., 1984, Development of homozygosity for chromosome 11p markers in Wilms' tumour, *Nature* **309**:172–174.

Parada, L. F., Land, H., Weinberg, R. A., Wolf, D., and Rotter, V., 1984, Cooperation between gene encoding p53 and ras in cellular transformation, *Nature* **312**:649–651.

Pelletier, J., Bruening, W., Kashtan, C. E., Mauer, S. M., Manivel, J. C., Striegel, J. E., Houghton, D. C., Junien, C., Habib, R., Fouser, L., Fine, R. N., Silverman, B. L., Haber, D. A., and Housman, D., 1991a, Germline mutations in the Wilms' tumor suppressor gene are associated with abnormal urogenital development in Denys–Drash syndrome, *Cell* **67**:437–447.

Pelletier, J., Bruening, W., Li, F. P., Haber, D. A., Glaser, T., and Housman, D. E., 1991b, WT1 mutations contribute to abnormal genital system development and hereditary Wilms' tumor, *Nature* **353**:431–434.

Peltomäki, P., Sistonen, P., Mecklin, J.-P., Pylkkänen, L., Järvinen, H., Simons, J. W., Cho, K. R., Vogelstein, B., and de la Chapelle, A., 1991, Evidence supporting exclusion of the *DCC* gene and a potion of chromosome 18q as the locus for susceptibility to hereditary nonpolyposis colorectal carcinoma in five kindreds, *Cancer Res.* **51**:4135–4140.

Ping, A. J., Reeve, A. E., Law, D. J., Young, M. R., Boehnke, M., and Feinberg, A. P., 1989, Genetic linkage of Beckwith–Weidemann syndrome to 11p15, *Am. J. Hum. Genet.* **44**:720–723.

Pinhasi-Kimhi, O., Michalovitz, D., Ben-Zeev, A., and Oren, M., 1986, Specific interaction between the p53 cellular tumour antigen and major heat shock proteins, *Nature* **320**:182–184.

Ponder, B., 1990, Neurofibromatosis gene cloned, *Nature* **346**:703–704.

Porteous, D. J., 1987, Chromosome mediated gene transfer: A functional assay for complex loci and an aid to human genome mapping, *Trends Genet.* **3**:177–182.

Pritchard, C. A., and Goodfellow, P. N., 1987, Investigation of chromosome-mediated gene transfer using the HPRT region of the human X chromosome as a model, *Genes Dev.* **1**:172–178.

Pritchard-Jones, K., Fleming, S., Davidson, D., Bickmore, W., Porteous, D., Gosden, C., Bard, J., Buckler, A., Pelletier, J., Housman, D., van Heyningen, V., and Hastie, N., 1990, The candidate Wilms' tumour gene is involved in genitourinary development, *Nature* **346**:194–197.

Rauscher, F. III, Morris, J. F., Tournay, O. E., Cook, D. M., and Curran, T., 1990, Binding of the Wilms' tumor locus zinc finger protein to the EGR-1 consensus sequence, *Science* **250**:1259–1262.

Raycroft, L., Wu, H. Y., and Lozano, G., 1990, Transcriptional activation by wild-type but not transforming mutants of the p53 anti-oncogene, *Science* **249**:1049–1051.

Rees, M., Leigh, S. E. A., Delhanty, J. D. A., and Jass, J. R., 1989, Chromosome 5 allele loss in familial and sporadic colorectal adenomas, *Br. J. Cancer* **59**:361–365.

Reeve, A. E., Housiaux, P. J., Gardner, R. J., Chewings, W. E., Grindley, R. M., and Millow, L. J., 1984, Loss of a Harvey ras allele in sporadic Wilms' tumour, *Nature* **309**:174–176.

Reeve, A. E., Sih, S. A., Raizis, A. M., and Feinberg, A. P., 1989, Loss of allelic heterozygosity at a second locus on chromosome 11 in sporadic Wilms' tumor cells, *Mol. Cell. Biol.* **9**:1799–1803.

Reis, A., Küster, W., Linss, G., Gebel, E., Hamm, H., Fuhrmann, W., Wolff, G., Groth, W., Gustafson, G., Kuklik, M., Bürger, J., Wegner, R. D., and Neitzel, H., 1992, Localisation of gene for the naevoid basal-cell carcinoma syndrome, *Lancet* **339**:617.

Rey, J. A., Bello, M. J., de Camps, J. M., Benitez, J., Sarasa, J. L., Boixados, J. R., and Sanchez, C. A., 1987, Cytogenetic clones in a recurrent neurofibroma, *Cancer Genet. Cytogenet.* **26**:157–163.

Rhim, J. S., 1988, Viruses, oncogenes, and cancer, *Cancer Detect. Prev.* **11**:139–149.

Riccardi, V. M., Sujansky, E., Smith, A. C., and Franke, U., 1978, Chromosomal imbalance in the aniridia–Wilms' tumor association: 11p interstitial deletion, *Pediatrics* **61**:604–610.

Ridley, A. J., Paterson, H. F., Noble, M., and Land, H., 1988, *ras*-mediated cell cycle arrest is altered by nuclear oncogenes to induce Schwann cell transformation, *EMBO J.* **7**:1635–1645.

Robbins, P. D., Horowitz, J. M., and Mulligan, R. C., 1990, Negative regulation of human c-fos expression by the retinoblastoma gene product, *Nature* **346**:668–671.

Rodrigues, N. R., Rowan, A., Smith, M. E. F., Kerr, I. B., Bodmer, W. F., Gannon, J. V., and Lane, D. P., 1990, p53 mutations in colorectal cancer, *Proc. Natl. Acad. Sci. USA* **87**:7555–7559.

Rose, E. A., Glaser, T., Jones, C., Smith, C. L., Lewis, W. H., Call, K. M., Minden, M., Champagne, E., Bonetta, L., Yeger, H., and Houseman, D. E., 1990, Complete physical map of the WAGR region of 11p13 localizes a candidate Wilms' tumor gene, *Cell* **60**:495–508.

Rustgi, A. K., Dyson, N., and Bernards, R., 1991, Amino-terminal domains of c-*myc* and N-*myc* proteins mediate binding to the retinoblastoma gene product, *Nature* **352**:541–544.

Sarnow, P., Ho, Y. S., Williams, J., and Levine, A. J., 1982, Adenovirus E1B-58kd tumor antigen and SV40 large tumor antigen are physically associated with the same 54kd cellular protein in transformed cells, *Cell* **28**:387–394.

Sato, T., Tanigami, A., Yamakawa, K., Akiyama, F., Kasumi, F., Sakamoto, G., and Nakamura, Y., 1990, Allelotype of breast cancer: Cumulative allele losses promote tumor progression in primary breast cancer, *Cancer Res.* **50**:7184–7189.

Sato, T., Saito, H., Swensen, H., Olifant, A., Wood, C., Danner, D., Sakamoto, T., Takita, K., Kasumi, F., Miki, Y., Skolnick, M., and Nakamura, Y., 1992, The human prohibitin gene located on chromosome 17q21 is mutated in sporadic breast cancer, *Cancer Res.* **52**:1643–1646.

Saxon, P. J., Srivatsan, E. S., and Stanbridge, E. J., 1986, Introduction of human chromosome 11 via microcell transfer controls tumorigenic expression of HeLa cells, *EMBO J.* **5**:3461–3466.

Schlessinger, D., 1990, Yeast artificial chromosomes: Tools for mapping and analysis of complex genomes, *Trends Genet.* **6**:248–258.

Schwartz, C. E., Haber, D. A., Stanton, V. P., Strong, L. C., Skolnick, M. H., and Housman, D. E., 1991, Familial predisposition to Wilms tumor does not segregate with the WT1 gene, *Genomics* **10**:927–930.

Seizinger, B. R., Rouleau, G. A., Lone, A. H., *et al.* (1987, Linkage analysis in von Recklinghausen neurofibromatosis (NF1) with DNA markers for chromosome 17, *Genomics* **1**:346–348.

Seizinger, B. R., Rouleau, G. A., Ozelius, L. J., *et al.* (1988, Von Hippel–Lindau disease maps to the region of chromosome 3 associated with renal cell carcinoma, *Nature* **332**:268–269.

Seizinger, B. R., Klinger, H. P., Junien, C., Nakamura, Y., Le Beau, M., Cavenee, W., Emanuel, B., Ponder, B., Naylor, S., Mitelman, F., Louis, D., Menon, A., Newsham, I., Decker, J., Kaelbling, M., Henry, I., and van Deimling, A., 1991, Report of the committee on chromosome and gene loss in human neoplasia, *Cytogenet. Cell. Genet.* **58**:1080–1096.

Simpson, N. E., Kidd, K. K., Goodfellow, P. J., McDermid, H., Myers, S., Kidd, J. R., Jackson, C. E., Duncan, A. M. V., Farrer, L. A., Brasch, K., Castiglione, C., Genel, M., Gertner, J., Greenberg, C. R., Gusell, J. F., Holden, J. J. A., and White, B. N., 1987, Assignment of multiple endocrine neoplasia type 2A to chromosome 10 by linkage, *Nature* **328**:528–530.

Skolnick, M. H., Cannon-Albright, L. A., Goldgar, D. E., Ward, J. H., Marshall, C. J., Schumann, G. B., Hogle, H., McWhorter, W. P., Wright, E. C., Tran, T. D., Bishop, D. T., Kushner, J. P., and Eyre, H. J., 1990, Inheritance of proliferative breast disease in breast cancer families, *Science* **250**:1715–1720.

Skuse, G. R., Kosciolek, B. A., and Rowley, P. T., 1991, The neurofibroma in von Recklinghausen neurofibromatosis has a unicellular origin, *Am. J. Hum. Genet.* **49**:600–607.

Slagle, B. L., Zhou, Y. Z., and Butel, J. S., 1991, Hepatitis B virus integration event in human chromosome 17p near the p53 gene identifies the region of the chromosome commonly deleted in virus-positive hepatocellular carcinomas, *Cancer Res.* **51**:49–54.

Smith, C. L., Lawrence, S. K., Gillepsie, G. A., Cantor, C. R., Weissman, S. M., and Collins, F. S., 1987, Strategies for mapping and cloning macroregions of mammalian genomes, *Meth. Enzymol.* **151**:461–489.

Solomon, E., and Bodmer, W. F., 1979, Evolution of sickle variant gene, *Lancet* **i**:923.

Solomon, E., Voss, R., Hall, V., Bodmer, W. F., Jass, J. R., Jeffreys, A. J., Lucibello, F. C., Patel, I., and Rider, S. H., 1987, Chromosome 5 allele loss in human colorectal carcinomas, *Nature* **328**:616–619.

Solomon, E., Borrow, J., and Goddard, A. D., 1991, Chromosome aberrations and cancer, *Science* **254**:1153–1160.

Sorensen, S. A., Mulvihill, J. J., and Nielsen, A., 1986, Long-term follow-up of von Recklinghausen neurofibromatosis. Survival and malignant neoplasms, *N. Engl. J. Med.* **314**:1010–1015.

Sotelo-Avila, C., and Gooch, W. M., 1976, Neoplasms associated with the Beckwith–Weidemann syndrome, *Perpect. Pediatr. Pathol.* **3**:255–272.

Soussi, T., Caron de Fromentel, C., Mechali, M., May, P., and Kress, M., 1987, Cloning and characterization of a cDNA from *Xenopus laevis* coding for a protein homologous to the human and murine p53, *Oncogene* **1**:71–78.

Sparkes, R. S., Sparkes, M. C., Wilson, M. G., Towner, J. W., Benedict, W., Murphree, A. L., and Yunis, J. J., 1980, Regional assignment of genes for human esterase D and retinoblastoma to chromosome band 13q14, *Science* **208**:1042–1044.

Sparkes, R. S., Murphree, A. L., Lingua, R. W., Sparkes, M. C., Field, L. L., Funderburk, S. J., and Benedict, W. F., 1983, Gene for hereditary retinoblastoma assigned to human chromosome 13 by linkage to Esterase D, *Science* **219**:971–973.

Srivastava, S., Zou, Z., Pirollo, K., Blattner, W., and Chang, E. H., 1990, Germ-line transmission of a mutated p53 gene in a cancer-prone family with Li–Fraumeni syndrome, *Nature* **348**:747–749.

Srivastava, S., Tong, Y. A., Devadas, K., Zou, Z.-Q., Sykes, V. W., Chen, Y., Blattner, W., Pirollo, K., and Chang, E. H., 1992, Detection of both mutant and wild-type p53 in normal skin fibroblasts and demonstration of a shared 'second hit' on p53 in diverse tumors from a cancer-prone family with Li–Fraumeni syndrome, *Oncogene* **7**:987–991.

Srivatsan, E. S., Benedict, W. F., and Stanbridge, E. J., 1986, Implication of chromosome 11 in suppression of neoplastic expression in human cell hybrids, *Cancer Res.* **46**:6174–6179.

Stanbridge, E. J., 1976, Suppression of malignancy in human cells, *Nature* **260**:17–20.

Stoler, A., and Bouck, N., 1985, Identification of a single chromosome in the normal human genome essential for suppression of hamster cell transformation, *Proc. Natl. Acad. Sci. USA* **82**:570–574.

Stürzbecher, H.-W., Chumakov, P., Welch, W. J., and Jenkins, J. R., 1987, Mutant p53 binds hsp 72/73 cellular heat shock-related proteins in SV40-transformed monkey cells, *Oncogene* **1**:201–211.

Su, L.-K., Kinzler, K. W., Vogelstein, B., Preisinger, A. C., Moser, A. R., Luongo, C., Gould, K. A., and Dove, W. F., 1992, A germline mutation of the murine homolog of the APC gene causes multiple intestinal neoplasia, *Science* **256**:668–669.

Takahashi, R., Hashimoto, T., Xu, H. J., Hu, S. X., Matsui, T., Miki, T., Bigo-Marshall, H., Aaronson, S. A., and Benedict, W. F., 1991, The retinoblastoma gene functions as a growth and tumor suppressor in human bladder carcinoma cells, *Proc. Natl. Acad. Sci. USA* **88**:5257–5261.

Tanaka, K., Oshimura, M., Kikuchi, R., Seki, M., Hayashi, T., and Miyaki, M., 1991, Suppression of tumorigenicity in human colon carcinoma cells by introduction of normal chromosome 5 or 18, *Nature* **349**:340–342.

T'Ang, A., Varley, J. M., Chakraborty, S., Murphree, A. L., and Fung, Y.-K. T., 1988, Structural rearrangement of the retinoblastoma gene in human breast cancer, *Science* **242**:263–266.

Templeton, D. J., Park, S. H., Lanier, L., and Weinberg, R. A., 1991, Nonfunctional mutants of the retinoblastoma protein are characterized by defects in phosphorylation, viral oncoprotein association, and nuclear tethering, *Proc. Natl. Acad. Sci. USA* **88**:3033–3037.

Toguchida, J., Ishizaki, K., Sasaki, M. S., Ikenaga, M., Sugimoto, M., Kotoura, Y., and Yamamuro, T., 1988, Chromosomal reorganization for the expression of recessive mutation of retinoblastoma susceptibility gene in the development of osteosarcoma, *Cancer Res.* **48**:3939–3943.

Ton, C. C., Huff, V., Call, K. M., Cohn, S., Strong, L. C., Housman, D. E., and Saunders, G. F., 1991, Smallest region of overlap in Wilms tumor deletions uniquely implicates an 11p13 zinc finger gene as the disease locus, *Genomics* **10**:293–297.

Tory, K., Brauch, H., Linehan, M., Barba, D., Oldfield, E., Filling-Katz, M., Nakamura, Y., White, R., Seizinger, B., Lerman, M. I., and Zbar, B., 1989, Specific genetic change in tumors associated with von Hippel–Lindau disease, *J. Natl. Cancer Inst.* **81**:1097–1101.

Trahey, M., and McCormick, F., 1987, A cytoplasmic protein stimulates normal N-*ras* p21 GTPase, but does not affect oncogenic mutants, *Science* **238**:542–545.

Tsai, Y. C., Nichols, P. W., Hiti, A. L., Williams, Z., Skinner, D. G., and Jones, P. A., 1990, Allelic losses of chromosomes 9, 11, and 17 in human bladder cancer, *Cancer Res.* **50**:44–47.

Van den Berghe, H., Vermaelen, K., Mecucci, C., Barbieri, D., and Tricot, G., 1985, The 5q–anomaly, *Cancer Genet. Cytogenet.* **17**:189–255.

Van Heyningen, V., Boyd, P. A., Seawright, A., Fletcher, J. M., Fantes, J. A., Buckton, K. E., Spowart, G., Porteous, D. J., Hill, R. E., Newton, M. S., and Hastie, N. D., 1985, Molecular analysis of chromosome 11 deletions in aniridia–Wilms tumor syndrome, *Proc. Natl. Acad. Sci. USA* **82**: 8592–8596.

Van Heyningen, V. Bickmore, W. A., Seawright, A. Fletcher, J. M., Maule, J., Fekete, G., Gessler, M., Bruns, G. A. P., Huerre-Jeanpierre, C., Junien, C., Williams, B. R. G., and Hastie, N. D., 1990, Role for the Wilms tumor gene in genital development? *Proc. Natl. Acad. Sci. USA* **87**:5383–5386.

Viskochil, D., Buchberg, A. M., Xu, G., Cawthon, R. M., Stevens, J., Wolff, R. K., Culver, M., Carey, J. C., Copeland, N. G., Jenkins, N. A., White, R., and O'Connell, P., 1990, Deletions and a translocation interrupt a cloned gene at the neurofibromatosis type 1 locus, *Cell* **62**:187–192.

Vogelstein, B., 1990, A deadly inheritance, *Nature* **348**:681–682.

Vogelstein, B., Fearon, E. R., Hamilton, S. R., Kern, S. E., Preisinger, A. C., Leppert, M., Nakamura, Y., White, R., Smits, A. M. M., and Bos, J. L., 1988, Genetic alterations during colorectal tumour development, *N. Engl. J. Med.* **319**:525–532.

Vogelstein, B., Fearon, E. R., Kern, S. E., Hamilton, S. R., Preisinger, A. C., Nakamura, Y.,and White, R., 1989, Allelotypes of colorectal carcinomas, *Science* **244**:207–211.

Wadey, R. B., Little, P. F., Pritchard, J., and Cowell, J. K., 1990, Isolation and regional localisation of DNA sequences from a human chromosome 11-specific cosmid library, *Hum. Genet.* **84**:417–423.

Wagner, S., and Green, M. R., 1991, A transcriptional tryst, *Nature* **352**:189–190.

Wallace, M. R., Andersen, L. B., Fountain, J. W., Odeh, H. M., Viskochil, D., Marchuk, D. A., O'Connell, P., White, R., and Collins, F. S., 1990a, A chromosome jump crosses a translocation breakpoint in the von Recklinghausen neurofibromatosis region, *Genes Chromosomes Cancer* **2**: 271–277.

Wallace, M. R., Marchuk, D. A., Andersen, L. B., Letcher, R. Odeh, H. M., Saulino, A. M., Fountain, J. W., Brereton, A., Nicholson, J., Mitchell, A. L., Brownstein, B. H., and Collins, F. S., 1990b, Type 1 neurofibromatosis gene: Identification of a large transcript disrupted in three NF1 patients, *Science* **249**:181–186.

Wang, N., and Perkins, K. L., 1984, Involvement of band 3p14 in t(3;8) hereditary renal carcinoma, *Cancer Genet. Cytogenet.* **11**:479–481.

Wasmuth, J. J., Park, C., and Ferrell, R. E., 1989, Report of the committee on the genetic constitution of chromosome 5, *Cytogenet. Cell. Genet.* **51**:137–148.

Weber, J. L., and May, P. E., 1989, Abundant class of human DNA polymorphisms which can be typed using the polymerase chain reaction, *Am. J. Hum. Genet.* **44**:388–396.

Weinberg, R. A., 1991, Tumor suppressor genes, *Science* **254:**1138–1146.

Weissman, B. E., and Stanbridge, E. J., 1983, Complementation of the tumorigenic phenotype in human cell hybrids, *J. Natl. Cancer Inst.* **70:**667–672.

Weissman, B. E., Saxon, P. J., Pasquale, S. R., Jones, G. R., Geiser, A. G., and Stanbridge, E. J., 1987, Introduction of a normal human chromosome 11 into a Wilms' tumor cell line controls its tumorigenic expression, *Science* **236:**175–180.

Whang-Peng, J., Bunn, P. A., Kao-Shan, C. S., Lee, E. C., Carney, D. N., Gazdar, A. F., and Minna, J. D., 1982a, A nonrandom chromosomal abnormality, del 3p (14–23) in human small cell lung cancer (SCLC), *Cancer Genet. Cytogenet.* **6:**119–134.

Whang-Peng, J., Kao-Shan, C. S., Lee, E. C., Bunn, P. A., Carney, D. N., Gazdar, A. F., and Minna, J. D., 1982b, Specific chromosomal defect associated with human small cell lung cancer: Deletion 3p(14–23), *Science* **215:**181–182.

White, T. J., Arnheim, N., and Erlich, H. A., 1989, The polymerase chain reaction, *Trends Genet.* **5:** 185–189.

Whyte, P., Buchkovich, K. J., Horowitz, J. M., Friend, S. H., Raybuck, M., Weinberg, R. A. and Harlow, E., 1988, Association between an oncogene and an antioncogene: The adenovirus E1A proteins bind to the retinoblastoma gene product, *Nature* **334:**124–129.

Wigler, M. H., 1990, GAPs in understanding ras, *Nature* **346:**696–697.

Wolf, D., and Rotter, V., 1984, Inactivation of p53 gene expression by an insertion of Moloney murine leukemia virus-like DNA sequences, *Mol. Cell. Biol.* **4:**1402–1410.

Xu, G. F., Lin, B., Tanaka, K., Dunn, D., Wood, D., Gesteland, R., White, R., Weiss, R., and Tamanoi, F., 1990a, The catalytic domain of the neurofibromatosis type 1 gene product stimulates ras GTPase and complements ira mutants of *S. cerevisiae*, *Cell* **63:**835–841.

Xu, G. F., O'Connell, P., Viskochil, D., Cawthon, R., Robertson, M., Culver, M., Dunn, D., Stevens, J., Gesteland, R., White, R., and Weiss, R., 1990b, The neurofibromatosis type 1 gene encodes a protein related to GAP, *Cell* **62:**599–608.

Xu, H. J., Sumegi, J., Hu, S. X., Banerjee, A., Uzvolgyi, E., Klein, G., and Benedict, W. F., 1991, Intraocular tumor formation of RB reconstituted retinoblastoma cells, *Cancer Res.* **51:**4481–4485.

Xu, W., Mulligan, L. M., Ponder, M. A., Liu, L., Smith, B. A., Mathew, C. G. P., and Ponder, B. A. J., 1992, Loss of NF1 alleles in phaeochromocytomas from patients with type 1 neurofibromatosis, *Genes Chromosomes Cancer* **4:**1–6.

Yagle, M. K., Parruti, G., Xu, W., Ponder, B. A., and Solomon, E., 1990, Genetic and physical map of the von Recklinghausen neurofibromatosis (NF1) region on chromosome 17, *Proc. Natl. Acad. Sci. USA* **87:**7255–7259.

Yatani, A., Okabe, K., Polakis, P., Halenback, R., McCormick, F., and Brown, A. M., 1991, *ras* p21 and GAP inhibit coupling of muscarinic receptors to atrial K^+ channels, *Cell* **61:**769–776.

Yokota, J., Akiyama, T., Fung, Y.-K. T., Benedict, W. F., Namba, Y., Hanaoka, M., Wade, M., Yokota, J., Wada, M., Shimosato, Y., Terada, M., and Sugimura, T., 1987, Loss of heterozygosity on chromosomes 3, 13, 17 in small-cell carcinoma and on chromosome 3 in adenocarcinoma of the lung, *Proc. Natl. Acad. Sci. USA* **84:**9252–9256.

Yokota, J., Tsukada, Y., Nakajima, T., Gotoh, M., Shimosato, Y., Mori, N., Tsunokawa, Y., Sugimura, T., and Terada, M., 1989, Loss of heterozygosity on the short arm of chromosome 3 in carcinoma of the uterine cervix, *Cancer Res.* **49:**3598–3601.

Yunis, E., Zuniga, R., and Ramirez, E., 1981, Retinoblastoma, gross internal malformations, and deletion 13q14–q31, *Hum. Genet.* **56:**283–286.

Zang, K. D., 1982, Cytological and cytogenetical studies on human meningioma, *Cancer Genet. Cytogenet.* **16:**249–274.

Zuppan, P., Hall, J. M., Lee, M. K., Ponglikitmongkol, M., and King, M.-C., 1991, Possible linkage of the estrogen receptor gene to breast cancer in a family with late-onset disease, *Am. J. Hum. Genet.* **48:**1065–1068.

Chapter 5

Gaucher Disease
Enzymology, Genetics, and Treatment

Gregory A. Grabowski

Division of Human Genetics
Children's Hospital Medical Center
Cincinnati, Ohio 45229-2899

INTRODUCTION

The eponym "Gaucher disease" designates the heterogeneous sets of signs and symptoms in patients with severely decreased intracellular hydrolysis of glucosylceramide (Fig. 1) and related glucosphingolipids. These autosomal recessively inherited disorders result almost exclusively from mutations in the gene encoding the lysosomal hydrolase, acid β-glucosidase or glucocerebrosidase (EC 3.2.1.45). Rare examples of mutations at the prosaposin or "activator" locus also lead to severe Gaucher disease-like phenotypes with glucosylceramide accumulation. At the clinical, biochemical, and molecular levels, marked heterogeneity has been delineated since the first description by Phillipe C. E. Gaucher in 1882 (Gaucher, 1882). Since this original description, massive splenomegaly and unusual, large storage cells in the spleen and bone marrow have become the hallmarks of Gaucher disease. In addition, this disease was the first lysosomal storage disease described (Gaucher, 1882), the second to have its enzymatic defect delineated (Brady *et al.*, 1965; Patrick, 1965), and the first to be successfully treated by enzyme therapy (Barton *et al.*, 1991; Beutler *et al.*, 1991b; Fallet *et al.*, 1992). The disease is the most frequent lysosomal storage disease and the most prevalent Jewish genetic disease ($q \sim 0.04$–0.05). This high frequency, very good genotype/phenotype

Advances in Human Genetics, Volume 21, edited by Henry Harris and Kurt Hirschhorn. Plenum Press, New York, 1993.

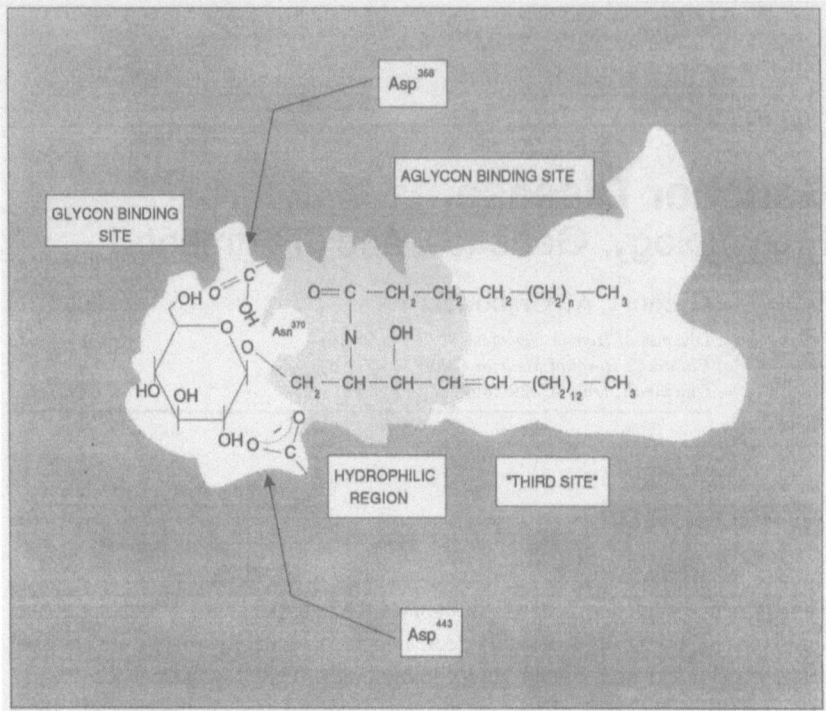

Fig. 1. Diagram of the active site of acid β-glucosidase containing glucosylceramide. Aspartates 358 and 443 are located near or in the active site and aspartate 443 has been identified as a binding site for Br-conduritol B epoxide, an active site-directed covalent inhibitor. The negatively charged residue near D443 and the functional group near D358 are the proposed locations of the nucleophile and proton donor necessary for catalysis. Asn 370 has been localized to the active-site region by site-directed mutagenesis and an allele encoding N370S is the most common Gaucher disease allele. The three "sites" and the hydrophilic region have been proposed by kinetic analyses to contain residues important for the interaction with substrate substituents. Continuity of the amino acid sequence is not implied in the figure.

correlations in the Ashkenazi Jewish population, and the occurrence of many asymptomatic to very mildly affected individuals make Gaucher disease a proto-type for genetic susceptibility testing, with the attendant counseling and social concerns.

CLINICAL CLASSIFICATION

Three major types of Gaucher disease have been delineated based on the absence (type 1) or presence and severity (types 2 and 3) of primary central

TABLE I.   Gaucher Disease: Clinical Types

Clinical feature	Type 1	Type 2	Type 3
Onset	Childhood/adulthood	Infancy	Childhood
Hepatosplenomegaly	+	+	+
Hypersplenism	+	+	+
Bone crises/fractures	+	−	+
Neurodegenerative course	−	+++	++
Survival	6–80+ years	<2 years	Second to fourth decade
Ethnic predilection	Ashkenazi Jewish	Panethnic	Swedish Norrbottnian
Frequency	~1/40,000–1/60,000, 1/400–1/600 (Ashkenazi Jewish)	<1/100,000	~1/50,000

nervous system involvement (Table I). To avoid the nonspecific descriptors "adult," "infantile," and "juvenile," the following nosology has been adopted: type 1—nonneuronopathic; type—acute neuronopathic; type 3—subacute neurono-pathic (Frederickson and Sloan, 1978). Within each type, however, even within the same ethnic and/or demographic groups, the phenotypes and genotypes can be markedly heterogeneous.

Gaucher Disease Type 1

The most common form of the disease is type 1, which is characterized by the lack of neuronopathic involvement. Estimates of the disease frequency range between 1/40,000 and 1/60,000 in the general population. These estimates reach 1/400 to 1/600 in the Ashkenazi population (Zimran et al., 1991; Beutler and Grabowski, 1993). The clinical manifestations of Gaucher disease type 1 result from hypertrophy and hyperplasia of macrophages in various organs (Desnick, 1982; Parkin and Brunning, 1982). This process leads to dysfunction of the liver and spleen, displacement of bone marrow elements by storage cells, and bony involvement leading to infarctions and fractures. Other organs of the monocyte/macrophage system also can be involved, including the lungs (pulmonary macro-phages), Peyer's patches, and lymph nodes. The heart and kidneys are rarely, if ever, involved by the disease process (Beutler and Grabowski, 1993).

The phenotypic variability of Gaucher disease type 1 is striking. Onset of clinical manifestations or age at diagnosis can occur from childhood to adult-hood. In 165 unrelated symptomatic patients, the mean age at onset was 21.1 years (range = 2–73 years) with a median age of 16 years (Sibille *et al.*, 1993). Also, asymptomatic or mildly affected, enzyme-deficient, "patients" have been diagnosed in the eighth and ninth decades of life (Berrebi *et al.*, 1984; Medoff and Bayrd, 1954; Beutler, 1977). Many of these individuals have been ascertained in the course of family studies or population surveys (Beutler, 1977; Zimran *et al.*, 1991). In comparison, affected children have been described with massive hepatosplenomegaly as well as liver function abnormalities, pancytopenias, and extensive skeletal abnormalities. The complications of Gaucher disease in the first or second decade of life may lead to demise in such severely involved patients.

In addition to the variation of the age at onset, there is great variability in the organ manifestations and degree of organ involvement among patients. Nearly all symptomatic patients present with splenomegaly. Some patients with severe visceral (liver and spleen) enlargement are largely spared skeletal involvement and vice versa. In other patients visceral involvement and skeletal involvement are approximately equal in severity.

Gaucher Disease Type 2

This variant is rare, with estimated frequencies of less than 1/100,000 live births. There is a moderate degree of heterogeneity in the onset and course of the acute neuronopathic form of the disease. Early and late onset varieties have been delineated (Kolodny *et al.*, 1982), but the overall clinical course is similar. The principal difference is in the time of onset, which varies between birth and 4–6 months. Indeed, Gaucher disease has been reported to manifest as hydrops fetalis (Sun *et al.*, 1984) and the author knows of additional cases where this has been the outcome. Also, severe neonatal ichthyosis ("collodion babies") has been described and resembles the skin abnormalities in the "knock-out" mouse model (Tybulewicz *et al.*, 1992).

Extensive visceral involvement with hepatosplenomegaly is the rule. Oculomotor abnormalities are often the first manifestations with the appearance of bilateral fixed strabismus (Kolodny *et al.*, 1982) or of oculomotor apraxia. This may result in the appearance of rapid head thrusts as an attempt to compensate when following a moving object (Conradi *et al.*, 1991). Hypertonia of the neck muscles with extreme fixed retroflection of the neck (opisthotonos), bulbar signs, limb rigidity, seizures, and, sometimes, choreoathetoid movements occur, although the latter are more common in type 3 disease. Most infants with type 2 disease die within the first 2 years of life. An exhaustive review of the clinical manifestations in type 2 disease is available (Frederickson and Sloan, 1978).

Gaucher Disease Type 3

In the general population Gaucher disease type 3 has an estimated frequency of 1/50,000 to 1/100,000 live births. The severity of type 3 disease is intermediate between that of types 1 and 2. Massive visceral involvement is usually present, although exceptions have been reported (Wenger *et al.*, 1983). Neurologic symptoms similar to those observed in type 2 disease are present but with a later onset and relatively lesser severity. The prototype of type 3 disease is the Norrbottnian form. The clinical manifestations have been described in detail (Tibblin *et al.*, 1982; Kyllerman *et al.*, 1990; Dreborg *et al.*, 1980; A. Erikson, 1986). The median age of onset of symptoms is 1 year, with a range in 22 patients of 1.4–14.2 years. The first symptoms are usually the result of visceral involvement, with neurologic findings developing in about half of the children during the first decade of life. As in type 2 disease, disorders of eye movement (Sidransky *et al.*, 1992) are the usual first symptoms with the subsequent development of other neurologic manifestations such as ataxia. Even in the Norrbottnian population with a homogeneous genetic mutation at the acid β-glucosidase locus, there is extensive phenotypic heterogeneity (A. Erikson *et al.*, 1987; A. Erikson and Wahlberg, 1985; Dreborg *et al.*, 1980). The variation in genetically heterogeneous populations would be expected to be greater.

SPECIFIC ORGAN SYSTEM INVOLVEMENT

Spleen

Although normal or nearly normal size spleens can be present in asymptomatic patients, splenomegaly is present in all symptomatic patients by physical examination or scanning procedures (Hill *et al.*, 1986). The degree of splenic enlargement is highly variable. Spleen sizes measured at autopsy (R. E. Lee, 1982) or surgery (Aufses and Salky, 1982), by computed tomography (Fallet *et al.*, 1992), or by magnetic imaging scans (Barton *et al.*, 1992; Beutler *et al.*, 1991b) range from about 5 to over 75 times normal size when adjusted for body weight [∼0.2% of body weight (Ludwig, 1979)]. The absolute sizes range from a few hundred grams to about 10 kg and can account for 15–25% of body weight. In a series of 72 patients (age 2–69 years) the mean splenic size as determined by quantitative CT scan was 17.9-fold greater than predicted (range 1.8–75, median 14-fold) (Sibille *et al.*, 1992). In proportion to body weight, the largest spleens were present in the younger severely affected patients with Gaucher disease types

1 and 3 (R. E. Lee, 1982; Stevens *et al.*, 1987; Sibille *et al.*, 1993). Exceptions have been observed with patients 40–70 years old having spleens 25- to 50-fold increased in weight (Fallet *et al.*, 1992). Although not systematically studied, the spleen appears to enlarge most rapidly in children. Acceleration of splenic enlargement in adults should prompt evaluations for other causes of blood dyscrasias: e.g., malignancies, immune-mediated thrombocytopenia (Lester *et al.*, 1984), and autoimmune hemolytic anemia (Haratz *et al.*, 1990) have been documented in patients with Gaucher disease.

Nodules on the surface of the spleen result from regions of extramedullary hematopoiesis, collections of Gaucher cells, or resolving fibrotic infarcts (R. E. Lee, 1982). Evidence of old splenic infarcts can be found in nearly all spleens that are 20-fold larger than normal. Most infarcts have been asymptomatic, but subcapsular infarcts can present as localized abdominal pain. Usually point tenderness can be elicited directly over the involved area of the spleen and the pain resolves within a few days to 1 week. The spleen is usually very hard. Ultrastructural studies have demonstrated active erythrophagocytosis by Gaucher cells (Elleder and Jirásek, 1981) as well as the extrusion of stored glucosylceramide tubular structures into the extracellular spaces in the spleen (Hibbs *et al.*, 1970).

Liver

Hepatomegaly occurs in over 50% of patients with Gaucher disease type 1 (James *et al.*, 1982). In a series of 88 patients, liver volumes ranged from 0.74 to 8.7 times the predicted normal (2.5% of body weight) with a mean value of 1.84 and a median value of 1.75 (Sibille *et al.*, 1983). Massively enlarged livers are usually hard and can have irregular surfaces. Frank hepatic failure and/or cirrhosis with portal hypertension and ascites are uncommon (R. E. Lee, 1982; James *et al.*, 1982), but do occur in a small percentage of patients, and bleeding from varices (Fellows *et al.*, 1975; Henderson *et al.*, 1991) may cause death. Minor abnormalities of liver function tests (increases in plasma transaminase and gamma-glutamyl transferase) are commonly present, even in mildly affected patients. Jaundice attributable to Gaucher disease is a very poor prognostic sign.

Hepatic involvement by Gaucher disease is always evident on biopsy by the presence of Gaucher cells, i.e., abnormal Kupffer cells, in the liver sinusoids. Importantly, even in the presence of extensive replacement of liver tissue by these Gaucher cells, the hepatocytes do not manifest grossly apparent glucosylceramide storage (R. E. Lee, 1982; James *et al.*, 1981). This sparing of the hepatocytes probably accounts for the low incidence of severe hepatic failure. Consistent with the monocyte/macrophage origin of Kupffer (Gaucher) cells was the recurrence of severe hepatic involvement following transplantation of a liver from a normal donor into a patient with Gaucher disease (Carlson *et al.*, 1990).

Hematologic Manifestations

Bleeding can occur from thrombocytopenia. Deficiencies of factor XI are prevalent, but may be coincidental because a deficiency of this clotting factor is common in Ashkenazi Jews (Seligsohn *et al.*, 1976; Berrebi *et al.*, 1992), the population that is most frequently affected by Gaucher disease.

Abnormalities of blood-formed elements are frequent, and thrombocytopenia is the most common in nonsplenectomized patients (Medoff and Bayrd, 1954). Early in the course of the disease it is usually due to splenic sequestration of platelets and responds to splenectomy (A. Zimran, unpublished). Later in the course of the disease, particularly following splenectomy, replacement of the marrow by Gaucher cells may lead to decreased production of megakaryocytes and other bone marrow elements. Anemia is usually mild to moderate, although severe anemias do occur. Leukopenia also occurs in some patients usually due to a combination of increased splenic sequestration, when the spleen is present, and decreased production because of replacement of the marrow by Gaucher cells. Isolated leukopenia is likely due to intercurrent illness and requires extensive evaluation. Bone marrow failure and myelofibrosis can occur in a small number of patients.

Bones

The skeletal manifestations of Gaucher disease can be totally debilitating (Stowens *et al.*, 1985). By radiographic (Stowens *et al.*, 1985; Hermann *et al.*, 1986; Beighton *et al.*, 1982; Goldblatt *et al.*, 1988; Myers *et al.*, 1975), scintigraphic, computed tomographic, or magnetic imaging scans (Hermann *et al.*, 1986; Rosenthal *et al.*, 1986; Cremin *et al.*, 1990) nearly all affected patients have bony lesions. The "Erlenmeyer flask deformity" of the distal femur is a common (~30%) radiographic finding, but is not universally present. Generalized bone loss has been documented (Hermann *et al.*, 1984, 1986).

Many patients with radiographically significant to severe Gaucher disease of bone have few or no bony symptoms. Episodic, excruciatingly painful "bone crises" occur in about 20–40% of patients (Goldblatt and Beighton, 1982). Fever occurs in patients with Gaucher disease, often (Amstutz, 1973; Draznin and

Singer, 1948), but not always (Billings *et al.*, 1973) in connection with bone crises. These are much more frequent in children and adolescents, but can occur for the first time in the third to seventh decades. Based on observations in several hundred type 1 and type 3 patients, the majority of severe bony lesions seem to appear during childhood and adolescence. This aggressive phase of bony involvement is followed in adults by a more slowly progressive process or a cessation of active bony destruction as assessed by current radiographic and MRI techniques. These observations provide a rationale for the institution of corrective therapy early in the course of the disease.

As summarized in Table II, marked variations have been reported in the degree and type of bone involvement (Stowens *et al.*, 1985; Hermann *et al.*, 1986; Cremin *et al.*, 1990). The first abnormality is the loss of calcification of trabecular bone or osteoporosis (Stowens *et al.*, 1985; Beighton *et al.*, 1982). Almost all bones have been involved, including, in approximate decreasing frequency, the femoral necks and heads, femoral shaft, humeri, vertebral bodies, tibias, ribs, pelvis, bones of the feet, calvarium, and mandible (Bildman *et al.*, 1972) [see Frederickson and Sloan (1978) for review]. Complete loss of bone marrow perfusion has been observed (Hermann *et al.*, 1986). The initiating event and pathophysiology of Gaucher disease of the bones remain poorly understood.

Lungs

Although it is uncommon, pulmonary involvement in Gaucher disease has a very severe prognosis. It may occur as a result of frank infiltration of lung by

TABLE II.   Radiologic Stages of Skeletal Lesions in Gaucher Disease Type 1[a]

Stage	Type of lesion/site involved	Radiographic appearance
1	Diffuse osteoporosis/tubular bones and vertebrae	Course trabecular pattern of osteoporosis
2	Medullary expansion/femora, long bones, ribs	Loss of normal concavity above femoral condyles; "Erlenmeyer flask" deformity
3	Localized destruction (osteolysis)/long bones	Small erosions (well-defined or moth-eaten); cortex rarefied and endosteal notching; ground-glass veiling
4	Ischemic necrosis, sclerosis, osteitis/long bones	Patch densities and erosions; serpiginous sclerotic streaks; layered periostitis; sequestra
5	Diffuse destruction; epiphyseal collapse; osteoarthrosis/hips, shoulders, vertebrae, sacroiliac joints	Flattening or irregular destruction of femoral or humeral heads; mixed lytic and sclerotic foci; larger "soap-bubble" pattern

[a]Adapted from Hermann *et al.* (1986) and Beighton *et al.* (1982).

Gaucher cells, particularly in children (Schneider *et al.*, 1977), but severe pulmonary disease also occurs in patients without demonstrable Gaucher cells in the lungs. In such patients right-to-left intrapulmonary shunting, probably secondary to liver disease, seems to be the cause.

Nervous System

The absence of primary central nervous system involvement defines Gaucher disease as type 1. However, occasionally central nervous symptoms can be observed as a secondary manifestation. Thus, massive systemic fat emboli involving the brain and lung have been reported (Melamed *et al.*, 1975). Compression of the spinal cord secondary to vertebral collapse has occurred (Markin and Skultety, 1984; Hermann *et al.*, 1989). In the author's experience with six cases and in the few autopsy reports (R. E. Lee, 1982) on type 1 patients, periadventitial accumulation of Gaucher cells has been the only consistent abnormality in the nervous system.

Microscopic abnormalities have been found in all patients dying of type 2 Gaucher disease (R. E. Lee, 1982; Banker *et al.*, 1962; Norman *et al.*, 1956; Espinas and Faris, 1969; Lacey and Terplan, 1984; Adachi *et al.*, 1967; Kaye *et al.*, 1986). Neuropathologic studies of Norrbottnian (Conradi *et al.*, 1988; Conradi *et al.*, 1984) and non-Swedish type 3 (R. E. Lee, 1982; Kaye *et al.*, 1986) patients have shown less severe involvement of the central nervous system. In contrast to other lipidoses, massive storage of material does not occur in the types 2 or 3 Gaucher disease brains. The most striking feature is the presence of "Gaucher cells" in the perivascular Virchow–Robin spaces in the cortex and deep white matter (Banker *et al.*, 1962; Adachi *et al.*, 1967; Norman *et al.*, 1956). Not infrequently these periadventitial cells also are found in the gray matter of the thalamus and subependymal tissues of the pons and medulla (Banker *et al.*, 1962; Adachi *et al.*, 1967). Neuronal loss is widespread in the brains of type 2 patients. Most consistently, large losses of neurons occur in the basal ganglia, nuclei of the midbrain, pons, and medulla and hypothalamus (Adachi *et al.*, 1967; Espinas and Faris, 1969; Kaye *et al.*, 1986). The dentate nucleus of the cerebellum always is severely involved. These findings are summarized in Table III. Severe neuronal loss in the cochlear nuclei and the superior olivary complex has been demonstrated by physiologic (BAER) studies and histologic correlation in the same patients (Lacey and Terplan, 1984). Free Gaucher cells are found within brain parenchyma of the cerebral cortex, with larger numbers in the occipital lobes (R. E. Lee, 1982; Kaye *et al.*, 1986; Banker *et al.*, 1962). The numbers of these cells in cerebral cortex appear to be greater in the more severe patients with type 2 disease and absent in type 3 patients (R. E. Lee, 1982; Conradi *et al.*, 1984). In either case,

TABLE III. Central Nervous System Abnormalities in Type 2 Gaucher Disease[a]

Region	Neuronal loss	Gliosis	Periadventitial Gaucher cells	Neuronophagia	Free Gaucher cells
Cerebral cortex					
Frontal	3+	2+	3+	2+	2→4+
Temporal	1→3+	2+	3+	0	3+
Parietal	4+	2+	3+	1+	2+
Occipital	1+	3+	3+	0	4+
White matter	4+	1+	4+	0	0
Thalamus	2→4+	2+	1+	4+	1+
Caudate nucleus	4+	2+	0	2→4+	1+
Claustrum	4+	2+	0	2+	2+
Putamen	3+	2+	0	1+	2+
Globus pallidus	4+	2+	0→1+	4+	0
Midbrain	2+4+	2+	1+	4+	0
Pons	3+	1+	1+	4+	1+
Medulla	3+	1+	1+	4+	0
Cerebellum					
Dentate nucleus	4→5+	0→1+	1→3+	4+	0
Purkinje layer	2+	0	0	0	0

[a]The severity of the abnormalities ranges from normal (0) to very extensive changes (5+).

they are few in number. The neuropathology has been reviewed in detail (Beutler and Grabowski, 1993).

Gaucher Cells

The presence in visceral tissues of lipid-engorged cells with a characteristic appearance (Gaucher, 1882), termed Gaucher cells, is the hallmark of this disease. These cells are derived from the monocyte/macrophage system through pre-Gaucher monocytoid cells in peripheral blood and bone marrow of affected patients. These cells contain small numbers of characteristic tubular cytoplasmic inclusions (Parkin and Brunning, 1982). This monocyte/macrophage origin of Gaucher cells and the evolution of monocytoid cells into end-stage Gaucher macrophages have had major implications for the understanding of the pathophysiology of Gaucher disease and toward the development of targeted enzyme augmentation therapy as well as gene and cellular replacement therapeutic strategies (see Therapeutic Prospects).

The Gaucher cell has a nearly pathognomonic appearance by light microscopy. In smears and imprints of affected tissues, Gaucher cells are about 20–100 μm in diameter (Parkin and Brunning, 1982), although larger cells can be isolated from affected tissues. These cells contain one or more nuclei and cytoplasm with a striated, fibrillary, or tubular pattern which has been likened to "wrinkled tissue paper" or "crumpled silk" (Parkin and Brunning, 1982). On Romanowsky-stained preparations these cells manifest a characteristic pale blue to gray cytoplasm. In comparison, the cells from patients with Niemann–Pick disease and other lipidoses are foam cells that contain discrete membrane-bound white or empty-appearing lipid inclusions.

By electron microscopy, the cytoplasm of pre-Gaucher cells and end-stage Gaucher cells has dilated membrane-bound vesicles (0.6–4 μm) containing characteristic twisted structures (R. E. Lee, 1968). These tubules are about 90% glucosylceramide and about 0.3% protein in composition (R. E. Lee, 1968). It has been suggested that the tubules result from the polymerization of glucosylceramide in a polar head-to-apolar tail manner. In addition to their distribution, the origin of the stored glucosylceramides in Gaucher cells has important implications for the pathophysiology and therapeutic interventions in Gaucher disease (Frederickson and Sloan, 1978). Evidence indicates that phagocytosis of blood-formed elements is the primary source of the accumulated intracellular lipid in the viscera. Erythrocyte fragments are present in Gaucher cells from spleen (Pennelli *et al.*, 1969; Hibbs *et al.*, 1970), suggesting the erythrocyte membrane as a source of stored glycolipid. Gaucher cells also are formed in culture by macrophage phagocytosis of exogenously supplied glucosylceramide (Christianson, 1941). The accumulation of β-galactose, N-acetylglucosamine, fucose, and sialic acid-containing glycopeptides and glycans has been demonstrated in delipidated Gaucher cells from splenic red pulp, liver, and the alveolar, perivascular, and interstitial areas of the lungs (Banker *et al.*, 1962; DeGasperi *et al.*, 1990). Since these glycoconjugates are not substrates for acid β-glucosidase, they probably represent digestive remnants of phagocytosed cells or cell membranes. From turnover calculations, leukocyte membranes have been suggested as the major source of stored glucosylceramide in visceral Gaucher cells (Kattlove *et al.*, 1969). The clear demonstration that previously stored intracellular glucosylceramide is extruded from lysing Gaucher cells into the extracellular spaces of the spleen (Pennelli *et al.*, 1969) indicates a source of the elevated plasma levels of this lipid.

PATHOPHYSIOLOGY

The Storage Substance: Glucosylceramide

The major natural substrate for acid β-glucosidase is N-acyl-sphingosyl-1-O-β-D-glucoside (Fig. 1). This compound also has been called glucosylceramide,

ceramide β-glucoside, or glucocerebroside and the enzyme has been referred to as glucocerebrosidase, glucosylceramidase, ceramide β-glucosidase, or acid β-glucosidase (EC 3.2.1.45). Although the chain length does vary, the usual sphingosyl moiety is (2**S**,3**R**,4**E**)-2-amino-4-octadecene-1,3-diol (Mislow, 1953; Carter and Fujino, 1956). Glucosylceramide is synthesized from ceramide and UDP-glucose by glucosylceramide synthase (Basu *et al.*, 1968, 1973; Coste *et al.*, 1985, 1986). Glucosylceramide is widely distributed in many mammalian tissues in small quantities as a metabolic intermediate in the synthesis and degradation of complex glycosphingolipids, such as the gangliosides or globoside. It is localized primarily in cellular membranes. Glucosylceramide is the penultimate intermediate in the degradation pathway of most complex glycosphingolipids (Fig. 2). Further degradation by the acid β-glucosidase produces ceramide, the latter being further

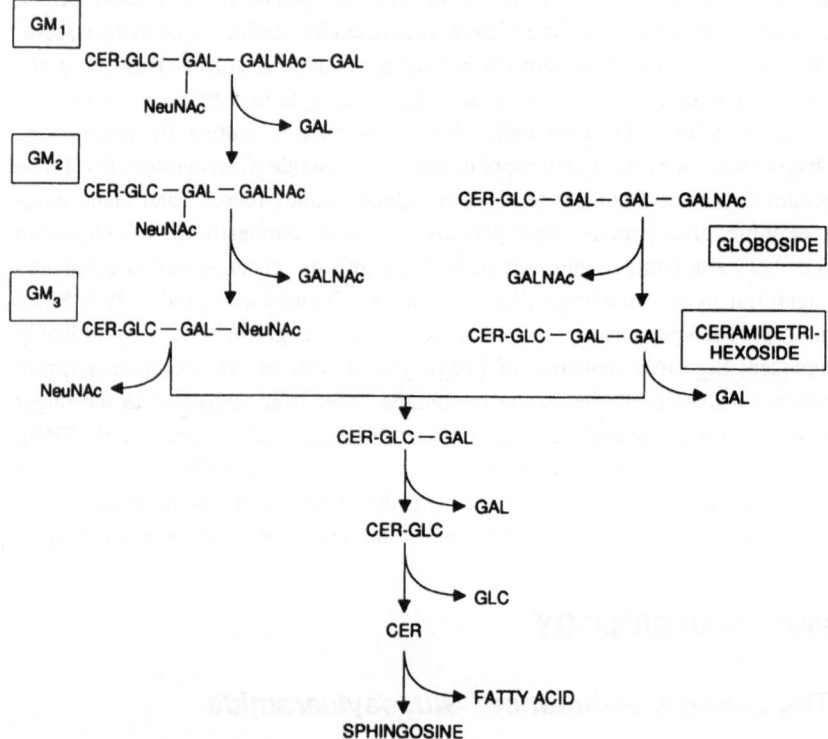

Fig. 2. Degradation pathway of lower-order gangliosides and neutral glycosphingolipids through the common pathway of glucosylceramide (CER-GLC).

degraded by acid ceramidase (Gatt, 1966; Yavin and Gatt, 1969) to sphingosine and fatty acid.

Plasma levels of glucosylceramide are elevated 2- to 20-fold in affected patients, but the levels do not correlate with the type of Gaucher disease nor the degree of involvement within a type (Dawson and Oh, 1977; Strasberg et al., 1983; Nilsson et al., 1982). Extensive lipid analyses of livers, spleens, and brains of Gaucher disease patients have shown large (20- to 100-fold) increases in glucosylceramide (Conradi et al., 1991; Kaye et al., 1986; Nilsson and Svennerholm, 1982a,b; Nilsson et al., 1985, 1982; Kennaway and Woolf, 1968; French et al., 1969). The fatty acid and sphingosyl base compositions of the accumulated glucosylceramides in liver and spleen are typical of visceral organs (Nilsson et al., 1985; Nilsson and Svennerholm, 1982a). There is a predominance of C_{20} to C_{24} fatty acids and shorter-chain sphingosyl bases in all Gaucher disease variants (Nilsson et al., 1985). Substantial concentrations of glucosylsphingosine are also present (Nilsson and Svennerholm, 1982b). Increased levels of glucosylceramide are found in brains from several disease variants. However, the distribution, amount, and type of this accumulated lipid vary with the Gaucher disease variant. The fatty acid and sphingosyl base compositions of the accumulated lipid in brain from one classical type 1 patient were consistent with the extraneural location and visceral origin of this lipid (Nilsson et al., 1985). In comparison, composition analyses of glucosylceramide from brains of nonsplenectomized type 2 and type 3 patients were consistent with a neural origin and location (Conradi et al., 1991; Nilsson and Svennerholm, 1982b; Nilsson et al., 1985). The levels of glucosylsphingosine accumulation also were severalfold increased over that found in the classical type I brain (Nilsson and Svennerholm, 1982b; Nilsson et al., 1985). In splenectomized type 3 patients much greater accumulations of glucosylceramide were found in liver and brain compared to their nonsplenectomized cohorts (Nilsson and Svennerholm, 1982b; Nilsson et al., 1982).

Pathogenesis

The earliest histologic abnormalities observed in several tissues are reticulum fibers surrounding Gaucher cells (R. E. Lee, 1982). Thus, visceral organ fibrosis and brain gliosis appear to occur prior to or coincident with secondary tissue events resulting in advanced disease. In visceral organs and bone, secondary disease findings include infarction, necrosis, and scarring. In brain of severe type 2 patients neuronal loss is the most apparent pathologic finding. It is possible that the regenerative capacities of visceral tissues could obscure the role of cell death as an early and/or unifying pathogenic event in these tissues. In addition, the tissue variation in pathology results from tissue-specific characteristics: i.e., the solid containment of bone marrow, the specialized extracellular environment of osteoclasts, and the sinusoidal structures of the liver and spleen. As a consequence, fibrosis can be considered the earliest pathologic event from which a unified pathophysiology of this disease should be developed. In addition, unlike most other lysosomal storage diseases, the storage material in Gaucher cells of visceral tissues derives from phagocytosis of components which are external to the cell. Neurons of the brain may be the only cell type in which endogenous synthesis plays an important pathogenic role in the neuronopathic Gaucher disease variants. These observtions suggest the presence of cytotoxic substances, i.e., glucosylceramide, its lysosphingolipid derivative, glucosylsphingosine, or other glucolipids. In visceral organs from patients with type 1, 2, or 3 disease glucosylsphingosine, if present, occurs only at exceedingly low concentrations, whereas in the CNS of type 2 and 3 patients significant amounts are present. The toxic role of this lysosphingolipid is consistent with the proposal of Hannun and Bell (1987), but experimental support in the lysosomal storage diseases is lacking. Glucosylceramide has been shown to have growth-promoting potential in animal models (Datta and Radin, 1988). In addition, normal macrophages release cytokines when exposed to solvent-solubilized glucosylceramide (Gery et al., 1981), although the levels of several interleukins and tumor necrosis factor α in plasma from Gaucher patients are not elevated (G. A. Grabowski and L. Meier, unpublished). As with most of the lysosomal storage diseases, the pathogenic mechanisms leading to the Gaucher disease phenotypes remain to be elucidated.

BIOCHEMISTRY

Acid β-Glucosidase

Acid β-glucosidase is a tightly associated peripheral membrane protein (Imai, 1985). Purification has been achieved from a variety of species, but most efforts have focused on the human enzyme (Pentchev et al., 1973; Furbish et al., 1977; Strasberg et al., 1982; Dinur et al., 1986; Grabowski and Dagan, 1984; Osiecki-Newman et al., 1986). The specific activities of the homogeneous acid β-glucosidase preparations varied by an order of magnitude ($\sim 5 \times 10^5$ to 5×10^6 nmole/hr per protein), primarily due to different assay conditions. Based on active site quantitation the human placental enzyme had a catalytic rate constant, k_{cat} of about 2350 min^{-1} for a variety of glucosylceramide and 4-alkylumbelliferyl-β-glucoside substrates (Grabowski et al., 1986). The human enzyme is a homomeric glycoprotein, and partial (Osiecki-Newman *et al.*, 1986; Dinur *et al.*, 1986) and complete (Tsuji *et al.*, 1986) amino acid sequences have been determined chemically and deduced from the nucleotide sequences of the cDNAs (Tsuji *et al.*, 1986; Sorge *et al.*, 1985; Graves *et al.*, 1988; Reiner *et al.*, 1987). The mature human polypeptide contains 497 amino acids with a calculated molecular weight M_r of 55,575. The pure glycosylated enzyme from placenta has a $M_r \sim 65,000$. Aerts *et al.* (1987) described a high-molecular-weight form of acid β-glucosidase in human cells and urine, termed glucocerebrosidase II, which was an enzyme/saposin C aggregate. Because acid β-glucosidase requires detergent solubilization, the native state of the enzyme in cells and tissues has not been resolved. Molecular weights, estimated by sedimentation and molecular exclusion chromatography, have varied between about 60,000 and 450,000 (Pentchev *et al.*, 1973; Prence *et al.*, 1985; Garrett *et al.*, 1985). In comparison, estimates of the native molecular weight in tissues by *in situ* radioinactivation of the enzymatic activity were consistent with either a monomeric or dimeric structure (Dawson and Ellory, 1985; Maret *et al.*, 1984, 1983). About 13% of the residues are basic (lysine, arginine, or histidine) and the calculated *pI* value is 7.2, which is consistent with experimental values (*pI* = 7.3–7.8) for placental enzyme preparations containing only the neutral mannosyl oligosaccharide core (Furbish *et al.*, 1981). The protein has about 11% leucine residues and 45% nonpolar amino acids, but transmembrane domains are not present in the mature polypeptide. Proteolytic digestion studies of acid β-glucosidase, obtained by *in vitro* translation in the presence of microsomal membranes, indicated the absence of a large cytoplasmic domain (A. H. Erickson *et al.*, 1985). No obvious basis for its tight membrane association is evident from the primary sequence.

The human and murine sequences have conserved placement of the five N-glycosylation sequons, but only those at the third and fifth positions were identical (O'Neill *et al.*, 1989). Takasaki *et al.* (1984) have shown typical bi- and tri-antennery complex N-linked oligosaccharides on the human placental acid β-glucosidase. The less common NeuAcα2→3Galβ1 linkages were found (Takasaki *et al.*, 1984). Endoglycosidase F digestion studies of human fibroblast acid β-glucosidase indicated that only four of the potential glycosylation sites are used (A. H. Erickson *et al.*, 1985). Expression of the human cDNA in prokaryotes indicated that glycosylation is essential for the development of a catalytically active enzyme (Grace *et al.*, 1992; Berg *et al.*, 1993), but it was not required to maintain activity in cells (Grace and Grabowski, 1990). Using site-directed mutagenesis, the first four glycosylation sequons were shown to be normally occupied and occupancy of the first sequon was needed to develop an active conformer (Berg *et al.*, 1993) (Fig. 3).

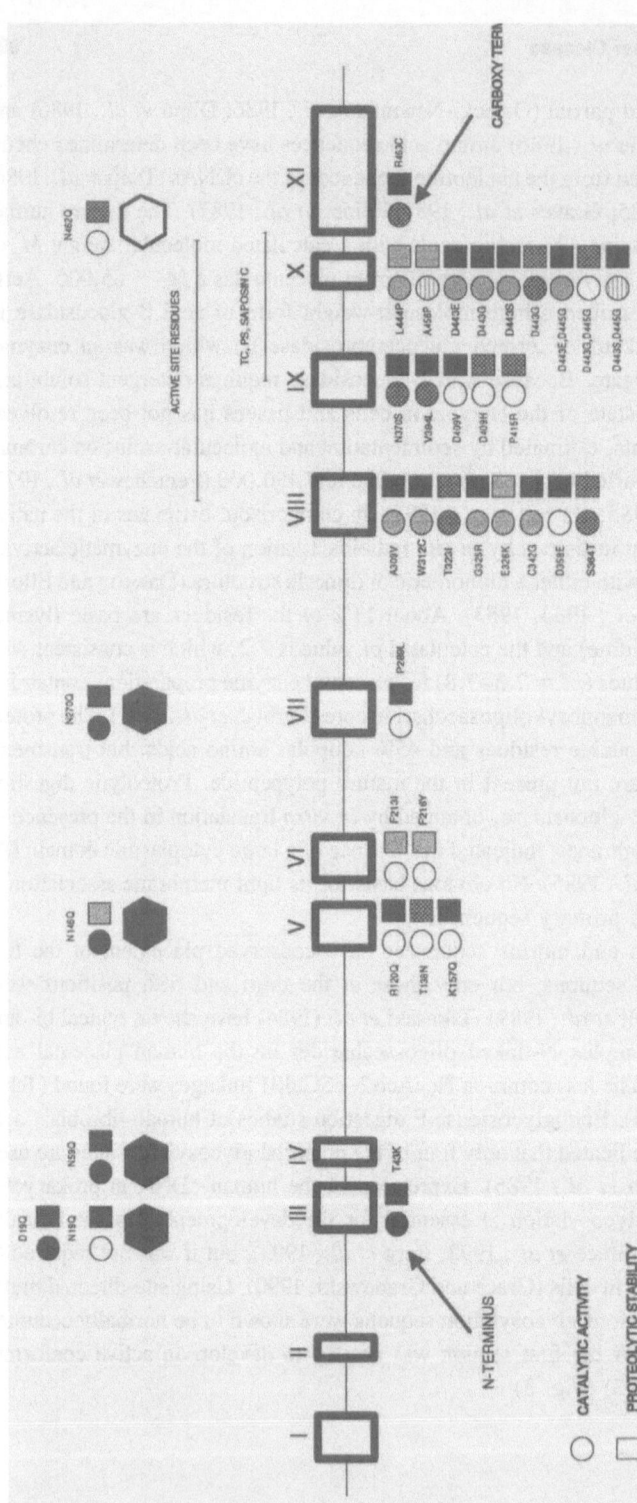

Fig. 3. The functional organization of the acid β-glucosidase gene. Exons are indicated in thickly outlined rectangles with roman numerals above. Introns are indicated by connecting thin lines. Open and filled symbols indicate lack or presence of function in mutations. The various fills represent degrees of abnormality. Circles and squares represent catalytic activity and stability to proteolysis, respectively, for each mutated enzyme. Filled or open hexagons represent occupied or unoccupied glycosylation sequons, respectively. Most mutant proteins were analyzed following expression of mutagenized cDNAs in insect cells.

Posttranslational Processing of Normal Acid β-Glucosidase

Biosynthesis of normal acid β-glucosidase has been characterized in cultured porcine kidney cells and human skin fibroblasts (Jonsson et al., 1987; Beutler and Kuhl, 1986; A. H. Erickson et al., 1985; Bergmann and Grabowski, 1989). Three molecular-weight forms of acid β-glucosidase were observed by immunoelectro-blotting of human skin fibroblast (Ginns et al., 1982; Jonsson et al., 1987; Fabbro et al., 1987) and brain extracts (Ginns et al., 1982). Total removal of N-linked oligosaccharides from acid β-glucosidase resulted in a single immunoreactive form (M_r = 56,000) of acid β-glucosidase (Fabbro et al., 1987). These findings demonstrated that the three forms of the enzyme were derived by differential posttranslational oligosaccharide remodeling (Fabbro et al., 1987; Bergmann and Grabowski, 1989). Use of ^{35}S-methionine metabolic labeling showed the first detectable forms of the human (Jonsson et al., 1987; Bergmann and Grabowski, 1989; A. H. Erickson et al., 1985) and porcine (A. H. Erickson et al., 1985) enzymes to contain high mannosyl chains (Jonsson et al., 1987; A. H. Erickson et al., 1985) which indicated residence in the endoplasmic reticulum. These initial forms were quantitatively remodeled by 24 hr into higher-molecular-weight forms with complex oligosaccharide chains (Jonsson et al., 1987; A. H. Erickson et al., 1985). Over about 72 hr a final glycosylated from (M_r = 59,000) was produced in the lysosomes. Acid β-glucosidase is not trafficked to the lysosome via the mannose-6-phosphate-mediated pathway (Kaplan et al., 1977; Varki and Korn-feld, 1981). Following signal peptide clipping, transport of the enzyme from the Golgi apparatus to the lysosomes was not associated with detectable proteolytic processing (A. H. Erickson et al., 1985) as had been observed with several other lysosomal hydrolases (Hasilik and Neufeld, 1984). This was confirmed by amino acid sequence analysis of the entire placental protein (Tsuji et al., 1986). The rate of transport of acid β-glucosidase from the endoplasmic reticulum to the Golgi apparatus is about 3 hr, which is long compared to that for other lysosomal enzymes (Bergmann and Grabowski, 1989). The half-life of acid β-glucosidase in cultured skin fibroblasts is estimated to be about 60 hr, which is short for lysosomal enzymes (Hasilik and Neufeld, 1984).

Posttranslational Processing of Acid β-Glucosidase in Cells from Gaucher Disease Patients

Ginns et al. (1982) suggested a differential processing of the mutant residual acid β-glucosidase in fibroblasts from variant patients. By immunoelectroblot-ting, the three normal steady-state forms of enzyme were found to be present in cultured skin fibroblasts from type 1 patients (Ginns *et al.*, 1982; Fabbro *et al.*, 1987). Remarkable heterogeneity of processing patterns was demonstrated (Fabbro *et al.*, 1987; Jonsson *et al.*, 1987) in ethnically heterogeneous populations of Gaucher disease patients. In retrospect, this would have been expected from the extensive genetic heterogeneity in this disease (Fig. 4) and (Table IV). The presence of two different mutant acid β-glucosidase alleles in cells from many patients makes assignment of differential stability or processing properties to a mutation difficult. However, in cells homozygous for the N370S or L444P alleles, normal or decreased stability, respectively, of the residual enzyme proteins has been observed (Bergmann and Grabowski, 1989; Jonsson *et al.*, 1987). By immunoelectronmicroscopic localization some mutations seem to preclude localization of the abnormal acid β-glucosidases to the lysosomes, particularly in type 2 disease (Willemsen *et al.*, 1988). Studies with protease inhibitors (Jonsson *et al.*, 1987; Tager *et al.*, 1986) and *in situ* hydrolysis of exogenously supplied glucosylceramide (Agmon *et al.*, 1993) indicate that some of the mutant enzymes do reach this compartment in most variants of Gaucher disease, so the block is not absolute. Recreation of the individual mutations and expression of each in heterologous systems has begun to indicate regions on the protein which modulate some of these properties (Grace and Grabowski, 1993; Grace *et al.*, 1990; 1992, 1993; Ohashi *et al.*, 1991) (Fig. 3) (see Molecular Enzymology).

Saposins C and A

In 1971, Ho and O'Brien (1971) characterized a factor from Gaucher disease spleen which enhanced acid β-glucosidase hydrolysis of 4-methylumbelliferyl-β-D-glucopyranoside (4MU-Glc). Subsequently, this "factor" has been designated heat-stable factor, co-β-glucosidase, saposin C, and SAP-2 (sphingolipid activator protein 2). Saposin C is now the term in common usage. The physiologic importance of saposin C was demonstrated by glucosylceramide storage and a Gaucher disease-like phenotype in two patients with a deficiency of saposin C cross-reacting immunological material and normal *in vitro* acid β-glucosidase activity (Christomanou *et al.*, 1987, 1989). Recently, a second protein activator of *in vitro* acid β-glucosidase activity has been described, saposin A (Morimoto *et al.*, 1989). Both of these activators derive from a single gene product by extensive proteolytic and glycosidic posttranslational processing (see Molecular Biology of the Activator Proteins).

The complete amino acid sequences of human (Kleinschmidt *et al.*, 1987) and guinea pig (Sano *et al.*, 1988) saposin C have been chemically determined and those from human (O'Brien *et al.*, 1988; Nakano *et al.*, 1989; Gavrieli-Rorman and Grabowski, 1989; Reiner *et al.*, 1989) and rat (Collard *et al.*, 1988) saposin C

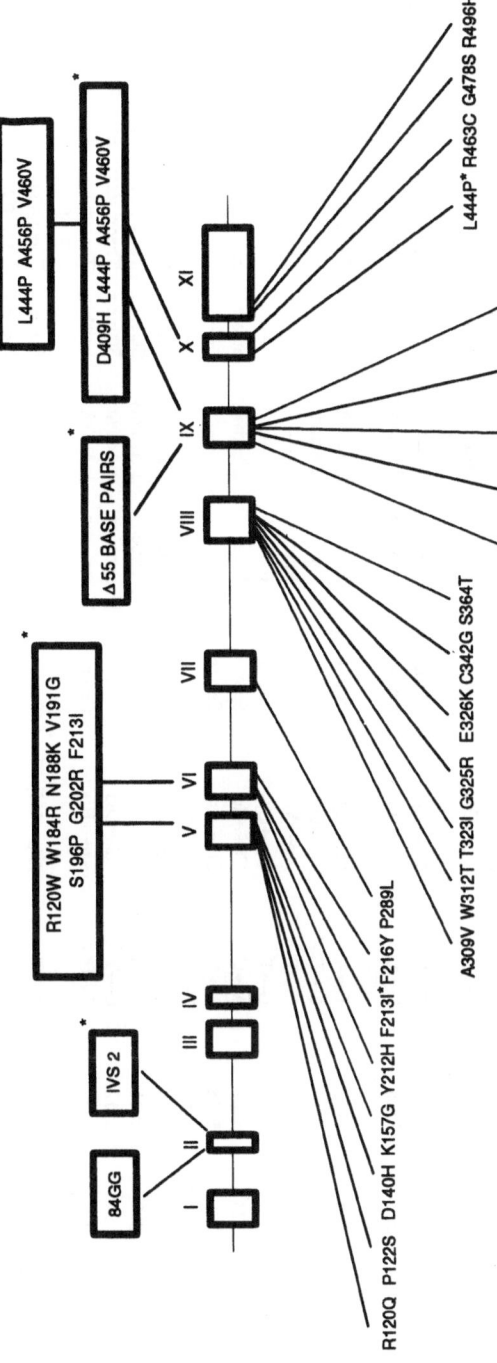

Fig. 4. Schematic representation of the distribution of Gaucher disease alleles in the acid β-glucosidase gene. The mutations below the schematic of the gene are point mutations and those above are nonsense or more complex alleles. The base substitutions in the cDNA and gene are presented in Table IV. Alleles marked with asterisks mimic base changes in the pseudogene (see Fig. 6). The alleles represented by N370S, 84GG, L444P, IVS2, and R463C are common alleles, whereas the other alleles are private or rare. The 84GG, Δ55, and IVS2 are, respectively, an insertion, a 55-bp deletion, and an exon2/intron2 splice junction mutation which are functionally null. The alleles containing multiple missense mutations (in boxes) probably arise from genetic rearrangements with the pseudogene.

TABLE IV. Gaucher Disease Alleles

cDNA #	Amino acid #	Genomic #	Base substitution	Amino acid substitution	Phenotypic effect	Frequency[a]	Enzymatic effect Catalytic	Enzymatic effect Stability	Ref.
A. Missense mutations that cause Gaucher disease									
476	120	3060	G→A	Arg→Gln	? Severe	Rare	Inactive	Stable	Graves et al. (1988)
481	122	3065	C→T	Pro→Ser	Mild	Rare	—	—	Beutler et al. (1992a)
535[b]	140	3119	G→C	Asp→His	?	Rare	—	—	Eyal et al. (1991)
1093[b]	326	5309	G→A	Glu→Lys	?		—	—	
586	157	3170	A→C	Lys→Gln	Severe	Rare	Inactive	Stable	Eyal et al. (1991), Latham et al. (1991)
751	212	3545	T→C	Tyr→His	?Mild	Rare	—	—	Beutler et al. (1992a)
754	213	3548	T→A[c]	Phe→Ile	Severe	Uncommon	↓↓ Activity	Unstable	Kawame and Eto (1991), He et al. (1993)
764	216	4113	T→A	Phe→Tyr	?Mild	Rare	↓↓ Activity	Unstable	Beutler and Gelbart (1990)
983	289	4332	C→T	Pro→Leu	?Mild	Rare	↓↓ Activity	Unstable	He et al. (1993)
1043	309	5259	C→T	Ala→Val	Mild	Rare	—	—	Latham et al. (1991)
1053	312	5269	G→T	Trp→Cys	?Mild	Rare	—	—	Latham et al. (1991)
1085	323	5301	A→C	Thr→Ile	Severe	Rare	↓ Activity	Stable	He et al. (1993)
1090	325	5306	G→A[c]	Gly→Arg	Severe	Rare	—	—	Eyal et al. (1990)
1141	342	5357	T→G	Cys→Gly	Severe	Rare	—	—	Eyal et al. (1990)
1208	364	5424	G→C	Ser→Thr	Mild	Rare	↓ Activity	Stable	Latham et al. (1991)
1226	370	5841	A→G	Asn→Ser	Mild	Common	↓ Activity	Stable	Tsuji et al. (1988a)
1297	394	5912	G→T	Val→Leu	Severe	Uncommon	↓ Activity	Stable	Theophilus et al. (1989b)
1342	409	5957	G→C[c]	Asp→His	Severe	Uncommon	↓↓ Activity	Unstable	Theophilus et al. (1989b), Eyal et al. (1990)
1343	409	5958	A→T	Asp→Val	Severe	Rare	↓↓ Activity	Stable	Theophilus et al. (1989b)
1361	415	5976	C→G	Pro→Arg	Severe	Rare	Inactive	Stable	Wigderson et al. (1989)
1448	444	6433	T→C[c]	Leu→Pro	Severe	Common	↓↓ Activity	Unstable	Tsuji et al. (1987)
1504	463	6489	C→T	Arg→Cys	?Severe	Uncommon	↓ Activity	Stable	Hong et al. (1990)
1604	496	6683	G→A	Arg→His	Mild	Uncommon	—	—	Beutler et al. (1992a)

B. Nonsense mutations that cause Gaucher disease

				Phenotypic effect			Ref.
84	1035	G→GG	Frameshift	Severe	Common	Null	Beutler et al. (1991a)
IVS2+1	1067	G→A	Aberrant splicing	Severe	Uncommon	Null	Beutler et al. (1992a), He and Grabowski (1992)
1263–1317	5879–5933[c]	Del	—	?Severe	Rare	Null	Beutler et al. (1992a)

C. Genetic rearrangements causal to Gaucher disease

Nucleotides involved	Location of crossover event(s)[d]	Phentypic effect	Frequency	Ref.
>1343 <1388	>5957 <6272[a]	Severe	Uncommon	Zimran et al. (1990)
>455 <475[f] >754	>3039 <5957 >3548	?Severe	Rare	Latham et al. (1990)
>1317 <1343	>5932 <5957	Severe	Uncommon	Eyal et al. (1990), Latham et al. (1990)
>1343 <1388	>5957 <6272	Severe	Uncommon	Latham et al. (1990), Hong et al. (1990)
>1225 <1263	>5588 <5878	Severe	Rare	Eyal et al., 1991, 1990

[a]Common = high frequency in at least one population; uncommon = found in a number of unrelated patients; rare = found in only one or two individuals.

[b]Both mutations are found in a single allele.

[c]Sequence mimicking that in the pseudogene.

[d]Only approximate ranges can be given, since the pseudogene and structural gene contain long identical sequences.

[e]Physical fusion with loss of intergenic segment.

[f]The first range represents crossover from gene to pseudogene; the second from pseudogene to functional gene. This region contains seven mutations; only six are identical with the corresponding pseudogene sequences. At genomic nt 3474, the nucleotide corresponds to that in the structural gene. Adapted from Beutler and Grabowski (1993)

have been deduced from their cDNA sequences. The 80-amino acid (M_r = 8950) human saposin C is a very acidic (calculated pI ~ 4.2) glycoprotein (San *et al.*, 1988), which is consistent with the presence of 20% acid (Asp and Glu) and 10% basic (Lys and His) amino acids. The single N-linked glycosylation sequon is occupied. The composition of the guinea pig saposin C oligosaccharides has been reported (Sano *et al.*, 1988; Kleinschmidt *et al.*, 1987). Similar to saposin C, saposin A is small (M_r = 9148), is very acidic (calculated pI = 4.14), and has six cysteine residues which align with those in saposin C (see below, Fig. 8A). Saposin A has two N-glycosylation sequons and both are occupied (Morimoto *et al.*, 1989). The human and murine proteins are nearly identical (Fig. 5).

Biosynthesis and Processing of Saposins C and A

Fujibayashi and Wenger (1986a), using rabbit anti-human saposin C anti-bodies in metabolic labeling studies, demonstrated that saposin C had extensive proteolytic and glycosidic processing in fibroblasts. In chase experiments follow-ing a short pulse (15 min), the major product was a 68-kilodalton (kD) glycosylated species which over 1 hr was converted to a 73 kD form. Subsequently, the total amount of these forms decreased coincident with the appearance of diffuse species with M_r ~ 12,000 and 9,000. By 5 hr of chase, the 12-kD species had nearly disappeared, while the 9-kD species was predominant. Endoglycosidase F diges-tion studies of ^{35}S-methionine-labeled saposin C immunoprecipitates demon-strated that the 73- and 68-kD or the 12- and 9-kD species were glycosylated species with polypeptides of M_r = 50,000 or 7600, respectively. Very similar results were reported for the biosynthesis of saposin B (Fujibayashi and Wenger, 1986b), which would now be expected since saposin C and saposin B are encoded by the same gene (see Molecular Biology of the Activator Proteins). Neither the

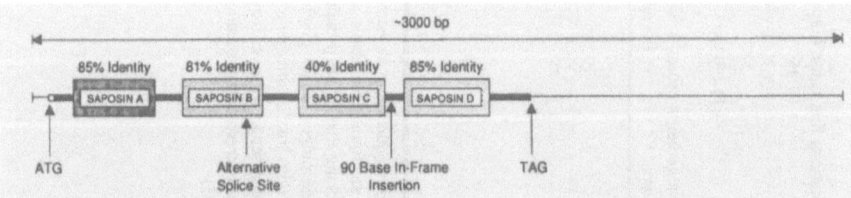

Fig. 5. Schematic diagram of the saposin A, B, C, and D regions of the rat and the human prosaposins. The amino acid identity between the rat and human sequences is shown above each region. The 90-base in-frame insertion occurs in the rat, but not the human, sequence. The alternative splice site (see text) has been demonstrated only in the human sequence.

sequence or events involved in proteolytic processing of the "prosaposin" precursor nor their potential tissue specificity has been elucidated. The highly homologous precursor produced in the rat analogue (SGP-1) (Collard et al., 1988) is not proteolytically processed prior to its secretion into the culture medium of Sertoli cells. Human prosaposin is found in serum, milk, and brain (Kishimoto et al., 1992), but its function is unknown.

Enzymology of Acid β-Glucosidase

Kinetic studies of acid β-glucosidase have been complicated by its membrane association and its requirements for detergents, delipidation for solubilization, and specific amphiphiles for reconstitution of enzymatic activity. The enzyme's natural substrates, glucosylceramides and its deacylated analogue (glucosylsphingosine), have been identified by their accumulation in the tissues of patients with the Gaucher disease variants (Nilsson and Svennerholm, 1982b; Nilsson et al., 1985). In the absence of detergents or negatively charged lipids, delipidated human (Mueller and Rosenberg, 1979; Dale et al., 1976; Basu and Glew, 1984; Glew et al., 1982), bovine (Legler and Liedtke, 1985), and murine (Basu and Glew, 1984) acid β-glucosidases have little or no hydrolytic activity toward natural lipid or more water-soluble synthetic substrates.

Because of the in vivo membrane association of acid β-glucosidase, investigators reasoned that specific membrane-derived lipids might be required for enzymatic activity. Dale et al. (1976) and Berent and Radin (1981) demonstrated that acid β-glucosidase activity toward 4MU-Glc was enhanced by negatively charged phospholipids in the series phosphatidylserine > phosphatidylinositol > phosphatidic acid, whereas phosphatidyl-ethanolamine or -choline had no effect on enzymatic activity. These results have been reproduced by numerous investigators in human, bovine, and murine sources (e.g., Mueller and Rosenberg, 1979; Osiecki-Newman et al., 1986; Legler and Liedtke, 1985; Grabowski et al., 1984; Glew et al., 1982), albeit with some variation in the composition of the assay mixtures and in the degree of activation. In artificial membrane systems, optimal activity also required the presence of negatively charged phospholipids to which acid β-glucosidase directly binds (Vaccaro et al., 1989). In a series of publications, Glew and co-workers delineated some structure/activity relationships between acid β-glucosidase and negatively charged phospholipids or glycosphingolipids (Basu and Glew, 1984, 1985; Basu et al., 1984, 1985, 1986; Prence et al., 1986a; Gonzales et al., 1988). Intriguingly, bis-(monoacylglyceryl)phosphate activated the enzyme. This lipid is present in liver lysosomal membranes, which suggest that the lipid milieu of β-glucosidase in the intact lysosome may play a role in modulating its activity in different tissues (Prence et al., 1986b). Characterization of the natural lipids associated with acid β-glucosidase from different tissues could provide an interesting insight into the enzyme's *in vivo* functional environment.

Saposin C has been shown to enhance acid β-glucosidase *in vitro* activity in the presence of negatively charged phospholipids (Brady et al., 1965; Berent and Radin, 1981; Grabowski et al., 1984; Prence et al., 1985). The direct interaction of saposin C with acid β-glucosidase has been extensively studied by many investigators (Ho and O'Brien, 1971; Berent and Radin, 1981) and it does not interact with glucosylceramide under equilibrium or nonequilibrium conditions (Berent and Radin, 1981). Consequently, saposin C enhances the hydrolysis of both glucosylceramide and 4MU-Glc substrates (Berent and Radin, 1981; Grabowski et al., 1984; Osiecki-Newman et al., 1987) in the presence of a variety of negatively charged lipids. Berent and Radin (1981) suggested that at low concentrations phosphatidylserine interacts with an anionic site on acid β-glucosidase, whereas at high concentrations it decreases the activation effects of saposin C by competing for this site on the enzyme. These effects are thought to be mediated by conformational changes in acid β-glucosidase so that the interaction of phosphatidylserine conforms the enzyme into a "poised" form for the interaction of saposin C (Berent and Radin, 1981). The use of inhibitory monoclonal antibodies has provided direct support for saposin C effecting a conformational change in acid β-glucosidase which enhances catalytic activity (Fabbro and Grabowski, 1991). It is not clear if saposins C and A bind to the same, overlapping (Van Weely et al., 1990), or different sites (Fabbro and Grabowski, 1991). No mutations in Gaucher disease have been shown to directly affect saposin C or A binding.

Using their results (Grabowski et al., 1984; Osiecki-Newman et al., 1987, 1988; Graves et al., 1986; Greenberg et al., 1990; Gatt et al., 1985; Grace et al., 1990) and previous work (J. S. Erickson and Radin, 1973; Hyun et al., 1975) on substrate specificity and inhibitor specificity, Grabowski and co-workers proposed a kinetic model of the active site of acid β-glucosidase which includes binding sites for the glycon head group, the sphingosyl moiety, and the fatty acid acyl chain of glucosylceramide (Fig. 1). Interaction at the glycon binding site requires a specific bonding with each hydroxyl group (Osiecki-Newman et al., 1988) on glucose, while the sphingosyl and fatty acid acyl binding sites are thought to accommodate hydrophobic chains up to about 16 carbon atoms in length. From the results of linear free-energy analyses, these latter sites have been proposed to be composed of multiple subsites which accommodate single methylene groups of substrates and inhibitors (Osiecki-Newman et al., 1987; Grabowski et al., 1984,

1986). The sphingosyl and fatty acid acyl chain lengths affect K_m and V_{max}, respectively. This model is similar to those proposed for the binding of globotriaosylceramide to saposin B (Wynn, 1986) and phospholipids to phospholipid transfer protein (Van Paridon et al., 1987). A nearly identical model has been proposed for a human neutral β-glucosidase (Glew et al., 1988). Although additional supporting structural data will be required for these models, they may serve as prototypes for binding sites on lipid transfer proteins and partial or complete active sites of glycosphingolipid hydrolases.

The specificity of glycon binding has been assessed by apparent binding constants for a series of epimers and substituted glucose derivatives (Osiecki-Newman et al., 1988). The enzyme binds β-glucose poorly, but has high affinity for 5-amino derivatives, i.e., the nojirimycins (Osiecki-Newman et al., 1988). Only the gluco derivatives of nojirimycin bound to the enzyme: i.e., the galacto-deoxynojirimycin and deoxymannojirimycin did not inhibit the enzyme. These results and the lack of inhibition by galactosylsphingosine (Grabowski et al., 1984) indicated the importance of the C2 and C4 hydroxyl configuration of the hexose. Alkyl-glycons were extremely powerful inhibitors of acid β-glucosidase (Grabowski et al., 1984; Osiecki-Newman et al., 1987, 1988). In particular, the N-alkyl-deoxynojirimycins and N-alkyl-β-glucosylamines had K_i values in the nanomolar and picomolar ranges, respectively (Osiecki-Newman et al., 1987; Greenberg et al., 1990). Since the additional free energies of binding were additive for the alkyl chain and the glycon head group, the existence of individual subsites for each methylene group up to about 16 was inferred to be within the active site (Osiecki-Newman et al., 1987). In addition, the N-alkyl-β-glucosylamines were slow-tight binding inhibitors (Morrison and Walsh, 1988) which induced a conformational change of the enzyme after formation of an initial binary complex (Greenberg et al., 1990). These results were consistent with the N-alkyl-glucosylamines being reaction intermediate analogues and with the catalytic mechanism requiring a conformational change in the enzyme during substrate hydrolysis. The catalytic mechanism is thought to involve the cleavage of the β-glucosidic linkage, the release of ceramide, donation of water to a glucose–enzyme complex, and the release of β-glucose (Grabowski et al., 1990). Recent use of site-directed mutagenesis and heterologous expression has provided some initial data to begin localization of portions of the enzyme which are responsible for catalytic activity (Fig. 3). Based on these studies and the degree of structural homology to the murine enzyme, the residues involved in catalysis and possibly saposin C binding appear to be located in the carboxy-terminal 40% of the amino acid sequence (Grabowski et al., 1990; Grace et al., 1990, 1991, 1993; O'Neill et al., 1989).

Molecular Enzymology of Acid β-Glucosidase

The existence of several different mutant alleles with different effects on enzymatic function was indicated by the phenomenological responses to the various modifiers of acid β-glucosidase activity (Osiecki-Newman *et al.*, 1986, 1987; Basu and Glew, 1984; Glew *et al.*, 1982; Grabowski *et al.*, 1985, 1986; Basu *et al.*, 1985; Wenger and Roth, 1982; Morimoto *et al.*, 1990) or differential stability of the protein (Bergmann and Grabowski, 1989) in cells from affected patients. These enzymes were separated into two general categories: (a) residual activities having blunted responses to negatively charged lipids and normal interaction with active site-directed inhibitors; (b) residual activities having enhanced response to negatively charged lipids and very abnormal interaction with active site-directed inhibitors (Osiecki-Newman *et al.*, 1987; Paulson *et al.*, 1989; Glew *et al.*, 1982; Grabowski *et al.*, 1984, 1985). Enzymes in the former category have normal or abnormal stability and processing, whereas those in the latter group have normal processing (Bergmann and Grabowski, 1989). Curiously, no mutant enzymes have been found which alter the K_M values for any substrate. This suggests a need to understand this parameter in its more fundamental aspects for this enzyme. With the extensive molecular heterogeneity (Table IV) of Gaucher disease, in retrospect, there was a remarkable consistency in the properties of the residual enzymes. Characterization of protein expressed from mutagenized cDNAs has begun to provide a functional map of the enzyme. As shown in Table V, several mutations lead to severely diminished catalytic rate constants. As expected, "disruptive" mutations (Bordo and Argos, 1991), e.g., D409V, lead to catalytic rate constants greater than 100-fold less than normal. However, some expected conservative substitutions, e.g., V394L or S364T, lead to very compromised enzymes and, in one case, D358E, an inactive enzyme (Grace *et al.*, 1990). Such mutations help define critical residues to catalytic activity. Figure 3 depicts the results from extensive analyses by Grace and co-workers (Grace and Grabowski, 1993; Berg *et al.*, 1993; Grace *et al.*, 1990, 1991, 1992, 1993) as a schematic of an initial function map. Using the transition-state analogues N-alkyl-β-glucosylamines, kinetic analyses of the N370S protein demonstrated no alterations in the required conformational change for catalysis and that the abnormal active function was due to diminished binding affinity (Greenberg *et al.*, 1990; Fabbro and Grabowski, 1991). Similarly, bromo-conduritol B epoxide (Br-CBE) was thought to be a mechanism-based covalent inhibitor whose attachment site to the enzyme defined the nucleophile in catalysis (Legler, 1977; Lalegerie *et al.*, 1982). Recent studies by Withers and co-workers (Gebler *et al.*, 1992; Trimbur *et al.*, 1992) cast doubt on the mechanistic generality of CBE's reaction with glycosidases. Mutagenesis studies of Asp[443] of the human enzyme were conducted

TABLE V. Specific Activity of Human Acid
β-Glucosidases Expressed in Sf9 Cells

cDNA	Glucosylceramide substrate nmole/hr per mg protein	1/CRIM SA[a]	Protein
Normal	2400	1	Stable
T138N	11–18	>100	Transient
F216Y	2.5–6.5	NA	Unstable
P289L	0.2–1.3	NA	Unstable
T3231	210	15	Stable
N370S	621	4	Stable
D358E	0.02	NA	Stable
S364T	306	~3	Unstable
V394L	450	5	Stable
D409H	4–5	NA	Unstable
D409V	0.3–10	>100	Stable
P415R	0.15–0.3	NA	Unstable
L444P	115	~20	Unstable
A456P	0.3–4	~100	Unstable
N462Q	13	>100	Stable
R463C	860	10	Stable
AcNPV[b]	0.02	NA	NA

[a]Relative CRIM specific activity referenced to the mature peptide.
[b]Sf9 cells infected with wild-type baculovirus.

to provide direct data on its role in catalysis. In addition, Asp^{445} was mutagenized to explore its potential role as an "opportunistic" nucleophile. All possible pairwise combinations of D, S, G, and E were substituted, expressed, and characterized. As shown in Figure 3, these results indicate the importance of aspartates 443 and 445 in catalytic function, but clearly demonstrate that neither is the catalytic nucleophile. These residues and a CBE binding site are within or near the active site. Clearly, additional analyses are required to elucidate further structure/function relationships.

MOLECULAR BIOLOGY

Genomic Organization

The complete sequences of the human acid β-glucosidase structural gene [7604 base pairs (bp)] and its nonprocessed pseudogene (5769 bp) have been reported (Horowitz et al., 1989). These sequences are contained within a 32-kilobase (kb) genomic DNA fragment with the pseudogene being about 16 kb 3' to the structural gene. Both sequences map to human chromosome 1(q21→31) (Devine *et al.*, 1982; Barneveld *et al.*, 1983; Ginns *et al.*, 1985) and each contains 11 exons (Fig. 6). In the regions present in both sequences, 96% nucleotide identity was found (Horowitz *et al.*, 1989). The pseudogene is smaller than the structural gene due to large deletions in introns 2 (313 bp), 4 (626 bp), 6 (320 bp), and 7 (277 bp) as well as two small exonic deletions (exon 9, 55 bp; and exon 4, 5 bp) (Fig. 6). The four large deletions in the pseudogene represent *Alu* sequences flanked by direct repeats within the structural gene. The structural gene also contains an inverted *Alu* sequence in intron 2 not found in the pseudogene (Horowitz *et al.*, 1989). In addition, the pseudogene contains numerous point substitutions and deletions in introns and exons scattered throughout the sequence. These differences and homologies have implications for the identification and characterization of mutations which result in Gaucher disease as well as the diagnostic accuracy of DNA-based tests for the disease variants (see Diagnosis by DNA Analysis).

Since *Alu* sequences are thought to be mobile elements (Hess *et al.*, 1983; Schmid and Jelinek, 1982) that may insert into DNA, the pseudogene may more closely resemble an ancestral acid β-glucosidase gene, with the current structural gene having been modified by insertion of *Alu* sequences. The proposed mechanism of insertion of *Alu* sequences (Jagadeeswaran *et al.*, 1981; Van Arsdell *et al.*, 1981) results in the duplication of single-copy sequences to form the direct repeats that flank the *Alu* sequences. This relationship is present in the appropriate corresponding regions of the pseudogene and structural gene. The mouse has only the structural gene (O'Neill *et al.*, 1989). In a study of 172 acid β-glucosidase alleles, eight intronic or 5′ flanking region polymorphisms were identified (Table VI). These base changes segregated with an ancient *Pvu*II polymorphic restriction site. Two of the eight intronic changes duplicate the sequence found in the pseudogene [nucleotides (nt) 2834 and 3297]. These appear to be localized alterations which do not involve large exchanges of structural gene and pseudogene material (Beutler *et al.*, 1992b). The base change at nucleotide 6144 marks the 5′ extent of a large 3′ exchange of structure and pseudogene material which includes the corresponding exons 10 and 11 (Beutler *et al.*, 1992b). Several disease alleles have also arisen by apparent rearrangements between the structural gene and pseudogene (Fig. 4). Only a single pseudogene sequence has been determined, however, and other forms may exist which might account for some rare alleles in isolated populations.

Analyses of the 5′ genomic sequences of the structural gene and the pseudogene indicated the presence of promoter elements (Reiner *et al.*, 1988a,b). The promoter of the structural gene contained two TATA boxes that lie between nucleotides (−23) to (−27) and (−33) to (−38) and two possible CAAT-like boxes between nucleotides (−90) to (−94) and (−96) to (−99) in relation to

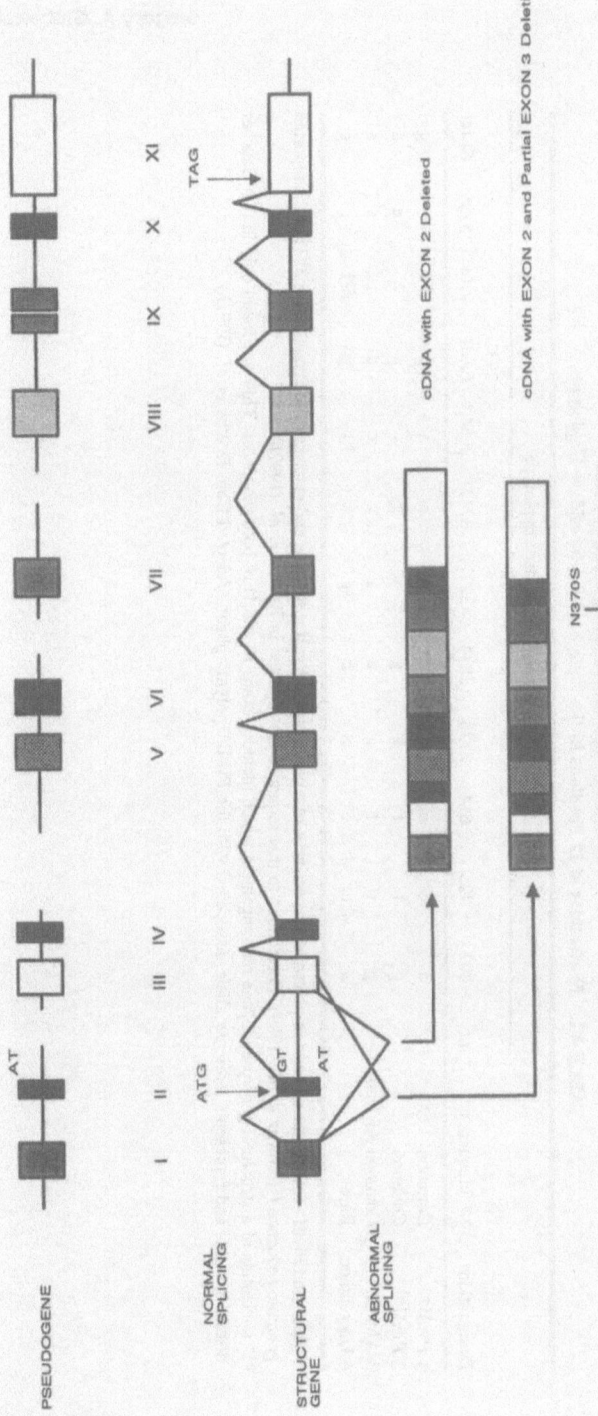

Fig. 6. Schematic diagram of the acid β-glucosidase pseudogene (top) and structural gene (bottom) and the depiction of the IVS2⁺¹ mutation leading to aberrant splicing. The homologous "exons" in the pseudogene and structural gene are indicated, as are the intronic deletions in the pseudogene. The pseudogene exon 9 contains a 55-bp deletion which corresponds to a rare disease allele designated Δ55. The G→A transition at the exon2/intron2 junction leads to the cDNA products shown at the bottom. The other allele in the patient encoded the N370S mutation.

TABLE VI. Nucleotides at 12 Positions in the Four Acid β-Glucosidase Haplotypes

Designation[a]	Frequency	nt	-802	-725	-614	2128	2834	3297	3747	3854	3931 PvuII	4644	5135	6144[b]
						Acid β-glucosidase haplotype								
1 PvuII+	Common		a	c	c	a	c	g	g	t	g	del	c	g
2 PvuII-	Common		g	t	t	g	g	a	g	c	a	a	a	a
3 African	Common African		g	t	t	g	a	a	g	c	a	a	a	a
4 Uncommon	Rare		a	c	c	a	c	g	a	t	g	del	c	g

[a] PvuII+ and PvuII- refer to the absence or presence, respectively, of a polymorphic PvuII restriction endonuclease site at nucleotide 3931 of the reported genomic sequence (Horowitz et al., 1989). Corrections to that sequence are reported in Beutler et al. (1992b).
[b] 5' nucleotide of a structural gene/pseudogene rearrangement which includes exons 10 and 11 of both sequences. This was found only in two Gaucher disease patients and represents a disease allele, associated with the PvuII+ polymorphism. Adapted from Beutler et al. (1992b).

the most 5' in-frame AUG of the mRNA (Horowitz et al., 1989; Reiner et al., 1988a,b). No Sp1 binding site, i.e., GGCGGG motif, was found. These results are interesting since the promoter elements that contain TATA and CAAT boxes have been identified most commonly in highly regulated genes (Dynan, 1986). The promoter activity was monitored using a plasmid pSVOCAT construct containing the appropriate acid β-glucosidase 5' sequences and the bacterial CAT gene (Horowitz et al., 1989; Reiner et al., 1988a,b). Following transfection of these constructs into different human cell lines, expression of CAT activity indicated that the promoter sequences from the structural gene were at least eight times more potent than the corresponding pseudogene sequences. The pseudogene contains several termination codons within its exons (Horowitz et al., 1989). Although the pseudogene has a very weak promoter (Dynan, 1986), transcription has been reported (Sorge et al., 1990).

No other primate or mammalian acid β-glucosidase sequence have been published, except that from the mouse (O'Neill et al., 1989). The mouse acid β-glucosidase gene has been mapped to mouse chromosome 3 and its genomic organization determined (O'Neill et al., 1989). The exons of the human and mouse genes have 84% nucleotide identity in the protein coding, 54% in the 5' noncoding, and 78% in the 3' noncoding regions. The positions of the 20 intron/exon junctions in the structural genes from each species are precisely conserved. Compared to the human sequence: (1) 64% of the nucleotide substitutions resulted from third position changes, (2) a single amino acid deletion, His^{273}, was predicted in the mouse sequence, (3) 100% of the cysteines, 91% of the prolines, and 89% of the glycines were preserved, and (4) the greatest variation in the predicted amino acid sequences occurred in the N-terminal 60%. All six uncommon amino acid substitutions as well as the His^{273} deletion occurred in this region of the predicted mature protein sequence. If analyzed on the basis of percentage of the mature protein sequence per exon compared to the percent variation per exon, the expected ratio of these parameters for random divergence would be about 1. When calculated from the published murine and human cDNA sequences, the 65 predicted nonconservative amino acid differences gave a ratio of about 1–1.5 for all exons except exons 8 and 9, where the ratio was 0.2 and 0.4, respectively. This result and the carboxy-terminal conservation suggested conserved functions for selected portions of these genes.

mRNA and cDNA Structure, Organization, and Expression

Northern analyses of poly (A)$^+$ RNAs from HeLa (epithelioid cells) and cultured skin fibroblasts from normal individuals and Gaucher disease patients identified three acid β-glucosidase-specific mRNAs of ~5.6, 2.5, and 2.0 kb in length (Reiner *et al.*, 1988a; Graves *et al.*, 1986). The longest mRNA represented an unspliced or partially spliced nuclear transcript. S1 nuclease analyses demonstrated that the shorter mRNAs arose from alternate transcription initiation and alternative polyadenylation (Graves *et al.*, 1986). Use of alternative polyadenylation signals about 450 bp apart was demonstrated by sequence analyses (Reiner *et al.*, 1988b). The full-length cDNA extends at least 237 bp 5' of the most 5' ATG (Horowitz *et al.*, 1989). Thus, the major full-length mRNA is about 2.5–2.6 kb in length and has about 250 kb of 5' and 550 bp of 3' untranslated sequence as well as a 100-bp poly (A) tail (Reiner *et al.*, 1988a; Wigderson *et al.*, 1989). Acid β-glucosidase is expressed constitutively in all tissues. The level of mRNA may be increased, however, by about 2–3 fold in cultured fibroblasts and lymphoid lines from Gaucher disease patients (Reiner *et al.*, 1988a; Wigderson *et al.*, 1989). Also, epithelial cells expressed high levels of acid β-glucosidase mRNA, while skin fibroblasts and promyelocytes had intermediate steady-state levels (Reiner *et al.*, 1987). In comparison, macrophages had low and B cells had barely detectable steady-state levels of acid β-glucosidase mRNA. These results were consistent with the parallel levels of CAT activity expressed from constructs containing the acid β-glucosidase promoter in the corresponding tissues (Reiner *et al.*, 1987). These results suggest the involvement of a feedback regulation of acid β-glucosidase expression.

An unusual feature of the human acid β-glucosidase cDNA was the presence of two 5' in-frame ATGs as possible translation initiation codons (Reiner *et al.*, 1987; Graves *et al.*, 1988; Wigderson *et al.*, 1989; Sorge *et al.*, 1985). A similar observation was made with acid sphingomyelinase (Schuchman *et al.*, 1991). These two ATGs begin 57 and 118 nucleotides 5' to the first nucleotide of the codons for the mature polypeptide. Although both can be used (Sorge *et al.*, 1987; Choudary *et al.*, 1986), the immediate translation products encoded by sequences downstream from these ATGs differed markedly. If translation is initiated from the most 5' ATG, the initial sequence of 19 amino acids is hydrophilic and charged. In comparison, a typical hydrophobic signal sequence of 20 amino acids in length would begin at the second ATG. The downstream ATG is preceded by a Kozak consensus sequence for translation initiation. The relative physiologic relevance, if any, of the most 5' ATG is unknown. Expression of functional acid β-glucosidase from the cDNAs in mammalian or insect cells (Grabowski *et al.*, 1988, 1989; Martin *et al.*, 1988) demonstrated their authenticity. However, the two originally reported cDNAs (Sorge *et al.*, 1985; Tsuji *et al.*, 1986) contain cloning artifacts when compared to genomic DNA sequences or to the chemically determined amino acid sequences. The reported Leu[259] was a cloning artifact which leads to an inactive enzyme and Phe[259] is the correct amino acid (Tsuji *et al.*, 1986; Choudary *et al.*, 1986). The His[495] for Arg[495] (Sorge *et al.*, 1985) also was a cloning artifact, but no functional abnormalities resulting from this substitution have been observed (G. A. Grabowski, unpublished).

Mutations in the Acid β-Glucosidase Gene Resulting in Gaucher Disease

Interest in the genetic mutations causal to Gaucher disease relates to the need to explain the marked heterogeneity of phenotypes within and among the variants. Also, the need for genetic markers of disease severity (Theophilus *et al.*, 1989a; Zimran *et al.*, 1989) has had increasing practical (i.e., prognostic) and basic (structure/function) import. Numerous disease alleles have been identified (Fig. 4 and Table IV). Several different nomenclatures used to designate the mutations are shown in Table IV. The genomic sequence numbering refers to the original report (Horowitz *et al.*, 1989), which has been corrected and extended (Beutler *et al.*, 1992b). The relationships of these mutant alleles to the disease have been established by family, population, and expression studies (see Disease Allele Frequencies and Genotype/Phenotype Correlations). A significant complicating problem for DNA diagnosis is the presence of the highly homologous pseudogene and the presence of disease alleles bearing multiple authentic, nonpolymorphic, point mutations. Since polymerase chain reaction (PCR) amplification of genomic sequences has supplanted the use of unamplified DNA, methods have exploited sequence differences to selectively amplify structural gene sequences (Theophilus *et al.*, 1989a; Zimran et al., 1989, 1990; Tsuji *et al.*, 1987, 1988b, 1990; Firon *et al.*, 1990). This is particularly important since several missense and nonsense mutations arise from introduction of "pseudogene" base changes into the structural gene (Table IV, Fig. 4). These substitutions can be single (i.e., 1448C, 754A) or multiple. The latter mutant alleles probably represent rearrangements between the homologous sequences by crossover events (Firon *et al.*, 1990) or other poorly defined genetic events. These types of alleles have been referred to as complex alleles (Latham *et al.*, 1990), a "pseudo" pattern (Hong *et al.*, 1990; Ohashi *et al.*, 1991), or a "rec" (recombinant) (Eyal *et al.*, 1990). In addition, several disease mutations in the same codon have occurred directly following a pseudogene difference (Theophilus *et al.*, 1989a,b; Graves *et al.*, 1986), suggesting the possibility of repair difficulties. The apparent imperfect exchange of exons 5 and 6 (Latham *et al.*, 1990) and other disease alleles with multiple exon substitutions or deletions indicate the participation of the pseudogene in the pathogenesis of this disease. The only splice junction abnormality mimics a pseudogene sequence at the exon2/intron2 splice donor site. This mutation was limited to a single base change and leads to the production of several alternatively spliced mRNA forms due to activation of a cryptic splice site (Fig. 6). There are several additional exonic differences between the structural gene and pseudogene, which are likely to be found in disease alleles in the future.

From studies on large populations of Gaucher disease patients, two mutations in exon 9 (N370S) and exon 10 (L444P) account for about 60% of all alleles present in patients with Gaucher disease. Table VII provides a compilation of frequency data on these two alleles in each of the three clinical variants. With the possible exception of one patient (Sidransky *et al.*, 1992), no patient with neuronopathic disease has had even a single copy of an N370S allele. In addition, with the exception of one Japanese patient (Masuno *et al.*, 1990), L444P homozygosity is associated with the neuronopathic variants or lethal disease early in childhood. Although these studies will require continuous updating, sufficient families have been studied to indicate the presence and prognosis of nonneuronopathic Gaucher disease from genotype analysis.

A puzzling finding among the neuronopathic phenotypes was apparent L444P homozygosity in both the very acute neuronopathic, type 2, and the subacute neuronopathic, type 3, patients (Theophilus *et al.*, 1989a; Tsuji *et al.*, 1987; Firon *et al.*, 1990). Complete sequencing of the cDNAs representing both alleles from such type 2 patients revealed the presence of multiply mutated alleles which contained the L444P missense error (Latham *et al.*, 1990). Consequently, these type 2 patients had two different alleles. The generality of this finding is not known.

TABLE VII.   Frequency of Mutant Alleles
among the Gaucher Disease Variants[a]

Variant	Mutant allele		
	N370S	L444P	Other
Type 1	439/704 (62%)	79/635 (12.4%)	107/352 (30%)
Type 2	0/30 (0%)	55/52 (48%)	12/18 (67%)
Type 3[b]	0/69 (0%)	55/80 (68.8%)	8/24 (33%)

[a]Adapted from Tsuji *et al.* (1987, 1988b), Sidransky *et al.* (1992), Theophilus *et al.* (1989a,b), Latham *et al.* (1990, 1991), Zimran *et al.* (1989), Firon *et al.* (1990), Choy *et al.* (1991), Sibille *et al.* (1993), Grace *et al.* (1991).
[b]Does not include the Norbottnian type 3 population.

Disease Allele Frequencies and Genotype/Phenotype Correlations

Of the numerous disease mutations only four occur with substantial frequencies in Gaucher disease populations. The N370S, 84GG, L444P, and IVS2^{+1} alleles account for about 60–70, 10, 5, and 2%, respectively, of the identified Gaucher disease alleles (He and Grabowski, 1992; Sibille *et al.*, 1993; Beutler *et al.*, 1992a) (Table VIII). The R463C and multiply mutated complex alleles also have been reported independently by several groups (see Table IV). In the Gaucher disease type 3 population from the Norrbottnian region of Sweden the L444P allele is found almost exclusively (Dahl *et al.*, 1990; Theophilus *et al.*, 1989a). The Ashkenazi Jewish Gaucher disease population has been the only other intensively studied population (Zimran *et al.*, 1991). In studies of over 300 symptomatic Jewish patients with type 1 disease the N370S, 84GG, L444P, and other (including the IVS2^{+1} and rare) alleles account for 71.2, 11.6, 4.44, and 2.12%, respectively, of the disease alleles in that population (He and Grabowski, 1992; Beutler *et al.*, 1992a). The largest discrepancies in allele frequencies are with the N370S allele, which varied by almost 10% in the studies by Beutler (1992a) (77.16%) and our analyses [68.7% (He and Grabowski, 1992)]. Thus, the allele identification rate in the Ashkenazi Jewish population using the above four alleles would vary between 89.34 and 95.9%. Our studies were strictly limited to symptomatic, unrelated patients and asymptomatic "patients" were excluded (Sibille *et al.*, 1993). This is a potential source of the frequency differences, since significant numbers of asymptomatic or very mildly affected family members and random Ashkenazi Jews are homozygous for the N370S mutation. Using molecular analyses, the true carrier frequencies for Gaucher disease alleles have been estimated by random screening in the healthy Ashkenazi Jewish population to be about 1/10 to 1/12

TABLE VIII.   Frequency of Gaucher Disease Mutations in 258 Jewish Patients

Mutation	Number of alleles	Percent
N370S	369	71.2
84GG	60	11.58
L444P and XOVR[a]	23	4.44
IVS2^{+1}	11	2.12
V394L	5	0.9

[a]XOVR is a fusion gene. Adapted from Beutler *et al.* (1992a) and Sibille *et al.* (1993).

individuals (Zimran *et al.*, 1991; Beutler *et al.*, 1992a). These results indicate a disease frequency of about 1/400 to 1/600. This is consistent with a large number of asymptomatic/mildly affected individuals in this population. In the author's experience with over 500 patients with documented acid β-glucosidase deficiency, only N370S homozygotes have been asymptomatic.

For effective delivery of medical care to patients affected with Gaucher disease, prognostic tests are required which predict the eventual degree of disease involvement in an individual or fetus. The simplest hypothesis is that various combinations of mutant and null alleles would lead to different phenotypic expressions of the disease. Genotype/phenotype correlations in 161 Ashkenazi Jewish patients demonstrate clear symptomatic prognostic import (Table IX). Symptomatic patients with N370S homozygosity had later onset and much less severe visceral and bony involvement than patients with the other genotypes. Such information has immediate genetic counseling and therapeutic impact.

Implicit in these data is the hypothesis that various thresholds of residual acid β-glucosidase activity levels exist in humans, as has been proposed in a more general manner by Conzelmann and Sandhoff (1983). One threshold prevents the development of symptomatic accumulation of acid β-glucosidase's major substrate, glucosylceramide. Obviously, this level is found in completely normal

TABLE IX. Genotype/Phenotype Correlations in Jewish Gaucher Disease

Genotype	Age of onset (years)	Bone (scale = 0– 5)	Liver enlargement[a]	Splenic enlargement[a]	Age at splenectomy (range) (years)
N370S/N370S	34.68	2.03	1.22	8.58	56
	(3–73)	(1–4)	(0.79–2.14)	(1.8–25.1)	(26–69)
	$n = 59$	$n = 37$	$n = 35$	$n = 33$	$n = 6$
	SD = 19.1	SD = 0.763	SD = 0.304	SD = 15.5	SD = 15.5
N370S/84GG	9.09	3.53	2.67	33.7	21.18
	(2–28)	(1–5)	(1.0–4.12)	(10.2–62)	(5–57)
	$n = 37$	$n = 26$	$n = 15$	$n = 9$	$n = 17$
	SD = 7.57	SD = 1.14	SD =0.93	SD = 13.4	SD = 13.4
N370S/L444P	16.9	3.11	2.11	22.23	25
	(2–35)	(1–5)	(1.7–2.69)	(6.89–43.3)	(2–44)
	$n = 17$	$n = 9$	$n = 10$	$n = 9$	$n = 4$
	SD = 11.3	SD = 1.23	SD = 0.33	SD = 12.1	SD = 19.1
N370S/IVS 2 or ? and ?/?	14.28	3.03	2.08	23.83	26.1
	(1–45)	(1–5)	(0.74–3.48)	(7.4–75)	(5–69)
	$n = 48$	$n = 38$	$n = 28$	$n = 7$	$n = 23$
	SD = 11.38	SD = 1.19	SD = 0.68	SD = 18.06	SD = 18.24

[a]Fold increase over normal.

individuals and carriers for Gaucher disease. The next level is in enzyme-deficient patients who are asymptomatic or mildly affected (e.g., N370S homozygotes). Another threshold of acid β-glucosidase activity is that which leads to the development of the severe and/or neuronopathic Gaucher disease phenotypes. This threshold may be bounded by type 1 patients with the N370S/"inactive allele" (e.g., 84GG, IVS2^{+1}) genotype and type 3 patients with homozygosity for L444P. The relative ethnic homogeneity of the Norrbottnian population of type 3 Gaucher disease patients provides a genetic background to assess the variation of the type 3 phenotype due to L444P homozygosity. Characterization of this population of patients indicated that all develop mental retardation, have severe visceral and/or bony disease, and have shortened life spans (Svennerholm et al., 1982). The severity and age of onset vary significantly among Norrbottnian type 3 patients (Dreborg et al., 1980). This may explain reports in the literature that homozygosity for L444P has been associated with very severely affected patients who were apparently free of neuronopathic manifestations (Theophilus et al., 1989a; Masuno et al., 1990). Potentially, if these young patients had survived longer, neurological manifestations could have developed. However, greater insight into the molecular determinants of this threshold should be provided by thorough characterization of the mutant acid β-glucosidase genes from other type 3 patients with horizontal gaze palsy and myoclonic seizures in the juvenile age group (Patterson et al., 1991; Yu et al., 1990), as well as from adult patients who develop the onset of mental deterioration due to Gaucher disease late in life or apparently nonneuronopathic teenage patients (Winkelman et al., 1983).

The distinction between types 2 and 3 is much clearer than those above and may be absolute, since type 2 disease has the onset of severe neuronopathic involvement in the first year of life and death occurs by about 2 years of age. In this regard the distinction between these two forms of Gaucher disease is similar to that between the infantile and the later onset forms of Tay–Sachs disease, Krabbe disease, and metachromatic leukodystrophy. For example, in Tay–Sachs disease, the infantile form has no residual β-hexosaminidase A activity (Meyerowitz and Proia, 1988), whereas the mutation in the later onset form of G_{M2}-gangliosidosis results in low levels (2–5% of normal) of this enzymatic activity (d'Azzo et al., 1984).

MOLECULAR BIOLOGY OF THE ACTIVATOR PROTEINS

Previous pulse-chasing processing studies of SAP-1 and SAP-2 indicated extensive posttranslational proteolytic processing of each from a large-molecular-weight precursor (Fujibayashi and Wenger, 1986a,b) and that the genes for both map to the same portion of chromosome 10 (Inui *et al.*, 1985; Fujibayashi *et al.*, 1985). The recent cloning of cDNAs encoding these activators has demonstrated that these two "activators" as well as two additional "activators" of sphingolipid hydrolases are encoded by a single gene in humans (Gavrieli-Rorman and Grabowski, 1989; O'Brien *et al.*, 1988; Nakano *et al.*, 1989; Reiner *et al.*, 1989) and rats (Collard *et al.*, 1988). As shown schematically in Fig. 7, the full-length human "proactivator" cDNA (Gavrieli-Rorman and Grabowski, 1989; Reiner *et al.*, 1989) contains regions of high amino acid similarity (>80%). The existence of these activators has been demonstrated by their partial or complete amino acid sequence determined from proteins isolated from human (Morimoto *et al.*, 1988, 1989; Kleinschmidt *et al.*, 1988; Dewji *et al.*, 1987) or guinea pig (Sano *et al.*, 1988) sources. The only major difference in the amino acid sequences predicted by the full-length cDNAs (Gavrieli-Rorman and Grabowski, 1989; Nakano *et al.*, 1989) was the presence of an in-frame 9-bp insertion (three amino acids) in the saposin B coding sequence of cDNAs isolated by Nakano *et al.* (1989) from lung and skin fibroblast libraries. Furthermore, the complete chemically determined sequence of saposin B did not contain the predicted three-amino acid insertion (Kleinschmidt *et al.*, 1988). These results suggest a potential for tissue-specific

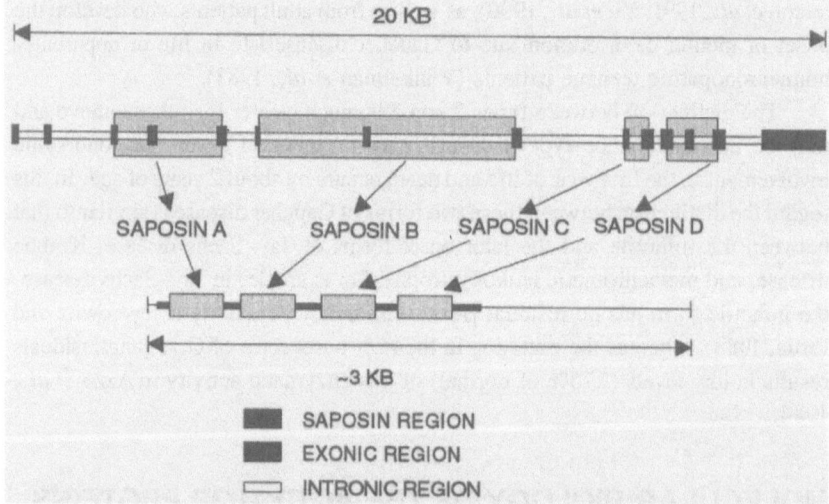

Fig. 7. Diagram of the human prosaposin genomic organization and the relationship to the cDNA. The 5' end of the gene has not been fully characterized, but is probably composed of several small exons and larger introns. No large contiguous regions of high homology are present in the intronic sequences. The exons are not numbered since the 5' organization is not complete. The shown structure accounts for all but about 100 bp of exonic sequence.

alternative splicing of the "proactivator" RNA in the saposin B region (Holt-schmidt et al., 1991b). A mutation resulting in aberrant splicing of the prosaposin mRNA leads to a 33-bp insertion in the saposin B region (Zhang et al., 1991) surrounding the 9-bp insertion described by Nakano et al. (1989). Similar to the acid β-glucosidase mRNA, two species of human proactivator mRNA arise from alternative polyadenylation (Holtschmidt et al., 1991a) and the levels of mRNA expression parallel those of acid β-glucosidase, i.e., the levels of proactivator mRNA were higher in normal skin fibroblasts than in B cells and the mRNA levels are higher in the corresponding cells from Gaucher disease patients. The genomic organization (Holtschmidt et al., 1991a; Rorman et al., 1992) and sequence (Rorman et al., 1992) have been determined for all but the most 5' region. The available sequence is about 21 kb long, which accounts for all but about 100 bp of cDNA sequence. The high homology of the various saposin regions suggested a duplication scheme for evolution. However, there were no significant intronic similarities to suggest regions of duplication prior to intron insertion. Based on the placement sites of intron insertion, Rorman et al. (1992) (Fig. 8) suggested a duplication and recombination model for the evolution of this gene from an ancestral saposin B gene. Several mutations in the prosaposin gene have led to clinical sphingolipid diseases mimicking isolated enzyme deficiencies, including: (1) saposin B mutations (Kretz et al., 1990; Rafi et al., 1990; Zhang et al., 1990, 1991) leading to metachromatic leukodystrophy-like diseases, (2) a saposin C mutation (Schnabel et al., 1991) leading to a Gaucher-like disease, and (3) a mutation in the prosaposin initiation codon leading to deficiencies of both sapo-sin B and C (Schnabel et al., 1992).

THERAPEUTIC PROSPECTS IN GAUCHER DISEASE

Gaucher disease has been considered a prime candidate for therapeutic interventions since it is the most prevalent lysosomal storage disease, its major forms are nonneuronopathic, and its visceral manifestations result from bone marrow-derived reticuloendothelial cells. Indeed, the monocyte/macrophage sys-tem involvement has led to the use of bone marrow transplantation for effective treatment of the nonneuronopathic variants (Rappeport and Ginns, 1984; Starer et al., 1987; Tsai et al., 1992). The morbidity and mortality of nonautologous bone marrow transplantation restricts its therapeutic usefulness to the most severely involved nonneuronopathic patients. Recent studies have focused on the use of large doses of enzyme which has been specifically modified to expose terminal α-mannose residues on its oligosaccharide for targeting to the reticuloendothelial

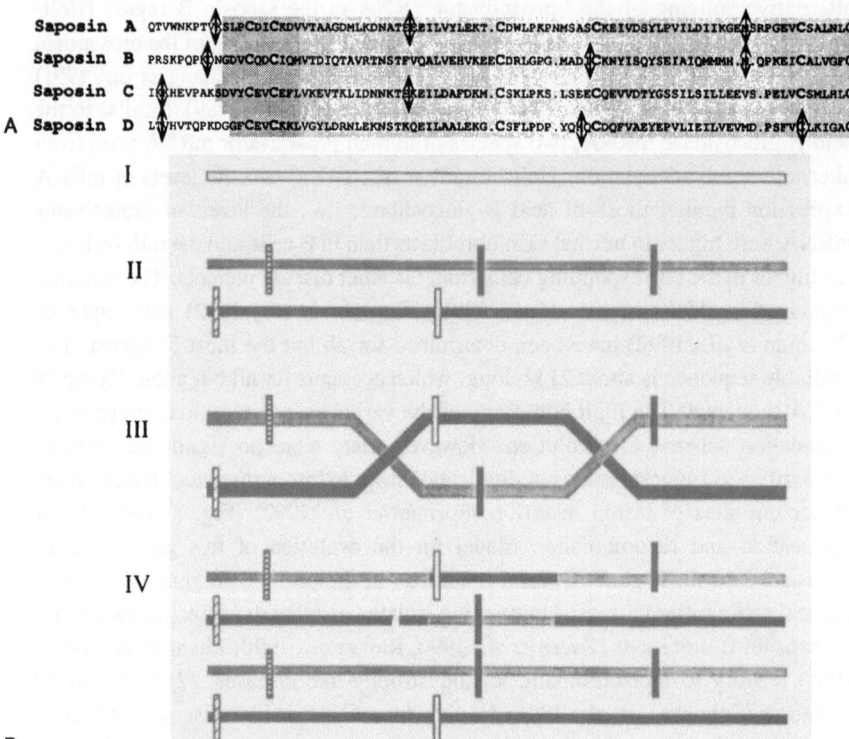

Fig. 8. (A) The alignment of the placement of the introns (double-ended arrows) in the saposin A, B, C, and D amino acid sequences. The stippled regions correspond to the mature saposin regions. (B) An evolutionary model from an ancestral saposin B (I) gene to account for the preservation of the intronic placement. In II a tandem duplicated version of I is shown with introns being inserted following the duplication of I. In III and IV are shown a proposed double crossover to lead to the intronic alignment shown in (A) and in IV. This mechanism required a tandem duplication of the structure in II.

organs (Barton *et al.*, 1991). Based on these initial results of bone marrow transplantation, enzyme replacement, and the recent successes in long-term expression of foreign genes in murine stem cells (Fink *et al.*, 1990; Nolta *et al.*, 1992; Correl *et al.*, 1992), the future development of genetic therapeutic approaches or more available enzyme replacement strategies portends a bright future for affected patients with Gaucher disease.

The available supportive therapies for Gaucher disease have been reviewed recently (Beutler and Grabowski, 1993) (Table X). Although many of these will

TABLE X. Available Intervention for Gaucher Disease

Hematologic care	Bone marrow transplantation
Iron and vitamin supplementation	Curative
Supportive intervention	Matched sibling donor
Total or partial splenectomy	High risk for nonlethal diseases
Androgen therapy	Prenatal diagnosis
Erythropoietin (not useful)	Heterozygote and homozygote detection
Bone crisis	Enzyme replacement
Pain management	
Core decompression	

continue to have supplementary roles in the total care of affected patients with type 1 disease, most have been supplanted by efficacious enzyme therapy.

Enzyme Therapy

In comprehensive reviews Desnick *et al.* (1976) and Desnick and Grabowski (1981) emphasized requisites for enzyme therapy:

1. The availability of large amounts of stable enzyme suitable for administration.
2. Delivery of the enzyme to the major cellular and organellar sites of pathology.
3. Development of practical serial evaluation procedures to monitor efficacy.

Early attempts at enzyme therapy for Gaucher disease (Brady *et al.*, 1974, 1975; Pentchev, 1977; Beutler *et al.*, 1977, 1980; Beutler and Dale, 1982) reemphasized these requisites. The macrophage and its tissue-specific counterparts (Kupffer cells, alveolar macrophages) have long been recognized as the principal cells involved in Gaucher disease type 1. The discovery of specific receptors for α-mannosyl-terminated oligosaccharides on macrophages provided the impetus to develop a modified acid β-glucosidase for targeted delivery (Achord *et al.*, 1978; Stahl *et al.*, 1978). Industrial-scale production of the human placental acid β-glucosidase with α-mannosyl residues exposed (Ceredase®; alglucerase) made possible the clinical and biochemical evaluation of enzyme therapy. The evaluation of this preparation in one child suggested clinical efficacy (Barranger *et al.*, 1989; Barton *et al.*, 1990). Larger clinical trials demonstrated (Barton *et al.*, 1990; Beutler *et al.*, 1991b; Fallet *et al.*, 1992; Figueroa *et al.*, 1992; Pastores *et al.*, 1993) clinical and biochemical efficacy in about 60 reported patients. Summaries of the visceral organ results are presented in Fig. 9A and 9B. In anemic patients the hemoglobin concentration of blood increases by about 1.5 gm% in the first 4–6 months of enzyme therapy. An additional 1 gm% increase is observed in the subsequent 9–18 months in persistently anemic patients (Pastores *et al.*, 1993). Platelet count increases are much slower and require at least 1 year to double. The most dramatic hemoglobin and platelet responses have been in splenectomized patients with pulmonary involvement, but in general the response to

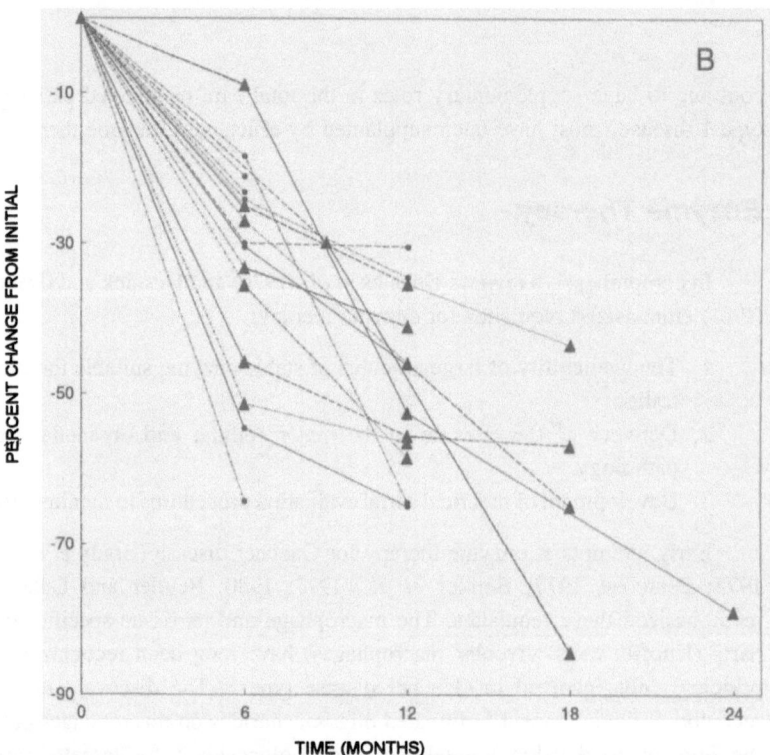

Fig. 9. (A) Hepatic and (B) splenic volume decreases following enzyme therapy (60–30 IU/kg q 2 weeks initial doses) with macrophage-targeted acid β-glucosidase (alglucerase). In (A) several different doses of alglucerase were used with an overall response of ~25% decrease in hepatic volume by 6 months. The squares represent patients who had or decreased to hepatic volumes of ≤150% of predicted normal by 6 months of therapy and demonstrate a nearly asymptotic approach to normal volumes. The presence or absence of a spleen did not influence the response to therapy. In (B) several different doses of enzyme demonstrated no overall trends in the effect of differing initial doses. The transitions to different types of lines for any patient indicate a decrease in dose by 50%. The patients with dosage reductions are indicated by filled triangles. Overall, no consistent differences in rates of response were observed with diminishing doses from 60–30 IU/kg to 20–10 IU/kg q 2 weeks.

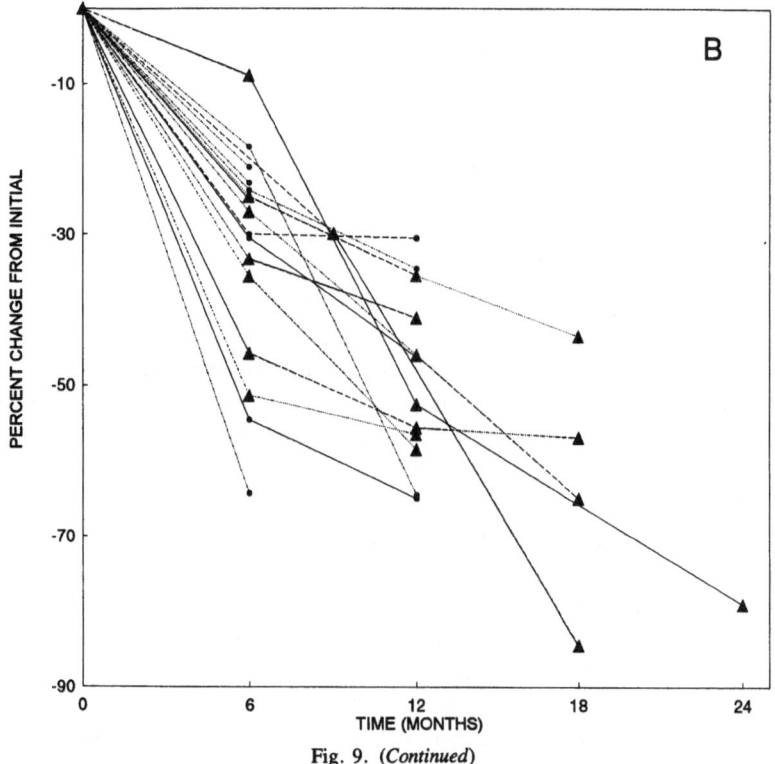

Fig. 9. (*Continued*)

enzyme therapy is the same in patients with or without intact spleens (Pastores *et al.*, 1993). Bone pain gradually decreases, but the X-ray abnormalities of bone appear to respond very slowly. Magnetic resonance imaging shows a decrease in bone marrow involvement (N. Barton, personal communication). About 10–15% (36/272) of patients treated develop antibodies against the enzyme protein (Richards *et al.*, 1993; Pastores *et al.*, 1993). A few of these develop pruritus with enzyme infusion, but it has been possible to continue therapy in such patients without the occurrence of serious reactions (Pastores *et al.*, 1993). One patient has had an anaphylactoid reaction with complement activation (G. A. Grabowski, unpublished observation). All antibodies reported to date have been IgG, predominantly IgG1 subclass (Richards *et al.*, 1993). The processing of the placental extract destroys known viruses, including the human immunodeficiency virus. Recombinant enzyme is not yet commercially available, but initial clinical trials have indicated efficacy (Grabowski *et al.*, 1993).

To monitor responses to therapy, some investigators have expressed the changes in visceral organs as a percentage decrease in excess liver or spleen volume (Figueroa et al., 1992). However, this method greatly exaggerates small decreases in livers that are only slightly enlarged. For example, a liver which is calculated to be 150% of normal size has an excess volume of 50%. If this decreases to 120% of normal, the percent decrease in excess liver is 60% (i.e., 0.3/0.5 × 100%), whereas the absolute change was 20% (30%/150% × 100%). If an additional 20% decrease is observed in the same patient, the liver is now normal in volume, so the absolute decrease of 17% corresponds to a 100% decrease in percent excess liver volume. This and the finding that the error of repeated measurement in volume determinations is from 4 to 12% indicate the inappropriateness of this calculation. In addition, nearly all patients receiving alglucerase gain substantial weight. Consequently, body weight normalization of hepatic volume calculations could suggest decreases when little or no absolute change actually occurs. This error coupled with the calculation of percent excess liver decrease could greatly exaggerate changes in visceral organ response. For adults, it is better to calculate the absolute percentage change referenced to an average body weight. In growing children this calculation is more problematic. The need to standardize the hepatic and splenic volumes to body weight is not trivial. For the liver volume, the response to alglucerase becomes less pronounced as this volume approaches normal (Pastores et al., 1993). At volumes less than about 150% of normal there is an asymptotic approach to the normal range. Thus, no or small changes could suggest a lack of therapeutic effect. If this occurred following a dose reduction, a misinterpretation would indicate a diminished response. Indeed, no difference in response rate was found with diminishing doses in patients with livers greater than 150% of normal and efficacy was demonstrated at lower doses. Currently, trials of decreasing doses of enzyme with differing schedules of administration (q 2 weeks or several times per week) are attempting to determine optimal doses and schedules of administration. This has not been established. To date, all doses and schedules have shown efficacy.

Very little data on the tissue, cellular, and subcellular distribution or the half-lives of the administered enzyme have been published. Only the single case report of enzyme analyses in tissues of a treated patient suggests persistence of exogenous enzyme for a significant period in particular organs (Fallet et al., 1992). The lack of such data is a severe limitation to optimizing doses and dosing schedules and acquiring such data is more relevant than current discussions limited to overall clinical responses. Indeed, ongoing clinical trials are being conducted without benefit of such formal tissue pharmacokinetic data.

In addition, several substantive and scientifically difficult issues need to be resolved in an effort to promote optimal use of this highly efficacious therapy or the development of a new generation of recombinant drugs (Grabowski *et al.*, 1993). Our zeal to decrease doses of alglucerase has been primarily focused on monitoring diminishing sizes of visceral organs and good hematologic responses. In focusing on these easily measured responses, are the effects on bone or lungs being more poorly documented? Is there a tissue-specific dose requirement which will determine overall efficacy of alglucerase therapy? Indeed, pulmonary involvement can be the most lethal and bony disease the most debilitating aspects of this disease. Is there a diminution in dose requirement for continued improvement as the mass of Gaucher cells and accumulated glucosylceramide decreases (i.e., how important is the distribution space of the enzyme to therapeutic response and what is that space?)? What is the minimal effective dose and at what degree of involvement? What is the maximal effective dose and who should receive it? To date, there is no clear relationship between the rate of improvement and initial doses between 30 and 120 IU/kg every month. Most importantly, is it more efficacious to prevent the disease manifestations than to attempt to return a very sick individual, with potentially irreversible tissue damage, to health? If very small amounts of enzyme are required to maintain health, then long-term alglucerase may be very affordable. In addition, the efficacy or lack thereof of enzyme therapy for the neuronopathic variants remains to be evaluated.

In view of the high cost of enzyme therapy with all current infusion protocols, the timing of the initiation of therapy must be carefully evaluated. In highly symptomatic patients therapy should begin immediately. In addition, other patients with major hematologic and visceral organ involvement should receive high priority. Although bone disease appears to be the slowest to respond, patients with substantial bone involvement are candidates for treatment to recover bone integrity and to prevent irreversibility of some bony damage. More difficult questions are posed by presymptomatic patients whose susceptibility to major disease manifestations can be anticipated by genotype analysis. Clearly, the issue of the timing and use of an apparently universally effective therapy must be carefully considered and the potential for preventive intervention should be evaluated. Although the drug appears safe, all risks of treatment may not, as yet, be appreciated.

Bone Marrow Transplantation

Since the manifestations of Gaucher disease type 1 are due to abnormalities of the macrophage, replacement of hematopoietic stem cells via allogeneic marrow transplantation should be curative. A number of patients with type 1 disease (Rappeport and Ginns, 1984; Starer *et al.*, 1987; Tsai *et al.*, 1992) have undergone bone marrow transplantation with therapeutic responses. There have been deaths secondary to the transplantation procedure. In patients with subacute neurono-pathic (type 3) disease, bone marrow transplantation has reversed the visceral disease and there are suggestions that the progression of neurologic disease may be arrested (Tsai *et al.*, 1992; Ringdén *et al.*, 1988; A. Erikson *et al.*, 1990). In one well-documented case, enzyme reconstitution to expected levels was achieved in visceral organs, but no increase in brain enzyme levels was found after 2½ years of full engraftment from an enzymatically normal sibling (Tsai *et al.*, 1992) (Table XI). The kinetics of cellular replacement were documented in liver and bone marrow of this patient. About 9 months was required to normalize the Kupffer cell population, whereas over 18 months was required for bone marrow normalization. This patient was uneventfully and fully engrafted for over 2 years. At autopsy of this patient, who died of pneumococcal sepsis, a few pockets of Gaucher cells were found in bone marrow, liver, lung, and lymph nodes. These results indicate a prolonged life span of tissue Gaucher cells.

The risks of the procedure impacts on the decision for bone marrow trans-plantation therapy. When successful, however, it provides the potential for perma-nent cure. In the short term, transplantation costs are less than those of enzyme augmentation, but the chronic care costs of complications may be very high. Recommendation of transplantation in type 1 patients is difficult due to the high (10%) mortality of the procedure, from long-term adverse effects (Sillivan and Reid, 1991). Indeed, the procedural risk is even higher in patients who most need treatment, i.e., those with advanced organ dysfunction. With the availability of safe enzyme therapy, there is likely little, if any, role for bone marrow transplanta-

TABLE XI.   Tissue Levels of Acid β-Glucosidase
in Normals and Gaucher Disease Patients[a]

Source	Liver	Lung	Lymph node	Brain
BMT Patient				
(4MUGlc)	35	14	19	1
(GC)	30	31	22	9
Normal ($n = 11$)				
(4MUGlc)	84 ± 25	23 ± 9	31 ± 11	61 ± 21
(GC)	71 ± 18	56 ± 12	37 ± 9	69 ± 25
Gaucher disease ($n = 5$)				
(4MUGlc)	12 ± 5	0.6 ± 0.5	4 ± 2	2 ± 1
(GC)	14 ± 7	6 ± 4	3 ± 2	11 ± 7

[a]Tissue samples were obtained at autopsy of normals between 2 and 24 hr of death. 4MUGlc and GC refer to the synthetic and glucosylceramide substrates, respec-tively. Levels are given as nmole/hr per mg protein. From Tsai *et al.* (1992).

tion in the treatment of Gaucher disease type 1. This modality should be systematically evaluated in type 3 disease, particularly since it is unknown if the neurologic disease could be prevented or arrested by the intravenous administration of an enzyme that is not known to cross the blood–brain barrier.

Gene Transfer

Since Gaucher disease type 1 can be cured by bone marrow transplantation, correction of the molecular defect in the patient's own hematopoietic stem cells by gene transfer should also prove effective. For such gene therapy to be effective, however, several special features of Gaucher disease must be considered. Acid β-glucosidase is an intracellular membrane protein which is not normally present or functional in the extracellular space. As such, cross correction between cells is not observed. Thus, genetic reconstitution would be required in 100%, or at least a large percentage, of bone marrow stem cells to avoid pathologic cellular mosaicism in tissues. Gaucher cells were demonstrated in liver, bone marrow, lungs, and lymph nodes of a patient receiving bone marrow transplantation (Tsai et al., 1992). This patient was fully engrafted with her homozygous normal brother's cells for 2½ years. Consequently, cellular replacement via turnover is the mechanism of therapy. These results also suggest that if only a portion of stem cells are corrected, Gaucher cells would develop in tissues at some level of macrophage chimerism and could have pathophysiologic effects. In addition, the persistence of Gaucher cells in the presence of a normal plasma glucosylceramide level for over 2 years in a bone marrow-transplanted patient indicates that efforts to directly lower plasma substrate levels will likely not decrease intracellular storage by a "diffusion" mechanism. Also, stem cells that produce active acid β-glucosidase do not have any obvious proliferative advantage over those that do not. Thus, therapeutic effects would be expected only if the patient's untransformed cells were ablated by chemotherapy or irradiation. Rational strategy for the treatment of Gaucher disease would be marrow ablation followed by autologous transplantation with transformed hematopoietic stem cells, and considerable effort has been expended to develop the required efficient gene transfer technology (Fink et al., 1990; Nolta et al., 1992; Correl et al., 1992; Ohashi et al., 1992).

The transfer of stable, functional acid β-glucosidase into cultured fibroblasts and transformed lymphoblasts using a retroviral vector was readily accomplished (Sorge et al., 1987), but transfer into hematopoietic stem cells has been difficult (Fink et al., 1990; Nolta et al., 1992; Correl et al., 1992). Recently, high-efficiency transfer of the human acid β-glucosidase cDNA into murine and human hematopoietic stem cells or progenitors has been accomplished (Kohn et al., 1991; Nolta *et al.*, 1992; Fink *et al.*, 1990; Ohashi *et al.*, 1992) with evidence of sustained gene expression. As gene therapy is developed for Gaucher disease and as the use of enzyme therapy is optimized, difficult decisions of "cure" with unknown risks and "treatment" will confront the treating physicians.

DIAGNOSIS

Morphological Diagnosis

The diagnosis of Gaucher disease has been most frequently suspected by finding Gaucher cells in tissues. Similar cells, pseudo-Gaucher cells, have been described in a variety of other disorders, including chronic granulocytic leukemia (Kattlove *et al.*, 1969; Keyserlingk *et al.*, 1972; Dosik *et al.*, 1972), thalassemia (Parkin and Brunning, 1982), multiple myeloma (Scullin *et al.*, 1979), Hodgkin disease (Zidar *et al.*, 1987; K. S. Lee *et al.*, 1982), plasmacytoid lymphomas (Parkin and Brunning, 1982), and in the acquired immunodeficiency syndrome (AIDS) in patients with *Mycobacterium avium* infections (Solis *et al.*, 1986). These "pseudo-Gaucher" cells do not contain the typical tubular structures of authentic Gaucher cells (R. E. Lee, 1982; Parkin and Brunning, 1982). Because of the need for an invasive procedure and its nonspecificity, histologic diagnosis of Gaucher disease may have selective use in monitoring the efficacy and progress in certain patients undergoing therapy for Gaucher disease.

Enzymatic Diagnosis

Numerous assay procedures have been developed to establish the diagnosis of Gaucher disease. Determinations of acid β-glucosidase activity levels in peripheral blood leukocytes, cultured skin fibroblasts, amniotic fluid cells, and chorionic villae are readily accomplished using the glucosylceramide or 4-methylumbelliferyl-β-D-glucopyranoside substrates. The absolute level of enzyme activity varies with the assay conditions and the type of substrate. A finding of <15% of mean normal activity is diagnostic of the disease [see Grabowski *et al.* (1990) for review]. It has been problematic to develop an *in vitro* assay system which mimics the native environment of the enzyme in the cell. Consequently, the determined *in vitro* acid β-glucosidase activity does not accurately reflect the available hydrolytic capacity toward glucosylceramide. As a result, the *in vitro* level of enzyme activity has not consistently differentiated the neuronopathic (types 2 and 3) and nonneuronopathic (type 1) variants.

In heterozygotes peripheral blood leukocytes and cultured skin fibroblasts have about one-half normal β-glucosidase activity. Most screening assays have used peripheral blood leukocyte components as the enzyme source. There is considerable overlap (up to 20%), however, between the normal and heterozygote ranges. This results in a large number of false positive and false negative results. Although many methods have been used to improve the accuracy of discrimination [see Grabowski *et al.* (1990) for review], a substantial overlap range remains. To establish 95% confidence ranges for inclusion of all obligate heterozygotes, about 25–30% of normal controls will be in the heterozygote range. Thus, enzyme assays for heterozygote detection are not suitable for mass screening and should be used with caution in at-risk families where there is an *a priori* probability of carrying a Gaucher disease gene.

Diagnosis by DNA Analysis

DNA-based technology has significant advantages over enzymatic diagnosis of Gaucher disease since the results are qualitative and the samples are very stable. Many Gaucher disease alleles have been characterized (Fig. 4, Table IV), but others have yet to be identified. As with all genetically heterogeneous traits, inability to detect a mutant allele does not exclude its presence. In the Ashkenazi Jewish population, the mutations at cDNA N370S, 84GG, L444P, and IVS2^{+1} account for 89–95% of the Gaucher disease alleles. Thus, Gaucher disease can be confirmed by DNA analysis in a high percentage of Jewish patients and heterozygote screening in this population is feasible by DNA analysis. Among non-Jewish patients, the mutations L444P and N370S account for 60–75% of the disease-producing alleles. In the non-Jewish populations the presence of many rare or private alleles decreases the utility of DNA testing for diagnosis and heterozygote screening. A variety of PCR-based techniques for facile detection of mutations have been described (Theophilus *et al.*, 1989a,b; Beutler *et al.*, 1990, 1991a; Eyal *et al.*, 1990; Firon *et al.*, 1990) and the amplification refractory mutation system (ARMS) has also been successfully employed (Mistry *et al.*, 1992).

CLINICAL GENETICS AND PREVENTION

Prenatal Diagnosis

Acid β-glucosidase activity is readily measured in amniotic fluid cells and chorionic villus samples and thus prenatal diagnosis is routine for all types of Gaucher disease (Schneider *et al.*, 1972; Svennerholm *et al.*, 1981; Heilbronner *et al.*, 1981). More recently, DNA-based technology has provided for identification of the genotype of the fetus where disease alleles are known. With the use of PCR, an ultrasensitive technique, one must be particularly concerned about contamination of fetal DNA with maternal DNA. However, in the absence of preferential amplification, maternal contamination, if small, should not present major difficulties.

Genetic Counseling

When a couple already has a child with Gaucher disease and both are heterozygotes, the recurrence risk, as in other autosomal recessive disorders, is 1 in 4. Because homozygotes for the N370S mutation often have very mild disease, there are many instances in which one of the parents is a homozygote. Indeed, this happens sufficiently frequently that Gaucher disease was once regarded as being transmitted as an autosomal dominant disorder (Greon, 1948). When one of the parents is a homozygote and the other heterozygous, the recurrence risk is 1 in 2.

Prediction of disease severity in an at-risk family is more straightforward if there is already an affected individual. There is a considerable concordance of disease involvement between sibs, but occasionally substantial differences can exist. When a child is at risk for L444P homozygosity the prognosis is very grave. The likelihood of eventual, if not early, neurologic involvement is high and in any case severe visceral and skeletal disease is to be expected. Counseling may be more difficult when one of the less common alleles is present.

In Gaucher disease type 1, genetic counseling and provision of medical care to affected patients has been particularly difficult due to the marked interfamilial phenotypic variation of the disease types (Beutler and Grabowski, 1993). With the advent of effective therapies, the genetic counseling and provision of care to families with Gaucher disease has become even more time consuming and difficult. In an effort to develop prognostic indicators, five objective parameters of Gaucher disease involvement (age of onset, bony severity, liver and splenic enlargement, and age at splenectomy) were determined and correlated with the genotypes in 161 symptomatic type 1 patients (Sibille *et al.*, 1993) (Table IX). These results demonstrate that symptomatic patients with genotypes other than N370S homozygosity will develop much more severe disease of their bone, liver, and spleen at a much earlier age. Based on these results in Gaucher disease type 1, genetic counseling and provision of medical care to at-risk families must now include informed discussions of fetal involvement and damage as well as the probability of future involvement by the disease process and the potential of early intervention therapy.

In spite of these clear differences in the clinical expression of different genotypes, there is tremendous intragroup variation. Some patients with N370S homozygosity manifest moderately severe disease and some patients with the N370S/"other mutant allele" genotypes have moderate disease (Beutler *et al.*, 1992a; Sidransky *et al.*, 1992; Sibille *et al.*, 1993). There are currently insufficient data to be certain how much of this variation is interfamilial, how much is intrafamilial, and how much may be due to modifer genes or environmental factors.

Clearly, the above discussion is predicated on the delivery of informed genetic counseling and genetic care to patients affected with Gaucher disease. The use of genotype/phenotype correlation data cannot be in a vacuum. They must be presented to patients and affected families in a highly controlled, carefully designed infrastructure, in centers staffed by knowledgeable, experienced medical genetic personnel. The lack of appropriate genetic infrastructure and treatment facilities is a disservice to the patients. In addition, knowledgeable physicians in the area of Gaucher disease must be available since broad variation in phenotypic manifestations, present even within the genotypes, must be addressed when providing care and counseling to affected patients and their families. Furthermore, as the base of well-studied patients expands, exceptions to the above genotype/phenotype correlations are likely to be detected. The rate of such exceptions will have significant impact on the information provided to at-risk and affected families. As a prototype for genetic susceptibility counseling, the complex personal and psychosocial issues will need to be addressed and require a restructuring of current genetic care delivery to accommodate such. The implications for the expanding list of other life-threatening or chronic, debilitating inherited diseases, e.g., cystic fibrosis, for which gene-based therapies and genotype/phenotype correlations will be developed necessitates the expansion and reeducation of genetic care providers.

REFERENCES

Achord, D. T., Brot, F. E., Bell, C. E., and Sly, W. S., 1978, Human β-glucuronidase: *In vivo* clearance and *in vitro* uptake by a glycoprotein recognition system on reticuloendothelial cells, *Cell* **15**:269–278.

Adachi, M., Wallace, B. J., Schneck, L., and Volk, B. W., 1967, Fine structure of central nervous system in early infantile Gaucher's disease, *Arch. Pathol. (Chicago)* **83**:513.

Aerts, J. M., Donker-Koopman, W. E., van Laar, C., Brul, S., Murray, G. J., Wenger, D. A., Barranger, J. A., Tager, J. M., and Schram, A. W., 1987, Relationship between the two immunologically distinguishable forms of glucocerebrosidase in tissue extracts, *Eur. J. Biochem.* **163**:583–589.

Agmon, V., Cherbu, S., Degan, A., Grace, M. E., Grabowski, G. A., and Gatt, S., 1993, Synthesis of novel fluorescent glycosphingolipids: Use in determining acid β-glucosidase activity *in situ* and correlation with genotype in Gaucher disease, in review.

Amstutz, H. C., 1973, The hip in Gaucher's disease, *Clin. Orthop.* **90**:83–89.

Aufses, A. H., Jr., and Salky, B. M., 1982, The surgical management of Gaucher's disease, *Prog. Clin. Biol. Res.* **95**:603–616.

Banker, B. Q., Miller, J. Q., and Crocker, A. C., 1962, The cerebral pathology of infantile Gaucher's disease, in: *Cerebral Sphingolipidoses* (S. M. Aronson and B. M. Volk, eds.), Academic Press, New York, pp. 73–99.

Barneveld, R. A., Keijzer, W., Tegelaers, F. P. W., Ginns, E. I., Geurts van Kessel, A., Brady, R. O., Barranger, J. A., Tager, J. M., Galjaard, H., Westerveld, A., and Reuser, A. J. J., 1983, Assignment of the gene coding for human β-glucocerebrosidase to the region of q21 to 31 of chromosome 1 using monoclonal antibodies, *Hum. Genet.* **64**:227.

Barranger, J. A., Ohashi, T., Hong, C. M., Tomich, J., Aerts, J. F. G. M., Tager, J. M., Nolta, J. A., Sender, L. S., Weiler, S., and Kohn, D. B., 1989, Molecular pathology and therapy of Gaucher disease, *Jpn. J. Inherit. Metab. Dis.* **51**:45–71.

Barton, N. W., Furbish, F. S., Murray, G. J., Garfield, M., and Brady, R. O., 1990, Therapeutic response to intravenous infusions of glucocerebrosidase in a patient with Gaucher disease, *Proc. Natl. Acad. Sci. USA* **87**:1913–1916.

Barton, N. W., Brady, R. O., Dambrosia, J. M., Di Bisceglie, A. M., Doppelt, S. H., Hill, S. C., Mankin, H. J., Murray, G. J., Parker, R. I., and Argoff, C. E., 1991, Replacement therapy for inherited enzyme deficiency—Macrophage-targeted glucocerebrosidase for Gaucher's disease, *N. Engl. J. Med.* **324**:1464–1470.

Basu, A., and Glew, R. H., 1984, Characterization of the phospholipid requirement of a rat liver β-glucosidase, *Biochem. J.* **224**:515–524.

Basu, A., and Glew, R. H., 1985, Characterization of the activation of rat liver β-glucosidase by sialosylgangliotetraosylceramide, *J. Biol. Chem.* **260**:13067–13073.

Basu, S., Kaufman, B., and Roseman, S., 1968, Enzymatic synthesis of ceramide-glucose and ceramide-lactose by glycosyltransferases from embryonic chicken brain, *J. Biol. Chem.* **243**:5802–5807.

Basu, S., Kaufman, B., and Roseman, S., 1973, Enzymatic synthesis of glucocerebroside by a glucosyltransferase from embryonic chicken brain, *J. Biol. Chem.* **248**:1388–1394.

Basu, A., Glew, R. H., Daniels, L. B., and Clark, L. S., 1984, Activators of spleen glucocerebrosidase from controls and patients with various forms of Gaucher's disease, *J. Biol. Chem.* **259**:1714–1719.

Basu, A., Prence, E., Garrett, K., Glew, R. H., and Ellingson, J. S., 1985, Comparison of N-acyl phosphatidylethanolamines with different N-acyl groups as activators of glucocerebrosidase in various forms of Gaucher's disease, *Arch. Biochem. Biophys.* **243**:28–34.

Basu, A., Glew, R. H., Wherrett, J. R., and Huterer, S., 1986, Comparison of the ability of phospholipids from rat liver lysosomes to reconstitute glucocerebrosidase, *Arch. Biochem. Biophys.* **245**:464–469.

Beighton, P., Goldblatt, J., and Sacks, S., 1982, Bone involvement in Gaucher disease, *Prog. Clin. Biol. Res.* **95**:107–129.

Berent, S. L., and Radin, N. S., 1981, β-glucosidase activator protein from bovine spleen ("coglucosidase"), *Arch. Biochem. Biophys.* **208**:248–260.

Berg, A., Grace, M. E., and Grabowski, G. A., 1993, Human acid β-glucosidase: N-Glycosylation site occupancy and role of glycosylation in enzyme activity, in review.

Bergmann, J. E., and Grabowski, G. A., 1989, Posttranslational processing of human lysosomal acid β-glucosidase: A continuum of defects of Gaucher disease type 1 and type 1 fibroblasts, *Am. J. Hum. Genet.* **44**:741–750.

Berrebi, A., Wishnitzer, R., and Von-der-Walde, U., 1984, Gaucher's disease: Unexpected diagnosis in three patients over seventy years old, *Nouv. Rev. Fr. Hematol.* **26**:201–203.

Berrebi, A., Malnick, S. D. H., Vorst, E. J., and Stein, D., 1992, High incidence of factor XI deficiency in Gaucher's disease, *Am. J. Hematol.* **40**:153.

Beutler, E., 1977, Gaucher's disease in an asymptomatic 72-year-old, *J. Am. Med. Assoc.* **237**:2529.

Beutler, E., and Dale, G. L., 1982, Gaucher disease: A century of delineation and research. Enzyme replacement therapy: Model and clinical studies, *Prog. Clin. Biol. Res.* **95**:703–716.

Beutler, E., and Gelbart, T., 1990, Gaucher disease associated with a unique *KpnI* restriction site: Identification of the amino-acid substitution, *Ann. Hum. Genet.* **54**:149–153.

Beutler, E., and Grabowski, G. A., 1993, Glucosylceramide lipidoses: Gaucher disease, in: *The Metabolic Basis of Inherited Disease*, 7th ed. (C. R. Scriver, A. L. Beaudet, W. S. Sly, and D. Valle, eds.), McGraw-Hill, New York, in press.

Beutler, E., and Kuhl, W., 1986, Glucocerebrosidase processing in normal fibroblasts and in fibroblasts from patients with type I, type II, and type III Gaucher disease, *Proc. Natl. Acad. Sci. USA* **83**: 7472–7474.

Beutler, E., Dale, G. L., Guinto, D. E., and Kuhl, W., 1977, Enzyme replacement therapy in Gaucher's disease: Preliminary clinical trial of a new enzyme preparation, *Proc. Natl. Acad. Sci. USA* **74**: 4620–4623.

Beutler, E., Dale, G. L., and Kuhl, W., 1980, Replacement therapy in Gaucher disease, *Birth Defects* **16**:369–381.

Beutler, E., Gelbart, T., and West, C., 1990, The facile detection of the nt 1226 mutation of glucocerebrosidase by 'mismatched' PCR, *Clin. Chim. Acta* **194**:161–166.

Beutler, E., Gelbart, T., Kuhl, W., Sorge, J., and West, C., 1991, Identification of the second common Jewish Gaucher disease mutation makes possible population-based screening for the heterozygous state, *Proc. Natl. Acad. Sci. USA* **88**:10544–10547.

Beutler, E., Kay, A., Saven, A., Garver, P., Thurston, D., Dawson, A., and Rosenbloom, B., 1991b, Enzyme replacement therapy for Gaucher disease, *Blood* **78**:1183–1189.

Beutler, E., Gelbart, T., Kuhl, W., Zimran, A., and West, C., 1992a, Mutations in Jewish patients with Gaucher disease, *Blood* **79**:1662–1666.

Beutler, E., West, C., and Gelbart, T., 1992b, Polymorphisms in the human glucocerebrosidase gene, *Genomics* **12**:795–800.

Bildman, B., Martinez, M. Jr., and Robinson, L. H., 1972, Gaucher's disease discovered by mandibular biopsy: Report of case, *J. Oral Surg.* **30**:510–512.

Billings, A. A., Post, M., and Shapiro, C. M., 1973, Febrile reaction of Gaucher's disease, *Ill. Med. J.* **144**:222–226.

Bordo, D., and Argos, P., 1991, Suggestions for "safe" residue substitutions in site-directed mutagenesis, *J. Mol. Biol.* **217**:721–729.

Brady, R. O., Kanfer, J. N., and Shapiro, D., 1965, Metabolism of glucocerebrosides. II. Evidence of an enzymatic deficiency in Gaucher's disease, *Biochem. Biophys. Res. Commun.* **18**:221–225.

Brady, R. O., Pentchev, P. G., Gal, A. E., Hibbert, S. R., and Dekaban, A. S., 1974, Replacement therapy for inherited enzyme deficiency. Use of purified glucocerebrosidase in Gaucher's disease, *N. Engl. J. Med.* **291**:989–993.

Brady, R. O., Pentchev, P. G., and Gal, A. G., 1975, Investigations in enzyme replacement therapy in lipid storage diseases, *Fed. Proc.* **34**:1310–1315.

Carlson, D. E., Busuttil, R. W., Giudici, T. A., and Barranger, J. A., 1990, Orthotopic liver transplantation in the treatment of complications of type 1 Gaucher disease, *Transplantation* **49**: 1192–1194.

Carter, H. E., and Fujino, Y., 1956, Biochemistry of the sphingolipids: IX. Configuration of the cerebrosidases, *J. Biol. Chem.* **21**:879.

Choudary, P. V., Barranger, J. A., Tsuji, S., Mayor, J., LaMarca, M. E., Cepko, C. L., Mulligan, R. C., and Ginns, E. I., 1986, Retrovirus-mediated transfer of the human glucocerebrosidase gene to Gaucher fibroblasts, *Mol. Biol. Med.* **3**:293–299.

Choy, F. Y., Woo, M., and Der Kaloustian, V. M., 1991, Molecular analysis of Gaucher disease: Screening of patients in the Montreal/Quebec region, *Am. J. Med. Genet.* **41**:469–474.

Christianson, O. O., 1941, Experimental lesions produced by cerebrosides, *Arch. Pathol. (Chicago)* **32**:369.

Christomanou, H., Kleinschmidt, T., and Braunitzer, G., 1987, N-terminal amino-acid sequence of a sphingolipid activator protein missing in a new human Gaucher disease variant, *Biol. Chem. Hoppe Seyler* **368**:1193–1196.

Christomanou, H., Chabás, A., Pämpols, T., and Guardiola, A., 1989, Activator protein deficient Gaucher's disease. A second patient with the newly identified lipid storage disorder, *Klin. Wochenschr.* **67**:999–1003.

Collard, M. W., Sylvester, S. R., Tsuruta, J. K., and Griswold, M. D., 1988, Biosynthesis and molecular cloning of sulfated glycoprotein 1 secreted by rat sertoli cells: Sequence similarity with the 70-kilodalton precursor to sulfatide/GM1 activator, *Biochemistry* **27**:4557–4564.

Conradi, N. G., Sourander, P., Nilsson, O., Svennerholm, L., and Erikson, A., 1984, Neuropathology of the Norrbottnian type of Gaucher disease. Morphological and biochemical studies, *Acta Neuropathol.* (Berl.) **65**:99–109.

Conradi, N. G., Kalimo, H., and Sourander, P., 1988, Reactions of vessel walls and brain parenchyma to the accumulation of Gaucher cells in the Norrbottnian type (type III) of Gaucher disease, *Acta Neuropathol.* (Berl.) **75**:385–390.

Conradi, N., Kyllerman, M., Månsson, J. E., Percy, A. K., and Svennerholm, L., 1991, Late-infantile Gaucher disease in a child with myoclonus and bulbar signs: Neuropathological and neurochemical findings, *Acta Neuropathol.* (Berl.) **82**:152–157.

Conzelmann, E., and Sandhoff, K., 1983, Partial enzyme deficiencies: Residual activities and the development of neurological disorders, *Dev. Neurosci.* **6**:58–71.

Correl, P. H., Colilla, S., Dave, H. P. G., and Karlsson, S., 1992, High levels of human glucocerebrosidase activity in macrophages of long-term reconstituted mice after retroviral infection of hematopoietic stem cells, *Blood* **80**:331–336.

Coste, H., Martel, M. B., Azzar, G., and Got, R., 1985, UDP-glucoseceramide glucosyltransferase from porcine submaxillary glands is associated with the Golgi apparatus, *Biochim. Biophys. Acta* **814**:1–7.

Coste, H., Martel, M. B., and Got, R., 1986, Topology of glucosylceramide synthesis in Golgi membranes from porcine submaxillary glands, *Biochim. Biophys. Acta* **858**:6–12.

Cremin, B. J., Davey, H., and Goldblatt, J., 1990, Skeletal complications of type I Gaucher disease: The magnetic resonance features, *Clin. Radiol.* **41**:244–247.

Dahl, N., Lagerström, M., Erickson, A., and Pettersson, U., 1990, Gaucher disease type III (Norrbottnian type) is caused by a single mutation in exon 10 of the glucocerebrosidase gene, *Am. J. Hum. Genet.* **47**:275–278.

Dale, G. L., Villacorte, D. G., and Beutler, E., 1976, Solubilization of glucocerebrosidase from human placenta and demonstration of a phospholipid requirement for its catalytic activity, *Biochem. Biophys. Res. Commun.* **71**:1048–1053.

Datta, S. C., and Radin, N. S., 1988, Normalization of liver glucosylceramide levels in the "Gaucher" mouse by phosphatidylserine injection, *Biochem. Biophys. Res. Commun.* **152**:155–160.

Dawson, G., and Ellory, J. C., 1985, Functional lysosomal hydrolase size as determined by radiation inactivation analysis, *Biochem. J.* **226**:283–288.

Dawson, G., and Oh, J. Y., 1977, Blood glucosylceramide levels in Gaucher's disease and its distribution amongst lipoprotein fractions, *Clin. Chim. Acta* **75**:149–153.

D'Azzo, A., Proia, R., Kolodny, E. H., Kaback, M., and Neufeld, E. F., 1984, Faulty association of α and β subunits in some forms of β-hexosaminidase A deficiency, *J. Biol. Chem.* **259**:11070–11074.

DeGasperi, R., Alroy, J., Richard, R., Goyal, V., Orgad, U., Lee, R. E., and Warren, C. D., 1990, Glycoprotein storage in Gaucher disease: Lectin histochemistry and biochemical studies, *Lab. Invest.* **63**:385–392.

Desnick, R. J., 1982, Gaucher disease: A century of delineation and understanding, *Prog. Clin. Biol. Res.* **95**:1–30.

Desnick, R. J., and Grabowski, G. A., 1981, Advances in the treatment of inherited metabolic diseases, *Ann. Hum. Genet.* **11**:281–369.

Desnick, R. J., Thorpe, S. R., and Fiddler, M. B., 1976, Toward enzyme therapy, *Physiol. Rev.* **56**: 57–99.

Devine, E. A., Smith, M., Arrendondo-Vega, F., Shafit-Zagardo, B., and Desnick, R. J., 1982, Regional assignment of the structural gene for human acid β-glucoside to q42→qtr on chromosome 1, *Cytogenet. Cell Genet.* **33**:340–344.

Dewji, N. N., Wenger, D. A., and O'Brien, J. S., 1987, Nucleotide sequence of cloned cDNA from human sphingolipid activator protein 1 precursor, *Proc. Natl. Acad. Sci. USA* **84**:8652–8656.

Dinur, T., Osiecki, K. M., Legler, G., Gatt, S., Desnick, R. J., and Grabowski, G. A., 1986, Human acid β-glucosidase: Isolation and amino acid sequence of a peptide containing the catalytic site, *Proc. Natl. Acad. Sci. USA* **83**:1660–1664.

Dosik, H., Rosner, F., and Sawitsky, A., 1972, Acquired lipidosis: Gaucher-like cells and "blue cells" in chronic granulocytic leukemia, *Semin. Hematol.* **9**:309–316.

Draznin, S. Z., and Singer, K., 1948, Legg-Perthes' disease: A syndrome of many etiologies? With clinical and roentgenographic findings in a case of Gaucher's disease, *Am. J. Roentgenol.* **60**:490.

Dreborg, S., Erikson, A., and Hagberg, B., 1980, Gaucher disease—Norrbottnian type. I. General clinical description, *Eur. J. Pediatr.* **133**:107–118.

Dynan, W. S., 1986, Promoters for housekeeping genes, *Trends Genet.* **2**:196–197.

Elleder, M., and Jiräsek, A., 1981, Histochemical and ultrastructural study of Gaucher cells, *Acta Neuropathol. Suppl.* (Berl.) **7**:208–210.

Erikson, A., 1986, Gaucher disease—Norrbottnian type (III). Neuropaediatric and neurobiological aspects of clinical patterns and treatment, *Acta Paediatr. Scand. Suppl.* **326**:1–42.

Erikson, A., and Wahlberg, I., 1985, Gaucher disease—Norrbottnian type. Ocular abnormalities, *Acta Ophthalmol.* (Copenh.) **63**:221–225.

Erikson, A., Karlberg, J., Skogman, A. L., and Dreborg, S., 1987, Gaucher disease (type III): Intellectual profile *Pediatr. Neurol.* **3**:87–91.

Erikson, A., Groth, C. G., Månsson, J. E., Percy, A., Ringdën, O., and Svennerholm, L., 1990, Clinical and biochemical outcome of marrow transplantation for Gaucher disease of the Norrbottnian type, *Acta Paediatr. Scand.* **79**:680–685.

Erickson, A. H., Ginns, E. I., and Barranger, J. A., 1985, Biosynthesis of the lysosomal enzyme glucocerebrosidase, *J. Biol. Chem.* **260**:14319–14324.

Erickson, J. S., and Radin, N. S., 1973, N-Hexyl-O-glucosyl sphingosine, an inhibitor of glucosyl ceramide β-glucosidase, *J. Lipid Res.* **14**:133–137.

Espinas, O. E., and Faris, A. A., 1969, Acute infantile Gaucher's disease in identical twins: An account of clinical and neuropathologic observations, *Neurology* **19**:133.

Eyal, N., Wilder, S., and Horowitz, M., 1990, Prevalent and rare mutations among Gaucher patients, *Gene* **96**:277–283.

Eyal, N., Firon, N., Wilder, S., Kolodny, E. H., and Horowitz, M., 1991, Three unique base pair changes in a family with Gaucher disease, *Hum. Genet.* **87**:328–332.

Fabbro, D., and Grabowski, G. A., 1991, Human acid β-glucosidase: Use of inhibitory and activating monoclonal antibodies to investigate the enzyme's catalytic mechanism and saposin A and C binding sites, *J. Biol. Chem.* **266**:15021–15027.

Fabbro, D., Desnick, R. J., and Grabowski, G. A., 1987, Gaucher disease: Genetic heterogeneity within and among the subtypes detected by immunoblotting, *Am. J. Hum. Genet.* **40**:15–31.

Fallet, S., Grace, M. E., Sibille, A., Mendelson, D. S., Shapiro, R. S., Hermann, G., and Grabowski, G. A., 1992, Enzyme augmentation in moderate to life-threatening Gaucher disease, *Pediatr. Res.* **31**:496–502.

Fellows, K. E., Grand, R. J., Colodny, A. H., Orsini, E. N., and Crocker, A. C., 1975, Combined portal and vena caval hypertension in Gaucher disease: The value of preoperative venography, *J. Pediatr.* **87**:739–743.

Figueroa, M. L., Rosenbloom, B. E., Kay, A., Garver, P., Thurston, D. W., Koziol, J. A., Gelbart, T., and Beutler, E., 1992, A less costly regimen of alglucerase to treat Gaucher's disease, *N. Engl. J. Med.* **327**:1632–1636.

Fink, J. K., Correll, P. H., Perry, L. K., Brady, R. O., and Karlsson, S., 1990, Correction of glucocerebrosidase deficiency after retroviral-mediated gene transfer into hematopoietic progenitor cells from patients with Gaucher disease, *Proc. Natl. Acad. Sci. USA* **87**:2334–2338.

Firon, N., Eyal, N., Kolodny, E. H., and Horowitz, M., 1990, Genotype assignment in Gaucher disease by selective amplification of the active glucocerebrosidase gene, *Am. J. Hum. Genet.* **46**: 527–532.

Frederickson, D. S., and Sloan, H. R., 1978, Glucosyl ceramide lipidoses: Gaucher's disease, in: *The Metabolic Basis of Inherited Disease*, 3rd ed. (J. B. Stanbury, J. B. Wyngaarden, and D. S. Frederickson, eds.), McGraw-Hill, New York, pp. 730–759.

French, J. H., Brotz, M., and Poser, C. M., 1969, Lipid composition of the brain in infantile Gaucher's disease, *Neurology* **19**:81–86.

Fujibayashi, S., and Wenger, D. A., 1986a, Synthesis and processing of sphingolipid activator protein-2 in cultured human fibroblasts, *J. Biol. Chem.* **261**:15339–15343.

Fujibayashi, S., and Wenger, D. A., 1986b, Biosynthesis of the sulfatide/G_{M1} activator protein in control and mutant cultured skin fibroblasts, *Biochim. Biophys. Acta* **875**:554–562.

Fujibayashi, S., Kao, R.-T., Jones, C., Morse, H., Law, M., and Wenger, D. A., 1985, Assignment of the gene for human sphingolipid activator protein-2 (SAP-2) to chromosome 10, *Am. J. Hum. Genet.* **37**:741–748.

Furbish, F. S., Blair, H. E., Shiloach, J., Pentchev, P. G., and Brady, R. O., 1977, Enzyme replacement therapy in Gaucher's disease: Large-scale purification of glucocerebrosidase suitable for human administration, *Proc. Natl. Acad. Sci. USA* **74**:3560–3563.

Furbish, F. S., Steer, C. J., Krett, N. L., and Barranger, J. A., 1981, Uptake and distribution of placental glucocerebrosidase in rat hepatic cells and effects of sequential deglycosylation, *Biochim. Biophys. Acta* **673**:425–435.

Garret, K. O., Prence, E. M., and Glew, R. H., 1985, Sucrose gradient analysis of phospholipid-activated β-glucosidase in type 1 and type 2 Gaucher's disease, *Arch. Biochem. Biophys.* **238**: 344–352.

Gatt, S., 1966, Enzymatic hydrolysis of sphingolipids. I. Hydrolysis and synthesis of ceramides by an enzyme from rat brain, *J. Biol. Chem.* **24**:2724.

Gatt, S., Dinur, T., Osiecki, K., Desnick, R. J., and Grabowski, G. A., 1985, Use of activators and inhibitors to define the properties of the active site of normal and Gaucher disease lysosomal β-glucosidase, *Enzyme* **33**:109–119.

Gaucher, P. C. E., 1882, De l'epithelioma primitif de la rate, hypertrophie idiopathique del la rate sans leucemie, M.D. Thesis, Paris.

Gavrieli-Rorman, E., and Grabowski, G. A., 1989, Molecular cloning of a human co-β-glucosidase cDNA: Evidence the four sphingolipid hydrolase activator proteins are encoded by single genes in humans and rats, *Genomics* **5**:486–492.

Gebler, J. C., Aebersold, R., and Withers, S. G., 1992, Glu-537, not Glu-461, is the nucleophile in the active site of (lacZ) β-galactosidase from *Escherichia coli*, *J. Biol. Chem.* **267**:11126–11130.

Gery, I., Zigler, J. S., Jr., Brady, R. O., and Barranger, J. A., 1981, Selective effects of glucocerebroside (Gaucher's storage material) on macrophage cultures, *J. Clin. Invest.* **68**:1182–1187.

Ginns, E. I., Brady, R. O., Pirruccello, S., Moore, C., Sorrell, S., Furbish, F. S., Murray, G. J., Tager, J., and Barranger, J. A., 1982, Mutations of glucocerebrosidase: Discrimination of neurologic and non-neurologic phenotypes of Gaucher disease, *Proc. Natl. Acad. Sci. USA* **79**:5607–5610.

Ginns, E. I., Choudary, P. V., Tsuji, S., Martin, B., Stubblefield, B., Sawyer, J., Hozier, J., and

Barranger, J. A., 1985, Gene mapping and leader polypeptide sequence of human gluco-cerebrosidase: Implications for Gaucher disease, *Proc. Natl. Acad. Sci. USA* **82**:7101–7105.

Glew, R. H., Daniels, L. B., Clark, L. S., and Hoyer, S. W., 1982, Enzymic differentiation of neurologic and nonneurologic forms of Gaucher's disease, *J. Neuropathol. Exp. Neurol.* **41**: 630–641.

Glew, R. H., Basu, A., LaMarco, K. L., and Prence, E. M., 1988, Mammalian glucocerebrosidase: Implications for Gaucher's disease, *Lab. Invest.* **58**:5–25.

Goldblatt, J., and Beighton, P., 1982, South African variants of Gaucher disease, *Prog. Clin. Biol. Res.* **95**:95–106.

Goldblatt, J., Sacks, S., and Beighton, P., 1978, The orthopedic aspects of Gaucher disease, *Clin. Orthop.* **137**:208–214.

Goldblatt, J., Sacks, S., Dall, D., and Beighton, P., 1988, Total hip arthroplasty in Gaucher's disease. Long-term prognosis, *Clin. Orthop.* **228**:94–98.

Gonzales, M. L., Basu, A., de Haas, G. H., Dijkman, R., van Oort, M. G., Okolo, A. A., and Glew, R. H., 1988, Activation of human spleen glucocerebrosidases by monoacylglycol sulfates and diacylglycerol sulfates, *Arch. Biochem. Biophys.* **262**:345–353.

Grabowski, G. A., and Dagan, A., 1984, Human lysosomal β-glucosidase: Purification by affinity chromatography, *Analyt. Biochem.* **141**:267–279.

Grabowski, G. A., Gatt, S., Kruse, J., and Desnick, R. J., 1984, Human lysosomal β-glucosidase: Kinetic characterization of the catalytic, aglycon, and hydrophobic binding sites, *Arch. Biochem. Biophys.* **231**:144–157.

Grabowski, G. A., Dinur, T., Osiecki, K. M., Kruse, J. R., Legler, G., and Gatt, S., 1985, Gaucher disease types 1, 2, and 3: Differential mutations of the acid β-glucosidase active site identified with conduritol B epoxide derivatives and sphingosine, *Am. J. Hum. Genet.* **37**:499–510.

Grabowski, G. A., Osiecki-Newman, K., Dinur, T., Fabbro, D., Legler, G., Gatt, S., and Desnick, R. J., 1986, Human acid β-glucosidase. Use of conduritol B epoxide derivatives to investigate the catalytically active normal and Gaucher disease enzymes, *J. Biol. Chem.* **261**:823–8269.

Grabowski, G. A., Graves, P. G., Grace, M., Bergmann, J. E., and Smith, F. I., 1988, Gaucher disease: Enzymatic and molecular studies, in: *Lipid Storage Disorders: Biological and Medical Aspects* (R. Salvayre, L. Douste-Blazy, and S. Gatt, eds.), Plenum Press, New York, pp. 793–803.

Grabowski, G. A., White, W. R., and Grace, M. E., 1989, Expression of functional human acid β-glucosidase in COS-1 and *Spodoptera frugiperda* cells, *Enzyme* **41**:131–142.

Grabowski, G. A., Gatt, S., and Horowitz, M., 1990, Acid β-glucosidase: Enzymology and molecular biology of Gaucher disease, *Crit. Rev. Biochem. Mol. Biol.* **25**:385–414.

Grabowski, G. A., Barton, N., Pastores, G., and Brady, R. O., 1993, Efficacy and safety of recombinant alglucerase for the therapy of Gaucher disease type 1, in preparation.

Grace, M. E., and Grabowski, G. A., 1990, Human acid β-glucosidase: Glycosylation is required for catalytic activity, *Biochem. Biophys. Res. Commun.* **168**:771–777.

Grace, M. E., and Grabowski, G. A., 1993, Molecular enzymology of human acid β-glucosidase, in: *The Biochemistry and Molecular Biology of β-Glucosidases* (A. Esen, ed.), ACS, Washington, D.C., in press.

Grace, M. E., Graves, P. N., Smith, F. I., and Grabowski, G. A., 1990, Analyses of catalytic activity and inhibitor binding of human acid β-glucosidase by site-directed mutagenesis. Identification of residues critical to catalysis and evidence for causality of two Ashkenazi Jewish Gaucher disease type 1 mutations, *J. Biol. Chem.* **265**:6827–6835.

Grace, M. E., Berg, A., He, G. S., Goldberg, L., Horowitz, M., and Grabowski, G. A., 1991, Gaucher disease: Heterologous expression of two alleles associated with neuronopathic phenotypes, *Am. J. Hum. Genet.* **49**:646–655.

Grace, M. E., Berg, A., He, G.-S., and Grabowski, G. A., 1992, Molecular enzymology of Gaucher disease, *Pediatr. Res.* **31**:133A.

Grace, M. E., Newman, K. M., Scheinker, V., He, G.-S., Berg, A., and Grabowski, G. A., 1993, Analysis of human acid β-glucosidase by site-directed mutagenesis and heterologous expression, in review.

Graves, P. N., Grabowski, G. A., Ludman, M. D., Palese, P., and Smith, F. I., 1986, Human acid β-glucosidase: Northern blot and S1 nuclease analysis of mRNA from HeLa cells and normal and Gaucher disease fibroblasts, *Am. J. Hum. Genet.* **39**:763–774.

Graves, P. N., Grabowski, G. A., Eisner, R., Palese, P., and Smith, F. I., 1988, Gaucher disease type 1: Cloning and characterization of a cDNA encoding acid β-glucosidase from an Ashkenazi Jewish patient, *DNA* **7**:521–528.

Greenberg, P., Merrill, A. H., Liotta, D. C., and Grabowski, G. A., 1990, Human acid β-glucosidase: Use of sphingosyl and *N*-alkyl-glucosylamine inhibitors to investigate the properties of the active site, *Biochim. Biophys. Acta* **1039**:12–20.

Greon, J., 1948, The hereditary mechanism of Gaucher's disease, *Blood* **3**:1238–1249.

Hannun, Y. A., and Bell, R. M., 1987, Lysosphingolipids inhibit protein kinase C: Implications for the sphingolipidoses, *Science* **235**:670–675.

Haratz, D., Manny, N., and Raz, I., 1990, Autoimmune hemolytic anemia in Gaucher's disease, *Klin. Wochenschr.* **68**:94–95.

Hasilik, A., and Neufeld, E. F., 1984, Biosynthesis of lysosomal enzymes in fibroblasts, *J. Biol. Chem.* **255**:4937–4945.

He, G.-S., and Grabowski, G. A., 1992, Gaucher disease: A G(+1) to A(+1) IVS2 splice donor site mutation causing exon 2 skipping in the acid β-glucosidase mRNA, *Am. J. Hum. Genet.* **51**:810–820.

He, G.-S., Grace, M. E., and Grabowski, G. A., 1993, Gaucher disease: Four rare alleles encoding F213I, P289L, T323I, and R463C in type 1 variants, *Hum. Mutation* **1**:423–427.

Heilbronner, H., Wurster, K. G., and Harzer, K., 1981, Prenatal diagnosis of Gaucher's disease, *Dtsch. Med. Wochenschr.* **106**:652–654 [in German].

Henderson, J. M., Gilinsky, N. H., Lee, E. Y., and Greenwood, M. F., 1991, Gaucher's disease complicated by bleeding esophageal varices and colonic infiltration by Gaucher cells, *Am. J. Gastroenterol.* **86**:346–348.

Hermann, G., Goldblatt, J., and Desnick, R. J., 1984, Kümmell disease: Delayed collapse of the trumatised spine in a patient with Gaucher type 1 disease, *Br. J. Radiol.* **57**:833–835.

Hermann, G., Goldblatt, J., Levy, R. N., Goldsmith, S. J., Desnick, R. J., and Grabowski, G. A., 1986, Gaucher's disease type 1: Assessment of bone involvement by CT and scintigraphy, *AJR (Am. J. Roentgenol.)* **147**:943–948.

Hermann, G., Wagner, L. D., Gendal, E. S., Ragland, R. L., and Ulin, R. I., 1989, Spinal cord compression in type I Gaucher disease, *Radiology* **170**:147–148.

Hess, J. F., Fox, M., Schmid, C., and Shen, C. K., 1983, Molecular evolution of the human adult α-globin-like gene region: Insertion and deletion of *Alu* family repeats and non-*Alu* DNA sequences, *Proc. Natl. Acad. Sci. USA* **80**:5970–5974.

Hibbs, R. G., Ferrans, V. J., Cipriano, P. R., and Tardiff, K. J., 1970, A histochemical and electron microscopic study of Gaucher cells, *Arch. Pathol.* **89**:137–153.

Hill, S. C., Reinig, J. W., Barranger, J. A., Fink, J., and Shawker, T. H., 1986, Gaucher disease: Sonographic appearance of the spleen, *Radiology* **160**:631–634.

Ho, M. W., and O'Brien, J. S., 1971, Gaucher's disease: Deficiency of "acid" β-glucosidase and reconstitution of enzyme activity *in vitro, Proc. Natl. Acad. Sci. USA* **68**:2810–2813.

Holtschmidt, H., Sandhoff, K., Furst, W., Kwon, H.-Y., Schnabel, D., and Suzuki, K., 1991a. The organization of the gene for the human cerebroside sulfate activator protein, *FEBS Lett.* **280**:267–270.

Holtschmidt, H., Sandhoff, K., Kwon, H. Y., Harzer, K., Nakano, T., and Suzuki, K., 1991b, Sulfatide activator protein: Alternative splicing that generates three mRNAs and a newly found mutation responsible for a clinical disease, *J. Biol. Chem.* **266**:7556–7560.

Hong, C. M., Ohashi, T., Yu, X. J., Weiler, S., and Barranger, J. A., 1990, Sequence of two alleles responsible for Gaucher disease, *DNA Cell Biol.* **9**:233–241.

Horowitz, M., Wilder, S., Horowitz, Z., Reiner, O., Gelbart, T., and Beutler, E., 1989, The human glucocerebrosidase gene and pseudogene: Structure and evolution, *Genomics* **4**:87–96.

Hyun, J. C., Misra, R. S., Greenblatt, D., and Radin, N. S., 1975, Synthetic inhibitors of glucocerebroside β-glucosidase, *Arch. Biochem. Biophys.* **166**:382–389.

Imai, K., 1985, Characterization of β-glucosidase as a peripheral enzyme of lysosomal membranes from mouse liver and purification, *J. Biochem.* (Tokyo) **98**:1405–1416.

Inui, K., Kao, R.-T., Fujibayashi, S., Jones, C., Morse, H. G., Law, M. L., and Wenger, D. A., 1985, The gene coding for a sphingolipid activator protein, SAP-2, is on chromosome 10, *Hum. Genet.* **69**:197–200.

Jagadeeswaran, P., Forget, B. G., and Weissman, S. M., 1981, Short interspersed repetitive DNA elements in eucaryotes: Transposable DNA elements generated by reverse transcription of RNA *pol* III transcripts, *Cell* **26**:141–142.

James, S. P., Stromeyer, F. W., Chang, C., and Barranger, J. A., 1981, Liver abnormalities in patients with Gaucher's disease, *Gastroenterology* **80**:126–133.

James, S. P., Stromeyer, F. W., Stowens, D. W., and Barranger, J. A., 1982, Gaucher disease: Hepatic abnormalities in 25 patients, *Prog. Clin. Biol. Res.* **95**:131–142.

Jonsson, L. M. V., Murray, G. J., Sorrell, S. H., Strijland, A., Aerts, J. F. G. M., Ginns, E. I., Barranger, J. A., Tager, J. M., and Schram, A. W., 1987, Biosynthesis and maturation of glucocerebrosidase in Gaucher fibroblasts, *Eur. J. Biochem.* **164**:171–179.

Kaplan, A., Achord, D. T., and Sly, W. S., 1977, Phosphohexosyl components of a lysosomal enzyme are recognized by pinocytosis receptors on human fibroblasts, *Proc. Natl. Acad. Sci. USA* **74**:2026–2030.

Kattlove, H. E., Williams, J. C., Gaynor, E., Spivack, M., Bradley, R. M., and Brady, R. O., 1969, Gaucher cells in chronic myelocytic leukemia: An acquired abnormality, *Blood* **33**:379–390.

Kawame, H., and Eto, Y., 1991, A new glucocerebrosidase-gene missense mutation responsible for neuronopathic Gaucher disease in Japanese patients, *Am. J. Hum. Genet.* **49**:1378–1380.

Kaye, E. M., Ullman, M. D., Wilson, E. R., and Barranger, J. A., 1986, Type 2 and type 3 Gaucher disease: A morphological and biochemical study, *Ann. Neurol.* **30**:223–230.

Kennaway, N. G., and Woolf, L. I., 1968, Splenic lipids in Gaucher's disease, *J. Lipid Res.* **9**:755–765.

Keyserlingk, D. G., Boll, I., and Albrecht, M., 1972, Electron microscopy and cytochemistry of Gaucher cells in chronic granulocytic leukaemia, *Klin. Wochenschr.* **50**:510–516 [in German].

Kishimoto, Y., Hiraiwa, M., and O'Brien, J. S., 1992, Saposins: Structure, function, distribution and molecular genetics, *J. Lipid Res.* **33**:1255–1267.

Kleinschmidt, T., Christomanou, H., and Braunitzer, G., 1987, Complete amino-acid sequence and carbohydrate content of the naturally occurring glucosylceramide activator protein (A1 activator) absent from a new human Gaucher disease variant, *Biol. Chem. Hoppe Seyler* **368**:1571–1578.

Kleinschmidt, T., Christomanou, H., and Braunitzer, G., 1988, Complete amino-acid sequence of the naturally occurring A2 activator protein for enzymic sphingomyelin degradation: Identity to the sulfatide activator protein (SAP-1), *Biol. Chem. Hoppe Seyler* **369**:1361–1365.

Kohn, D. B., Nolta, J. A., Weinthal, J., Bahner, I., Yu, X. J., Lilley, J., and Crooks, G. M., 1991, Toward gene therapy for Gaucher disease, *Hum. Gene Ther.* **2**:101–105.

Kolodny, E. H., Ullman, M. D., Mankin, H. J., Raghavan, S. S., Topol, J., and Sullivan, J. L., 1982, Phenotypic manifestations of Gaucher disease: Clinical features in 48 biochemically verified type 1 patients and comment on type 2 patients, *Prog. Clin. Biol. Res.* **95**:33–65.

Kretz, K. A., Carson, S., Morimoto, S., Kishimoto, Y., Fluharty, A. L., and O'Brien, J. S., 1990, Characterization of a mutation in a family with saposin B deficiency: A glycosylation site defect, *Proc. Natl. Acad. Sci. USA* **87**:2541–2544.

Kyllerman, M., Conradi, N., Mánsson, J. E., Percy, A. K., and Svennerholm, L., 1990, Rapidly progressive type III Gaucher disease: Deterioration following partial splenectomy, *Acta Paediatr. Scand.* **79**:448–453.

Lacey, D. J., and Terplan, K., 1984, Correlating auditory evoked and brainstem histologic abnormalities in infantile Gaucher's disease, *Neurology* **34**:539–541.

Lalegerie, P., Legler, G., and Yon, J. M., 1982, The use of inhibitors in the study of glycosidases, *Biochemie* **64**:977.

Latham, T., Grabowski, G. A., Theophilus, B. D., and Smith, F. I., 1990, Complex alleles of the acid β-glucosidase gene in Gaucher disease, *Am. J. Hum. Genet.* **47**:79–86.

Latham, T. E., Theophilus, B. D., Grabowski, G. A., and Smith, F. I., 1991, Heterogeneity of mutations in the acid β-glucosidase gene of Gaucher disease patients, *DNA Cell Biol.* **10**:15–21.

Lee, K. S., Tobin, M. S., Chen, K. T., Ahmed, F., and Gomez-Leon, G., 1982, Acquired Gaucher's cells in Hodgkin's disease, *Am. J. Med.* **73**:290–294.

Lee, R. E., 1968, The fine structure of the cerebroside occurring in Gaucher's disease, *Proc. Natl. Acad. Sci. USA* **61**:484–489.

Lee, R. E., 1982, The pathology of Gaucher disease, *Prog. Clin. Biol. Res.* **95**:177–217.

Legler, G., 1977, Glucosidases, *Meth. Enzymol.* **46**:368–381.

Legler, G., and Bieberich, E., 1988, Active site directed inhibition of a cytosolic β-glucosidase from calf liver by bromoconduritol B epoxide and bromoconduritol F, *Arch. Biochem. Biophys.* **260**: 437–442.

Legler, G., and Liedtke, H., 1985, Glucosylceramidase from calf spleen. Characterization of its active site with 4-alkylumbelliferyl-β-glucosides and *N*-alkyl derivatives of 1-deoxynojirimycin, *Biol. Chem. Hoppe Seyler* **366**:1113–1122.

Lester, T. J., Grabowski, G. A., Goldblatt, J., Leiderman, I. Z., and Zaroulis, C. G., 1984, Immune thrombocytopenia and Gaucher's disease, *Am. J. Med.* **77**:569–571.

Ludwig, J., 1979, *Current Methods of Autopsy Practice*, 2nd ed., W. B. Saunders, Philadelphia.

Maret, A., Potier, M., Salvayre, R., and Douste-Blazy, L., 1983, Modification of subunit interaction in membrane-bound acid β-glucosidase from Gaucher disease, *FEBS Lett.* **160**:93–97.

Maret, A., Potier, M., Salvayre, R., and Douste-Blazy, L., 1984, Modifications of the molecular weight of membrane-bound nonspecific β-glucosidase in type 1 Gaucher disease determined *in situ* by the radiation inactivation method, *Biochim. Biophys. Acta* **799**:91–94.

Markin, R. S., and Skultety, F. M., 1984, Spinal cord compression secondary to Gaucher's disease, *Surg. Neurol.* **21**:341–346.

Martin, B. M., Tsuji, S., LaMarca, M. E., Maysak, K., Eliason, W., and Ginns, E. I., 1988, Glycosylation and processing of high levels of active human glucocerebrosidase in invertebrate cells using a baculovirus expression vector, *DNA* **7**:99–106.

Masuno, M., Tomatsu, S., Sukegawa, K., and Orii, T., 1990, Non-existence of a tight association between a 444leucine to proline mutation and phenotypes of Gaucher disease: High frequency of a *Nci*I polymorphism in the non-neuronopathic form, *Hum. Genet.* **84**:203–206.

Medoff, A. S., and Bayrd, E. D., 1954, Gaucher's disease in 29 cases: Hematologic complications and effect of splenectomy, *Isr. J. Med. Sci.* **40**:481–492.

Melamed, E., Cohen, C., Soffer, D., and Lavy, S., 1975, Central nervous system complication in a patient with chronic Gaucher's disease, *Eur. Neurol.* **13**:167–175.

Meyerowitz, R., and Proia, R., 1988, cDNA clone for the α-chain of human β-hexosaminidase: Deficiency of α-chain mRNA in Ashkenazi Tay–Sachs fibroblasts, *Proc. Natl. Acad. Sci. USA* **81**: 5394–5398.

Mislow, K., 1953, The geometry of sphingosine, *Am. Chem. Soc.* **74**:155.

Mistry, P. K., Smith, S. J., Ali, M., Hatton, C. S., McIntyre, N., and Cox, T. M., 1992, Genetic diagnosis of Gaucher's disease, *Lancet* **339**:889–892.

Morimoto, S., Martin, B. M., Kishimoto, Y., and O'Brien, J. S., 1988, Saposin D: A sphingomyelinase activator, *Biochem. Biophys. Res. Commun.* **156**:403–410.

Morimoto, S., Martin, B. M., Yamamoto, Y., Kretz, K. A., O'Brien, J. S., and Kishimoto, Y., 1989, Saposin A: Second cerebrosidase activator protein, *Proc. Natl. Acad. Sci. USA* **86**:3389–3393.

Morimoto, S., Yamamoto, Y., O'Brien, J. S., and Kishimoto, Y., 1990, Distribution of saposin proteins (sphingolipid activator proteins) in lysosomal storage and other diseases, *Proc. Natl. Acad. Sci. USA* **87**:3493–3497.

Morrison, J. F., and Walsh, C. T., 1988, The behavior and significance of slow-binding inhibitors, *Adv. Enzymol. Mol. Biol.* **61**:201.

Mueller, O. T., and Rosenberg, A., 1979, Activation of membrane-bound glucosylceramide: β-Glucosidase in fibroblasts cultured from normal and glucosylceramidotic human skin, *J. Biol. Chem.* **254**:3521–3525.

Myers, H. S., Cremin, B. J., Beighton, P., and Sacks, S., 1975, Chronic Gaucher's disease: Radiological findings in 17 South African cases, *Br. J. Radiol.* **48**:465–469.

Nakano, T., Sandhoff, K., Stumper, J., Christomanou, H., and Suzuki, K., 1989, Structure of full-length cDNA coding of sulfatide activator, a co-β-glucosidase and two other homologous proteins: Two alternate forms of the sulfatide activator, *J. Biochem.* **105**:152–154.

Nilsson, O., and Svennerholm, L., 1982a, Characterization and quantitative determination of gangliosides and neutral glycosphingolipids in human liver, *J. Lipid Res.* **23**:327–334.

Nilsson, O., and Svennerholm, L., 1982b, Accumulation of glucosylceramide and glucosylsphingosine (psychosine) in cerebrum and cerebellum in infantile and juvenile Gaucher disease, *J. Neurochem.* **39**:709–718.

Nilsson, O., Håkansson, G., Dreborg, S., Groth, C. G., and Svennerholm, L., 1982, Increased cerebroside concentration in plasma and erythrocytes in Gaucher disease: Significant differences between type I and type III, *Clin. Genet.* **22**:274–279.

Nilsson, O., Grabowski, G. A., Ludman, M. D., Desnick, R. J., and Svennerholm, L., 1985, Glycosphingolipid studies of visceral tissues and brain from type 1 Gaucher disease variants, *Clin. Genet.* **27**:443–450.

Nolta, J. A., Yu, X. J., Bahner, I., and Kohn, D. B., 1992, Retroviral-mediated transfer of the human glucocerebrosidase gene into cultured Gaucher bone marrow, *J. Clin. Invest.* **90**:342–348.

Norman, R. M., Urich, J., and Lloyd, O. C., 1956, The neuropathology of infantile Gaucher's disease, *J. Pathol. Bacteriol.* **72**:121.

O'Brien, J. S., Kretz, K. A., Dewji, N., Wenger, D. A., Esch, F., and Fluharty, A. L., 1988, Coding of two sphingolipid activator proteins (SAP-1 and SAP-2) by same genetic locus, *Science* **241**:1098–1101.

Ohashi, T., Hong, C. M., Weiler, S., Tomich, J. M., Aerts, J. M. F. G., Tager, J. M., and Barranger, J. A., 1991, Characterization of human glucocerebrosidase from different mutant alleles, *J. Biol. Chem.* **266**:3661–3669.

Ohashi, T., Boggs, S., Robbins, P., Bahnson, A., Patrene, K., Wei, F-S., Wei, H-F., Li, J., Lucht, Y., Fei, Y., Clark, S., Kimak, M., He, H., Mowery-Rushton, P., and Barranger, J. A., 1992, Efficient transfer and sustained high expression of the human glucocerebrosidase gene in mice and their functional macrophages following transplantation of bone marrow transduced by a retroviral vector, *Proc. Natl. Acad. Sci. USA* **89**:11332–11336.

O'Neill, R. R., Tokoro, T., Kozak, C. A., and Brady, R. O., 1989, Comparison of the chromosomal localization of murine and human glucocerebrosidase genes and of the deduced amino acid sequences, *Proc. Natl. Acad. Sci. USA* **86**:5049–5053.

Osiecki-Newman, K. M., Fabbro, D., Dinur, T., Boas, S., Gatt, S., Legler, G., Desnick, R. J., and Grabowski, G. A., 1986, Human acid β-glucosidase: Affinity purification of the normal placental and Gaucher disease splenic enzymes on *N*-alkyl-deoxynojirimycin-sepharose, *Enzyme* **35**:147–153.

Osiecki-Newman, K., Fabbro, D., Legler, G., Desnick, R. J., and Grabowski, G. A., 1987, Human acid β-glucosidase: Use of inhibitors, alternative substrates and amphiphiles to investigate the properties of the normal and Gaucher disease active sites, *Biochim. Biophys. Acta* **915**:87–100.

Osiecki-Newman, K., Legler, G., Grace, M., Dinur, T., Gatt, S., Desnick, R. J., and Grabowski, G. A., 1988, Human acid β-glucosidase: Inhibition studies using glucose analogues and pH variation to characterize the normal and Gaucher disease glycon binding sites, *Enzyme* **40**:173–188.

Parkin, J. L., and Brunning, R. D., 1982, Pathology of the Gaucher cell, *Prog. Clin. Biol. Res.* **95**: 151–175.

Pastores, G., Sibille, A., and Grabowski, G. A., 1993, Enzyme therapy in Gaucher disease type 1: Dosage efficacy and adverse effects in thirty-three patients treated for six to twenty-four months, *Blood*, in review.

Patrick, A. D., 1965, Short communications: A deficiency of glucocerebrosidase in Gaucher's disease, *Biochem. J.* **97**:17C–18C.

Patterson, M. C., Higgins, J. J., Fedio, P., Brady, R. O., and Baron, N. W., 1991, Markers of type 3 neuronopathic Gaucher's disease, *Neurology* **41**:332–333.

Paulson, J. A., Marti, G. E., Fink, J. K., Sato, N., Schoen, M., and Karcher, D. S., 1989, Richter's transformation of lymphoma complicating Gaucher's disease, *Hematol. Pathol.* **3**:91–96.

Pennelli, N., Scarvilli, F., and Zacchello, F., 1969, The morphogenesis of Gaucher cells investigated by electron microscopy, *Blood* **34**:331–347.

Pentchev, P. G., 1977, Enzyme replacement therapy in Gaucher's and Fabry's disease, *Ann. Clin. Lab. Sci.* **7**:251–253.

Pentchev, P. G., Brady, R. O., Hibbert, S. R., Gal, A. E., and Shapiro, D., 1973, Isolation and characterization of glucocerebrosidase from human placental tissue, *J. Biol. Chem.* **248**:5256–5261.

Prence, E., Chakravorti, S., Basu, A., Clark, L. S., Glew, R. H., and Chambers, J. A., 1985, Further studies on the activation of glucocerebrosidase by a heat-stable factor from Gaucher spleen, *Arch. Biochem. Biophys.* **236**:98–109.

Prence, E. M., Garrett, K. O., and Glew, R. H., 1986a, A kinetic study of the effects of galacto-cerebroside 3-sulphate on human spleen glucocerebrosidase. Evidence for two activator-binding sites, *Biochem. J.* **237**:655–662.

Prence, E. M., Garrett, K. O., Panitch, H., Basu, A., Glew, R. H., Wherrett, J. R., and Huterer, S., 1986b, Sulfogalactocerebroside and bis-(monoacylglyceryl)-phosphate as activators of spleen glucocerebrosidase, *Clin. Chim. Acta* **156**:179–189.

Rafi, M. A., Zhang, X.-L., DeGala, G., and Wenger, D. A., 1990, Detection of a point mutation in sphingolipid activator protein-1 mRNA in patients with a variant from metachromatic leuko-dystrophy, *Biochem. Biophys. Res. Commun.* **166**:1017–1023.

Rappeport, J. M., and Ginns, E. I., 1984, Bone-marrow transplantation in severe Gaucher's disease, *N. Engl. J. Med.* **311**:84–88.

Reiner, O., Wilder, S., Givol, D., and Horowitz, M., 1987, Efficient *in vitro* and *in vivo* expression of human glucocerebrosidase cDNA, *DNA* **6**:101–108.

Reiner, O., Wigderson, M., and Horowitz, M., 1988a, Structural analysis of the human gluco-cerebrosidase gene, *DNA* **7**:107.

Reiner, O., Wigderson, M., and Horowitz, M., 1988b, Characterization of the normal human glucocerebrosidase genes and a mutated form in Gaucher's patients, in: *Lipid Storage Disorders. Biological and Medical Aspects* (R. Salvayre, L. Douste-Blazy, and S. Gatt, eds.), Plenum Press, New York, pp. 29–39.

Reiner, O., Dagan, O., and Horowitz, M., 1989, Human sphingolipid activator protein-1 and sphingo-lipid activator protein-2 are encoded by the same gene, *J. Mol. Neurosci.* **1**:225.

Richards, S., Olson, T. A., and McPherson, J. M., 1993, Immunologic events in 332 patients receiving Ceredase, in review.

Ringdén, O., Groth, C. G., Erickson, A., Bäckman, L., Granqvist, S., Månsson, J. E., and Svennerholm, L., 1988, Long-term follow-up of the first successful bone marrow transplantation in Gaucher disease, *Transplantation* **46**:66–70.

Rorman, E. G., Scheinker, V., and Grabowski, G. A., 1992, Structure and evolution of the human prosaposin chromosomal gene, *Genomics* **13**:312–318.

Rosenthal, D. I., Scott, J. A., Barranger, J., Mankin, H. J., Saini, S., Brady, T. J., Osier, L. K., and Doppelt, S., 1986, Evaluation of Gaucher disease using magnetic resonance imaging, *J. Bone Joint Surg.* (Am. Vol.) **68**:802–808.

Sano, A., Radin, N. S., Johnson, L. I., and Tarr, G. E., 1988, The activator protein for glycosylceramide β-glucosidase from guinea pig liver: Improved isolation method and complete amino acid sequence, *J. Biol. Chem.* **263**:19597–19601.

Schmid, C. W., and Jelinek, W. R., 1982, The *Alu* family of dispersed repetitive sequences, *Science* **216**:1065–1070.

Schnabel, D., Schröder, M., and Sandhoff, K., 1991, Mutation in the sphingolipid activator protein 2 in a patient with a variant of Gaucher disease, *FEBS Lett.* **284**:57–59.

Schnabel, D., Schreoder, M., Furst, W., Klein, A., Hurwitz, R., Zenk, T., Weber, J., Harzer, K., Paton, B. C., Poulos, A., Suzuki, K., and Sandhoff, K., 1992, Simultaneous deficiency of sphingolipid activator proteins 1 and 2 is caused by a mutation in the initiation codon of their common gene, *J. Biol. Chem.* **267**:3312–3315.

Schneider, E. L., Ellis, W. G., Brady, R. O., McCulloch, J. R., and Epstein, C. J., 1972, Infantile (type II) Gaucher's disease: *In utero* diagnosis and fetal pathology, *J. Pediatr.* **81**:1134–1139.

Schneider, E. L., Epstein, C. J., Kaback, M. J., and Brandes, D., 1977, Severe pulmonary involvement in adult Gaucher's disease. Report of three cases and review of the literature, *Am. J. Med.* **63**:475–480.

Schuchman, E. H., Suchi, M., Takahashi, T., Sandhoff, K., and Desnick, R. J., 1991, Human acid sphingomyelinase: Isolation, nucleotide sequence and expression of the full-length and alternatively spliced cDNAs, *J. Biol. Chem.* **266**:8531–8539.

Scullin, D. C., Jr., Shelburne, J. D., and Cohen, H. J., 1979, Pseudo-Gaucher cells in multiple myeloma, *Am. J. Med.* **67**:347–352.

Seligsohn, U., Zitman, D., Many, A., and Klibansky, C., 1976, Coexistence of factor XI (plasma thromboplastin antecedent) deficiency and Gaucher's disease, *Isr. J. Med. Sci.* **12**:1448–1452.

Sibille, A., Eng, C., Kim, S.-G., Pastores, G., and Grabowski, G. A., 1993, Phenotype/genotype correlations in Gaucher disease type 1: Clinical and therapeutic implications, *Am. J. Hum. Genet.*, in press.

Sidransky, E., Tsuji, S., Martin, B. M., Stubblefield, B., and Ginns, E. I., 1992, DNA mutation analysis of Gaucher patients, *Am. J. Med. Genet.* **42**:331–336.

Sillivan, K. M., and Reid, C. D., 1991, Introduction to a symposium on sickle cell anemia: Current results of comprehensive care and the evolving role of bone marrow transplantation, *Semin. Hematol.* **28**:177.

Solis, O. G., Belmonte, A. H., Ramaswamy, G., and Tchertkoff, V., 1986, Pseudogaucher cells in *Mycobacterium avium* intracellulare infections in acquired immune deficiency syndrome (AIDS), *Am. J. Clin. Pathol.* **85**:233–235.

Sorge, J., West, C., Westwood, B., and Beutler, E., 1985, Molecular cloning and nucleotide sequence of the human glucocerebrosidase gene, *Proc. Natl. Acad. Sci. USA* **82**:7289–7293.

Sorge, J., Kuhl, W., West, C., and Beutler, E., 1987, Complete correction of the enzymatic defect on type I Gaucher disease fibroblasts by retroviral-mediated gene transfer, *Proc. Natl. Acad. Sci. USA* **84**:906–909.

Sorge, J., Gross, E., West, C., and Beutler, E., 1990, High level transcription of the glucocerebrosidase pseudogene in normal subjects and patients with Gaucher disease, *J. Clin. Invest.* **86**:1137–1141.

Stahl, P. D., Rodman, J. S., Miller, M. J., and Schlesinger, P. H., 1978, Evidence for receptor-mediated binding of glycoproteins, glycoconjugates, and lysosomal glycosidases by alveolar macrophages, *Proc. Natl. Acad. Sci. USA* **75**:1399–1403.

Starer, F., Sargent, J. D., and Hobbs, J. R., 1987, Regression of the radiological changes of Gaucher's disease following bone marrow transplantation, *Br. J. Radiol.* **60**:1189–1195.

Stevens, P. G., Kumari-Subaiya, S. S ., and Kahn, L. B., 1987, Splenic involvement in Gaucher's disease: Sonographic findings, *J. Clin. Ultrasound* **15**:397–400.

Stowens, D. W., Teitelbaum, S. L., Kahn, A. J., and Barranger, J. A., 1985, Skeletal complications of Gaucher disease, *Medicine* (Baltimore) **64**:310–322.

Strasberg, P. M., Lowden, J. A., and Mahuran, D., 1982, Purification of glucosylceramidase by affinity chromatography, *Can. J. Biochem.* **60**:1025–1031.

Strasberg, P. M., Warren, I., Skomorowski, M. A., and Lowden, J. A., 1983, HPLC analysis of neutral glycolipids: An aid in the diagnosis of lysosomal storage disease, *Clin. Chim. Acta* **132**:29–41.

Sun, C. C., Panny, S., Combs, J., and Gutberlett, R., 1984, Hydrops fetalis associated with Gaucher disease, *Pathol. Res. Pract.* **179**:101–104.

Svennerholm, L., Håkansson, G., Lindsten, J., Wahlström, J., and Dreborg, S., 1981, Prenatal diagnosis of Gaucher disease. Assay of the β-glucosidase activity in amniotic fluid cells cultivated in two laboratories with different cultivation conditions, *Clin. Genet.* **19**:16–22.

Svennerholm, L., Dreborg, S., Erikson, A., Groth, C. G., Hillborg, P. O., Håkansson, G., Nilsson, O., and Tibblin, E., 1982, Gaucher disease of the Norrbottnian type (type III). Phenotypic manifestations, *Prog. Clin. Biol. Res.* **95**:67–94.

Tager, J. M., Aerts, J. M., Jonsson, M. V., Murray, G. J., Van Weely, S., Strijland, A., Ginns, E. I., Reuser, J. J., Schram, A. W., and Barranger, J. A., 1986, Molecular forms, biosynthesis and maturation of glucocerebrosidase, a membrane associated lysosomal enzyme deficient in Gaucher disease, in: *Enzymes of Lipid Metabolism II* (L. Freysz, H. Dreyfus, R. Massarelli, and S. Gatt, eds.), Plenum Press, New York, pp. 735–745.

Takasaki, S., Murray, G. J., Furbish, F. S., Brady, R. O., Barranger, J. A., and Kobata, A., 1984, Structure of the N-asparagine-linked oligosaccharide units of human placental β-glucocerebrosidase, *J. Biol. Chem.* **259**:10112–10117.

Theophilus, B., Latham, T., Grabowski, G. A., and Smith, F. I., 1989a, Gaucher disease: Molecular heterogeneity and phenotype–genotype correlations, *Am. J. Hum. Genet.* **45**:212–225.

Theophilus, B. D., Latham, T., Grabowski, G. A., and Smith, F. I., 1989b, Comparison of RNase A, a chemical cleavage and GC-clamped denaturing gradient gel electrophoresis for the detection of mutations in exon 9 of the human acid β-glucosidase gene, *Nucleic Acids. Res.* **17**:7707–7722.

Tibblin, E., Dreborg, S., Erikson, A., Håkansson, G., and Svennerholm, L., 1982, Hematological findings in the Norrbottnian type of Gaucher disease, *Eur. J. Pediatr.* **139**:187–191.

Trimbur, D. E., Warren, R. A., and Withers, S. G., 1992, Region-directed mutagenesis of residues surrounding the active site nucleophile in β-glucosidase from *Agrobacterium faecalis*, *J. Biol. Chem.* **267**:10248–10251.

Tsai, P., Lipton, J. M., Sahdev, I., Najfeld, V., Rankin, L. R., Slyper, A. H., Ludman, M. D., and Grabowski, G. A., 1992, Allogenic bone marrow transplantation in severe Gaucher disease, *Pediatr. Res.* **31**:503–507.

Tsuji, S., Choudary, P. V., Martin, B. M., Winfield, S., Barranger, J. A., and Ginns, E. I., 1986, Nucleotide sequence of cDNA containing the complete coding nsequence for human lysosomal glucocerebrosidase, *J. Biol. Chem.* **261**: 50–53.

Tsuji, S., Choudary, P. V., Martin, B. M., Stubblefield, B. K., Mayor, J. A., Barranger, J. A., and Ginns, E. I., 1987, A mutation in the human glucocerebrosidase gene in neuronopathic Gaucher's disease, *N. Engl. J. Med.* **316**:570–575.

Tsuji, S., Martin, B. M., Barranger, J. A., Stubblefield, B. K., LaMarca, M. E., and Ginns, E. I., 1988a, Genetic heterogeneity in type 1 Gaucher disease: Multiple genotypes in Ashkenazic and non-Ashkenazic individuals, *Proc. Natl. Acad. Sci. USA* **85**:2349–2352; Erratum, *Proc. Natl. Acad. Sci. USA* **85**(15):5708 (1988).

Tsuji, S., Martin, B. M., Barranger, J. A., Stubblefield, B. K., LaMarca, M. E., and Ginns, E. I., 1988b, Genetic heterogeneity in type 1 Gaucher disease: Multiple genotypes in Ashkenazic and non-Ashkenazic individuals, *Proc. Natl. Acad. Sci. USA* **85**:2349–2352, 5708.

Tybulewicz, V. L. J., Tremblay, M. L., LaMarca, M. E., Willemsen, R., Stubblefield, B. K., Winfield, S., Zablocka, B., Sidransky, E., Martin, B. M., Huang, S. P., Mintzer, K. A., Westphal, H., Mulligan, R. C., and Ginns, E. I., 1992, Animal model of Gaucher's disease from targeted disruption of the mouse glucocerebrosidase gene, *Nature* **357**:407–412.

Vaccaro, A. M., Tatti, M., Ciaffoni, F., and Salvioli, R., 1989, Factors affecting the binding of glucosylceramidase to its natural substrate dispersion, *Enzyme* **42**:87–97.

Van Arsdell, S. W., Denison, R. A., Bernstein, L. B., and Weiner, A. M., 1981, Direct repeats flank snRNA pseudogenes, *Cell* **26**:11–17.

Van Paridon, P. A., Visser, A. J., and Wirtz, K. W., 1987, Binding of phospholipids to the phosphatidylinositol transfer protein from bovine brain as studied by steady-state and time-resolved fluorescence spectroscopy, *Biochim. Biophys. Acta* **898**:172–180.

Van Weely, S., Aerts, J. M., Van Leeuwen, M. B., Heikoop, J. C., Donker-Koopman, W. E., Barranger, J. A., Tager, J. M., and Schram, A. W., 1990, Function of oligosaccharide modification in glucocerebrosidase, a membrane-associated lysosomal hydrolase, *Eur. J. Biochem.* **191**:669–677.

Varki, A., and Kornfeld, A., 1981, Lysosomal enzyme targeting: N-Acetylglucosaminyl-phosphotransferase selectively phosphorylated native lysosomal enzymes, *J. Biol. Chem.* **256**:11977–11980.

Wenger, D. A., and Roth, S., 1982, Homozygote and heterozygote identification, *Prog. Clin. Biol. Res.* **95**:551–572.

Wenger, D. A., Roth, S., Kudoh, T., Grover, W. D., Tucker, S. H., Kaye, E. M., and Ullman, M. D., 1983, Biochemical studies in a patient with subacute neuropathic Gaucher disease without visceral glucosylceramide storage, *Pediatr. Res.* **17**:344–348.

Wigderson, M., Firon, N., Horowitz, Z., Wilder, S., Frishberg, Y., Reiner, O., and Horowitz, M., 1989, Characterization of mutations in Gaucher patients by cDNA cloning, *Am. J. Hum. Genet.* **44**:365–377.

Willemsen, R., van Dongen, J. M., Aerts, J. M., Schram, A. W., Tager, J. M., Goudsmit, R., and Reuser, A. J., 1988, An immunoelectron microscopic study of glucocerebrosidase in type 1 Gaucher's disease spleen, *Ultrastruct. Pathol.* **12**:471–478.

Winkelman, M. D., Banker, B. Q., Victor, M., and Moser, H. W., 1983, Non-infantile neuronopathic Gaucher's disease: A clinicopathologic study, *Neurology* **33**:994–1008.

Wynn, C. H., 1986, A triple-binding-domain model explains the specificity of the interaction of a sphingolipid activator protein (SAP-1) with sulfatide, G_{M1}-ganglioside and globotriaosylceramide, *Biochem. J.* **240**:921–924.

Yavin, Y., and Gatt, S., 1969, Further purification and properties of rat brain ceramidase, *Biochemistry* **8**:1692–1698.

Yu, C., Merrick, H. F. W., Verderese, C., Brady, R. O., Currie, J. N., and Barton, N. W., 1990, Horizontal supranuclear gaze palsy: A marker for severe systemic involvement in type III Gaucher's disease, *Neurology* **40**(Suppl. 1):357.

Zhang, X.-L., Rafi, M. A., DeGala, G., and Wenger, D. A., 1990, Insertion in the mRNA of a metachromatic leukodystrophy patient with sphingolipid activator-1 deficiency, *Proc. Natl. Acad. Sci. USA* **87**:1426–1430.

Zhang, X.-L., Rafi, M. A., DeGala, G., and Wenger, D. A., 1991, The mechanism for a 33-nucleotide insertion in mRNA causing sphingolipid activator protein (SAP-1) deficient metachromatic leukodystrophy, *Hum. Genet.* **87**:211–215.

Zidar, B. L., Hartsock, R. J., Lee, R. E., Glew, R. H., LaMarco, K. L., Pugh, R. P., Raju, R. N., and Shackney, S. E., 1987, Pseudo-Gaucher cells in the bone marrow of a patient with Hodgkin's disease, *Am. J. Clin. Pathol.* **87**:533–536.

Zimran, A., Sorge, J., Gross, E., Kubitz, M., West, C., and Beutler, E., 1989, Prediction of severity of Gaucher's disease by identification of mutations at DNA level, *Lancet* **2**:349–352.

Zimran, A., Sorge, J., Gross, E., Kubitz, M., West, C., and Beutler, E., 1990, A glucocerebrosidase fusion gene in Gaucher disease. Implications for the molecular anatomy, pathogenesis, and diagnosis of this disorder, *J. Clin. Invest.* **85**:219–222.

Zimran, A., Gelbart, T., Westwood, B., Grabowski, G. A., and Beutler, E., 1991, High frequency of the Gaucher disease mutation at nucleotide 1226 among Ashkenazi Jews, *Am. J. Hum. Genet.* **49**:855–859.

Addendum

CHAPTER 1: PEROXISOMAL DISEASES

Hugo W. Moser

Peroxisome Functions

A recent article by van den Bosch *et al.* (1992) provides an excellent review of peroxisome biochemistry.

Van Veldoven *et al.* (1992) have studied the substrate specificities of three peroxisomal acyl-CoA oxidases. Palmitoyl-CoA oxidase (long-chain acyl-CoA ligase) also oxidizes prostaglandin CoA esters. Pristanoyl-CoA oxidase can also oxidize very long-chain acyl-CoA.

Synthesis of Docosahexaenoic Acid

Docosahexaenoic acid (DHA) is a prominent normal constituent of brain and retina (Bazan, 1990). Voss *et al.* (1992) have shown that it is synthesized via a pathway that involves the formation of tetracosahexaenoic acid (24:6), which then is chain-shortened to DHA (22:6), and they suggest that this last reaction takes place in the peroxisome. As discussed in the addendum on therapy, DHA levels are reduced in the tissues and body fluids of patients with disorders of peroxisome biogenesis, and DHA administration may be of clinical benefit.

Advances in Human Genetics, Volume 21, edited by Henry Harris and Kurt Hirschhorn. Plenum Press, New York, 1993.

Presence of Superoxide Dismutase

Wanders *et al.* (1992b) and Dhaunsi *et al.* (1992) have shown that peroxisomes contain Cu-Zn superoxide dismutase, an enzyme that is involved in the defense against O_2^- radicals, such as those produced by xanthine oxidase, which has also been localized to the peroxisome (Augermüller, 1982). These findings indicate that the peroxisome also acts as a defense against the damaging effects of O_2^- radicals, such as those produced by environmental agents or during hypoxia-reperfusion.

Activation and Oxidation of Pristanic Acid in the Peroxisome

Wanders *et al.* (1992a) have shown that pristanic acid is activated to its CoA-ester in rat liver. The pristanoyl-CoA synthetase is present in peroxisomes, mitochondria, and microsomes. Competition and immunoprecipitation studies suggest that it is the same enzyme as long-chain fatty acyl-CoA synthetase (EC 6.2.1.3) (Miyazawa *et al.*, 1987). Studies with rat liver indicate that pristanoyl-CoA oxidase is localized exclusively in the peroxisome, and that, unlike peroxisomal acyl-CoA oxidase, this enzyme is not induced by clofibrate. The localization of pristanoyl-CoA oxidase to the peroxisome supports the formulation that defective function of this enzyme is the cause of the phytanic and pristanic acid accumulations that are observed in the peroxisomal disorders.

D-Amino Acid Oxidase

Konno and Yasumura (1992) have recently reviewed the distribution and role of D-amino acid oxidase. In mammals, this peroxisomal enzyme is present in highest concentration in the proximal convoluted tubules of the kidney, the hepatocytes, and the cerebellar Bergman glial cells and astrocytes. While this enzyme has been conserved through evolution, its physiological role remains unclear.

Targeting Mechanisms for Peroxisomal Proteins

Swinkels *et al.* (1991) have shown that the targeting signal for peroxisomal 3-ketoacyl-CoA thiolases A and B in rat liver resides in the cleavable amino-terminus. It is not present at the extreme amino-terminus but is preceded by ten or seven amino acids. This peroxisomal targeting sequence is distinct from other known amino-terminal signals that direct other proteins to mitochondria, endoplasmic reticulum, or chloroplasts.

Direct Demonstration That Import of Peroxisomal Proteins Is Defective In Zellweger Syndrome

Up to now, evidence for the hypothesis that the underlying defect in the disorders of peroxisome biogenesis was due to defective import mechanisms was circumstantial. Walton *et al.* (1992) have now provided positive evidence. They performed microinjection studies with two proteins (Luciferase and albumen) that had been conjugated to the peroxisomal targeting sequence Ser-Lys-Leu. They demonstrated that in normal human fibroblast cell lines these proteins were transported into peroxisomes, but that this did not occur in cell lines derived from patients with Zellweger syndrome.

A Peroxisome "Precursor Particle"

Aikawa *et al.* (1991) have isolated a catalase-containing particle with a density that is lower than that of peroxisomes from human fibroblasts. The particle also contained other peroxisomal enzymes, such as L-alpha-hydroxyacid oxidase, and comparison with the results of Balfe *et al.* (1990) suggests that it also contains 3-ketoacyl-CoA thiolase in the unprocessed form. Cells from patients with Zellweger syndrome also contained this particle. Santos *et al.* (1988) speculate that it represents an immature or incomplete form of peroxisomes, and that it is distinct from the previously described peroxisome ghosts.

Further Observations on Peroxisome Ghosts

A recent report by Suzuki *et al.* (1992) provides additional information about peroxisome ghosts and their contents. With the aid of indirect immunofluorescence staining and cell fractionation techniques, they confirmed that fibroblast cells from patients with Zellweger or NALD lacked particulate catalase and also the nonspecific transfer protein, confirming the hypothesis that these proteins are not imported into the organelle. They also confirmed that there were membranous structures that contain the PMP70 integral membrane proteins, i.e., the peroxisome "ghosts" described previously by Santos *et al*. (1988). A new finding was their observation that a fraction of 3-ketoacyl-CoA thiolase was also present in the particulate fraction, thus indicating that the "ghosts" are not totally lacking in enzymatic activity. The fact that the protein is present in the unprocessed form may indicate that the precursor protein is recognized by the membrane but cannot enter the matrix, or that it is not processed due to the absence of proteolytic enzyme.

Abnormal Plasma Lipoproteins in Disorders of Peroxisome Biogenesis

Mandel *et al*. (1992) have studied in detail the plasma lipoprotein changes in a 7-year-old patient with infantile Refsum syndrome. Cholesterol levels in the LDL and HDL fractions were reduced to 26 to 29% of control, respectively. The LDL demonstrated a significant reduction in its lipid-to-protein ratio and a decreased susceptibility to *in vitro* oxidation. The interactions with macrophages were altered, as reflected by decreased uptake and increased cholesterol esterification rate.

Complementation Analysis: Additional studies

Yajima *et al*. (1992) performed complementation analyses in 17 cell lines from patients with disorders of peroxisome biogenesis. They identified five complementation groups, which they designated as groups A–E. Cross-correction studies showed that their group E, which included Zellweger, NALD, and infantile Refsum syndrome phenotypes, was identical to Roscher group 1 (Roscher *et al*., 1989), and that the Yajima group C (Zellweger phenotype) was equivalent to Roscher group 4. Yajima groups A, B, and D were distinct from Roscher groups 1, 2, 3, 4, and 6 (Roscher group 5 was not available for comparison). This study, therefore, indicates that there exist at least two complementation groups in addition to the six that had been identified by Roscher *et al*. (1989).

Molecular Basis of Disorders of Peroxisome Biogenesis

Two recent studies provide the first insights into the molecular biology of the disorders of peroxisome biogenesis. Shimozawa *et al*. (1992) demonstrated the role of peroxisomal assembly factor-1. This is a 35-kDa peroxisomal integral protein that restored the assembly of peroxisomes in one peroxisome-deficient Chinese hamster ovary cell mutant (Tsukamoto, 1991). This same protein restored peroxisome assembly in fibroblasts derived from a Zellweger disease patient. The mutant cell lines had a point mutation that resulted in premature termination of PAF-1. Addition of normal PAF-1 restored normal peroxisome assembly. The patients' parents were heterozygous for this mutation. The patient belonged to a unique complementation group, i.e., it did not match any of the other groups so far identified.

Gärtner et al. (1992) have studied the molecular biology of the 70-kDa peroxisomal membrane protein, which is a member of the superfamily of ATP-binding proteins that are involved in transport systems. They cloned and sequenced cDNAs for human PMP70 and mapped the gene to chromosome 1. Study of cell lines from 32 patients with Zellweger syndrome revealed two mutant PMP70s. Both of the patients belonged to the most common complementation groups (Roscher group 1) (Roscher et al., 1989). One allele had a splice site mutation and the second a missense mutation.

Therapy of Disorders of Peroxisome Biogenesis

Setchell et al. (1992) reported that oral administration of cholic and chenode-oxycholic acid improved liver function, growth and neurological status in a 6-month old boy with Zellweger syndrome. Martinez (1992) administered DHA by mouth to two patients with neonatal adrenoleukodystrophy. This therapy restored red blood DHA levels to normal and unexpectedly also appeared to have a favorable effect on very-long-chain fatty acid and plasmalogen levels. Of greatest interest was the observation that one of the NALD patients showed substantial improvement in neurological status. While the results of these two studies are provocative, the observations are anecdotal. Additional controlled studies are needed.

Pathogenesis of X-Linked ALD

Powers et al. (1992) have proposed that cytokines play a role in the inflammatory response. They have found an excess of tumor necrosis factor alpha in macrophages and astrocytes at the advancing margin of the lesion. This finding may have implications for therapy, since it is possible to modulate tumor necrosis factor activity with pharmacological agents such as pentoxifylline.

Theda *et al.* (1992) have compared the biochemical abnormalities in post-morten ALD brain white matter and myelin with the degree of histological abnormality. Areas that were undergoing demyelination showed the expected marked excess of C26:0 in the cholesterol ester fractions. In contrast, in areas in which myelin was histologically intact, the cholesterol ester fraction was normal in amount and fatty acid composition, suggesting that biochemical changes in this fraction were a secondary event. The most striking abnormality in this zone was a 39-fold excess of C26:0 in the phosphatidylcholine fraction. The authors speculated that the alterations in this fraction play a role in the immunologically or cytokine-mediated inflammatory response that is responsible for the rapid progression of the cerebral forms of the illness.

Segregation analysis has been performed to determine the cause of the striking intrafamilial phenotypic variability in adrenoleukodystrophy, specifically the frequent co-occurrence of the childhood cerebral forms and adrenomyeloneuropathy within the same kindred and nuclear families (Moser *et al.*, 1980). This study involved 3862 members of 89 kindreds and included 147 patients with childhood cerebral ALD and 83 men with adrenomyeloneuropathy. Segregation analysis indicated that there was a 20:1 likelihood for the existence of an autosomal modifier gene. This postulated gene may act by determining the severity of the cerebral inflammatory response, with the persons with the milder form of the illness adrenomyeloneuropathy displaying little or no inflammation.

Additional Disorders due to Deficient Function of a Single Peroxisomal Enzyme

Human Dihydroxyacetone Phosphate Acyltransferase Deficiency

Wanders *et al.* (1992b) have described a patient with an isolated deficiency of peroxisomal dihydroxyacetone phosphate acyltransferase deficiency. It is of interest that the clinical presentation was similar to that of rhizomelic chondrodysplasia punctata (PCDP), indicating that the isolated abnormality in plasmalogen metabolism is sufficient to produce the skeletal eye and neurological abnormalities associated with RCDP.

Therapy of X-Linked Adrenoleukodystrophy

Ten boys in the early stages of childhood cerebral adrenoleukodystrophy have received bone marrow transplantation. Early results in carefully selected patients are encouraging, but longer follow-up is required for definitive evaluation.

Rhizomelic Chondrodysplasia Punctata (RCDP): Heterogeneity and Additional Observations about Pathogenesis

Genetic Heterogeneity

Heikoop *et al.* (1992) have utilized complementation analysis to demonstrate genetic heterogeneity in patients with RCDP. The newly identified patients showed the biochemical triad characteristic of RCDP, but complemented cell lines from patients with RCDP.

Smeitink *et al.* (1992), Pike *et al.* (1990), and Poll-The *et al.* (1991) demonstrated phenotypic heterogeneity. They reported two patients who showed the biochemical features characteristic of RCDP but with limbs of normal length. The patients did have cataracts, dysmorphic features and progressive neurological deficit. One of the Dutch patients (Smeitink *et al.*, 1992) was a cousin of the Canadian patient (Pike *et al.*, 1990).

Heikoop *et al.* (1990) have added significant new information relevant to the pathogenesis of RCDP. They examined peroxisomes in cultured skin fibroblasts of RCDP patients with immunofluorescence techniques. Anticatalase and anti-69-kDa peroxisomal membrane protein antibodies showed that peroxisomes were normal in number and size. In contrast, no staining was obtained with a monoclonal antibody to 3-oxoacyl-CoA thiolase. The most plausible interpretation is that this protein did not enter the peroxisome. RCDP thus may be a subcategory of the disorders of peroxisome biogenesis, with import defects that are more restricted than that in Zellweger syndrome, NALD, or infantile Refsum disease.

References

Aikawa, J., Chen, W. W., Kelley, R. I., Tada, K., Moser, H. W., and Chen, G. L., 1991, Low-density particles (W-particles) containing catalase in Zellweger syndrome and normal fibroblasts, *Proc. Natl. Acad. Sci.* **88**:10084–10088.

Augermüller, S., Bruder, G., Volkl, A., Wesch, H., and Fahimi, H. D., 1987, Localization of xanthine oxidase in crystalline cores of peroxisomes. A cytochemical and biochemical study, *Eur. J. Cell Biol.* **45**(1):137–144.

Balfe, A., Hoefler, G., Chen, W. W., and Watkins, P. A., 1990, Aberrant subcellular localization of peroxisomal 3-ketoacyl-CoA thiolase in the Zellweger syndrome and rhizomelic chondrodysplasia punctata, *Pediatr. Res.* **27**(3):304–310.

Bazan, N. G., 1990, Supply of n-3 polyunsaturated fatty acids and their significance in the central nervous system, in: *Nutrition and the Brain*, Volume 8, Raven Press, New York, pp. 1–22.

Dhaunsi, G. S., Gulati, S., Singh, A. K., Orak, J. K., Asayama, K., and Singh, I., 1992, Demonstration of Cu-Zn superoxide dismutase in rat liver peroxisomes: Biochemical and immunological evidence, *J. Biol. Chem.* **267**(10):6870–6873.

Gartner, J., Moser, H., and Valle, D., 1992, Mutations in the 70K peroxisomal membrane protein gene in Zellweger syndrome, *Nature Genet.* **1**:16–23.

Heikoop, J. C., Vanroermund, C. W. T., Just, W. V., Ofman, R., Schutgens, R. B. H., Heymans, H. S. A., Wanders, R. J. A., and Tager, J. M., 1990, Rhizomelic chondrodysplasia punctata— Deficiency of 3-oxoacyl-coenzyme a thiolase in peroxisomes and impaired processing of the enzyme, *J. Clin. Invest.* **86**(1):126–130.

Heikoop, J. C., Wanders, R. J. A., Strijland, A., Purvis, R., Schutgens, R. B. H., and Tager, J. M., 1992, Genetic and biochemical heterogeneity in patients with the rhizomelic form of chondrodysplasia punctata—a complementation study, *Hum. Genet.* **89**:439–444.

Konno, R., and Yasumura, Y., 1992, D-amino-acid oxidase and its physiological function, *Int. J. Biochem.* **24**(4):519–524.

Mandel, H., Berant, M., Meiron, D., Aizin, A., Oiknine, J., Brook, J. G., and Aviram, M., 1992, Plasma lipoproteins and monocyte-macrophages in a peroxisome-deficient system: Study of a patient with infantile Refsum disease, *J. Inher. Metab. Dis.* **15**:774–784.

Martinez, M., 1992, Treatment with docosahexaenoic acid favorably modifies the fatty acid composition of erythrocytes in peroxisomal patients, in: *Proceedings of the Second International Symposium on Clinical, Biochemical, and Molecular Aspects of Fatty Acid Oxidation*, Wiley-Liss, New York, pp. 389–397.

Miyazawa, W., Hayashi, H., Hijikata, M., Ishii, N., Furuta, S., Kagamiyama, H., Osumi, T., and Hashimoto,T., 1987, Complete nucleotide sequence of cDNA and predicted amino acid sequence of rat acyl-CoA oxidase, *J. Biol. Chem.* **262**(17):8131–8137.

Moser, H. W., Moser, A. B., Kawamura, N., Migeon, B., O'Neill, B. P., Fenselau, C., and Kishimoto, Y., 1980, Adrenoleukodystrophy: Studies of the phenotype, genetics and biochemistry, *Johns Hopkins Med. J.* **147**:217–224.

Pike, M. G., Applegarth, D. A., Dunn, H. G., Bamforth, S. J., Tingle, A. J., Wood, B. J., Dimmick, J. E., Harris, H., Chantler, J. K., and Hall, J. G., 1990, Congenital rubella syndrome associated with calcific epiphyseal stippling and peroxisomal dysfunction, *J. Pediatr.* **116**(1):88–94.

Poll-The, B. T., Maroteaux, P., Narcy, C., Quetin, P., Guesnu, M., Wanders R. J. A., Schutgens, R. B. H., and Saudubray, J. M., 1991, A new type of chondrodysplasia punctata associated with peroxisomal dysfunction, *J. Inher. Metab. Dis.* **14**:361–363.

Powers, J. M., Liu, Y., Moser, A., and Moser, H., 1992, The inflammatory myelinopathy of adreno-leukodystrophy: Cells, effector molecules, and pathogenetic implications, *J. Neuropathol. Exp. Neurol.* **51**(6):630–643.

Roscher, A. A., Hoefler, S., Hoefler, G., Paschke, E., Paltauf, F., Moser, A., and Moser, H. W., 1989, Genetic and phenotypic heterogeneity in disorders of peroxisome biogenesis—A complementation study involving cell lines from 19 patients, *Pediatr. Res.* **26**:67–72.

Santos, M. J., Imanaka, T., Shio, H., Small, G. M., and Lazarow, P. B., 1988, Peroxisomal membrane ghosts in Zellweger syndrome—Aberrant organelle assembly, *Science* **239**:1536–1538.

Setchell, K. D. R., Bragetti, P., Zimmer-Nechemias, L., Daugherty, C., Pelli, M. A., Vaccaro, R., Gentili, G., Distrutti, E., Dozzini, G., Morelli, A., and Clerici, C., 1992, Oral bile acid treatment and the patient with Zellweger syndrome, *Hepatology* **15**:198–207.

Shimozawa, N., Tsukamoto, T., Suzuki, Y., Orii, T., Shirayoshi, Y., Mori, T., and Fujiki, Y., 1992, A human gene responsible for Zellweger syndrome that affects peroxisome assembly, *Science* **255**:1132–1134.

Smeitink, J. A. M., Beemer, F. A., Espeel, M., Conckerwolcke, R. A. M.G., Jakobs, C., Wanders, R. J. A., Schutgens, R. B. H., Roels, F., Duran, M., Dorland, L., Berger, R., and Poll-The, B. T., 1992, Bone dysplasia associated with phytanic acid accumulation and deficient plasmalogen synthesis: A peroxisomal entity amenable to plasmapheresis, *J. Inher. Metab. Dis.* **15**:377–380.

Suzuki, Y., Shimozawa, N., Yajima, S., Orii, T., Yokota, S., Tashiro, Y., Osumi, T., and Hashimoto,

T., 1992, Different intracellular localization of peroxisomal proteins in fibroblasts from patients with aberrant peroxisome assembly, *Cell Struct. Funct.* **17**:1–8.

Swinkels, B. W., Gould, S. J., Bodnar, A.G., Rachubinski, R. A., and Subramani, S., 1991, A novel, cleavable peroxisomal targeting signal at the amino-terminus of the rat 3-ketoacyl-CoA thiolase, *EMBO J.* **10**(11):3255–3262.

Theda, C., Moser, A. B., Powers, J. M., and Moser, H. W., 1992, Phospholipids in X-linked adrenoleukodystrophy white matter—Fatty acid abnormalities before the onset of demyelination, *J. Neurol. Sci.* **110**:195–204.

Tsukamoto, T., Miura, S., and Fujiki, Y., 1991, Restoration by a 35K membrane protein of peroxisome assembly in a peroxisome-deficient mammalian cell mutant, *Nature* **350**:77–81.

van den Bosch, H., Schutgens, R. B. H., Wanders, R. J. A., and Tager, J. M., 1992, Biochemistry of peroxisomes, *Annu. Rev. Biochem.* **61**:157–197.

Van Veldhoven, P. P., Vanhove, G., Assselberghs, S., Eyssen, H. J., and Mannaerts, G. P., 1992, Substrate specificities of rat liver peroxisomal acyl-coA oxidases: Palmitoyl-CoA oxidase (inducible acyl-CoA oxidase), pristanoyl-CoA oxidase (non-inducible acyl-CoA oxidase), and trihydroxycoprostanoyl-CoA oxidase, *J. Biol. Chem.* **267**(28):20065–20074.

Voss, A., Reinhart, M., Sankarappa, S., and Sprecher, H., 1992, The metabolism of 7, 10, 13, 16, 19-docosapentaenoic acid to 4, 7, 10, 13, 16, 19-docosahexaenoic acid in rat liver is independent of a 4-desaturase, *J. Biol. Chem.* **266**:19995–20000.

Walton, P. A., Gould, S. J., Feramisco, J. R., and Subramani, S., 1992, Transport of microinjected proteins into peroxisomes of mammalian cells: Inability of Zellweger cell lines to import proteins with the SKL tripeptide peroxisomal targeting signal, *Mol. Cell. Biol.* **12**(2):531–541.

Wanders, R. J. A., and Denis, S., 1992, Identification of superoxide dismutase in rat liver peroxisomes, *Biochim. Biophys. Acta.* **1115**:259–262.

Wanders, R. J. A., Denis, S., Van Roermund, C. W. T., Jakobs, C., and ten Brink, H. J., 1992a, Characteristics and subcellular localization of pristanoly-CoA synthetase in rat liver, *Biochim. Biophys. Acta.* **1125**:274–279.

Wanders, R. J. A., Schumacher, H., Heikoop, J., Schutgens, R. B. H., and Tager, J. M., 1992b, Human dihydroxyacetonephosphate acyltransferase deficiency: A new peroxisomal disorder, *J. Inher. Metab. Dis.* **15**:389–391.

Yajima, S., Suzuki, T., Shimozawa, N., Yamaguchi, S., Orii, T., Fujiki, Y., Osumi, T., Hashimoto, T., and Moser, H. W., 1992, Complementation study of peroxisome-deficient disorders by immunofluorescence staining and characterization of fused cells, *Hum. Genet.* **88**(5):491–499.

Index